Genetic Engineering of Symbiotic Nitrogen Fixation and Conservation of Fixed Nitrogen

BASIC LIFE SCIENCES

Alexander Hollaender, General Editor

Associated Universities, Inc.
Washington, D.C.

Recent volumes in the series

Volume 11 • PHOTOSYNTHETIC CARBON ASSIMILATION
Edited by Harold W. Siegelman and Geoffrey Hind

Volume 12 • GENETIC MOSAICS AND CHIMERAS IN MAMMALS
Edited by Liane B. Russell

Volume 13 • POLYPLOIDY: Biological Relevance
Edited by Walter H. Lewis

Volume 14 • GENETIC ENGINEERING OF OSMOREGULATION
Impact on Plant Productivity for Food, Chemicals, and Energy
Edited by D. W. Rains, R. C. Valentine, and Alexander Hollaender

Volume 15 • DNA REPAIR AND MUTAGENESIS IN EUKARYOTES
Edited by W. M. Generoso, M. D. Shelby, and F. J. deSerres

Volume 16 • DEVELOPMENT AND NEUROBIOLOGY OF *DROSOPHILA*
Edited by O. Siddiqi, P. Babu, Linda M. Hall, and Jeffrey C. Hall

Volume 17 • GENETIC ENGINEERING OF SYMBIOTIC NITROGEN FIXATION AND CONSERVATION OF FIXED NITROGEN
Edited by J. M. Lyons, R. C. Valentine, D. A. Phillips, D. W. Rains and R. C. Huffaker

Volume 18 • TRENDS IN THE BIOLOGY OF FERMENTATIONS FOR FUELS AND CHEMICALS
Edited by Alexander Hollaender, Robert Rabson, Palmer Rogers, Anthony San Pietro, Raymond Valentine, and Ralph Wolfe

A Continuation Order Plan is available for this series. A contintutation order will bring delivery of each new volume immediately upon publication. Volumes are billed only upon actual shipment. For further information please contact the publisher.

Genetic Engineering of
Symbiotic Nitrogen Fixation
and Conservation of Fixed Nitrogen

Edited by

J. M. Lyons

R. C. Valentine

D. A. Phillips

D. W. Rains

R. C. Huffaker

University of California, Davis, California

PLENUM PRESS • NEW YORK AND LONDON

Library of Congress Cataloging in Publication Data

Symposium on Enhancing Biological Production of Ammonia From Atmospheric Nitrogen and Soil Nitrate (1980 : Tahoe City, Calif.)
Genetic engineering of symbiotic nitrogen fixation and conservation of fixed nitrogen

(Basic life sciences; v. 17)
Co-sponsored by NSF Grant PFR 77-07301 and the College of Agricultural and Environmental Sciences, University of California, Davis.
Bibliography: p.
Includes index.
1. Nitrogen—Fixation—Congresses. 2. Genetic engineering—Congress. 3. Micro-organisms, Nitrogenfixing—Congresses. I. Lyons, James M. II. National Science Foundation (U.S.) III. University of California, Davis. College of Agricultural and Environmental Sciences. IV. Title. V. Series.

QR 89.7.S95 1980 589 81-4683
ISBN 0-306-40730-2 AACR2

Proceedings of a symposium of Enhancing Biological Production of Ammonia From Atmospheric Nitrogen and Soil Nitrate, held June 29-July 3, 1980, at Granlibakken, Lake Tahoe, California, co-sponsored by NSF Grant PFR 77-07301 and the College of Agricultural and Environmental Sciences, University of California, Davis.

ORGANIZING COMMITTEE

D. W. Rains, Chair
Suzanne Epperly
Nelinia Henry
R. C. Valentine
D. A. Phillips
R. C. Huffaker
J. M. Lyons

Any opinions, findings, conclusions, or recommendations herein are those of the speakers and do not necessarily reflect the views of the National Science Foundation

A Division of Plenum Publishing Corporation
233 Spring Street, New York, N.Y. 10013

Printed in the United States of America

A very special thanks to
Suzanne Epperly and Nelinia Henry
for their exceptional contributions
to the organization and conduct
of the conference.

We greatly appreciate the effort of
Bill Parsons and the staff of
Granlibakken, Lake Tahoe.

FOREWORD

The present volume developed from a symposium entitled "Enhancing Biological Production of Ammonia From Atmospheric Nitrogen and Soil Nitrate" that was held at Lake Tahoe, California in June, 1980. The meeting was supported by the National Science Foundation, Division of Engineering and Applied Sciences and by the College of Agricultural and Environmental Sciences, University of California, Davis. A total of 99 scientists from 41 institutions participated.

Plants capture solar energy in photosynthesis and use mineral nutrients to produce human food and fiber products. The extent to which such materials are removed from agricultural production sites represents a permanent drain of mineral nutrients. Some plants of agronomic importance such as alfalfa, soybean, and clover associate with soil bacteria and use photosynthetic energy to reduce N_2 to NH_3. Many other free-living bacteria and some symbioses involving procaryotes and eucaryotes also reduce N_2. Such processes represent one natural mechanism by which Man can augment soil N for agronomic purposes without using fossil fuel to synthesize and distribute N fertilizer. Other metabolic conversions in the N cycle and physical leaching processes remove N made available through N_2 fixation. Thus nitrification, denitrification, and utilization of soil N by plants are processes that must be considered if one is to conserve N captured by N_2 fixation.

The meeting at Lake Tahoe united scientists from many disciplines to review the literature and to discuss current research directed toward the goal stated in the symposium title. One recurring theme in those discussions was the thought that solutions to many problems mentioned might be provided by genetic manipulation of key metabolic processes in the N cycle. This volume outlines many traditional genetic techniques being used to solve problems related to N_2 fixation, NO_3^- utilization, and NO_3^- conservation. More importantly, it reports some of the first efforts to provide solutions through methods of genetic engineering.

The application of genetic engineering techniques to problems of plant production provides vast potential for contributions.

The major limitation of such methods could be the identification of appropriate problems and the provision of an interpretive framework in which to use this new technology. This volume drawn from the symposium presentations highlights various aspects of the biological processes discussed and indicates some problems that might be amenable to solutions through genetic engineering.

CONTENTS

I. INTRODUCTION

Alexander Hollaender

Associated Universities, Inc.
Council for Research Planning
in Biological Sciences

PREFATORY CHAPTER* - ENHANCING BIOLOGICAL PRODUCTION OF AMMONIA FROM ATMOSPHERIC NITROGEN AND SOIL NITRATE

Alexander Hollaender

Associated Universities, Inc.
Council for Research Planning in Biological Sciences
1717 Massachusetts Avenue, N.W.
Washington, D.C. 20036

My friends from Davis asked me to describe how this physical chemist-biologist-geneticist got so interested in nitrogen fixation. This is not difficult to understand when one considers my background and the events that were taking shape in scientific history during my early education.

During the late 1920's and early 1930's, many interesting scientific developments took place, particularly in the fields of physics and biology. In the physical sciences, this marked the beginning of the modern development and structural understanding of the nucleus, electrons and other heretofore unidentified fundamental particles. My real interests were in the more biological sciences and as my academic career began at the University of Wisconsin at Madison in 1927, I became especially conscious of the contemporary developments in this sphere of study, particularly the interrelation of physical chemistry and biology. I was compelled to comprehend the effects of radiation on living cells and the important observation of Herman Muller -- that mutations could be induced in Drosophila by ioinizing radiation -- was one of the many significant connections revealed between radiation and biology at that time. Of course we were all very aware at Madison of the work on provitamin activation by ultraviolet radiation which leads

*This work was supported in part by U. S. Department of Energy Contract EY-76-C-02-0016 with Associated Universities, Inc., Brookhaven National Laboratory.

to the production of vitamin D. Quantum efficiency of photosynthesis was another very prominent topic at that time, enlarged by the contrary opinions of Warburg and the group in the United States studying the efficiency of photosynthesis.

When my major study changed from physical chemistry to biology after I received my Ph.D., I became very much involved in plant physiology, and research in nitrogen fixation became a logical step from the study of photosynthetic processes. At that time, Prof. Farrington Daniels was teaching in the Physical Chemistry Department of the University. His interest in nitrogen oxides carried over from his war experience in the Washington Fixed Nitrogen Laboratory which was headed up during and immediately following World War I by Dr. Frederick Cottrell. The many interesting approaches developed and patented by Cottrell, a creative physical chemist, led to the formation of the Research Corporation.

One could not help being involved in the lively discussions of nitrogen fixation at Madison. Research accelerated as the Frasch Foundation gave the University of Wisconsin a substantial grant to continue the innovative leads that were developing there at that time. E. B. Fred, I. L. Baldwin and E. McCoy who were joined by Perry Wilson and later by Robert Burris increased the activity and also my interest in this fascinating field. Their work led to many publications by the University faculty, many of which became most significant in this area.

During the war, interest in nitrogen fixation became somewhat dormant as more urgent problems demanded priority. The establishment of the National Science Foundation represented a major encouragement to basic research and its potential application. As a result, nitrogen fixation research resumed in this country and much important experimentation was taking place abroad as well.

I remained in close contact with members of the National Science Foundation and my continued interest in the mechanisms of mutation production has initiated many discussions concerning the improvement of assimilation methodologies and the link with hydrogen production as a possible competitor for N_2 fixation. Theories evolved about 'genetic engineering' by means of hybridization, plasmid transfer and various other approaches to open up new possibilities for the improvement of nitrogen fixation processes.

This led to the informative symposium on 'Genetic Engineering for Nitrogen Fixation' held at the Brookhaven National Laboratory in 1977 sponsored by the National Science Foundation to which there was a most impressive response. Three years later, the conference on 'Enhancing Biological Production of Ammonia from Atmospheric Nitrogen and Soil Nitrate' took place at Tahoe City.

There are many possibilities for broadening the application of N_2 fixation by newly developing techniques that accept not only a wider range of microorganisms but also extend the range of plants which can live in symbiosis with these organisms.

These are goals which may well open new areas of agriculture to help increase our food supplies without compromising existing energy resources. Although my close association with this field has not been exercised through direct laboratory experimentation, it is not difficult to recognize that the development of the basic problems which underlie N_2 fixation will disclose the horizons of our agricultural potential. This symposium demonstrates that considerable progress has been made in certain sections of the fundamental aspects of N_2 fixation research since the 1977 conference. Through such presentations we see that basic biology, genetics, biochemistry and genetic engineering are closely interrelated and that, in the long run, these may offer yet undiscovered applications of biological research to agriculture.

SUPPORT OF PROBLEM FOCUSED NITROGEN FIXATION RESEARCH AT THE NATIONAL SCIENCE FOUNDATION

H. T. Huang

National Science Foundation
1800 G Street, NW
Washington, D.C. 20050

On behalf of the National Science Foundation it is my pleasure and privilege to welcome all of the participants to this symposium on "Enhancing Biological Production of Ammonia from Atmospheric Nitrogen and Soil Nitrate."

Since 1975, the National Science Foundation has funded a focused research effort on biological nitrogen fixation under its statutory authority to "sustain applied scientific research relevant to national problems of public interest." The effort was initiated under the RANN (Research Applied to National Needs) program, and maintained under RANN's successors, ASRA (Applied Science and Research Application) and EAS (Engineering and Applied Science) Directorates.

Looking back over the past six years, we can readily discern the causes which have helped to establish and sustain biological nitrogen fixation as a focused element in the applied research activity at NSF. First, there was the growing recognition of the importance of biological nitrogen fixation to the future welfare of the nation in the aftermath of the oil embargo and quadrupling of petroleum prices of 1973. These events had led to drastic price increases and occasional scarcities of materials such as chemical feedstock, synthetic polymers and nitrogen fertilizer, all essential to the smooth functioning of a modern industrial society. They underscored the need for the United States to develop alternative sources of critical materials which would not have to depend on petroleum and natural gas for their manufacture. Enhancing biological nitrogen fixation in the field would appear to be an ideal, energy-conserving way to reduce the nations' reliance on chemical nitrogen fertilizer for maintaining its agricultural productivity.

Secondly, at about the same time, there was increasing realization that in recombinant DNA research, a new technology was emerging which would greatly expand the capability of microbial geneticists to manipulate organisms so as to enhance their ability to fix molecular nitrogen, or to endow non-fixing organisms with this ability. Recombinant DNA was attracting an unusual amount of interest in the scientific community, in government and among public interest groups, as a tool of enormous power to genetically engineer new forms of life. Proponents point with enthusiasm to the many benefits that could accrue through the application of recombinant DNA technology, while opponents expressed great apprehension about the potential disasters that could befall us from the escape of artificially created life forms into the natural environment; an apprehension that has been greatly reduced with accumulation of knowledge concerning this technology.

The recognition of a national need for enhanced biological nitrogen fixation, and the emergence of a new, powerful technology for genetic engineering, converged to provide a basis for continued support of the nitrogen fixation effort in the applied science activity of the National Science Foundation. To assess the potential benefits and environmental consequences of this research, NSF funded two conferences on "Genetic Engineering for Nitrogen Fixation" in 1977. One was held at the Brookhaven National Laboratory, New York, March 13-17[1], to define the current status and future needs of research in this field, and the other at the National Academy of Sciences, Washington, D.C., October 5-6[2], to discuss the societal impacts of the application of the new technology with both scientific and non-scientific segments of the public.

The establishment of biological nitrogen fixation as a line item in NSF's budget for applied science was formalized in Fiscal Year 1977 although support for this research had already begun as part of the Enzyme Technology program under RANN. So far, annual expenditures on this topic are as follows:

---

[1] Genetic Engineering for Nitrogen Fixation. Edited by Alexander Hollaender et al. Plenum Publishing Corporation, New York, 1977.

[2] Report of the Public Meeting on Genetic Engineering for Nitrogen Fixation. Edited by Alexander Hollaender. NSF/RA 780004.

FY 1975	$0.207 million
FY 1976	0.424
FY 1977	1.412
FY 1978	1.581
FY 1979	1.228
FY 1980	0.970

The above tabulation signifies a substantial increase in the level of support in Fiscal Year 1977. Further increases in subsequent years were actually planned, but new, unexpected budgetary constraints which soon developed in FY 79 and FY 80, virtually erased all possibility of their realization. At any rate, in the climate of 1977, it was felt at NSF that the time was ripe to entertain proposals for interdisciplinary multicomponent projects which could hasten the application of the new techniques for genetic engineering to enhance the ability of natural systems to fix atmospheric nitrogen. One result of this attitude was the award in FY 1977 of a five year continuing grant to the University of California, Davis, PFR 77-07301, entitled "Enhancing Biological Production of Ammonia from Atmospheric Nitrogen and Soil Nitrate," with James M. Lyons, as the principal investigator, and Raymond C. Valentine, Donald A. Phillips, D. William Rains and Ray C. Huffaker as co-principal investigators. The research consists of four interconnected and complementary elements:

1. Genetic enhancement of nitrogen fixation
2. Increasing symbiotic nitrogen fixation in legumes
3. Maximizing nitrogen fixation in rice fields
4. Enhancement of biological conversion of nitrate to ammonia

As part of the obligations associated with the award, the investigators are encouraged to consult annually with an advisory committee consisting of experts in various aspects of research and applications in biological nitrogen fixation, and to hold a conference, once every other year, to disseminate results of the research to the community of potential users. Advisory committee meetings were held in November 1978, and March 1979, and a users' conference was held in May 1978.

In the summer of 1979 the investigators suggested to this program manager that the next user's conference scheduled for 1980, should be expanded into an international symposium open to a wide audience of researchers and users and to include contributions on aspects of research related to the Davis project from other laboratories both within and outside the United States. In order to make the conference truly international in scope, the investigators found it necessary to apply for an NSF grant which would support travel and expenses of invited speakers, especially those from abroad.

At NSF we agreed in principle with the idea of an enlarged users' conference. We further agreed that funds alloted in the FY 1979 incremental budget for the advisory committee meeting and the users' conference could be transferred to help defray the cost of the expanded conference. Thus fortified, the organizers sent out announcements for the symposium in March 1980. As it turned out, the application for a conference grant from NSF had to be declined due to lack of funds, and for a time there was serious doubt as to whether the proposed conference would be held. Fortunately, through the efforts of Dean Charles E. Hess, the organizers were able to obtain a special grant from the College of Agricultural and Environmental Sciences, University of California, Davis, and the proposed symposium at Granlibakken became a reality.

Such then were the circumstances which attended the birth of this symposium. It is our hope that the presentations and discussions will benefit not only the NSF supported project at Davis, but also biological nitrogen fixation research in general around the world. The proceedings will undoubtedly influence the course of future support at the National Science Foundation and other funding agencies for research in biological nitrogen fixation.

Finally, may I take this opportunity to thank the organizers, Drs. Lyons, Valentine, Phillips, Rains and Huffaker, without whose persistence and dedication, the conference could not have been held, all the participants, especially those from abroad, for their contributions to the program and informal discussions, the College of Agricultural and Environmental Sciences, University of California, Davis, for its sponsorship of the symposium, and Dr. Alexander Hollaender for his help in the publication of the proceedings.

BIOLOGICAL NITROGEN FIXATION RESEARCH AT THE UNIVERSITY OF CALIFORNIA, DAVIS

C. E. Hess, Dean

College of Agricultural & Environmental Sciences
University of California
Davis, CA 95616

We are particularly proud to encourage and support the program in biological nitrogen fixation. It is an ideal project to be conducted in the environment of a College of Agricultural and Environmental Sciences and an Agricultural Experiment Station. First, the project has tremendous potential significance to agricultural productivity using an environmentally sensitive approach. A significant component of increased productivity in agriculture has been due to the use of nitrogen fertilizers. But as the costs of nitrogen fertilizers have increased with the increased cost of fossil fuels and concerns have been expressed about nitrogen contamination of ground water, the enhancement and possible expansion of biological nitrogen fixation becomes particularly attractive and could have significant economic and environmental value.

Second, and equally important, is that the project represents a research approach that we all seek but seldom accomplish. The interaction among basic biologists and scientists interested in applying knowledge to the solution of agricultural problems is truly synergistic.

We appreciate the support of the National Science Foundation and the Science Board in providing the resources which greatly facilitates the interaction of basic and applied research on a sustained basis. I believe that the model established by the project Enhancing Biological Production of Ammonia from Atmospheric Nitrogen and Soil Nitrate could be used in other areas of scientific pursuit (e.g., biosalinity and genetic engineering in agriculture) and hope that the National Science Foundation will encourage the submission and funding of similar projects in the future.

II. GENETICS AND REGULATION OF NITROGEN FIXATION

MOLECULAR CLONING OF NITROGEN FIXATION GENES FROM *RHIZOBIUM MELILOTI*

Frederick M. Ausubel and Gary B. Ruvkun

Cellular and Developmental Biology Group
Department of Biology, Biological Laboratories
Harvard University
Cambridge, MA 02138

Until recently, the focus of interest in our laboratory has been on cloning *nif* genes from the free-living nitrogen-fixing species *Klebsiella pneumoniae* (Cannon et al., 1977; Cannon et al., 1979; Riedel et al., 1979). *K. pneumoniae* was chosen for the first attempts to clone *nif* genes because its close resemblance to *Escherichia coli* allowed it to be directly studied by *E. coli* genetic techniques.

This article describes the molecular cloning of nitrogen fixation (*nif*) genes from *Rhizobium meliloti*, the nitrogen-fixing symbiont of alfalfa (*Medicago sativa* L.). In contrast to *K. pneumoniae*, *R. meliloti* does not express *nif* genes in the free-living state, apart from its host plant. Therefore, the isolation and characterization of *nif* mutants using standard bacterial genetic techniques would require screening thousands of potential bacterial mutants on individual plants. Even when identified, *nif* mutations would be difficult to study in *R. meliloti* because of the lack of sophisticated genetic techniques to manipulate mutations in *Rhizobium* species. To circumvent these problems, we have adapted a variety of recombinant DNA technologies to the study of *nif* genes in *R. meliloti*. In particular, we have been able to identify *R. meliloti nif* genes on the basis of their homology with cloned *K. pneumoniae nif* genes (Ruvkun and Ausubel, 1980a). In addition, we have developed a general method for the site directed mutagenesis of *R. meliloti* DNA sequences cloned in *E. coli* and the subsequent replacement of the wild-type parental DNA sequences with the mutated sequences in the *R. meliloti* genome (Ruvkun and Ausubel, manuscript submitted). Following a brief review of *K.*

pneumoniae nif molecular genetics, we describe our recent progress in cloning and manipulating R. meliloti nif genes.

NITROGEN FIXATION GENES IN KLEBSIELLA PNEUMONIAE

A cluster of at least 17 nif genes maps near the operator end of the histidine (his) biosynthesis operon in K. pneumoniae. Fifteen genes in the cluster (nifQ, B, A, L, F, M, V, S, U, N, E, K, D, H, and J) have been ordered unambiguously on the basis of deletion mapping and have been grouped into seven operons on the basis of polarity induced by phage Mu or by transposons Tn5, Tn7 and Tn10 (MacNeil et al., 1978; Elmerich et al., 1978; Merrick et al., 1978, 1980). The two remaining nif genes in the cluster (X and Y) were identified on the basis of the elimination of a polypeptide by Tn5 insertion when a recombinant plasmid carrying nif sequences was used to infect mini cells (W. Klipp and A. Pühler, personal communication). A genetic map of the his-nif region of the K. pneumoniae chromosome is illustrated diagrammatically at the top of Figure 1.

The molecular cloning of Klebsiella pneumoniae nif genes was facilitated by the clustering of nif genes in K. pneumoniae because cloning a series of overlapping DNA fragments from the his-nif region guaranteed inclusion of all nif genes on cloned sequences. The major strategy employed for cloning nif DNA fragments was to construct a shotgun clone bank of K. pneumoniae DNA in various plasmid vectors, to transform $hisD^-$ and/or nif^- strains of K. pneumoniae, and to select His^+ and/or Nif^+ transformants. A series of EcoRI fragments from the his-nif region cloned in recombinant plasmids are depicted in Figure 1. Pühler et al. (1979) have also constructed clones carrying all the genes in the nif cluster and MacNeil and Brill (1980) have constructed a series of λ nif transducing phages in vivo which collectively cover the entire nif cluster. One plasmid, pWK25, constructed by Klipp and Pühler (personal communication) is shown in Figure 1.

The cloned DNA fragments shown in Figure 1 were used to obtain a detailed restriction map of the nif gene cluster (Riedel et al., 1979). In order to correlate the restriction map with the genetic map, Riedel et al. (1979) devised a method for determining the precise location of mutations caused by the insertion of transposable genetic elements into nif genes. The overall strategy was to use the Southern gel transfer procedure (Southern, 1975) to monitor alterations of specific nif DNA restriction fragments in strains carrying TN5, TN10, or phage Mu insertions in various nif genes. Each insertion had been assigned a location within one of 15 genes in the nif cluster (MacNeil et al., 1978; Elmerich et al., 1978; Merrick et al., 1978, 1980). Cloned nif DNA fragments were used as hybridization probes. Using this procedure, Riedel et al. (1979) determined the location of 86 nif insertion mutations.

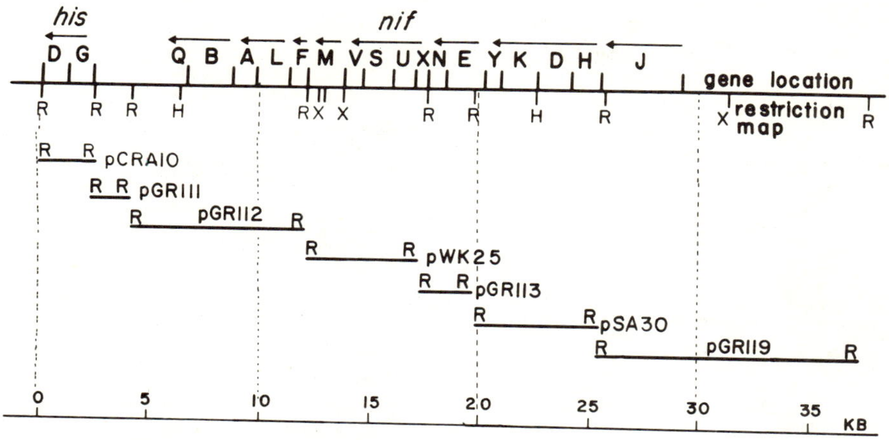

Fig. 1. Genetic and physical map of the hif-nif cluster in *K. pneumoniae*. The horizontal arrows indicate *nif* operons and their direction of transcription (see text). The approximate locations of the *nif* and *his* genes were determined by physical mapping of insertion mutations (Riedel et al., 1979). Accurate locations (±200 bp) for the boundaries between *nifL* and *nifF*, *nifN* and *nifE*, and *nifD* and *nifH* were determined because, in these cases, insertions in the two adjacent genes mapped closely to each other. pCRA10 (Cannon et al., 1977); pGR111, pGR112, pGR113, pGR119 (Riedel, 1980); pWK25 (W. Klipp and A. Pühler, personal communication); and pSA30 (Cannon et al., 1979) are small amplifiable plasmids carrying the *his* or *nif* DNA restriction fragments indicated. R = *Eco*RI, H = *Hind*III, X = *Xho*I.

Some of the results of the physical mapping studies are shown in Figure 1 which shows: 1) the approximate location of each *nif* gene based on transposon mapping; and 2) the restriction map of the *his-nif* region of the *K. pneumoniae* chromosome for selected restriction endonucleases.

INTERSPECIES HOMOLOGY OF NITROGENASE GENES

As stated above, because *R. meliloti nif* genes are only expressed after the bacteria have invaded their host plants, it is difficult to isolate and characterize *nif* mutants in this species. For these reasons, we sought to identify *R. meliloti* nitrogenase genes by using cloned *K. pneumoniae nif* DNA as hybridization probes to identify particular DNA restriction fragments from *R. meliloti* which contain sequences homologous to

K. pneumoniae nif genes (Ruvkun and Ausubel, 1980a). The success of this approach depended on the evolutionary conservation of nif gene sequences.

To test whether nif genes from K. pneumoniae are homologous to DNA from R. meliloti and from other nitrogen-fixing species, Ruvkun and Ausubel (1980a) labeled cloned K. pneumoniae nif DNA with ^{32}P and hybridized this labeled DNA to "Southern" blots of restriction endonuclease digested DNAs from a variety of prokaryotic species. DNA isolated from 13 nitrogen-fixing species including a Gram positive bacterium, an actinomycetous bacterium, and three species of cyanobacteria hybridized to a 6.3 kb EcoRI K. pneumoniae DNA fragment, cloned in pSA30, which carries the nitrogenase structural genes nifK, D and H (see Fig. 1). In contrast, no hybridization was observed when the same ^{32}P-labeled pSA30 was used to probe DNA from 12 different species which do not fix nitrogen. From these data it was concluded that the restriction fragments from the nitrogen-fixing species which hybridize to the K. pneumoniae nif DNA fragment carry genes which code for nitrogenase polypeptides ("nitrogenase" DNA). Nuti et al. (1979) and Mazur et al. (1980) also found that at least a portion of the structural gene sequences for K. pneumoniae nitrogenase hybridized to R. leguminosarum and Anabaena 7120 DNA, respectively.

CLONING OF R. MELILOTI NIF HOMOLOGOUS DNA

In order to clone nitrogenase structural gene sequences from R. meliloti, Ruvkun and Ausubel (1980a) constructed a gene library of R. meliloti EcoRI fragments derived from a partial EcoRI digestion in the cloning plasmid vector pBR322 (Bolivar et al., 1977) and then screened the gene library for clones containing a DNA fragment homologous to the 6.3 kb EcoRI fragment cloned in plasmid pSA30 (Ruvkun and Ausubel, 1980a). Out of approximately 4000 clones screened by the method of Hanahan and Meselson (1980), one was found to contain a plasmid, pRmR1, which contained the 3.9 kb R. meliloti fragments that hybridized to the K. pneumoniae probe.

Restriction analysis of purified pRmR1 DNA revealed that pRmR1 contained 6 different R. meliloti fragments ranging in size from 1.2 kb to 5.0 kb in addition to the 3.9 kb EcoRI fragment homologous to pSA30. Analysis of R. meliloti DNA by the Southern gel transfer and hybridization technique using ^{32}P-labeled pRmR1 as a hybridization probe, showed that the 5.0 kb R. meliloti EcoRI fragment in pRmR1 is adjacent to the 3.9 kb EcoRI fragment on the R. meliloti genome. In order to facilitate the analysis of these two EcoRI fragments, the 3.9 kb and 5.0 kb fragments were recloned individually in the vector pACYC184 (Chang and Cohen, 1978) to yield plasmids pRmR2 and pRmR3, respectively (Ruvkun and Ausubel,

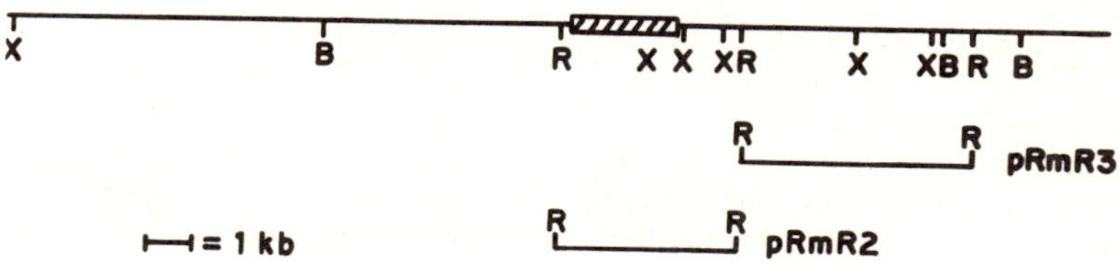

Fig. 2. Restriction map of the R. meliloti nif region. The restriction map of the nif region of the R. meliloti genome was constructed as follows: total DNA was isolated from R. meliloti strain Rm1021 (Str^r wild type), as described by Marmur (1961); approximately 1 μg of DNA was digested with restriction endonucleases BamHI, EcoRI, and/or XhoI (5 units, Bethesda Research Laboratories, Rockville, MD) and restriction fragments were separated on 0.7% agarose gels and transferred to nitrocellulose by the method of Southern (1975) as described (Ruvkun and Ausubel, 1980a). Plasmids pRmR2 and pRmR3 were constructed by ligating EcoRI cleaved pRmR1 DNA and EcoRI cleaved pACYC184 DNA with T4 ligase as described (Cannon et al., 1977); transforming E. coli strain HB101 (Cohen et al., 1977); screening Tc transformants for ones that were ampicillin (Ap) and chloramphenicol (Cm) sensitive, and analyzing individual $Tc^rCm^sAp^s$ transformants with a small scale plasmid isolation procedure (Klein et al., 1980) for ones containing the 3.9 and the 5.0 kb EcoRI fragments. Plasmids pRmR2 and pRmR3 were isolated by the cleared lysate technique (Clewell and Helinski, 1969) and labeled by nick translation (Maniatis et al., 1975; Rigby et al., 1977) with α^{32}P-TPP (350 Ci/mmole, Amersham) and DNA polymerase I (Boehringer Mannheim) to a specific activity of 5 x 10^7 cpm/μg of DNA, and 10^6 cpm of labeled DNA probe was hybridized to each 10 x 15 cm nitrocellulose filter and exposed to Kodak XR-5 film as described (Ruvkun and Ausubel, 1980a). The region of homology in pRmR2 to K. pneumoniae is cross-hatched and was localized as described in the text and in Figure 5. X = XhoI; B = BamHI; R = EcoRI.

1980b). A restriction map of a portion of the *R. meliloti* genome containing the 3.9 kb and 5.0 kb *EcoRI* fragments is shown in Figure 2.

SITE DIRECTED MUTAGENESIS OF *R. MELILOTI NIF* GENES

In order to obtain genetic verification that the 3.9 kb fragment cloned in pRmR2 contains essential *nif* genes, and to initiate a detailed genetic analysis of *R. meliloti nif* genes, we developed a method to replace the *R. meliloti* wild-type genomic *nif* genes with homologous sequences altered by transposon Tn5 mutagenesis. Transposon Tn5 was chosen for this mutagenesis because it confers resistance to kanamycin (Km) and neomycin (Nm) in *E. coli* and *R. meliloti*, exhibits little insertion site specificity, and in general causes polar insertion mutations (Berg, 1977). The experimental strategy we adopted involved the series of steps summarized below and diagrammed in Figures 3 and 4.

Step I (see Fig. 3): Plasmid pRmR2 was mutagenized with Tn5 by infecting E. coli strain HB101 containing pRmR2 with λ::Tn5. Pladmids carrying Tn5 were collected by isolation of plasmid DNA from Km^r survivors of λ::Tn5 infected cells, transformation of strain HB101, and selection of tetracycline and kanamycin resistant transformants. Plasmid DNA was prepared from individual Tc^rKm^r clones using a small scale purification procedure (Klein et al., 1980), and the plasmid DNA was digested with *EcoRI*. Because Tn5 is 5.7 kb and does not contain an *EcoRI* site, insertions of Tn5 into the 3.9 kb *nif* homologous *R. meliloti EcoRI* restriction fragment were monitored (using agarose gel electrophoresis) by the appearance of a 9.6 kb *EcoRI* restriction fragment and disappearance of the 3.9 kb *EcoRI* restriction fragment. Approximately 50% of the plasmids contained Tn5 insertions in the 3.9 kb *EcoRI* fragment. Because Tn5 contains mapped *XhoI* and *HindIII* sites (Jorgenson et al., 1979), the sites of these Tn5 insertions along the 3.9 kb *EcoRI* restriction fragment were determined by double restriction endonuclease digestions of each pRMR2::Tn5 plasmid with *XhoI* and *EcoRI* or *HindIII* and *EcoRI*. Six plasmids with Tn5 insertions distributed along the entire length of the 3.9 kb *EcoRI* fragment were chosen for further analysis. Figure 5 shows the insertion sites of these 6 Tn5 mutations.

Step II (see Figs. 3 and 4): A DNA transformation system has not been developed for *R. meliloti*. Therefore, in order to introduce the *nif*-region::Tn5 insertions into *R. meliloti*, each *nif*-region::Tn5 *EcoRI* fragment was recloned into the conjugative cloning vector pRK290. (Plasmid pRK290, constructed by Ditta et al., manuscript submitted, also see paper in this volume, is a 20 kb derivative of the P-group conjugative plasmid RK2, confers tetracycline (Tc) resistance, and contains a single *EcoRI* site suitable for cloning. pRK290 is *tra*$^-$ but can be complemented

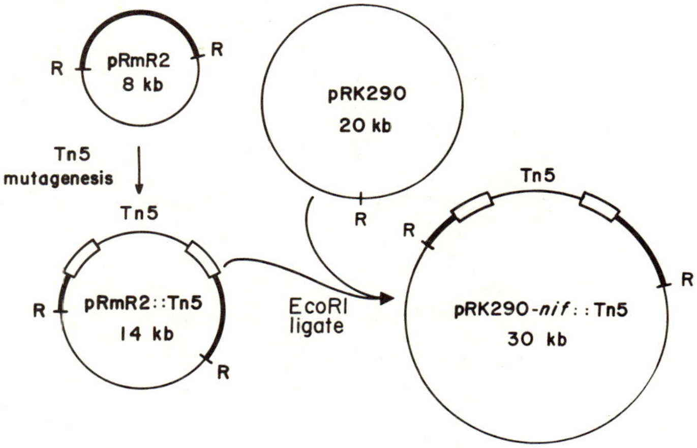

Fig. 3. Experimental strategy used to mutagenize pRmR2 with Tn5 and transfer the mutated R. meliloti DNA to R. meliloti. E. coli strain HB101/pRmR2 (thr leu thi recA hsdR hsdM pro str^r Tc^r) was grown at 37 C in LB medium containing 0.2% maltose and 10 μg/ml Tc to a density of 5 x 10^8 cells per ml and infected with λb221rex::Tn5 cI857 O_{am8} P_{am29} (Berg, 1977) at a multiplicity of 0.1 particles per cell. Approximately 5 x 10^3 kanamycin resistant survivors were obtained per plate when 0.2 ml of infected cells were plated on LB agar containing 20 μg/ml Km at 32 C. Colonies from 3 plates were resuspended in 10 ml of 10 mM $MgSO_4$, plasmid DNA was isolated (Clewell and Helinski, 1969), and used to transform $CaCl_2$ treated strain HB101 (Cohen et al., 1972) and Tc^r and Tc^rKm^r transformants were selected on LB agar containing 10 μg/ml Tc and 20 μg/ml Km. Approximately 20 Tc^rKm^r transformants were obtained per 10^4 Tc^r transformants. Plasmid DNA was prepared from individual Tc^rKm^r clones (Klein et al., 1980), and the plasmid DNA was digested with EcoRI, with XhoI and EcoRI, or HindIII and EcoRI. Six plasmids with Tn5 insertions distributed along the entire length of the 3.9 kb EcoRI fragment were chosen for further analysis. Figure 5 shows the insertion sites of these 6 Tn5 mutations. A 0.5 μg aliquot of each of the six pRmR2::Tn5 plasmids was digested with EcoRI and mixed with a 0.5 μg aliquot of pRK290 DNA which also had been digested with EcoRI. The mixed DNAs were ligated with one unit of T4 DNA ligase, as described (Cannon et al., 1977), and the ligated DNA was used to transform E. coli strain MM294 (endoI hsdR hsdM pro). pRK290-nif-region::Tn5 plasmids were selected as described in the legend to Figure 4.

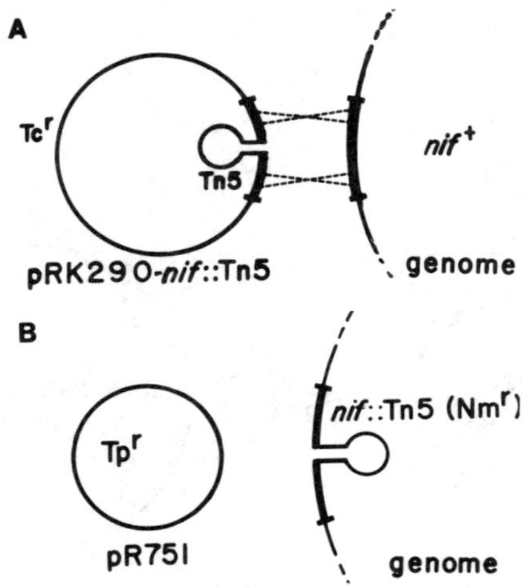

Fig. 4. Experimental strategy used to introduce specific Tn5 mutations into the R. meliloti genome. Panel A: E. coli strain MM294 transformed with a ligated mixture of EcoRI cut pRK290 and pRmR2::Tn5 DNAs (see legend to Fig. 3) was grown nonselectively in LB broth for 2 hrs at 32 C and then a 0.1 ml aliquot was diluted into 5 ml LB containing 10 μg/ml Tc and 20 μg/ml Km and grown to saturation at 30 C. 0.1 ml of this culture was mated simultaneously with 0.1 ml R. meliloti strain Rm1021 (Nif^+Str^r) and 0.1 ml MM294/pRK2013 and incubated on an LB agar plate overnight at 32 C. (Plasmid pRK2013 (Figurski and Helinski, 1979) contains the tra genes of RK2 cloned in plasmid ColE1 and complements pRK290 in trans for transfer of pRK290.) $Str^rTc^rNm^r$ R. meliloti exconjugants were selected on LB agar containing 250 μg/ml streptomycin (Str), 5 μg/ml Tc, and 20 μg/ml Nm. Because pRK2013 (which confers Nm^r) cannot replicate in R. meliloti (Ditta et al., 1980), this selection yielded R. meliloti exconjugants containing pRK290-nif-region::Tn5 plasmids which arose at a frequency of 10^{-3}. Panel B: E. coli strain J53/pR751 ((Jobanputra and Datta, 1974); pR751 confers trimethoprim (Tp) resistance) was conjugated with each of the R. meliloti M1021/pRK290-nif-region::Tn5 strains as described above and $Str^rTp^rNm^r$ exconjugants were selected on LB agar containing 250 μg/ml Str, 700 μg/ml Tp, and 30 μg/ml Nm. Exconjugants arose at a frequency of 10^{-6} and several were tested by the Southern gel transfer and hybridization procedure in order to verify that Tn5 had recombined into the R. meliloti genome by means of flanking nif DNA sequency homology.

in trans for conjugation by pRK2013 (Figurski and Helinski, 1979)). In this step, we took advantage of the fact that Tn5 is not cleaved by *Eco*RI which enabled us to remove the *nif*-region::Tn5 insertions intact from pRmR2::Tn5 plasmids by digestion with *Eco*RI. The resulting pRK290-*nif*-region::Tn5 hybrid plasmids were conjugated into *R. meliloti*. Total DNA was isolated from the *R. meliloti* exconjugants and analyzed by the Southern gel transfer and hybridization procedure (Southern, 1975). The pRK290-*nif*-region::Tn5 hybrid plasmids had the expected restriction map and replicated stably in *R. meliloti*.

Step III (see Fig. 4): Our strategy for isolating *R. meliloti* strains in which the plasmid-borne *nif*-region::Tn5 fragment had replaced the genomic wild-type *nif* region was based on the asumption that homologous recombination between pRK290-*nif*-region::Tn5 and the *R. meliloti* genome would occur in a small subpopulation of cells and result in the "transfer" of the Tn5 insertion mutant from the plasmid to the genome. Cells in which this recombination event had occurred were detected by conjugation of a second P-group plasmid, pR751 (Jobanputra and Datta, 1974), which is incompatible with pRK290, into the *R. meliloti* strains containing the pRK290-*nif*-region::Tn5 plasmid and simultaneous selection for neomycin resistance (retention of Tn5) and trimethoprim (Tp) resistance (conferred by pR751). All of the Tp^rNm^r exconjugants were Tc^s, indicating loss of pRK290. These *R. meliloti* $Tp^rNm^rTc^s$ exconjugants could have resulted from Tn5 transpositions to pR751 or to the *R. meliloti* genome as well as from homologous recombination between pRK290-*nif*-region::Tn5 and the *R. meliloti* genome. To demonstrate that Tn5 had recombined into the *R. meliloti* genome by means of flanking *nif* DNA sequence homology, DNA was isolated from the $Tp^rNm^rTc^s$ exconjugants, cleaved with *Bam*HI, and analyzed by the Southern gel transfer and hybridization technique using ^{32}P-labeled pRmR2 DNA as a hybridization probe. Homologous recombination of each *nif*-region::Tn5 insertion from the plasmid into the *R. meliloti* genome *nif*-region would result in a different predictable change in the restriction map of that region whereas transposition of Tn5 would not be expected to change the restriction map of the genomic *nif* region. The results, not shown, indicated that in all the $Tp^rNm^rTc^s$ strains examined, each *nif*-region ::Tn5 had replaced the normal 3.9 kb *nif* region in the *R. meliloti* genome via a homologous recombination event. Among 24 $Tp^rNm^rTc^s$ colonies analyzed none contained Tn5 transpositions to new locations in the genome or to pR751, and none contained replicon fusions between pR751 and pRK290-*nif*-region::Tn5.

Step IV: Each *R. meliloti* strain containing a verified Tn5 insertion in the *nif* region of the genome was inoculated onto a sterile alfalfa seedling and nitrogenase activity was determined 3-4 weeks later using the acetylene reduction technique as described in the legend to Table 1. Among the six Tn5 insertions in

Table 1. Level of nitrogenase activity in whole plants with nodules containing various *Rhizobium meliloti* strains.

	nanomoles/ml of ethylene produced after 24 hrs in acetylene gas
Alfalfa plant alone	<1
+ Rm 1021	500±100
+ Rm 1021/pRK290	400±100
+ Rm 1021/pR751	400±100
+ Rm 1021/pRK290-*nif*-region::Tn5	500±100
+ Rm 1021 *nif*-region::Tn5 #2	500±100
+ Rm 1021 *nif*-region::Tn5 #12	<1
+ Rm 1021 *nif*-region::Tn5 #7	<1
+ Rm 1021 *nif*-region::Tn5 #10	<1
+ Rm 1021 *nif*-region::Tn5 #22	<1
+ Rm 1021 *nif*-region::Tn5 #20	500±100
+ Rm 1021 *met*::Tn5	300±100

Legend to Table 1: Alfalfa seeds (variety Iroquois) were sterilized for 30 minutes in 70% ethanol then 30 minutes in 50% commercial bleach and then germinated in the dark in distilled water for 3 days. Individual seedlings were planted on 6 ml slants of nitrogen-free agar (0.1% $CaHPO_4$, 0.02% K_2HPO_4, 0.02% $MgSO_4 \cdot 7H_2O$, 0.02% NaCl, 0.01% $FeCl_3$, 1.0% agar) (Vincent, 1970) in 18 x 150 mm test tubes and grown at 27 C with overhead illumination of 200 foot candles on 16 hour days for 7 days prior to bacterial inoculation. A freshly grown colony of *R. meliloti* grown on LB agar plus appropriate drugs was picked with a flat toothpick, suspended in 5 ml distilled water, and used to inoculate 5 alfalfa seedlings. After 4 weeks of growth under the conditions described above, the tubes were sealed with serum stoppers, injected with 1 ml acetylene gas per tube, and incubated at room temperature in the dark for 24 hours. Ethylene gas production was monitored by gas chromatography as described (Cannon et al., 1977). Control plants not inoculated with bacteria did not form nodules and no acetylene reduction was detected. All plants inoculated with *R. meliloti* formed nodules but those inoculated with strains containing *nif*$^-$::Tn5 insertions did not exhibit detectable levels of nitrogenase activity and had approximately 5 to 10 times the normal number of nodules per plant. Both the control plants and the plants inoculated with Nif$^-$ bacterial strains were stunted and yellow after 4 weeks of growth compared to plants inoculated with Nif$^+$ bacterial strains which were green and visibly larger.

the nif region, four strains showed undetectable levels of nitrogenase activity (Nif$^-$) upon establishment of symbiosis, whereas two had normal nitrogenase activity (Nif$^+$) (see Fig. 5 and Table 1). The construction of two Nif$^+$ strains with genomic Tn5 insertions in the 3.9 kb EcoRI fragment demonstrates that the construction of the four Nif$^-$::Tn5 mutations is unlikely to be an artifact of the cloning and recombinational techniques used in our gene replacement strategy. All of the strains with a normal genomic nif region and in addition plasmid pRK290-nif-region::Tn5 were Nif$^+$ indicating that none of the nif-region::Tn5 insertions were dominant nif$^-$ mutations. All six strains containing Tn5 recombined into the genome were prototrophic, indicating that the Nif$^-$ phenotype of four of the strains was not an indirect effect of auxotrophy.

LOCALIZATION OF SPECIFIC NITROGENASE GENES ON pRmR2

In conjunction with the data presented in Table 1 and Figure 5 which correlates Tn5 map position with Nif phenotype, we have utilized pRmR2::Tn5 plasmids to map the approximate positions of R. meliloti nif genes on the 3.9 kb EcoRI fragment. This was accomplished by hybridizing specific K. pneumoniae restriction fragments carrying regions of genes nifD (nitrogenase) and nifH (nitrogenase reductase) to various restriction digests of pRmR2 and pRmR2::Tn5 plasmids. The pRmR2::Tn5 plasmids were useful in this procedure because the insertion of Tn5 introduced new restriction sites into the 3.9 kb EcoRI fragment which could be used to subdivide the fragment into a variety of pieces with known end points. We found that sequences within both K. pneumoniae nifD and nifH are conserved in R. meliloti. The conserved R. meliloti "nifD" sequences are localized to 1.5 kb between Tn5 insertions #2 and #10, and the conserved R. meliloti "nifH" sequences are localized to 1.0 kb between Tn5 insertion #10 and XhoI site B indicated in Figure 5.

As expected on the basis of the interspecies DNA homology with K. pneumoniae nif genes, those Tn5 insertions which cause a Nif$^-$ phenotype in R. meliloti are localized near or in these regions of homology, whereas the two Tn5 insertions which have no effect on nif gene expression are on the two ends of the 3.9 kb R. meliloti EcoRI fragment and not in the region of interspecies nif gene homology. Because Tn5 insertion #2 does not cause a Nif$^-$ phenotype, the nif gene D equivalent in R. meliloti probably starts to the right of this Tn5 insertion (see Fig. 5). The protein subunits for nitrogenase and nitrogenase reductase are of average molecular weights 55K and 28K, respectively, in other Rhizobium species (Israel et al., 1974; Whiting and Dilworth, 1974); therefore the minimum length of DNA necessary to code for these proteins should be 2.2 kb. Thus it is likely that no more than two nif genes map in the 1.8 kb of DNA between TN5 insertions #2 and #7 and that either the Nif$^-$::Tn5 insertion #22, 2.6 kb from

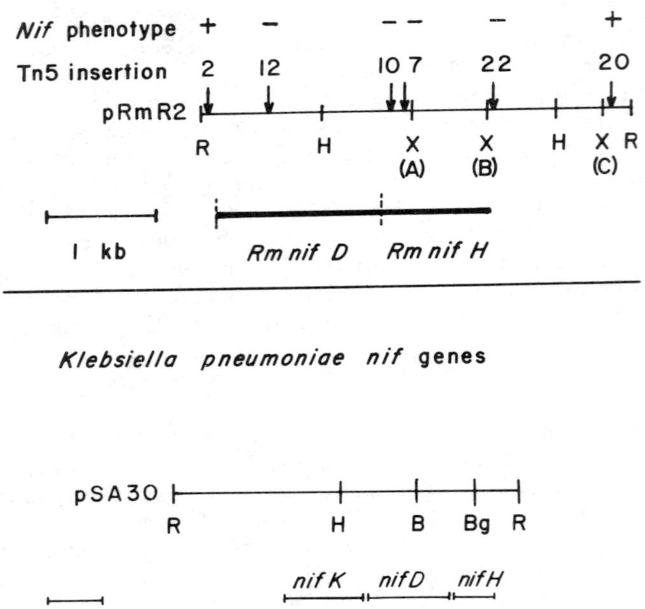

Fig. 5. Physical genetic map of R. meliloti nif genes on pRmR2. Restriction maps of plasmid pRmR2 and each pRmR2::Tn5 plasmid were constructed using a combination of digestions with restriction endonucleases EcoRI, HindIII, XhoI, and BamHI followed by gel electrophoresis in 1% to 1.5% agarose gels. These digested DNAs were transferred to nitrocellulose by the method of Southern (1975) and hybridized with probes derived from the K. pneumoniae 6.3 kb EcoRI fragment cloned in pSA30. As shown in panel B, the 1.4 kb HindIII BamHI fragment of pSA30 contains the C terminal coding region of nifD and the 0.7 kb BglII-EcoRI fragment contains the N terminal coding region of nifH (Riedel et al., 1979; V. Sundaresan, personal communication). These fragments were eluted from preparative agarose gels as described (Ruvkun and Ausubel, 1980a) and labeled with α^{32}P-TTP (350 Ci/mmole, Amersham) using the nick translation method to a specific activity of 5 x 10^7 cpm/μg of DNA. Approximately 10^6 cpm were hybridized to the nitrocellulose sheets. Regions of homology in pRmR2 to K. pneumoniae nifD and nifH genes (shown in heavy lines) were localized by a combination of restriction endonuclease sites present in the 3.9 kb EcoRI fragment and new restriction sites generated by various Tn5 insertions. R = EcoRI, H = HindIII, X = XhoI, B = BamHI, Bg = BglII.

Tn5 insertion #2, is in nifH or its promoter, or defines an adjacent nif gene.

DISCUSSION AND CONCLUSIONS

An important conclusion to be drawn from the data presented here is that cloned R. meliloti DNA sequences homologous to K. pneumonaie nifD and nifH play an essential role in symbiotic nitrogen fixation. The data rule out the possibility that the DNA homology between K. pneumoniae nif genes and R. meliloti DNA is due to cryptic nif genes not functional in symbiotic nitrogen fixation or due to homology with genes evolutionarily (or fortuitously) related to K. pneumoniae nif genes. Based on the interspecies homology with K. pneumoniae nifD and nifH and on the mutational analysis described here, we conclude that R. meliloti contains genes equivalent to K. pneumoniae nifD and nifH, though we cannot as yet assign these homologous R. meliloti nif genes to separate complementation groups.

The ability to rapidly construct mutations on cloned DNA fragments and then recombine these mutations into the R. meliloti genome enables us to scan the R. meliloti genome in the vicinity of the 3.9 kb EcoRI fragment for other nif genes. Preliminary results indicate that other genes affecting symbiotic nitrogen fixation are clustered in this region of the R. meliloti genome (G. Ruvkun and S. Long, unpublished results).

The techniques described in this paper should be generally applicable to the production of specific mutations in a variety of Gram negative prokaryotic species (see papers by Ditta et al., and Hennecke and Mielenz, in this volume for further applications). These techniques should also be useful in the initiation of genetic studies of DNA fragments cloned from Gram negative prokaryotes by non-genetic criteria, e.g., homology to inducible RNA, or homology to specific DNA sequences. Finally, DNA fragments identified and cloned on the basis of transposon insertional inactivation can be further mutagenized using these techniques.

ACKNOWLEDGMENTS

We are pleased to acknowledge G. Ditta for providing plasmids pRK290 and pRK2013 prior to publication, G. Jacoby for providing plasmid pR751, and R. Hyde for preparing this manuscript. This work was supported by USDA/SEA/CRGO grant 7900-481 and NSF grant PCM 78-06834 awarded to F.M.A.

REFERENCES

Berg, D. E., 1977, Insertion and excision of the transposable kanamycin resistance determinant Tn5, in: "DNA Insertion Elements, Plasmids and Episomes," A. I. Bukhari et al., eds., Cold Spring Harbor Laboratory, New York.

Bolivar, F., Rodriguez, R., Green, R., Betlach, M., Heyneker, H., and Boyer, H., 1977, Construction and characterization of new cloning vehicles. II. A multipurpose cloning system, Gene, 2:95.

Cannon, F. C., Riedel, G. E., and Ausubel, F. M., 1977, Recombinant plasmid that carries part of the nitrogen fixation (nif) gene cluster of Klebsiella pneumoniae, Proc. Natl. Acad. Sci. USA, 74:2963.

Cannon, F. C., Riedel, G. E., and Ausubel, F. M., 1979, Overlapping sequences of Klebsiella pneumoniae nif DNA cloned and characterized, Molec. Gen. Genet., 174:59.

Chang, A. C. Y., and Cohen, S. N., 1978, Construction and characterization of amplifiable multicopy DNA cloning vehicles derived from the P15A cryptic mini plasmid, J. Bacteriol., 134:1141.

Clewell, D. B., and Helinski, D. R., 1969, Supercoiled circular DNA-protein complex in Escherichia coli: purification and induced conversion to an open circular DNA form, Proc. Natl. Acad. Sci. USA, 62:1159.

Cohen, S. N., Chang, A. C. Y., and Hsu, C. L., 1972, Non-chromosomal antibiotic resistance in bacteria: Genetic transformation of E. coli by R factor DNA, Proc. Natl. Acad. Sci. USA, 69:2110.

Ditta, G., Stanfield, S., Corbin, D., and Helinski, D. R., 1980, A new broad host range cloning system for gram-negative bacteria and its use in constructing a gene bank of Rhizobium meliloti, manuscript submitted for publication.

Elmerich, C., Houmard, J., Sibold, L., Manheimer, J., and Charpin, N., 1978, Genetic and biochemical analysis of mutants induced by bacteriophage Mu DNA integration into Klebsiella pneumoniae nitrogen fixation genes, Mol. Gen. Genet., 165:181.

Figurski, D. H., and Helinski, D. R., 1979, Replication of an origin-containing derivative of plasmid RK2 dependent on a plasmid function provided in trans, Proc. Natl. Acad. Sci. USA, 76:1648.

Hanahan, D., and Meselson, M., 1980, Plasmid screening at high colony density, Gene, 10:63.

Israel, D. W., Howard, R. L., Evans, H. J., and Russell, S. A., 1974, Purification and characterization of the molybdenum-iron protein component of nitrogenase from soybean nodule bacteroids, J. Biol. Chem., 249:500.

Jobanputra, R. S., and Datta, N., 1974, Trimethoprim resistant R factors in enterobacteria from clinical specimens, J. Med. Microbiol., 7:169.

Jorgenson, R. A., Rothstein, S. J., and Reznikoff, W. S., 1979, A restriction enzyme cleavage map of Tn5 and location of a region encoding neomycin resistance, Mol. Gen. Genet., 177:65.
Klein, R. D., Selsing, E., and Wells, R. D., 1980, A rapid micro-scale technique for isolation of recombinant plasmid DNA suitable for restriction enzyme analysis, Plasmid, 3:88.
MacNeil, D., and Brill, W. J., 1980, Isolation and characterization of lambda specialized transducing bacteriophages carrying Klebsiella pneumoniae nif genes, J. Bacteriol., 141:1264.
MacNeil, T., MacNeil, D., Roberts, G., Supiano, M. A., and Brill, W. J., 1978, Fine structure mapping and complementation analysis of nif (nitrogen fixation) genes in Klebsiella pneumoniae, J. Bacteriol., 136:253.
Maniatis, T., Jeffrey, A., and Kleid, D., 1975, Nucleotide sequence of the rightward operator of phage lambda, Proc. Natl. Acad. Sci. USA, 72:1184.
Marmur, J., 1961, A procedure for the isolation of deoxyribonucleic acid from micro-organisms, J. Mol. Biol., 3:208.
Mazur, B. J., Rice, D., and Haselkorn, R., 1980, Identification of blue-green algal nitrogen fixation genes by using heterologous DNA hybridization probes, Proc. Natl. Acad. Sci. USA, 77:186.
Merrick, M., Filser, M., Kennedy, C., and Dixon, R. A., 1978, Polarity of mutations induced by insertion of transposons Tn5, Tn7 and Tn10 into the nif gene cluster of Klebsiella pneumoniae, Mol. Gen. Genet., 165:103.
Merrick, M., Filser, M., Dixon, R., Elmerich, C., Sibold, L., and Houmard, J., 1980, The use of translocatable genetic elements to construct a fine-structure map of the Klebsiella pneumoniae nitrogen fixation (nif) gene cluster, J. Gen. Microbiol., 117:509.
Nuti, M. P., Lepidi, A. A., Prakash, R. K., Schiliperoort, R. A., and Cannon, F. C., 1979, Evidence for nitrogen fixation (nif) genes on indigenous Rhizobium plasmids, Nature, 282:533.
Pühler, A., Burkhardt, H. J., and Klipp, W., 1979, Cloning of the entire region for nitrogen fixation from Klebsiella pneumoniae in a multicopy plasmid vehicle in Escherichia coli, Mol. Gen. Genet., 176:17.
Riedel, G. E., Ausubel, F. M., and Cannon, F. C., 1979, A physical map of chromosomal nitrogen fixation (nif) genes of Klebsiella pneumoniae, Proc. Natl. Acad. Sci. USA, 76:2866.
Riedel, G. E., 1980, The use of molecular cloning to study the organization and expression of the nitrogen fixation (nif) genes of Klebsiella pneumoniae, Ph.D. Thesis, Harvard University.
Rigby, P. W. J., Dieckmann, M., Rhodes, C., and Berg, P., 1977, Labeling deoxyribonucleic acid to high specific activity in vitro by nick translation with DNA polymerase I, J. Mol. Biol., 113:237.
Ruvkun, G. B., and Ausubel, F. M., 1980a, Interspecies homology of nitrogenase genes, Proc. Natl. Acad. Sci. USA, 77:191.

Ruvkun, G. B., and Ausubel, F. M., 1980b, A general method for site directed mutagenesis in prokaryotes: Construction of mutations in symbiotic nitrogen fixation genes of _Rhizobium meliloti_, manuscript submitted for publication.

Southern, E. M., 1975, Detection of specific sequences among DNA fragments separated by gel electrophoresis, _J. Mol. Biol._, 98:503.

Vincent, J. M., 1970, "A Manual for Practical Study of Root-Nodule Bacteria," Blackwell Scientific Press, Oxford.

Whiting, M. J., and Dilworth, M. J., 1974, Legume root nodule nitrogenase purification, properties, and studies on its genetic control, _Biochim. Biophys. Acta_, 371:337.

CLONING DNA FROM RHIZOBIUM MELILOTI USING A NEW BROAD HOST RANGE, BINARY VEHICLE SYSTEM

G. Ditta, S. Stanfield, D. Corbin, and D. R. Helinski

University of California, San Diego
La Jolla, CA 92093

Molecular cloning techniques now permit the dissection of DNA to almost any desired degree in order to study both structure and function of individual genes. An important strategy for analyzing cloned DNAs involves reintroducing them into their host of origin under various conditions designed to affect gene function. This procedure is greatly facilitated if the primary cloning vector is capable of autonomous replication in multiple hosts. Double vectors, capable of replication in more than one host, have been developed for use in Escherichia coli/yeast (Struhl et al., 1979) and Bacillus subtilis/Staphylococcus aureus (Lofdahl et al., 1978). There are many other organisms of medical, agricultural, and economic importance, however, for which no such dual purpose vectors exits. We have therefore developed a broad host range cloning system for gram-negative bacteria from the naturally-occurring drug resistance plasmid RK2. This system, whose possible development was first suggested some time ago (Meyer et al., 1977), has proved exceedingly valuable to our laboratory in working with the nitrogen-fixing alfalfa symbiont, Rhizobium meliloti.

BINARY VEHICLE SYSTEM

RK2 is a 56 Kilobase pair (Kbp) drug resistance plasmid very closely related to, if not identical with, plasmids bearing the designation RP1, RP4 and R68 (Burkhardt et al., 1979). It confers resistance to the antibiotics tetracycline, kanamycin, and ampicillin. Like other members of the P-1 incompatibility group, RK2 has a very wide conjugative host range. It will transfer to a variety of different gram-negative bacteria and be stably maintained in such hosts. While RK2 has been used directly as a recombinant cloning vector, its large size is a serious drawback.

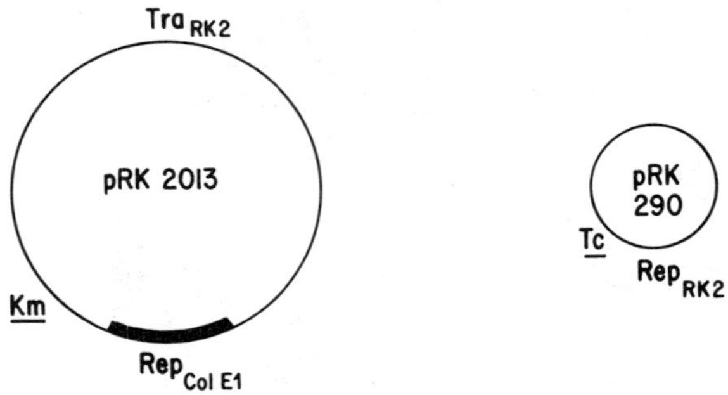

Fig. 1. Binary vehicle system. pRK290 is the cloning vector. pRK2013 is a helper plasmid essential for conjugal transfer of pRK290. Rep and Tra refer to generalized replication and conjugal transfer capability, respectively, and are not intended to designate specific genes. Tc and Km refer to tetracycline resistance and kanamycin resistance, respectively.

In order to reduce the size of RK2 and yet still retain enough of its valuable broad host range characteristics, a two-plasmid, binary vehicle system has been developed (Fig. 1). The actual cloning vector in this system is a 20 Kbp deletion derivative of RK2 called pRK290. This plasmid codes for tetracycline resistance and has two single restriction enzyme sites where DNA can be inserted without loss of plasmid function. Since the regions of DNA necessary for conjugal transfer have been deleted, pRK290 is incapable of self-transfer. On the other hand, it can be mobilized at high frequency by a second plasmid, pRK2013, which is composed of the entire complement of RK2 transfer genes cloned onto the small *E. coli*-specific plasmid Col El. To use this system, DNA from a particular gram-negative bacteria is first cloned into *E. coli* using pRK290, and pRK2013 is subsequently introduced in order to move the cloned DNA back into its original host. In actual practice, the mobilization step is simple and quick (see Transfer Properties). Also, since the cloned DNA is non-transmissible in the absence of mobilizing helper, a high degree of biological containment is achieved.

Although one could clone directly into the original bacterium with pRK290, there are many advantages to using E. coli as an intermediate host. In the first place, procedures for transforming E. coli are well established, while developing a high-frequency transformation system for some bacteria could prove quite difficult. Another advantage is that it is possible to isolate large quantities of plasmid DNA for analysis, again using well-characterized procedures. A most important consideration is that in E. coli, the cloned DNA exists in a heterologous chromosomal background. Hence radiolabeled probes can be used for identification of particular clones via colony hybridization (Grunstein and Hogness, 1975). Finally, transposon insertion, a powerful technique for both genetic and physical analysis, is easily carried out in E. coli.

pRK290

The essential regions of pRK290 are those coding for its broad host range replication properties, its mobilizability, and its resistance to tetracycline. In RK2, these functions are located on widely separated parts of the molecule, and include trfA and trfB, trans-acting replication functions (Thomas and Helinski, 1979), ori RK2, the origin of replication (Meyer and Helinski, 1977), and rlx, a cis-acting site necessary for mobilizability (Guiney and Helinski, 1977). In order to design a strategy for constructing pRK290, it was necessary to know the location of these functions relative to available restriction enzyme sites. Details of the final plan chosen are described elsewhere (Ditta et al., 1980). At one stage, use was made of a previously constructed RK2 deletion derivative, pRK2501 (Kahn et al., 1980). A map of pRK290 is shown in Figure 2.

There are two available cloning sites in pRK290, Bgl II and EcoRI. Since insertion at either of these sites does not inactivate any known plasmid function, it is necessary to pretreat the enzyme-cleaved vector with bacterial alkaline phosphatase (BAP) before ligation. This removes the terminal 5' phosphates and renders the molecule incapable of being covalently recircularized by DNA ligase. Transformable molecules can be generated, however, when exogenous DNA is combined with BAP-treated vector.

Despite its moderately large size, pRK290 specifically lacks restriction enzyme sites for BamHI, HindIII, PstI, KpnI, HpaI and XhoI. This feature is often useful in the physical analysis of large cloned DNAs.

It frequently happens that deletion derivatives of wild-type plasmids show reduced stability in the absence of positive selective pressure. pRK290 was therefore tested in both E. coli and R. meliloti for stability during non-selective log-phase growth

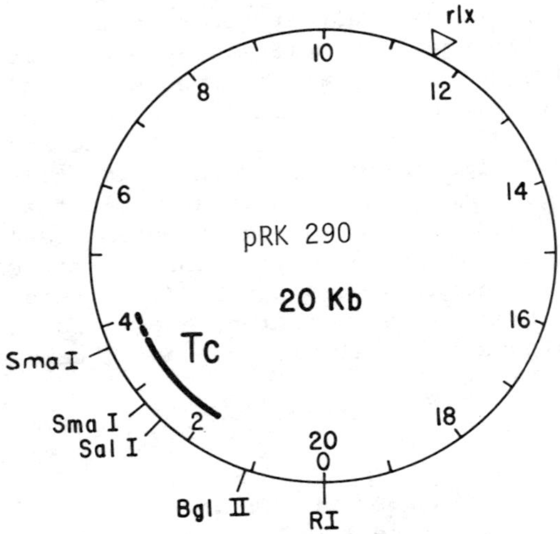

Fig. 2. Physical map of pRK290. Coordinates are in kilobase pairs. rlx is the relaxation complex site (Ditta et al., 1980).

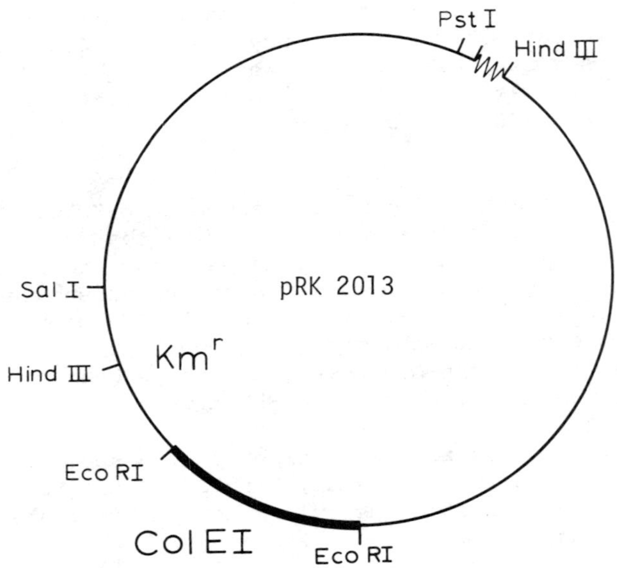

Fig. 3. Physical map of pRK2013. Coordinates are in kilobase pairs. The wavy line represents approximately 1 kilobase of phage Mu DNA (taken from Figurski and Helinski, 1979).

Table 1. Plasmid stability.

Strain	Generations without selection	% Loss per generation
E.c. HB101 (pRK290)	25	0
E.c. HB101 (pRK290 R2)	25	0.3
E.c. HB101 (pRK290 R9)	25	0.1
E.c. HB101 (pRK290 B3)	25	0.1
E.c. HB101 (pRK290 B9)	25	3.0
R.m. 2011 (pRK290)	27	0.2
R.m. 2011 (pRK290 R2)	26	0.4
R.m. 2011 (pRK290 R9)	26	0.3
R.m. 2011 (pRK290 B3)	26	0.3
R.m. 2011 (pRK290 B9)	26	1.6

R2 and R9 are cloned EcoR1 fragments; B3 and B9 are cloned Bgl II fragments.

(Table 1). Plasmid loss was considerably less than 1% per generation in either host. Derivatives carrying cloned DNAs at either the Bgl II or EcoRI sites were also tested. Only a slight decrease in stability was observed. The copy number of pRK290 in *E. coli* is approximately 5. This is well within the reported range of 5 to 8 copies per chromosomal equivalent for RK2 (Figurski et al., 1979).

pRK2013

The kanamycin-resistant helper plasmid pRK2013 is shown in Figure 3. It was constructed some time ago in connection with experiments designed to determine the essential replication regions of RK2 (Figurski and Helinski, 1979). Since pRK2013 carries the entire complement of RK2 transfer genes, it is self-transmissible to hosts that permit replication of Col El. The relatively narrow host range of the Col El region is an important factor in binary system mates with some gram-negative recipients (see below). pRK2013 lacks the RK2 origin of replication but retains both *trfA* and *trfB*, as well as much of the rest of RK2. It is probably for this reason that moderate incompatibility exist between helper and vector when both are present in the same cell. In the absence of selective pressure, pRK2013 is preferentially lost from cells carrying pRK290. This is also a valuable factor in some binary system matings (see below).

TRANSFER PROPERTIES

Transfer data for binary vehicle system matings from *E. coli* to a number of different gram-negative bacteria are shown in Table 2. Matings were performed by mixing equal numbers of donors and recipients together, collecting the cells on nitrocellulose filters, and incubating a minimum of three hours on non-selective agar plates at the optimum temperature for the recipient. HB101 was used as the *E. coli* host for two reasons. First, it is recombination deficient since vector and helper share large regions of homology. Second, as the intermediate host in cloning experiments, HB101 lacks a restriction system which could potentially inactivate cloned DNAs. All matings were done as triparental matings (i.e., *E. coli* (pRK2013) x *E. coli* (pRK290 x Recipient) since the transfer frequency under these conditions is comparable to that observed when both plasmids are together initially in the same cell. This is an important facet of the binary vehicle system, since it greatly facilitates manipulations as well as increasing the overall level of biological containability. In Table 2, it can be seen that a high rate of vector transfer was observed for all the strains surveyed, including some such as *R. meliloti* and *Acinetobacter calcoaceticus*, which are quite distantly related to *E. coli*. The situation with respect to the helper plasmid varies somewhat. For some hosts, such as *Rhizobium* and *Pseudomonas*, the transfer frequency of pRK2013 is considerably less than for pRK290. Most of the kanamycin-resistant *Rhizobium* exconjugants are probably generated via recombinational rescue of pRK2013 and pRK290, since the transfer frequency of pRK2013 alone to *Rhizobium* is several orders of magnitude less than that observed in the binary system matings. For other hosts, transfer of pRK2013 is quite efficient and the majority of conjugants selected on tetracycline contain both helper and vector. However, since there is a certain degree of incompatibility between these two plasmids, kanamycin-sensitive clones lacking pRK2013 are easily segregated by a single restreaking on tetracycline. Thus, in all cases it is easy to achieve vector-specific transfer to the host of interest.

RHIZOBIUM GENE BANK

One of our primary motivations for developing the binary vehicle system was to use it to study the molecular genetics of symbiotic nitrogen fixation in *Rhizobium*. The nodulation of agricultural legumes has been estimated to account for nearly half of the annual global production of fixed nitrogen (Evans and Barber,

Table 2. Plasmid transfer frequencies.

Recipient	Tc^R conjugants / recipient	Km^R/Nm^R conjugants / recipient
Rhizobium meliloti 104B5 nal	$8.3x10^{-2}$	$5.6x10^{-4}$
Serratia marcescens nal	$6.6x10^{-2}$	$2.2x10^{-1}$
Klebsiella pneumoniae M5A1	$1.4x10^{-1}$	$8.8x10^{-1}$
Pseudomonas aeruginosa PAO nal	$2.6x10^{-1}$	$8.4x10^{-7}$
Acinetobacter calcoaceticus rif	$8.3x10^{-4}$	$3.0x10^{-4}$

Tc^R = tetracycline resistance; Km^R = kanamycin resistance; Nm^R = neomycin resistance; nal = nalidixic acid resistance; rif = rifampicin resistance (take from Ditta et al., 1980).

1977), and is a matter of great agricultural importance. Nevertheless, little is known about the molecular details of infection, nodulation, and fixation by rhizobia. In an attempt to understand more about these processes, we have represented the total DNA of a typical fast-growing Rhizobium species, Rhizobium meliloti, as a collection of random DNA fragments cloned onto pRK290. This so-called "gene bank" is a readily available source of cloned Rhizobium DNA. It can be examined in a variety of ways to discover which DNA fragments carry symbiotically important genes. The use of a broad host range cloning vector such as pRK290 greatly extends the types of analyses which can be carried out to find such important genes.

Construction of the gene bank was accomplished as follows. Total Rhizobium DNA was partially digested with the enzyme Bgl II and size-fractionated on sucrose gradients. DNA in the range of 15-30 Kbp was combined with Bgl II-digested, BAP-treated PK290 in a series of ligations. A total of 1285 transformants were obtained. Using a rapid small-scale isolation procedure, approximately 300 of these were analyzed for plasmid DNA content. In Figure 4, plasmid DNA from 15 such transformants is shown following Bgl II digestion and agarose gel electrophoresis. The uppermost band in each lane except lanes 2 and 5 is the vector pRK290. The cloned insert, if present, is displayed as a set of Bgl II fragments. From this analysis, it was concluded that 72% or 929 of the original transformants carried cloned Rhizobium DNA inserts. The

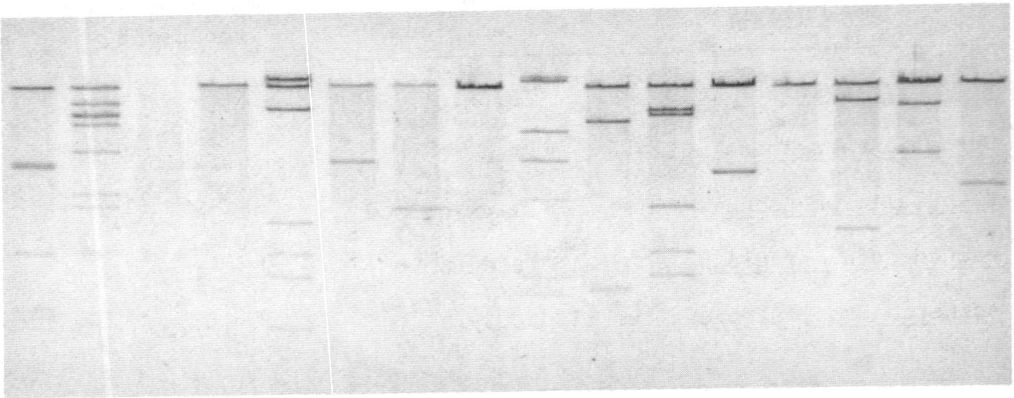

Fig. 4. Bg II digests of cloned R. meliloti DNA. pRK290 is the uppermost band in each lane except lane 5. Lane 9 is a set of DNA standards (taken from Ditta et al., 1980).

average size insert was estimated to be approximately 19 Kbp. An effort was specifically made to clone large inserts for several reasons. Most importantly, it increases the probability that functional gene clusters will be preserved intact. Further, it reduces the number of samples required to provide statistical representation. Using the formula of Clarke and Carbon (Clarke and Carbon, 1976) and assuming the size of the Rhizobium chromosome to be comparable to that of E. coli, there is a 98% probability that a given gene will be represented somewhere within the bank. Finally, cloning large fragments reduces any non-random bias caused by the use of a site-specific restriction endonuclease.

NIF GENES

One of the most obvious symbiotically important set of genes in Rhizobium is that coding for the structural components of the enzyme nitrogenase. In those organisms which have been studied, the enzyme complex directly responsible for reducing atmospheric nitrogen has been found to consist of two components. Component I, referred to as the molyldenum-iron protein, is a tetramer composed of two pairs of non-identical subunits whose molecular weights are in the range of 50,000 to 60,000 daltons. Component II, a dimer, contains two identical subunits of approximately 35,000 daltons. The minimum amount of DNA necessary to code for three such proteins is on the order of 3-4 Kbp. The structural organization and mode of regulation of these genes in Rhizobium is unknown. Ruvkun et al. (Ruvkun and Ausubel, 1976) have shown that cloned

DNA representing the nitrogenase structural genes (*nif* genes) of the free-living nitrogen-fixing bacteria *Klebsiella pneumoniae* will hybridize to total cellular DNA from *R. meliloti*, as well as to eighteen other nitrogen-fixing organisms, including *Clostridium pasteurianum*, a gram-positive bacterium, *Anabaena*, a blue-green algae, and *Frankia*, an *Actinomycete*. It will not, however, hybridize to the DNA of organisms which do not fix nitrogen. The actual region of homology was estimated to be less than 1.6 Kbp in the case of *R. meliloti*, and to correspond to no more than two of the three *Klebsiella* nitrogenase genes (i.e., gene(s) D and/or H). We have used colony hybridization (Grunstein and Hogness, 1975) to identify those clones in the gene bank which carry homology to the same *nif* DNA probe. Three such clones have been identified, the largest of which carries a 26 Kbp insert. All three contain a 3.6 Kbp Bgl II fragment containing the entire region of homology. When total *R. meliloti* DNA is examined by Southern blot analysis (Southern, 1975), the same 3.6 Kbp Bgl II fragment is the only fragment which is detected. We therefore conclude that at least a portion of the *R. meliloti nif* gene set has been identified as a component of the gene bank. Further studies are in progress to substantiate this point and to address various questions concerning organization and control of this important region of DNA.

SUMMARY

A two-plasmid, binary vehicle system has been developed for cloning DNA from Gram-negative bacteria. After initial isolation in *E. coli*, cloned DNA can be reintroduced at high frequency into the host of origin using a single triparental bacterial mating procedure. The vector component of this system has been used to clone the entire genome of *Rhizobium meliloti* as a collection of inserts averaging 19 Kbp in size. A portion of the nitrogenase structural gene set has been identified on three of these inserts.

REFERENCES

Burkardt, H. J., Reiss, G., and Puhler, A., 1979, Relationship of group P-1 plasmids revealed by heteroduplex experiments: RP1, RP4, R68 and RK2 are identical, *J. Gen. Microbiol.*, 114:341.

Clarke, L., and Carbon, J., 1976, A colony bank containing synthetic Col El hybrid plasmids representative of the entire *E. coli* genome, *Cell*, 9:91.

Ditta, G., Stanfield, S., Corbin. D., and Helinski, D. R., 1980, A new broad host range cloning system for Gram-negative bacteria and its use in constructing a gene bank of *Rhizobium meliloti*,

Evans, H. J., and Barber, L. E., 1976, Biological nitrogen fixation for food and fiber production, *Science*, 197:332.

Figurski, D., and Helinski, D. R., 1979, Replication of an origin-containing derivative of plasmid RK2 dependent on a plasmid function provided in trans, Proc. Natl. Acad. Sci. USA, 74: 1658.

Figurski, D. H., Meyer, R. J., and Helinski, D. R., 1979, Suppression of Col El replication properties by the Inc P-1 plasmid RK2 in hybrid plasmids constructed in vitro, J. Mol. Biol., 133:295.

Grunstein, M., and Hogness, D. S., 1975, Colony hybridization: A method for the isolation of cloned DNA that contain a specific gene, Proc. Natl. Acad. Sci. USA 72:3961.

Guiney, D. G., and Helinski, D. R., 1979, The DNA-protein relaxation complex of the plasmid RK2: location of the site-specific nick in the region of the proposed origin of transfer, Mol. Gen. Genet., 1976:183.

Kahn, M., Kolter, R., Thomas, C., Figurski, D., Meyer, R., Remaut, E., and Helinski, D. R., 1980, Plasmid cloning vehicles derived from plasmids Col El, F, R6K and RK2, in: "Methods in Enzymology, Vol. 68, Recombinant DNA," R. Wu, ed., Academic Press, New York.

Lofdahl, S., Sjostrom, J.-E., and Philipson, L., 1978, A vector for recombinant DNA in Staphylococcus aureus, Gene, 3:161.

Meyer, R. J., Figurski, D., and Helinski, D. R., 1977, Properties of the plasmid RK2 as a cloning vehicle, in: "DNA Insertion Elements, Plasmids, and Episomes," A. I. Bukhari, J. A. Shapiro, and S. L. Adhya, eds., Cold Springs Harbor Laboratory.

Meyer, R. J., and Helinski, D. R., 1977, Unidirectional replication of the P-group plasmid RK2, Biochim. Biophys. Acta, 478:109.

Ruvkun, G. B., and Ausubel, F. M., 1980, Interspecies homology of nitrogenase genes, Proc. Natl. Acad. Sci. USA, 77:191.

Southern, E. M., 1975, Detection of specific sequences among DNA fragments separated by gel electrophoresis, J. Mol. Biol., 98:503.

Struhl, K., Stinchcomb, D. T., Scherer, S., and Davis, R. W., 1979, High-frequency transformation of yeast: Autonomous replication of hybrid DNA molecules, Proc. Natl. Acad. Sci. USA, 76:1035.

Thomas, C. M., and Helinski, D. R., 1979, Regions of the broad host range plasmid RK2 that are required for replication and maintenance, J. Bact., 141:213.

MOLECULAR CLONING OF *RHIZOBIUM JAPONICUM* DNA IN *E. COLI* AND IDENTIFICATION OF NITROGEN FIXATION (*NIF*) GENES

Hauke Hennecke[1,2] and Jonathan R. Mielenz[2,3]

[1]Institute for Microbiology, University of Regensburg, D-8400 Regensburg, W. Germany, and [2]Plant Growth Laboratory, University of California, Davis, CA 95616

INTRODUCTION

In the last few years considerable interest has been devoted to studies of the genetics of root nodule bacteria, *Rhizobium* spp., which are the nitrogen fixing symbionts of legumes. Scientists in this field hope to eventually genetically engineer new *Rhizobium* strains with improved properties in symbiosis.

Recently, circular linkage maps have been established for three strains of *R. meliloti* (Meade and Signer, 1977; Kondorosi et al., 1977; Casadesus and Olivares, 1979) and for *R. leguminosarum* strain 300 (Beringer et al., 1978), using plasmids of the Pl incompatibility group for the mobilization of the chromosome. However, no genetic analysis of the rhizobial nitrogen fixation (*nif*) genes has been completed so far. Screening for *nif*⁻ mutants and scoring for the Nif phenotypes requires time consuming plant nodulation assays. *R. japonicum*, the soybean symbiont, offers the advantage that one can study the expression of the *nif* genes under free-living, i. e., asymbiotic, conditions (Keister, 1975; Kurz and LaRue, 1975; Pagan et al., 1975). In order to approach the genetic organization and regulation of the *nif* genes of *R. japonicum* in more detail, we have attempted to isolate *nif* genes from this organism utilizing recombinant DNA technology.

---

[3]Present address: Biotechnology Department, Moffett Technical Center, CPC International, Argo, Illinois 60501

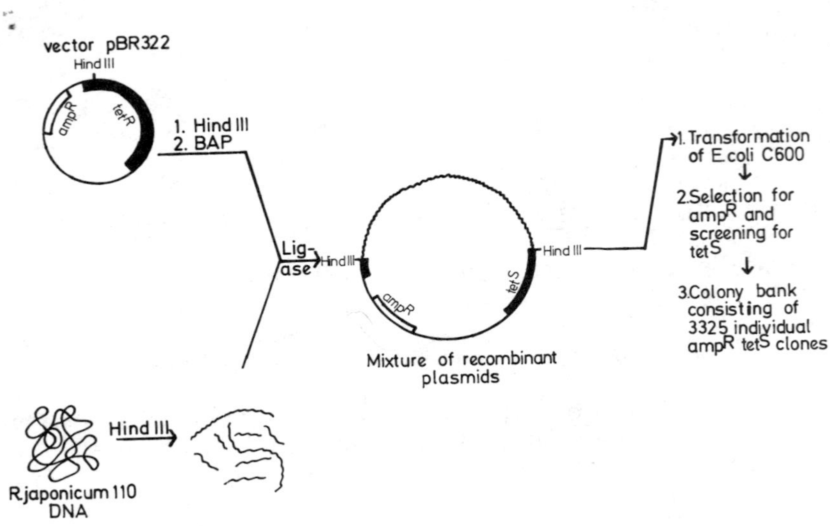

Fig. 1. Cloning scheme.

CONSTRUCTION OF A GENE BANK OF *R. JAPONICUM* 110 DNA IN *E. COLI*

Figure 1 shows an outline of the cloning procedure. Total DNA from *R. japonicum* strain 110 was isolated and digested with the restriction enzyme *Hind*III (Mielenz et al., 1979). The plasmid cloning vehicle was pBR322 (Bolivar et al., 1977) which has a single *Hind*III site in the gene region coding for tetracycline resistance. Before annealing and ligating these two types of DNA, the vector was treated with the enzyme bacterial alkaline phosphatase (BAP) (Ulrich et al., 1977). BAP removes the terminal 5'-phosphate from DNA. Since T4 DNA ligase requires both a 5'phosphate and a 3'hydroxyl group for activity, BAP-treated DNA cannot be ligated together. Therefore, treatment of the cloning plasmid pBR322 with BAP prevented self-ligation and allowed the plasmid to be ligated almost exclusively to the non-BAP treated *R. japonicum* DNA. Following ligation, this mixture of recombinant plasmids was introduced into *E. coli* strain C600R^-M^+ by transformation (Mandel and Higa, 1970). Since the *Hind*III site on pBR322 is in the promoter of the genes for tetracycline resistance, insertion of foreign DNA at that site should eliminate or reduce tetracyline resistance. Analysis of the transformed *E. coli* cells for the two drug resistance phenotypes carried on pBR322, ampicillin (amp^R > 100 μg/ml) and tetracycline (tet^R, 75 μg/ml), showed that > 99%

Table 1. The effect of treatment of pBR322 with bacterial alkaline phosphatase before ligation.

Type and source of DNA ligated and transformed into E. coli C600	Transformants per microgram pBR322 selected for	
	Amp^R	Tet^R
R. japonicum 110 DNA (HindIII) plus pBR322 (HindIII, BAP)	9,600	44
pBR322 (HindIII, BAP)	220	180
pBR322 (closed circular)	54,000	22,000
No DNA	0	0

of the cells were amp^R and tetracycline sensitive (tet^S) (Table 1). Therefore, the BAP treatment allowed the creation of a population of E. coli containing recombinant plasmids in a single step without any involved enrichment procedure. Putative plasmid-containing E. coli cells were plated on a rich medium (LB) containing ampicillin. A total of 3325 amp^R colonies were transferred to microtiter plates. Each well contained 0.2 ml LB ampicillin plus 10% dimethylsulfoxide, a microbial antifreeze. This colony bank is maintained frozen at -20°C.

In order to evaluate the size distribution of the cloned R. japonicum DNA inserted into pBR322, the recombinant plasmids from the total pool of selected clones were isolated and analyzed on agarose gels after digestion with HindIII (Fig. 2). The cloning plasmid is the predominant band in the middle of the pattern while the fragments of cloned DNA range in size from about 15 x 10^6 to less than 1 x 10^6 daltons. A similar size distribution was found when the plasmids from 12 randomly chosen colonies were analyzed (not shown). Assuming an average size of 3 x 10^6 daltons for the inserts and 3 x 10^9 daltons for the R. japonicum genome, the colony bank theoretically could represent three times the total genome.

The question of expression of Rhizobium genes in E. coli has been addressed by attempts to complement a variety of auxotrophs of E. coli with the recombinant plasmids. For example, all 3325 clones have been checked for complementation of the leucine requirement of the E. coli $C600R^-M^+$ host cell. No clones were found to be Leu^+. Furthermore four additional genes in E. coli have been

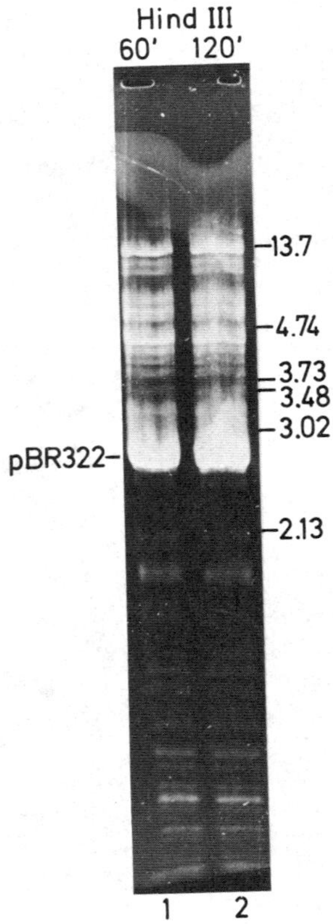

Fig. 2. Analysis of cloned Rhizobium japonicum DNA on 0.7% agarose gels. Total plasmid DNA was extracted from transformed E. coli cells containing the HindIII "gene bank" of R. japonicum strain 110, and digested for 1 and 2 hrs with HindIII (lanes 1 and 2, respectively). The numbers on the right give the molecular weight markers (in Md.) of DNA fragments obtained from EcoRl digested phage λ DNA.

checked for complementation by the gene bank DNA: His^-, Arg^-, Thy^-, Pro^- mutants were not complemented by cloned rhizobial DNA, suggesting an apparent biological block in the expression of R. japonicum DNA by E. coli.

SEARCH FOR CLONED R. JAPONICUM NIF GENES

The primary structure of the enzyme nitrogenase is highly conserved over a wide variety of nitrogen fixing microorganisms. For example, the constituent components of the nitrogenase enzyme complex of Klebsiella pneumoniae can form an active hybrid enzyme by mixing them with the constituent components of the R. japonicum nitrogenase complex (Emerich and Burris, 1978). Furthermore, immunological cross-reaction was found between the nitrogenase of K. pneumoniae and R. japonicum (Maier and Brill, 1976; Scott et al., 1979). More recently, Ruvkun and Ausubel demonstrated that part of the nitrogenase structural genes from K. pneumoniae hybridizes to total R. japonicum DNA (Ruvkun and Ausubel, 1980). Therefore, the presumptive sequence homology between nif genes from K. pneumoniae and R. japonicum should make it possible to detect R. japonicum nif genes by screening the colony bank for the presence of any cross-hybridizing material with the help of the colony hybridization technique (Grunstein and Hogness, 1975).

The molecular probe used for the hybridization experiments was the EcoRl fragment from the recombinant plasmid pSA30 (Cannon et al., 1979) which carries almost exclusively the structural genes, nifKDH, of the nitrogenase complex of Klebsiella pneumoniae strain M5Al (see Fig. 3). After enzymatic digestion of pSA30 with EcoRl, the nif DNA fragment was purified by two consecutive runs on a preparative agarose gel. The DNA fragment was "nick-translated" (Maniatis et al., 1975) with DNA polymerase I (Boehringer Mannheim) in the presence of thymidine 5-(α-32P)triphosphate (400 Ci/mmol) to yield a specific radioactivity of 0.5 to 1 x 10^8 cpm/µg DNA. The 3325 clones of the colony bank were grown in groups of 47-48 on nitrocellulose filters layered on top of LB agar containing ampicillin (100 µg/ml). When the colony size reached about 2-3 mm, the filters were layered on LB agar containing ampicillin and chloramphenicol (100 µg/ml) for 24 hrs, and then further treated as described (Chinault and Carbon, 1979). This plasmid amplification step dramatically increased the hybridization signal, as shown in Figure 4. The colony hybridization experiment was done in duplicate employing about 5 to 7 x 10^6 cpm per filter. Upon examination of the autoradiograms, 21 colonies gave a positive response. The recombinant plasmids were isolated from all of these colonies using standard techniques (Clewell and Helinski, 1969), digested with Hind III and analyzed on agarose gels. The result of this analysis is shown in Table 2 and Figure 5. All plasmids have a common 5.8 megadalton (Md) insert. Using the Southern hybridization technique (Southern, 1975), we confirmed that this DNA fragment carries the nif homology region. Plasmids from 15 colonies have only this single insert, while 5 others contain an additional 2.0 Md fragment. One plasmid was found to have 3 fragments. It is unknown as yet whether the additional pieces are natural "neighbors" of the 5.8 Md DNA sequence or just random cointegrates produced during the cloning experiment.

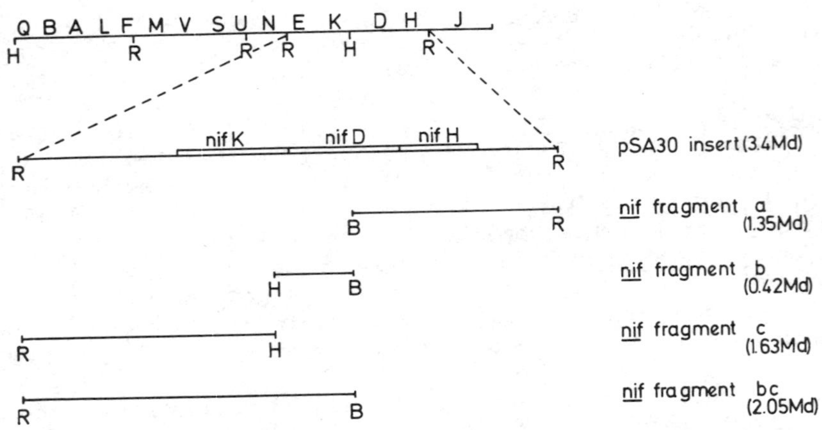

Fig. 3. Klebsiella pneumoniae nif DNA fragments used for hybridization experiments. The top part of the figure depicts the nif gene region of K. pneumoniae strain M5A1. The EcoRI fragment containing almost exclusively the structural genes of the nitrogenase complex, nifHDK, had been cloned by Cannon et al. (1979) into the vector plasmid pACYC184 (Chang and Cohen, 1978) to yield the recombinant plasmid pSA30. Subfragments of this region were obtained after preparative agarose gel electrophoresis of double digests with EcoRI/BamHI (R, B), HindIII/BamHI (B, H), and HindIII/EcoRI (H, R). All subfragments were made radioactive by "nick translation" (Maniatis et al., 1975).

Table 2. Types of recombinant plasmids isolated.

Colony number	Size(s) of insert(s)	Fragment hybridizing to (^{32}P) nifKDH DNA
6G12 19B4 62F10 11A2 20B8 63C4 12F9 22B11 67A5 15C2 26B12 68G11 19H2 59H4 69F3	5.8 Md	5.8
13A5 34E10 47B3 50D12 64D7	5.8, 2.0 Md	5.8
67D6	5.8, 1.6, 1.35 Md	5.8

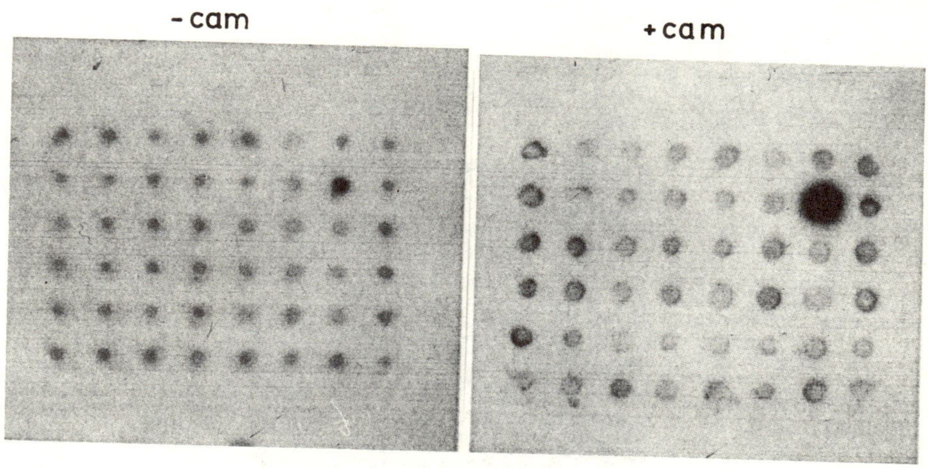

Fig. 4. Section from the colony hybridization experiments showing an example of 48 out of the 3325 colonies screened. One colony on this plate gave a positive hybridization signal. The signal could be dramatically enhanced when the nitrocellulose filters with the grown colonies were transferred to LB agar plates containing chloramphenicol (cam) in order to allow amplification of the plasmids within the cells.

Next we focussed our attention on the following two questions: 1) How large is the actual region of homology within the cloned 5.8 Md fragment, and where is this region located on a physical restriction map of the plasmid?; 2) Which of the structural genes of the K. pneumoniae nitrogenase complex, nifH, nifD, or nifK, carries the homology region?

First a physical map has been established for one representative plasmid (isolated from colony number 15C2), using the restriction enzymes HindIII, EcoRI, SalI and BamHI. When p15C2 was digested with each of these enzymes in separate assays, the following numbers of bands were detected on agarose gels: HindIII (2) EcoRI (4), SalI (3), and BamHI (3). In addition, double digestions were performed in every possible combination. This data, in combination with the available information on restriction sites within pBR322 (Bolivar et al., 1977) allowed a rough map of p15C2 to be established (Fig. 8). Part of the experimental detail is shown in Figure 6 (left part), with the digestion patterns of p15C2 with (from left to right) BamHI, BamHI plus HindIII, HindIII, EcoRI plus HindIII, and EcoRI. When this pattern was transferred to nitrocellulose filters via the Southern blotting technique (Southern, 1975) and hybridized with radioactive nifKDH DNA from K. pneumoniae (see Fig. 3), the restriction fragments carrying the homology region

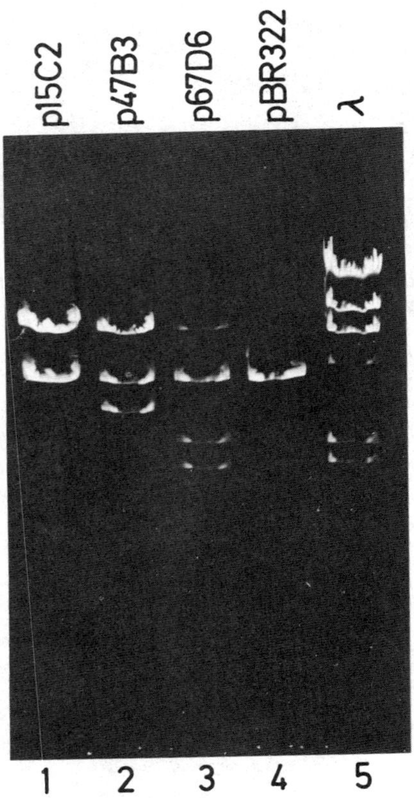

Fig. 5. Agarose gel electrophoresis of representative recombinant plasmids after digestion with HindIII. Plasmids are shown from colony No. 15C2 (1), 47B3 (2) and 67D6 (3). Lane 4: vector plasmid pBR322 as control; lane 5: λ DNA digested with HindIII.

could be identified (Fig. 6, right part). Hence the maximal regeion of homology was located on the fragments designated as 1 and 2 in Figure 8. In order to determine which of the *K. pneumoniae nif* genes carries the homology region, the following experiment was carried out. After enzymatic digestion of p15C2 with EcoRI, the fragments of DNA were separated on an agarose gel. This DNA-pattern was transferred to nitrocellulose filters and hybridized to three different subfragments of the *nifDH* region (Fig. 3). All these subfragments were isolated by preparative agarose gel electrophoresis and then labeled with (^{32}P) by "nick-translation." Subfragment a (see Fig. 3) contains *nifH* and part of *nifD*, subfragment b contains *nifD* and part of *nifK*, and subfragment c contains the rest of *nifK*, part of *nifE* and *nifY*

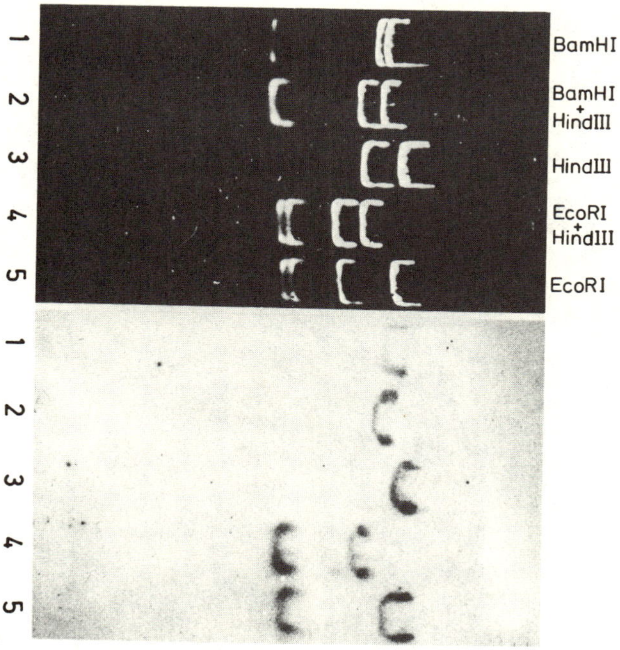

Fig. 6. Gel analysis of the restriction digest of the recombinant plasmid p15C2 and "Southern" hybridization with (^{32}P) nifKDH DNA from pSA30. Figure 6A shows the ethidium bromide stained bands produced after digestion with (1) BamHI, (2) BamHI plus HindIII, (3) HindIII, (4) HindIII plus EcoRI, and (5) EcoRI. This pattern was transferred to nitrocellulose filters and hybridized to the "nick-translated" nifKDH fragment. Figure 6B shows the autoradiogram of the filter after hybridization.

(A. Pühler, personal communication). As shown in Figure 7, no hybridization was observed with subfragment c, while weak hybridization to fragment 1 of p15C2 was obtained with subfragment a. By far the strongest hybridization was exerted by subfragment b, specifically to the EcoRI fragment no. 2 of p15C2.

As summarized in Figure 8, the results of the experiments performed in Figures 6 and 7 imply that 1) the region of homology is less than 1 MD in size, and 2) that only the nifD gene of *K. pneumoniae* is homologous to one of the *R. japonicum nif* genes, although a weak homology to nifH cannot be excluded. These findings are in good agreement with the results of Ruvkun and Ausubel

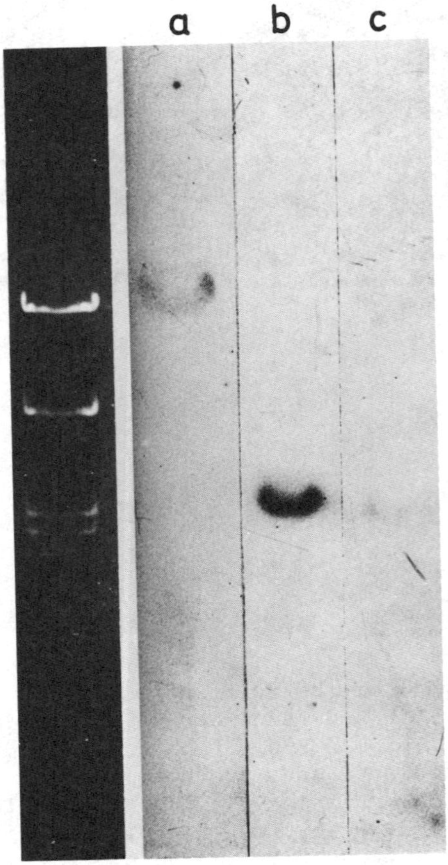

Fig. 7. Southern hybridization of the "nick-translated" nif subfragments a, b, and c (see Fig. 3) to the EcoRI fragments of plasmid p15C2. The first lane shows the ethidium bromide stained bands of EcoRI digested p15C2. Three such patterns were blotted individually to three nitrocellulose filters and then one of each was hybridized to either subfragment a, b, or c as shown in autoradiograms a, b, and c, respectively.

(1980). They determined the homology region between *K. pneumoniae* and *R. meliloti* strain M2011 to be less than one megadalton in size.

The availability of purified nitrogen fixation genes from *Rhizobium japonicum* now makes it possible to study their organization and mode of expression.

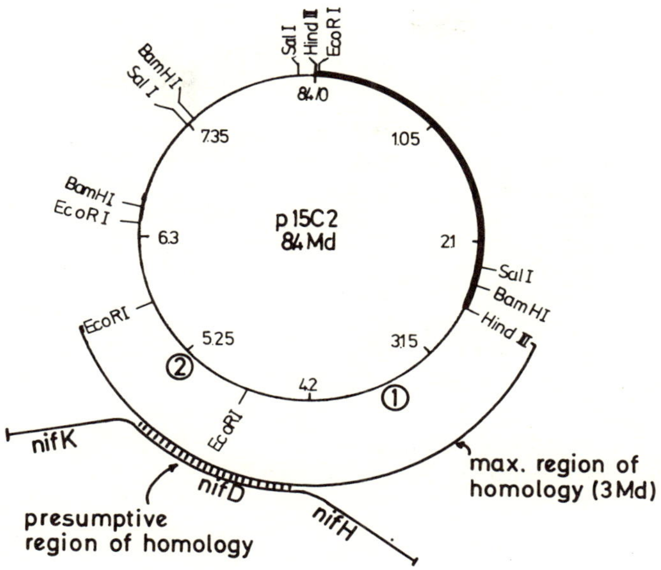

Fig. 8. Restriction map of one of the recombinant plasmids, p15C2, and the approximate location of the homology region with K. pneumoniae nif DNA.

ACKNOWLEDGMENTS

We would like to thank R. Valentine and A. Böck for their support. The help of C. Hershberger, Eli Lilly & Co., in the initial cloning experiments is greatly appreciated. We also thank F. Ausubel and F. Cannon for providing bacterial strains and helpful suggestions, and G. Tondl and P. Frischeisen for expert technical assistance. This work was supported by grants from the National Science Foundation (NSF PFR 77-07301) and the Continental Grain Company to R. Valentine and from the Deutsche Forschungsgemeinschaft to H. Hennecke.

REFERENCES

Beringer, J. E., Hoggen, S. A., and Johnston, A. W. B., 1978, Linkage mapping in Rhizobium leguminosarum by means of R plasmid-mediated recombination, J. Gen. Microbiol., 104:201.

Bolivar, F., Rodriguez, R. L., Greene, P. J., Betlach, M. C., Heyneker, H. L., Boyer, H. W., Crosa, J. H., and Falkow, S., 1977, Construction and characterization of new cloning vehicles. II. A multipurpose cloning system, Gene 2:95.

Casadesus, J. and Olivares, J., 1979, Rough and fine linkage mapping of the *Rhizobium meliloti* chromosome, Mol. Gen. Genet., 174: 203.

Cannon, F. C., Riedel, G. E., and Ausubel, F. M., 1979, Recombinant plasmid that carries part of the nitrogen fixation (*nif*) gene cluster of *Klebsiella pneumoniae*, Mol. Gen. Genet., 174:59.

Chang, A. C. Y., and Cohen, S. N., 1978, Construction and characterization of amplifiable multicopy DNA cloning vehicles derived from the P15A cryptic miniplasmid, J. Bacteriol., 134: 1141.

Chinault, A. C., and Carbon, J., 1979, Overlap hybridization screening: Isolation and characterization of overlapping DNA fragments surrounding the leu22 gene on yeast chromosome III, Gene, 5:111.

Clewell, D. B., and Helinski, D. R., 1969, Supercoiled circular DNA protein complex in *E. coli*: purification and induced conversion to an open circular DNA form, Proc. Natl. Acad. Sci. USA, 62:1159.

Emerich, D, W., and Burris, R. H., 1978, Complementary functioning of the component proteins of nitrogenase from several bacteria, J. Bacteriol., 134:936.

Grunstein, M., and Hogness, D. S., 1975, Colony hybridization: A method for the isolation of cloned DNAs that contain a specific gene, Proc. Natl. Acad. Sci. USA, 72:3961.

Keister, D. L., 1975, Acetylene reduction by pure cultures of Rhizobia, J. Bacteriol., 123:1265.

Kondorosi, A., Kiss, G. B., Forrai, T., Vincze, E., and Banfalvi, Z., 1977, Circular linkage map of *Rhizobium meliloti* chromosome, Nature, 265:525.

Kurz, W. G. W., and LaRue, T. A., 1975, Nitrogenase activity in Rhizobia in absence of host plant, Nature, 256:407.

Maier, R. J., and Brill, W. J., 1976, Ineffective and non-nodulating mutant strains of *Rhizobium japonicum*, J. Bacteriol., 134:936.

Mandel, M., and Higa, A., 1970, Calcium-dependent bacteriophage DNA infection, J. Mol. Biol., 53:159.

Maniatis, T., Jeffrey, A., and Kleid, D. G., 1975, Nucleotide sequence of the rightward operator of phage λ, Proc. Natl. Acad. Sci. USA, 72:1184.

Meade, H. M., and Signer, E. R., 1977, Genetic mapping of *Rhizobium meliloti*, Proc. Natl. Acad. Sci. USA, 74:2076.

Mielenz, J. R., Jackson, L. E., O'Gara, F., and Shanmugam, K. T., 1979, Fingerprinting bacterial chromosomal DNA with restriction endonuclease *Eco*RI: Comparison of *Rhizobium* spp. and identification of mutants, Can. J. Microbiol., 25:803.

Pagan, J. D., Child, J. J., Scowcroft, W. R., and Gibson, A. H., 1975, Nitrogen fixation by Rhizobium cultured on a defined medium, Nature, 256:406.

Ruvkun, G. B., and Ausubel, F. M., 1980, Interspecies homology of nitrogenase genes, Proc. Natl. Acad. Sci. USA, 77:191.

Scott, D. B., Hennecke, H., and Lim, S. T., 1979, The biosynthesis of nitrogenase MoFe protein polypeptides in free-living cultures of Rhizobium japonicum, Biochim. Biophys. Acta, 565:365.

Southern, E. M., 1975, Detection of specific sequences among DNA fragments separated by gel electrophoresis, J. Mol. Biol., 98:503.

Ullrich, A., Shine, J., Chirgwin, J., Pictet, R., Tischer, E., Rutter, W. J., and Goodman, H. M., 1977, Rat insulin genes: Construction of plasmids containing the coding sequences, Science, 196:1313.

THE IDENTIFICATION, LOCATION AND MANIPULATION OF GENES IN *RHIZOBIUM*

John E. Beringer

Soil Microbiology Department
Rothamsted Experimental Station
Harpenden, Hertfordshire AL5 2JQ, UK

INTRODUCTION

The importance of nitrogen fixation by rhizobia in legume root nodules has been recognized for a long time and has led to the establishment of inoculation industries in many countries. However, despite the extensive use of *Rhizobium* strains in agriculture since the beginning of this century, the strains in use are 'wild-type' isolates and no genetically-manipulated strains are presently available. This is partly due to the fact that *Rhizobium* genetic studies have only flourished over the last few years. A much more important impediment has been a lack of understanding of the specific attributes of *Rhizobium* strains which are required. These involve properties such as competitiveness, efficiency of fixation and an ability to become established in particular soil conditions. Attempts to understand the genetic basis of these properties, so that improved strains can be constructed by genetic means, have only just started. The aim of this brief review is to summarize our present knowledge of *Rhizobium* genetics and indicate the methods that are available to produce improved strains in the future.

THE IDENTIFICATION OF GENES

Genetic studies depend upon the isolation of strains which carry mutations. This enables one to determine how many genes are involved in the normal functioning of, for example, a metabolic pathway and what the function of each gene is. The isolation of mutants requires suitable mutagenic procedures and an efficient method for screening survivors of the metagenic treatment for the class of mutant which is required. Rhizobia have been reported to be susceptible to mutagenesis by procedures involving treatments with radiation, chemicals and the introduction of transposons (see Beringer et al., 1980a). Interestingly there is considerable variation in the efficiency of mutagenesis between and within *Rhizobium* species (Cunningham, 1980). N-methyl-N'-nitro-N-nitrosoguanidine has been used for fast- and slow-growing rhizobia and is probably the only mutagen which has been effective with all the strains of *Rhizobium* tested. Why some mutagens are so variable in their ability to induce mutation in different strains and species of *Rhizobium* is unknown. For the slow-growing rhizobia most workers have found it difficult to induce mutations. This may be due to the intrinsic metabolic differences of these bacteria or possibly to the difficulty in separating bacteria so that a colony is formed by a single bacterium, rather than a clump. This difficulty has yet to be resolved so that adequate numbers of mutants can be isolated for genetic studies.

Recently transposon mutagenesis has been used for *R. leguminosarum*, *R. trifolii*, *R. phaseoli*, and *R. meliloti*. This procedure involves the introduction of a transposon, which confers drug resistance on the recipient, and selection for those bacteria which have stably inherited the drug resistance (Beringer et al., 1978a). Because transposons are unique sequences of DNA that cannot replicate themselves, stable inheritance of the drug resistance is a result of the insertion of the transposon into the DNA of the host (Kleckner et al., 1977). This inactivates the gene into which it is inserted. If that gene is part of an operon this blocks the function of the genes in the operon distal to the insertion. An advantage of transposon mutagenesis is that only one mutagenic lesion is formed, so that a clone derived from an initial transposition event consists of a collection of bacteria which have undergone the same mutagenic event. This is quite unlike other treatments where the expectation is that more than one mutagenic lesion has occurred and therefore the analysis of the properties of the mutant must take this into account.

A far more important feature of transposon-induced mutation is that the site of mutation is defined by the presence of the drug resistance. Therefore the mutation can be readily mapped and the defective gene can be transferred to other bacteria simply by selecting for the drug resistance. The transposon also provides a relatively simple way for identifying the DNA which carries the mutant gene. This can be done in two ways, by looking for hybridization with labeled transposon DNA and by looking at restriction enzyme digests to determine which fragment is altered in size.

For the fast-growing rhizobia, at least, the range of mutagenic techniques available is adequate. Providing that a number of different mutagens are tested there would appear to be no problem in inducing mutations. A more serious problem with

Rhizobium species is in the screening of survivors of mutagen treatments for specific classes of mutants. The isolation of auxotrophic and drug-resistant mutants is no more difficult than for other bacteria. However the particularly interesting mutants that one wishes to isolate for Rhizobium strains are those which are defective in symbiotic properties. The problem here is that each colony must be tested independently on a suitable legume host plant for the establishment of normal nitrogen-fixing nodules. These tests are time consuming, expensive and can only be used to handle a limited number of colonies (more than 250 plants tested per day per person would be unusual). This obviously limits the number of strains that can be tested and also the ease with which genetic mapping and strain construction can be attempted.

Despite this limitation a number of symbiotically-defective mutants of Rhizobium have been isolated after chemical (Maier and Brill, 1976; Beringer et al., 1977) and transposon mutagenesis (Buchanan-Wollaston et al., 1980). Some of these mutants have been mapped and shown to be either plasmid-borne or chromosomal.

THE LOCATION OF GENES

The mapping of genes is important because it facilitates their manipulation during strain construction experiments. The first essential is to know whether a mutation maps to the chromosome or to extrachromosomal DNA, such as plasmids. If the gene is chromosomal its location on the chromosome can be determined by mapping techniques utilizing DNA transfer between bacteria. For the fast-growing species of Rhizobium we know that there is a single circular chromosome and large numbers of genetically marked strains of R. leguminosarum (Beringer et al., 1978b) and R. meliloti (Meade and Signer, 1977; Kondorosi et al., 1977; Julliot and Boistard, 1979; Casadesus and Olivares, 1979a) are available for mapping purposes. The order of genes on the chromosomes of R. phaseoli and R. trifolii is probably identical to that of R. leguminosarum (Johnston and Beringer, 1977) and shows striking similarities to that of R. meliloti strains (Kondorosi et al., 1980). The similarity of the chromosomes of R. trifolii and R. phaseoli with R. leguminosarum is so great that mapping of genes in these 'species' can be achieved by crossing the bacteria with suitable marked strains of R. leguminosarum.

Plasmid-borne genes can be mapped by transduction if the plasmid is sufficiently large that only part is packaged in a transducing phage. This has been used to map genes on a large plasmid in R. leguminosarum (Buchanan-Wollaston et al., 1980). Another, usually easier and more accurate method is to isolate pure plasmid DNA, digest it with appropriate restriction enzymes and analyse the fragmentation pattern. If the mutation is caused by a DNA insertion (such as occurs during transposon mutation) the fragment carrying the inactivated gene will have an altered molecular weight in the mutants as compared to the control. Likewise if mutation is due to deletion or invertion, alterations in restriction patterns will be evident. Obviously such simple physical methods are not suitable for mapping mutations arising from base-pair substitutions, etc., where there is no significant alteration in the molecular weight of the plasmid. This accentuates the value of a mutagenic technique, such as transposon mutation, which alters the DNA in a way which enables physical studies to be done.

The initial problem of deciding whether a mutation is chromosomal or plasmid-borne can be resolved by two main methods. The most general one is to attempt to map the mutation to the chromosome. The absence of linkage to genes from any part of the chromosome is useful genetic evidence for an extrachromosomal location for the mutation. Further genetic evidence can be obtained by introducing plasmids from other strains and looking for suppression of the mutant phenotype. Both these procedures have been used to show that genes are plasmid-borne in R. leguminosarum (Hirsch, 1978; Brewin et al., 1980; Buchanan-Wollaston et al., 1980).

THE MANIPULATION OF GENES

The three main methods of gene transfer in bacteria (transformation, transduction and conjugation) have been reported for many Rhizobium species (see reviews by Dénarié and Truchet, 1976; Schwinghamer, 1977; Beringer, 1980; Beringer et al., 1980a,b). Unfortunately, like all aspects of Rhizobium genetics, very little success has been achieved with slow-growing species of Rhizobium and, as yet, we are probably only able to perform extensive genetic manipulations with strains of R. leguminosarum, R. meliloti, R. phaseoli and R. trifolii. Transformation has been reported for some strains of these species but has not yet been important for mapping or strain construction experiments. However the great value of transformation for introducing purified plasmid DNA into strains of bacteria suggests that it will play an important role in genetic studies in the future.

Gene transfer mediated by plasmids has been reported extensively for Rhizobium species. R plasmids belonging to the P-1 incompatibility group have very wide host ranges among Gram-negative bacteria and have probably been transferred to and between strains of all the species of Rhizobium. They appear to be stably inherited and can therefore be utilized to introduce 'foreign' DNA into Rhizobium from other species or genera of bacteria. R primes can be constructed using restriction enzyme methods with purified DNA (Jacob et al., 1976; Julliot and Boistard, 1979). When transformation is not available, as with strain 300 of *R. leguminosarum*, the constructed plasmids can be re-introduced into the *Rhizobium* strain by transforming them into *Escherichia coli* and then crossing them into *R. leguminosarum* by conjugation. This was the procedure used by Jacob et al. (1976) for *R. leguminosarum* and by Julliot and Boistard (1979) for *R. meliloti*. These plasmids can then be transferred between strains and species of *Rhizobium* by conjugation. The efficiency with which this occurs will vary with the size of the plasmid and whether or not the particular strains used carry modification and restriction systems which destroy the foreign DNA (see Johnston et al., 1978a).

R primes carrying *Rhizoibum* DNA can be constructed *in vivo* by crossing different *Rhizobium* strains. This can either be done by crossing different species which show low levels of chromosomal homology, for example, *R. meliloti* and *R. leguminosarum* (Johnston et al., 1978b) or by selecting for high frequency transfer of a specific marker in crosses within a species (Kiss et al., 1980). R primes selected in this manner have been very useful for 'mapping' new mutations by looking for suppression of the defective phenotype after the introduction of R primes carrying different alleles (Johnston et al., 1978a; Kondorosi et al., 1980). Because they are transmissible to other bacterial genera R primes can be used to look at the expression of *Rhizobium* genes in different genera and to determine which genes in a metabolic pathway map at that point on the chromosome. For example by transferring different Trp^+ R primes to *Pseudomonas aeruginosa* it was a relatively simple matter to demonstrate that the tryptophan genes of *R. leguminosarum*, which were located at three different regions, composed of *trp F,A*, and *B*, *trp C* and *D* and *trp E* though the order of F, A and B could not be determined (Johnston et al., 1978a).

A further use for R primes is to promote increased frequencies of chromosome transfer in crosses between strains. The chromosomal DNA in the R primes is homologous with the same region in the chromosome and therefore when an R prime donor is crossed there is enhanced possibility that chromosomal DNA from that region will also be transferred. This has been demonstrated in *R. meliloti* by Julliot and Boistard (1979) using R primes constructed *in vitro* and by Kondorosi et al. (1980) using R primes constructed *in vivo*. Jacob et al. (1976) did not observe enhanced chromosome mobilization with two R primes carrying *R. leguminosarum* DNA, and not all of the R primes constructed by Julliot and Boistard (1979) had this property. This absence of enhanced sex factor ability could have been due to a number of factors including the possibility that *Rhizobium* plasmid DNA was carried by some of the R primes.

It appears that all species of *Agrobacterium* and probably *Rhizobium* contain large plasmids with molecular weights greater than 10^8 (see Beringer et al., 1980b). The ability to produce bacteriocins (Hirsch, 1979; Brewin et al., 1980), host range determinants (Higashi, 1967; Johnston et al., 1978c; Brewin et al., 1980), genes required for plasmid transfer (Buchanan-Wollaston et al., 1980) and at least one of the nitrogenase structural genes (Nuti et al., 1979; Ruvkun and Ausubel, 1980) have been reported to be plasmid borne in fast-growing species of Rhizobium. All of these properties have been reported to be transferred between Rhizobium strains (see Beringer et al., 1980b), usually through conjugation mediated by the plasmid itself. Thus it appears that some strains of Rhizobium (and perhaps all) have at least part of the genes required for the symbiosis with legumes on plasmids and are able to exchange these genes by conjugation. One of the R. leguminosarum plasmids involved in determining host range is transmissible at a frequency of about 10^{-2} to R. trifolii or R. phaseoli (Johnston et al., 1978c), which implies that a considerable exchange of genes between rhizobia is occurring in nature. However, strains which have received a plasmid carrying host-range genes for another species initially appear to be less efficient in nodulating the appropriate host than the wild-type donor (Johnston et al., 1978c).

Transduction has, and will, play an important role in the construction of new Rhizobium strains and in the mapping of genes. As previously stated it can be used to transfer fragments of plasmid DNA and to establish linkage between plasmid-borne genes. It has the advantage that it can be used to transfer fairly small fragments of DNA which facilitates strain construction because one can be reasonably sure that only a single reasonably short fragment of DNA is integrated, avoiding the introduction of many unknown and probably unwanted genes. Transducing phages are available for strains of R. meliloti (Kowalski, 1971; Casadesus and Olivares, 1979b) and R. leguminosarum and R. trifolii (Buchanan-Wollaston, 1979). The relative ease with which these phages have been isolated suggests that, for the fast-growing species, it should not be difficult to obtain transduction in a given strain, especially if virulent phage are used with procedures designed to reduce the killing which would normally occur, as was done by Casadesus and Olivares (1979b) and Buchanan-Wollaston (1979).

No discussion of genetic manipulation would be complete without reference to the cloning of genes. This has been alluded to in the discussion of R primes, but not discussed in terms of the isolation and analysis of symbiotically-important genes. The fragmentation and cloning of DNA using restriction enzymes is now a routine procedure and a number of clone banks of Rhizobium DNA have been made. The isolation and characterization of those fragments which carry symbiotically-important genes largely depends upon the use of hybridization probes to find the clone carrying the gene whose DNA is homologous with that of the gene being studied. For the nitrogenase structural gene(s) this involved the use of purified Klebsiella nif DNA as a probe of the Rhizobium

DNA (Nuti et al., 1979; Ruvkun and Ausubel, 1980). For other genes the most suitable method presently available is to use homology with a transposon. The idea is to induce mutations in Rhizobium using transposon mutagenesis and then select out those clones which are symbiotically defective. By definition the DNA flanking the transposon should be that of the symbiotically-important gene and hence isolation of a fragment of DNA carrying the transposon will isolate the gene, or part of it. The gene can be found by hybridization of the purified fragment containing the transposon with DNA from the clone bank for that or other closely related species. This procedure is already being used in a number of laboratories to isolate genes, to study their function in different strains and species and for sequencing studies.

SUMMARY

For fast-growing species of Rhizobium we have the necessary techniques for inducing mutations and mapping them. The ease with which this can be done is largely governed by the type of mutant that is required. For most classes of symbiotically defective mutants this will always be limited by the necessity to carry-out plant tests to determine the defective phenotype. A major limiting problem is that we do not know in detail what is involved in the establishment of a normal nitrogen fixing nodule. For breeding purposes we would like strains which are competitive and hence will form the majority of nodules and we also require these strains to be efficient at nitrogen fixation. The number of genes involved and their functions is quite unknown.

Despite this reservation we are well equipped to carry out fairly extensive programmes of genetic manipulation of Rhizobium strains, initially using a pragmatic approach of crossing two good strains and selecting for a better one. As the description of the important genes improves it will be possible to carry-out more refined manipulations, probably using DNA from Rhizobium clone banks. While we are at present unable to contemplate this range of genetic manipulations with the slow-growing rhizobia, it is likely that most of the immediate problems will be resolved over the next few years. The concepts and methods for the genetic manipulation of their genes will have been resolved with the fast-growing Rhizobium species and thus progress should be rapid.

REFERENCES

Beringer, J. E., 1980, The development of Rhizobium genetics, J. Gen. Microbiol., 116:1.

Beringer, J. E., Johnston, A. W. B., and Wells, B., 1974, The isolation of conditional ineffective mutants of Rhizobium leguminosarum, J. Gen. Microbiol., 98:339.

Beringer, J. E., Beynon, J. L., Buchanan-Wollaston, A. V., and Johnston, A. W. B., 1978a, Transfer of the drug resistance transposon Tn5 to Rhizobium, Nature, Lond., 276:633.

Beringer, J. E., Hoggan, S. A., and Johnston, A. W. B., 1978b, Linkage mapping in Rhizobium leguminosarum by means of R plasmid-mediated recombination, J. Gen. Microbiol., 104:201.

Beringer, J. E., Brewin, N. J., and Johnston, A. W. B., 1980a, Genetics, in: "Ecology of Nitrogen Fixation: Rhizobium," Vol. II, W. J. Broughton, ed., Oxford University Press, Oxford.

Beringer, J. E., Brewin, N. J., and Johnston, A. W. B., 1980b, The genetic analysis of Rhizobium in relation to symbiotic nitrogen fixation, Heredity, in press.

Brewin, N. J., Beringer, J. E., Buchanan-Wollaston, A. V., Johnston, A. W. B., and Hirsch, P. R., 1980, Transfer of symbiotic genes with bacteriocinogenic plasmids in Rhizobium leguminosarum, J. Gen. Microbiol., 116:261.

Buchanan-Wollaston, A. V., 1979, Generalized transduction in Rhizobium leguminosarum, J. Gen. Microbiol., 112:135.

Buchanan-Wollaston, A. V., Beringer, J. E., Brewin, N. J., Hirsch, P. R., and Johnston, A. W. B., 1980, Isolation of symbiotically defective mutants in Rhizobium leguminosarum by insertion of the transposon Tn5 into a transmissible plasmid, Molec. Gen. Genet., 178:185.

Casadesus, J., and Olivares, J., 1979a, Rough and fine linkage mapping of the Rhizobium meliloti chromosome, Molec. Gen. Genet., 174:203.

Casadesus, J., and Olivares, J., 1979b, General transduction in Rhizobium meliloti by a thermosensitive mutant of bacteriophage DF2, J. Bacteriol., 139:316.

Cunningham, D. A., 1980, Genetic studies on the nitrogen fixing bacterium Rhizobium trifolii, Ph.D. Thesis, University of Edinburgh, Edinburgh.

Dénarié, J., and Truchet, G., 1976, Genetics of Rhizobium: a short survey, in: "Proceedings of the First International Symposium on Nitrogen Fixation," Vol. 2, W. E. Newton and C. J. Nyman, eds., Washington State University Press, Pullman.

Higashi, S., 1967, Transfer of clover infectivity of Rhizobium trifolii to Rhizobium phaseoli as mediated by an episomic factor, J. Gen. Appl. Microbiol., 13:391.

Hirsch, P. R., 1978, Studies of plasmids in Rhizobium leguminosarum, Ph.D. Thesis, University of East Anglia, Norwich.

Hirsch, P. R., 1979, Plasmid-determined bacteriocin production by Rhizobium leguminosarum, J. Gen. Microbiol., 113:219.

Jacob, A. E., Cresswell, J. M., Hedges, R. W., Coetzee, J. N., and Beringer, J. E., 1976, Properties of plasmids constructed by the in vitro insertion of DNA from Rhizobium leguminosarum or Proteus mirabilis into RP4, Molec. Gen. Genet., 147:315.

Johnston, A. W. B., and Beringer, J. E., 1977, Chromosomal recombination between Rhizobium species, Nature, Lond., 267:611.

Johnston, A. W. B., Bibb, M. J., and Beringer, J. E., 1978a, Tryptophan genes in Rhizobium - their organization and their transfer to other bacterial genera, Molec. Gen. Genet., 165:323.

Johnston, A. W. B., Setchell, S. M., and Beringer, J. E., 1978b, Interspecific crosses between Rhizobium leguminosarum and R. meliloti: formation of haploid recombinants and of R-primes, J. Gen. Microbiol., 104:209.

Johnston, A. W. B., Beynon, J. L., Buchanan-Wollaston, A. V., Setchell, S. M., Hirsch, P. R., and Beringer, J. E., 1978c, High frequency transfer of nodulating ability between strains and species of Rhizobium, Nature, Lond., 276:634.

Julliot, J. S., and Boistard, P., 1979, Use of RP4-prime plasmids constructed in vivo to promote a polarized transfer of the chromosome in Escherichia coli and Rhizobium meliloti, Molec. Gen. Genet., 173:289.

Kiss, G. B., Dobo, K., Dusha, I., Breznovits, A., Orosz, L., Vincze, E., and Kondorosi, A., 1980, Isolation and characterization of an R-prime plasmid from Rhizobium meliloti, J. Bacteriol., 141:121.

Kleckner, N., Roth, J., and Botstein, D., 1977, Genetic engineering in vivo using translocatable drug resistance elements; New methods in bacterial genetics, J. Molec. Biol., 116:125.

Kondorosi, A., Kiss, G. B., Forrai, T., Vineze, E., and Banfalvi, Z., 1977, Circular linkage map of Rhizobium meliloti chromosome, Nature, Lond., 268:525.

Kondorosi, A., Vincze, E., Johnston, A. W. B., and Beringer, J. E., 1980, A comparison of three Rhizobium linkage maps, Molec. Gen. Genet., 178:403.

Kowalski, M., 1971, Transduction in Rhizobium meliloti, Pl. Soil Special Vol., 63.

Maier, R. J., and Brill, W. J., 1976, Ineffective and non-nodulating mutant strains of Rhizobium japonicum, J. Bacteriol., 127:763.

Meade, H. M., and Signer, E. R., 1977, Genetic mapping of Rhizobium meliloti, Proc. Natl. Acad. Sci. USA, 74:2076.

Nuti, M. P., Lepidi, A. A., Prakesh, R. K., Schilperoort, R. A., and Cannon, F. C., 1979, Evidence for nitrogen fixation (nif) genes on indigenous Rhizobium plasmids, Nature, Lond., 282:533.

Ruvkun, G. B., and Ausubel, F. M., 1980, Interspecies homology of nitrogenase genes, Proc. Natl. Acad. Sci. USA, 77:191.

Schwinghamer, E. A., 1977, Genetic aspects of nodulation and dinitrogen fixation by legumes: the microsymbiont, in: "A Treatise on Dinitrogen Fixation," Section 3, Biology, R. W. F. Hardy and W. S. Silver, eds., John Wiley and Sons, New York.

THE ROLE OF *RHIZOBIUM* PLASMIDS IN HOST SPECIFICITY

N. J. Brewin, J. L. Beynon, and A. W. B. Johnston

John Innes Institute
Colney Lane
Norwich, NR4 7UH, UK

HOST RANGE AND *RHIZOBIUM* TAXONOMY

There is no such thing as a non-nodulating field isolate of *Rhizobium*. It is a contradiction in terms because *Rhizobium* is defined by its ability to induce root nodules on a legume host. Indeed, the species to which a particular *Rhizobium* field isolate is assigned depends exclusively on the genus of the host legume from which it was originally isolated. Thus *R. japonicum* would have been isolated from a soybean root nodule and *R. meliloti* from an alfalfa root nodule. This form of classification works well provided that there is very little overlap between the host-ranges of the various *Rhizobium* species (Wilson, 1944). However it would appear to be unsatisfactory in three major areas. Firstly, the large and heterogeneous 'cowpea miscellany' is a collection of *Rhizobium* isolates with broad and diverse host ranges. Secondly, it is doubtful whether nodulation *per se* (rather than the formation of nitrogen-fixing nodules) can be regarded as an adequate criterion for species assignment, because it is known for example that *R. leguminosarum* strain 300, which forms effective nitrogen-fixing nodules on peas, will also form nodules on *Phaseolus* and *Trifolium* (Hepper and Lee, 1979; J. E. Beringer, personal communication; J. L. Beynon, unpublished results) although on these last two hosts the induced nodules fail to fix nitrogen. Thirdly, and perhaps more fundamentally, we shall argue in the ensuing presentation that strains of *R. leguminosarum*, *R. trifolii* and *R. phaseoli* are almost indistinguishable by all taxonomic criteria (including chromosome homology) except host-range itself, and that host-range characteristics represent a plasmid-borne trait that can be transferred from strain to strain across the 'species barrier'.

CHROMOSOME EXCHANGE BETWEEN RHIZOBIUM SPECIES

For the fast growing *Rhizobium* species *R. meliloti* and *R. leguminosarum* a single circular genetic linkage group has been established using Pl group R plasmids to mobilise chromosomal genes (Meade and Signer, 1977; Kondorosi et al., 1977; Beringer et al., 1978). These maps have recently been compared by Kondorosi et al. (1980). It is clear from Figure 1 that the gene order of the two linkage maps is basically similar. Furthermore, the apparent colinearity of the linkage data has been confirmed by the use of eight R-primes carrying small fragments of the *R. meliloti* chromosome to study the pattern of suppression of mutant phenotypes in *R. leguminosarum* and *R. meliloti* (Fig. 1).

Although the order of chromosomal alleles is similar, the extent of DNA homology between *R. meliloti* and *R. leguminosarum* is apparently insufficient to allow normal chromosome recombination (Johnston et al., 1978b) in conjugational crosses which used the Pl group plasmid R68.45. By contrast, interspecific crosses between *R. phaseoli* or *R. trifolii* and *R. leguminosarum* resulted in the transfer of genetic markers at frequencies that were no different from those obtained between different field isolates of *R. leguminosarum* itself, and the introduced chromosomal alleles were integrated into the recipient chromosome to yield haploid recombinants (Johnston and Beringer, 1977). Similarly, recombination between two different strains of *R. meliloti* (Rm2011 and Rm41) occurred at high frequency, yielding haploid recombinants (Kondorosi et al., 1980). Thus there appear to be at least two distinct groups of fast growing rhizobia, one represented by *R. leguminosarum*, *R. trifolii* and *R. phaseoli* and the other by *R. meliloti*; this is consistent with the taxonomic relationships between these species (Graham, 1964).

In an attempt to map the genes concerned with host range specificity, an extensive series of genetic crosses was undertaken involving *R. phaseoli*, *R. trifolii* and *R. leguminosarum*. These yielded a set of haploid recombinants in which the chromosome of *R. leguminosarum* had been replaced, section by section by the corresponding part of the genome from either *R. phaseoli* or *R. trifolii* (J. L. Beynon and A. W. B. Johnston, unpublished results). In no case did any of these interspecific recombinants lose the ability to nodulate peas, let alone acquire the ability to nodulate one of the other legume hosts. These genetic data provide strong evidence that the chromosomes of *R. leguminosarum* *R. trifolii* and *R. phaseoli* are essentially identical and that the genes determining host range are to be found elsewhere, presumably on a plasmid.

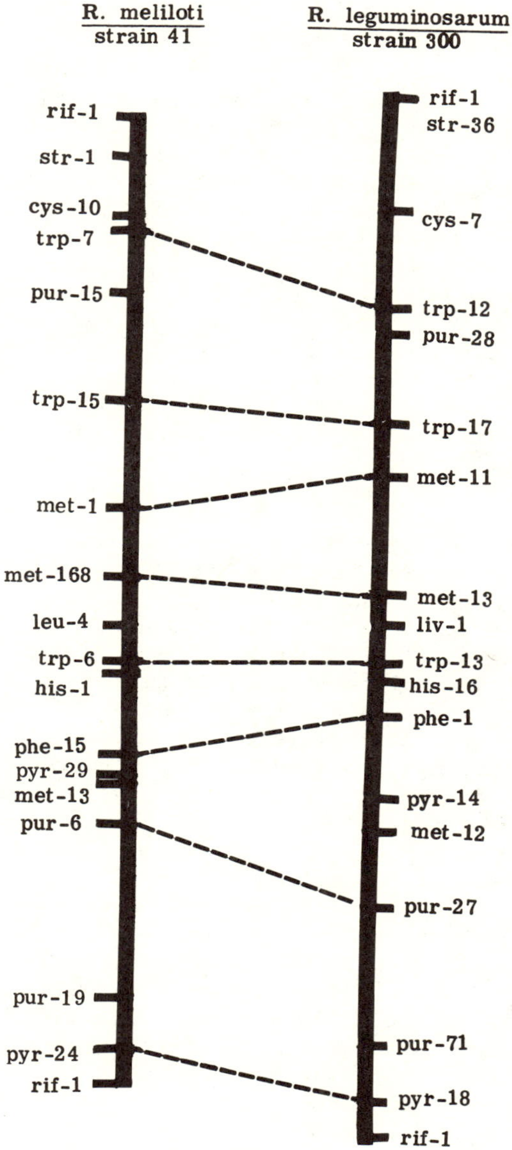

Fig. 1. Comparison of the linkage maps of R. meliloti strain 41 and R. leguminosarum strain 300 (adapted from Kondorosi et al., 1980) showing only those alleles with apparently similar map locations. Alleles connected by solid lines are suppressed by the same R-prime carrying fragments of the R. meliloti chromosome.

THE PLASMIDS OF R. LEGUMINOSARUM STRAIN 300

Genetic evidence that Rhizobium plasmids might determine gene functions concerned with symbiotic nitrogen fixation (Higashi, 1967; Johnston et al., 1978a; Zurkowski and Lorkiewicz, 1978) has recently been supported by the observation that some of the genes concerned with nitrogenase synthesis (nif) are plasmid borne (Nuti et al., 1979; Ruvkun and Ausubel, 1980). In an attempt to correlate plasmids and symbiotic phenotypes of Rhizobium strains, plasmids recovered from lysates of Rhizobium have been visualized following electrophoresis on agarose gels (Casse et al., 1979; Gross et al., 1979; Hirsch et al., 1980. These plasmids are very large, having molecular weights in the range 60-250 million, or perhaps larger. In the case of R. leguminosarum every strain was found to contain more than one plasmid. Among 14 different field isolates, each recovered in the same region of England, each had a unique pattern of plasmids (G. Hombrecher, personal communication.

We have investigated the plasmids of R. leguminosarum strain 300 because this strain has been genetically well characterised and various plasmids have been transferred into and out of it. Although three plasmid bands were originally detected in this strain (Prakash et al., 1979), it now seems likely that there are six distinct plasmids (Hirsch et al., 1980). The two largest plasmids, which have molecular weights in excess of 220 million, are often not recovered in plasmid preparations because their size makes them very susceptible to mechanical and nucleolytic breakage during plasmid isolation. (For this reason we have no idea whether other even larger plasmids might exist which have never been isolated intact.)

The third largest plasmid of strain 300 has a molecular weight of c. 205 million. This plasmid carries determinants for nodule formation (Nod$^+$) and nodule function (Fix$^+$) (Buchanan-Wollaston et al., 1980) and nitrogenase (P. R. Hirsch, personal communication). Two derivatives of this plasmid exist (Fig. 2). There is a partial deletion mutant (molecular weight c. 180 million) which lacks Nod$^+$, Fix$^+$ and nitrogenase functions: this is present in strains 6015 and 16015. There is also a derivative of this plasmid containing the transposon Tn5 as an insertion (A. V. Buchanan-Wollaston and G. Hombrecher, unpublished results). The plasmid is not self-transmissible (Brewin et al., 1980a). However Nod$^+$, Fix$^+$ and the introduced Kan-r determinant can be mobilised following genetic recombination with introduced transmissible plasmids such as pRL3JI and pRL4JI (Brewin et al., 1980; N. J. Brewin, unpublished observations).

The next plasmid in the series from strain 300 has a molecular weight of c. 165 million. Nothing further is known about

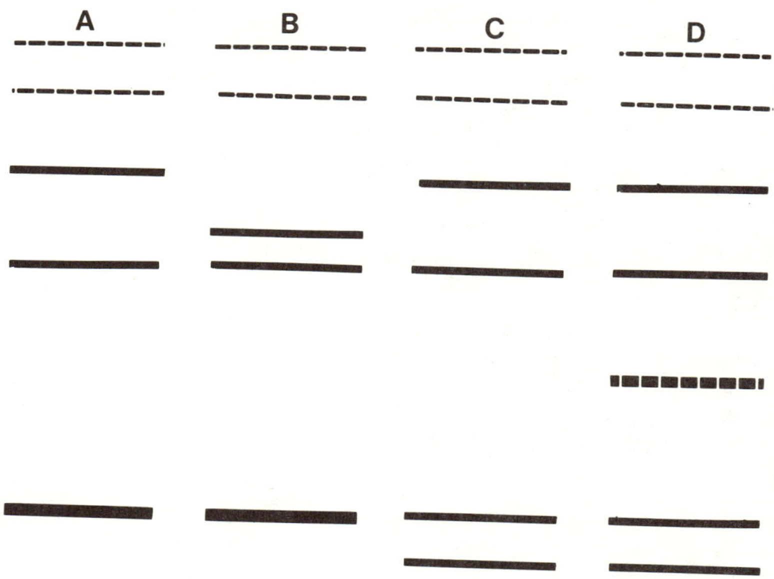

Fig. 2. Plasmids of R. leguminosarum strain 300 and its derivatives following electrophoresis in agarose gels (0.7%) containing Tris-borate buffer, represented as line drawings of these plasmids. The direction of electrophoresis was from top to bottom. A) Strain 300 field isolate. B) Strain 6015 (Nod^- Fix^- plasmid deletion mutant). C) Strain 1062 (carries small deletion in one of the small comigrating plasmids of strain 300). D) Strain 1062 containing pRL1JI.

this plasmid. Finally, the fastest migrating plasmid band of strain 300 probably comprises two co-migrating plasmids (molecular weight c. 100 million). In one derivative of strain 300 (strain 1062) one of these plasmids may have suffered a small deletion so that the single bright band is resolved into a doublet (Fig. 2 and Hirsch et al., 1980). Nothing is known of the functions determined by these two small plasmids except that one of them apparently belongs to the same incompatibility group as the nodulation plasmid of a strain of R. phaseoli (see below).

TRANSMISSIBLE PLASMIDS FROM VARIOUS R. LEGUMINOSARUM STRAINS

Several transmissible plasmids derived from other field isolates of *R. leguminosarum* have been introduced into derivatives of *R. leguminosarum* strain 300 (Table 1). Of these, the best studied is the bacteriocinogenic plasmid pRL1JI and its derivative, pJB5JI, which contains the transposon Tn5 (Johnston et al., 1978a; Hirsch, 1979; Brewin et al., 1980a; Buchanan-Wollaston et al., 1980; Hirsch et al., 1980; Beynon et al., 1980). This plasmid, which has the same mobility on gels as the Ti plasmid of *Agrobacterium tumefaciens* strain C58 (molecular weight c. 130 million), is transmissible between strains of *R. leguminosarum* at high frequency (10^{-2} per recipient). When transferred to the Nod$^-$ Fix$^-$ mutant (strain 6015 - see above) of *R. leguminosarum*, it restored the wild-type phenotype (Johnston et al., 1978a; Buchanan-Wollaston et al., 1980). It was also able to suppress the phenotypes of several other symbiotically-defective strains (Brewin et al., 1980a). Perhaps the most surprising property of this plasmid is that, following transfer to strains of *R. trifolii* and *R. phaseoli*, it conferred on these strains the ability to nodulate peas in addition to their normal hosts, clover and *Phaseolus* beans, respectively (Johnston et al., 1978a; Beynon et al., 1980). Interspecific transfer of pRL1JI from *R. leguminosarum* to *R. trifolii* and *R. phaseoli* occurred at the same frequency as when *R. leguminosarum* was used as recipient. However, establishment of plasmid-linked markers in *R. meliloti* was very much reduced (c. 10^{-4} per recipient) and none of these transconjugant clones had acquired the ability to nodulate peas (A. W. B. Johnston, unpublished observations). Preliminary results suggest that pRL1JI also carries genes for nitrogenase (*nif*) in addition to the genes concerned with nodule formation and nodule fixation (Fix$^+$) which have already been identified (P. R. Hirsch, personal communication).

The plasmid pRL1JI was first identified (Hirsch, 1979) because it codes for the production of a non-dialysable *medium* sized bacteriocin (Med$^+$). It was also noted that strains such as 300 which received pRL1JI lost the ability to produce a dialysable *small* sized bacteriocin. This phenomenon could signify that the introduction of pRL1JI had resulted in the loss of a resident plasmid (coding for *small* bacteriocin) from the recipient strain. However, two lines of evidence suggest that this interpretation is incorrect. Firstly, when pRL1JI was introduced into strain 300 a single new band was acquired and none of the resident plasmid bands was displaced (Hirsch et al., 1980), suggesting that pRL1JI is compatible with all the resident plasmids of strain 300. Secondly, when a derivative of strain 300 carrying pRL1JI was cured of this plasmid the production of *small* bacteriocin was restored (P. R. Hirsch, personal communication), arguing strongly that the presence of pRL1JI had repressed the production of *small* bacteriocin.

The fact that the introduction of pRL1JI did not result in the elimination of any plasmids from strain 300 implies that 300 strains containing pRL1JI may be diploid for any gene functions that are common to the new plasmid and the resident plasmids. This is apparently the case because a series of transposon insertions into transmissible plasmids from strain 300 containing pRL1JI did not result in any symbiotically defective phenotypes, although when transferred into the plasmid deletion mutant 6015, 14/165 of the Tn5-containing transmissible plasmids induced a phenotype that was symbiotically defective (Buchanan-Wollaston, et al., 1980).

Two other transmissible plasmids appear to have very similar bacteriocinogenic properties to pRL1JI. These are pRL3JI which has a molecular weight of 125-160 x 10^6 (different transconjugants appear to contain plasmids of different sizes presumably as a result of genetic recombination in the new host) and pRL4JI (molecular weight 160 x 10^6) (Hirsch et al., 1980). However, unlike pRL1JI, these two plasmids lack the Nod^+ and Fix^+ genes necessary to restore several symbiotically defective R. leguminosarum mutants to the wild-type phenotype (Brewin et al., 1980).

All three bacteriocinogenic plasmids, pRL1JI, pRL3JI and pRL4JI appear to be related and they cannot coexist within the same cell without undergoing recombination. Genetic evidence is consistent with the possibility that these plasmids can recombine (Brewin et al., 1980a).

Another nodulation plasmid (termed pRL5JI) has been isolated from R. leguminosarum strain TOM (Brewin et al., 1980b). This strain (unlike strain 300) is capable of nodulating the primitive pea line cv. Afghanistan (Winarno and Lie, 1979) and other pea lines from the Afghanistan area. When the nodulation plasmid pRL5JI was transferred to the Nod^- Fix^- mutant strain 16015 (a derivative of strain 300), the transconjugants were Nod^+ Fix^+ and now capable of nodulating cv. Afghanistan. A similar result was obtained when a strain of R. phaseoli was used as recipient (N. J. Brewin, unpublished results). The transferred plasmid has a molecular weight of c. 160 million. It is apparently compatible with all the resident plasmids of strain 300 and also with any of the bacteriocinogenic plasmids pRL1JI, pRL3JI and pRL4JI. Unlike these plasmids, pRL5JI does not code for a medium bacteriocin and its transfer frequency is rather low (c. 10^{-6} per recipient).

The large majority of the R. leguminosarum strains examined have been unable to transfer nodulation ability to a Nod^- plasmid deletion mutant (N. J. Brewin, unpublished results). R. leguminosarum strain 300 was an example of such a strain (Brewin et al., 1980a), and in this case it is known that the nodulation plasmid is not transmissible (because a derivative exists which contains a Tn-5 insertion that provides a strong selection for plasmid transfer). However, the Nod^+ determinants of this non-transmissible plasmid can be mobilised at low frequency following the introduction into strain 300 of a transmissible plasmid, pRL3JI or pRL4JI, which is itself Nod^- (Brewin et al., 1980). Presumbaly these transmissible plasmids either recombine with or cointegrate with the nodulation plasmid of strain 300, since in subsequent crosses nodulation ability can often be transferred at high frequencies (Brewin et al., 1980a).

We have also attempted to transfer the nodulation plasmid from R. leguminosarum strain 128C53 (Bethlenfalvay and Phillips, 1979) in the hope that the determinants for hydrogenase uptake (Hup^+) and nodulation (Nod^+) might be carried on the same plasmid. (This strain contains only two plasmids.) Unfortunately the nodulation plasmid did not appear to be transmissible to the Nod^- mutant 16015. We therefore attempted to mobilise Nod^+ by the introduction of Kan-r derivatives of pRL3JI and pRL4JI (termed pVW3JI and pVW5JI, respectively). Following the introduction of either of these two plasmids into strain 128C53 the smaller of the two resident plasmids (MW c. 190 million on agarose gels) always disappeared (N. J. Brewin, unpublished results). Although no new band could be seen, it was probable that the resident plasmid (termed pRL6JI) had recombined to form a cointegrate with the introduced plasmid which would then have been too large to recover in normal plasmid preparations. There was, however, genetic evidence for the existence of such cointegrates. Kanamycin resistance from these derivatives could be transferred to strain 300 at low frequencies (10^{-7} per recipient) and then at high frequencies (10^{-3} per recipient) in subsequent crosses.

When a derivative of 128C53 carrying determinants from pVW3JI was used as the donor and a Nod^- Fix^- mutant (strain 16015) of R. leguminosarum was used as the recipient, the majority of the Kan-r transconjugants obtained were Nod^+ Fix^+ Hup^+. Neither Nod^+ nor Hup^+ determinants was carried on pVW3JI (the original derivative of pRL3JI) and hence these characters must have been acquired as a result of recombination between the Kan-r marker carried by pVW3JI and a resident Nod^+ Hup^+ plasmid of strain 128C53 (namely pRL5JI). Similarly, when a derivative of strain 128C53 containing pVW5JI (i.e., pRL4JI::Tn5) was used in crosses, the majority of Kan-r 16015 transconjugants were Nod^+ Hup^+. (The only other class obtained, Nod^-, being presumably the result of fragmentation of the pVW5JI/pRL6JI cointegrate.) Although pRL6JI does not appear to be self-transmissible it is interesting to note that the strain which carries this plasmid produces a medium bacteriocin similar to those specified by pRL1JI and pRL4JI and the strain produces no small bacteriocin.

Table 1. Transmissible plasmids of *Rhizobium leguminosarum*.

Group I	Mol. Wt. x10^{-6}	Transfer frequency (per recipient)	Bacteriocin	Symbiotic Determinants			Original field isolate	Reference
			Med	Nod	Fix	Hup		
pRL1JI	130	10^{-2}	Med	Nod	Fix	-	248	Brewin et al. (1980a)
pRL3JI	125/160	10^{-2}	Med	-	-	-	306	Brewin et al. (1980a)
pRL4JI	160	10^{-2}	Med	-	-	-	309	Brewin et al. (1980a)
pRL6JI	190	10^{-7}	?Med	Nod	Fix	Hup	128C53	N. J. Brewin (unpublished results) Bethlenfalvay & Phillips (1979)
Group II								
pRL5JI	160	10^{-6}	-	Nod	Fix	-	TOM	Brewin et al. (1980b)

In Table 1, the Nod$^+$ Hup$^+$ plasmid from strain 128C53 which we term pRL6JI has been grouped with the bacteriocinogenic plasmids pRL1JI, pRL3JI and pRL4JI, despite the fact that it is not itself transmissible and we have as yet no direct evidence that it codes for the medium bacteriocin produced by strain 128C53. Nevertheless, the observed recombination between pRL6JI and introduced derivatives of pRL1JI, pRL3JI and pRL4JI suggests that these plasmids are in some way related.

TRANSFER OF R. LEGUMINOSARUM PLASMIDS TO R. PHASEOLI

In order to investigate the transfer of large plasmids from R. leguminosarum to closely related species, a strain of R. phaseoli (1233) was used as a recipient (Beynon et al., 1980). This strain has two plasmids of molecular weight c. 200 million. The smaller of these two plasmids carries determinants for nodulation ability on Phaseolus beans as well as determining the production of a dark brown (melanin) pigment by mature rhizobial colonies growing on plates of tryptone-yeast extract. (The significance of this pigment is unknown, but nearly all strains of R. phaseoli examined produced it, whereas out of 20 strains of R. leguminosarum and 20 strains of R. trifolii examined, none produced the pigment.) Spontaneous plasmid deletions occurred at a frequency of 0.5%. Introduction of the smallest plasmid from R. leguminosarum strain 300, always resulted in elimination of the smallest plasmid from R. phaseoli. Hence this plasmid, of unknown function, from R. leguminosarum apparently belongs to the same incompatibility group as the plasmid from R. phaseoli which determines nodulation and pigment production.

The introduction of the nodulation plasmid pJB5JI (a derivative of pRL1JI) from R. leguminosarum into R. phaseoli did not result in the loss of either plasmid from R. phaseoli (provided that the smallest plasmid from R. leguminosarum strain 300 was not also cotransferred). The plasmid pJB5JI conferred to the transconjugant derivatives of R. phaseoli the ability to nodulate peas effectively (Johnston et al., 1978a). However, unless the resident nodulation plasmid of R. phaseoli had already been eliminated, the nodulation of peas was considerably delayed compared to R. leguminosarum itself. This phenomenon might be due to some form of functional interference occurring between two plasmids specifying nodulation for different host legumes. This possibility is consistent with the observation that, when an inoculant strain was used which contained both the R. phaseoli nodulation plasmid and pJB5JI, all the clones recovered from pea nodules had either suffered a substantial deletion in the R. phaseoli nodulation plasmid or else lost it completely (Beynon et al., 1980), suggesting that genetic information on this plasmid had to be eliminated (spontaneously) before nodulation specified by the R. leguminosarum plasmid pJB5JI could proceed normally.

Similar phenomena may well occur in transconjugants of *R. trifolii* that have received pJB5JI (Johnston et al., 1978a) because here too nodulation of peas appeared to be somewhat impaired. However this system has not yet been studied in any detail.

CONCLUSIONS

The correlation of plasmid genetics with a physical analysis of plasmid composition is still in its infancy but a number of generalisations can now usefully be made.

1) Host range is a plasmid-determined trait, at least for the three closely related species *R. leguminosarum*, *R. trifolii* and *R. phaseoli*.

2) Other functions such as bacteriocin production, pigment production (in *R. phaseoli*) and an uptake hydrogenase may also be carried on nodulation plasmids.

3) Compatible plasmids can carry the same or similar genetic information, leading in some cases studied to the presence of two (or perhaps even three) copies of Nod^+ and Fix^+ determinants co-existing within the same cell.

4) When determinants for the nodulation of two different legume hosts co-exist within the same cell nodulation of either host may be impaired.

5) Plasmids may be either transmissible or non-transmissible. Genetic determinants carried on non-transmissible plasmids can sometimes be mobilised following the introduction of a transmissible plasmid which recombines (or cointegrates) with the non-transmissible plasmid.

6) When a new plasmid is introduced into a *Rhizobium* strain, various kinds of inter-plasmid rearrangements can occur. Sometimes a resident plasmid is eliminated; sometimes a cointegrate is formed with a resident plasmid; sometimes there is recombination with a resident plasmid resulting in two new plasmids of different sizes; sometimes there is apparently no interaction at all.

These observations raise many questions concerning the multifarious *Rhizobium* plasmids and their ecological importance in competition both within and without the legume root nodule.

ACKNOWLEDGMENTS

We are grateful to Professor D. A. Hopwood for his criticisms of this manuscript.

REFERENCES

Beringer, J. E., Hoggan, S. A., and Johnston, A. W. B., 1978, Linkage mapping in Rhizobium leguminosarum by means of R-plasmid mediated recombination, J. Gen. Microbiol., 104:201.

Bethlenfalvay, G. J., and Phillips, D. A., 1979, Variation in nitrogenase and hydrogenase activity of Alaska pea root nodules, Plant Physiol., 63:816.

Brewin, N. J., Beringer, J. E., Buchanan-Wollaston, A. V., Johnston, A. W. B., and Hirsch, P. R., 1980a, Transfer of symbiotic genes with bacteriocinogenic plasmids in Rhizobium leguminosarum, J. Gen. Microbiol., 116:261.

Brewin, N. J., Beringer, J. E., and Johnston, A. W. B., 1980b, Plasmid-mediated transfer of host-range specificity between two strains of Rhizobium leguminosarum, J. Gen. Microbiol., in press.

Buchanan-Wollaston, A. V., Beringer, J. E., Brewin, N. J., Hirsch, P. R., and Johnston, A. W. B., 1980, Isolation of symbiotically defective mutants of Rhizobium leguminosarum by insertion of the transposon Tn5 into a transmissible plasmid, Molec. Gen. Genet., 178:185.

Casse, F., Boucher, C., Julliot, J. S., Michel, M., and Dénarié, J., 1979, Identification and characteristics of large plasmids in Rhizobium meliloti using agarose gel electrophoresis, J. Gen. Microbiol., 113:229.

Graham, P. H., 1964, The application of computer techniques to the taxonomy of the root nodule bacteria of legumes, J. Gen. Microbiol., 35:511.

Gross, D. C., Vidaver, A. K., and Klucas, R. V., 1979, Plasmids, biological properties and efficacy of nitrogen fixation in Rhizobium japonicum strains indigenous to alkaline soils, J. Gen. Microbiol., 114:257.

Hepper, C. M., and Lee, L., 1979, Nodulation of Trifolium subterraneum by Rhizobium leguminosarum, Plant and Soil, 51:441.

Higashi, S., 1967, Transfer of clover infectivity of Rhizobium trifolii to Rhizobium phaseoli as mediated by an episomic factor, J. Gen. and Appl. Microbiol., 13:391.

Hirsch, P. R., 1979, Plasmid-determined bacteriocin production by Rhizobium leguminosarum, J. Gen. Microbiol., 113:219.

Hirsch, P. R., Van Montagu, M., Johnston, A. W. B., Brewin, N. J., and Schell, J., 1980, Physical identification of bacteriocinogenic, nodulation and other plasmids in strains of Rhizobium leguminosarum, J. Gen. Microbiol., in press.

Johnston, A. W. B., and Beringer, J. E., 1977, Chromosomal recombination between Rhizobium species, Nature, 267:611.

Johnston, A. W. B., Beynon, J. L., Buchanan-Wollaston, A. V., Setchell, S. M., Hirsch, P. R., and Beringer, J. E., 1978a, High frequency transfer of nodulating ability between strain and species of Rhizobium, Nature, 276:635.

Johnston, A. W. B., Setchell, S. M., and Beringer, J. E., 1978b, Interspecific crosses between Rhizobium leguminosarum and R. meliloti: formation of haploid recombinants and of R primes, J. Gen. Microbiol., 104:209.

Kondorosi, A., Kiss, K. B., Forrai, T., Vincze, L. E., and Banfalvi, Z., 1977, Circular linkage map of the Rhizobium meliloti chromosome, Nature, 268:525.

Kondorosi, A., Vincze, E., Johnston, A. W. B., and Beringer, J. E., 1980, A comparison of three Rhizobium linkage maps, Molec. Gen. Genet., 178:403.

Meade, H. M., and Signer, E. R., 1977, Genetic mapping of Rhizobium meliloti, Proc. Natl. Acad. Sci., USA, 74:2076.

Nuti, M. P., Lepidi, A. A., Prakash, R. K., Schilperoort, R. A., and Cannon, F. C., 1979, Evidence for nitrogen fixation (nif) genes on indigenous Rhizobium plasmids, Nature, 282:533.

Prakash, P. K., Hooykaas, P. J. J., Ledeboer, A. M., Kijne, J., Schilperoort, R. A., Nuti, M. P., Lepidi, A. A., Casse, F., Boucher, C., Julliot, J. S., and Denarie, J., 1980, Detection isolation and characterisation of large plasmids in Rhizobium, in: "Proceedings of the 3rd International Symposium on Nitrogen Fixation," W. E. Newton and W. H. Orme-Johnson, eds., University Park Press, Madison.

Ruvkun, G. B., and Ausubel, F. M., 1980, Interspecies homology of nitrogenase genes, Proc. Natl. Acad. Sci. USA, 77:191.

Wilson, J. K., 1944, Over five hundred reasons for abandoning the cross-inoculation groups of the legumes, Soil Sci., 58:61.

Winarno, R., and Lie, T. A., 1979, Competition between Rhizobium strains in nodule formation: interactions between nodulating and non-nodulating strains, Plant and Soil., 51:135.

Zurkowski, W., and Lorkiewicz, Z., 1978, Effective method for the isolation of non-nodulating mutants of Rhizobium trifolii, Genet. Res. (Camb.), 32:311.

"REDOX CONTROL" OF NITROGEN FIXATION: AN OVERVIEW

K. T. Shanmugam*, S. T. Lim, Sherman S. M. Hom, D. B. Scott, and H. Hennecke

*Dept. of Microbiology and Cell Science, University of Florida, Gainesville, Fl 32611 and Plant Growth Laboratory, University of California, Davis, CA 95616

Biological nitrogen fixation, an energy intensive process, is subject to several control processes in the cell (Brill, 1975; Ludden and Burris, 1976; Carithers and Yoch, 1979; Shanmugam and Hennecke, 1979). These include both genetic regulatory mechanisms and protein modification (in some organisms like Rhodospirillum rubrum). Addition of either NH_4^+ or O_2 to a nitrogen-fixing culture immediately leads to repression of nitrogenase biosynthesis. In Klebsiella pneumoniae, nitrogenase synthesis is also regulated by growth temperature (Hennecke and Shanmugam, 1979). Nitrogenase is undetectable at elevated temperatures (> 37 C) under growth conditions that allow the production of nitrogenase at permissive temperatures. Although NH_4^+ and high temperature (39 C) repress nitrogenase biosynthesis, they have no detectable effect on preformed nitrogenase protein and catalytic activity. Ammonium ion mediated regulation of nitrogenase (NH_4^+-control) is believed to be mediated through the assimilation of NH_4^+ to the level of organic compounds. The mechanism by which temperature affects the production of nitrogenase is unknown. These two control systems will not be discussed in this communication, with emphasis here on the role of O_2 as a modulator of synthesis of nitrogenase. The term 'redox control' used in the title refers to the fact that molecular oxygen (and other compounds such as nitrate) have the dual role as terminal electron acceptors as well as genetic modulators of cellular activity. A lucid description of the role of O_2 as a biochemical regulator is presented by Schlegel and coworkers in this volume.

REGULATION OF NITROGENASE SYNTHESIS BY O_2

Nitrogenase is a non-heme iron-sulfur molybdenum protein operating at a low mid-point potential (E_o' about -0.69 v) and as such is extremely sensitive to O_2 (Watt and Bulen, 1976; Mortenson, 1978). Exposing nitrogenase to molecular O_2 leads to inactivation of the enzyme and also to decay of the protein tertiary structure (Eady et al., 1972). All nitrogen fixing organisms have evolved some unique and specific mode of protection of nitrogenase from oxygen (Shanmugam and Valentine, 1975b; Yates, 1977). These include high respiration rate as in _A. vinelandii_, induction of nitrogenase only under anaerobic conditions as in facultative anaerobes like _Klebsiella_, induction of nitrogenase only under low partial pressure of O_2 as in _Rhizobium_, or induction of nitrogenase only in a specialized organelle like heterocysts in _Anabaena_ or the root nodule in a plant-_Rhizobium_ symbiosis. In no case is a nitrogen fixing organism described in which nitrogenase is not O_2 sensitive. Even in _A. vinelandii_, an aerobic nitrogen fixer, nitrogen fixation decreases with increasing oxygen tension (Drozd and Postgate, 1970).

Brill and his co-workers (St. John et al., 1974) observed that _K. pneumoniae_ failed to produce nitrogenase in the presence of O_2. Using _K. pneumoniae_ strains that are derepressed for nitrogenase biosynthesis in the presence of NH_4^+ (Shanmugam and Valentine, 1975a), Eady and his co-workers (Eady et al., 1978) observed that the regulatory properties observed in the presence of O_2 is independent of the control exerted by NH_4^+. It is also known that slow growing _Rhizobium_ spp. produce nitrogenase activity only at low partial pressures of O_2 (Keister and Ranga Rao, 1977). Experiments from our laboratory show that this is due to the absence of nitrogenase polypeptide synthesis at high partial pressures of oxygen (Scott et al., 1979). Unfortunately, similar studies were not carried out with other nitrogen-fixing organisms. These studies clearly demonstrate that synthesis of nitrogenase is repressed by O_2 although the mechanism by which this is effected is largely unknown.

REGULATION OF SYNTHESIS OF ELECTRON TRANSPORT PATHWAYS BY O_2

Klebsiella, Escherichia

In bacteria, unlike higher organisms, there are at least 4 major electron transport pathways, each utilizing a different terminal electron acceptor (Wimpenny, 1969; Harrison, 1972; Payne, 1973; Morris, 1975; Cole, 1976; Stouthamer, 1976; Haddock and Jones, 1977). These electron transport pathways are produced in response to the availability of the terminal electron acceptor.

1) Aerobic respiration-terminal e^- acceptor is O_2 - E_o' of O_2/H_2O is +0.82 V.

2) Anaerobic respiration-terminal e^- acceptor is NO_3^- - E_o' of NO_3^-/NO_2^- is +0.42 V.

3) Anaerobic respiration-denitrification-terminal e^- acceptor is NO_2^- - E_o' of $2NO_2^-/N_2O$ is +0.77 V, E_o' of $2NO_2^-/N_2$ is +0.97 V.

4) Anaerobic fermentation-terminatl e^- acceptors are organic compounds and H^+ - E_o' of H^+/H_2 is -0.42 V.

The anaerobic e^- transport pathway employs several terminal dehydrogenases and hydrogenase. The accepted role for these proteins is the removal of the excess electrons (reducing power) generated during sugar degradation by the glycolytic enzyme systems. Typical examples are lactate dehydrogenase, alcohol dehydrogenase, fumarate reductase and formate hydrogen lyase (formate dehydrogenase and hydrogenase).

In a typical case, electrons originating from formate are transferred through a series of electron transport proteins to hydrogenase (formate-hydrogen lyase reaction). Hydrogenase uses this reductant during catalytic reduction of H^+ to H_2 and this reaction is generally used as an indicator of anaerobic fermentative metabolism.

The above mentioned other three electron transport systems, utilizing several cytochromes, are known to produce energy during the oxidation of reductant (NADH). Although NO_3^- respiration and denitrification may coexist in the same cell, the electron transport components belonging to pathway 1, 2, 3 and 4 are not known to coexist in the same cell. However, a given organism may have the capacity to carry out all 4 reactions under appropriate conditions.

It is well known that in several organisms, electron transport pathways are subject to control by environmental factors (Harrison, 1972). For example, in both E. coli and K. aerogenes anaerobiosis leads to repression of the synthesis of components of aerobic electron transport pathway and induction of anaerobic electron transport proteins (Cole, 1976; Stouthamer, 1976).

Addition of oxygen to an anaerobic culture represses the production of anaerobic electron transport proteins and terminal dehydrogenases. In the presence of oxygen the cell produces aerobic electron transport proteins that couple the available reductant (NADH) to molecular oxygen. This process also yields energy for growth of the cell. If NO_3^- is added to the anaerobic cell, instead of O_2, as the terminal electron acceptor, the cell reduces it to the level of NO_2^- and accumulates it in the medium. Excess electrons produced during anaerobic growth of the culture are thus effectively removed during NO_3^--dependent respiration. Anaerobic electron transport proteins like formate dehydrogenase-2 are not produced during NO_3^- respiration. Addition of oxygen to a culture respiring on NO_3^- also leads to repression of NO_3^- reductase. Similarly, denitrification in the presence of NO_2^- is also repressed by the presence of O_2. In turn NO_2^- also represses the production of typical anaerobic proteins.

These experimental results suggest a hierarchy among the several electron transport systems. The cell prefers oxygen as the terminal electron acceptor over NO_3^- or NO_2^-. Nitrate and nitrite are favored electron acceptors over organic compounds or H^+. It seems essential that the electron transport to terminal electron acceptor (O_2 or NO_3^-) occurs (leading to the reduction of the terminal electron acceptor and oxidation of NADH) for the above mentioned regulation to be effective. Failure to do so abolishes the terminal oxidant mediated repression of the electron transport pathways. For example, an E. coli mutant strain with defects in ubiquinone biosynthesis and thus incapable of aerobic respiration and growth at the expense of succinate or lactate was found to produce nitrate reductase even under aerobic conditions (Giordano et al., 1978). Chlorate-resistant mutants that are defective in NO_3^- respiration produced both hydrogenase and nitrogenase activities in the presence of NO_3^- (Stouthamer, 1976; Hom et al., 1980). Mutant strains of E. coli with pleiotropic defects in NO_2^- metabolism (nirA) produced hydrogenase in the presence of NO_2^- (Douglas et al., 1974). NirA mutants were found to be defective in the production of N_2O (Satoh, personal communication). These studies show that flow of electrons to a given acceptor (O_2, NO_3^-, or NO_2^-) is essential for the repression of electron transport to other electron acceptors whose E_0' is more negative (O_2/H_2O, +0.82 V; NO_3^-/NO_2^-, +0.42 V; NO_2^-/N_2O, +0.77 V; fumarate/succinate, +0.03 V; acetaldehyde/ ethanol, -0.20 V; H^+/H_2, -0.42 V).

Besides the components of the electron transport proteins, levels of several TCA cycle enzymes also respond to the availability of oxidants (Wimpenny and Cole, 1967; Yamamoto and Ishimoto, 1975). Under anaerobic conditions, aconitase, isocitrate dehydrogenase and fumarase were found to be low, compared to aerobic growth conditions. This effectively prevents the production of NAD(P)H, the main product of TCA-cycle, under anaerobic conditions in which reoxidation of NAD(P)H is met with considerable difficulty.

Pichinoty (1962) proposed that this regulatory process described above is mediated by the oxidation-reduction state of a particular electron carrier in the electron transport pathway. Wimpenny and Necklen (1971) demonstrated a correlation between the redox potential of the culture and presence of or absence of anaerobic proteins (for example, hydrogenase). It was later shown that the E_h of the culture alone is not sufficient for this control (O'Brien and Morris, 1971). The presence or absence of terminal e^- acceptors is an integral part of the control system. As later studies demonstrated, the reduction of the terminal e^-

acceptor is also an essential requirement. This regulatory system has been termed as 'redox control' and studied in E. coli and K. aerogenes but the mechanism by which this is effected is still unknown.

Rhizobium

It has been observed that several wild-type Rhizobium species induce oxygen-reactive cytochromes different from those of free-living cells when the organisms are in association with the plant or are respiring under anaerobic conditions, using nitrate as the terminal oxidant (see Table 1). These results can be summarized as follows.

The cytochrome pattern of aerobic cells of slow growing species, R. japonicum and R. lupini, and of the fast-growing species, R. leguminosarum, resemble each other quantitatively and qualitatively. Bacteroids of these species completely lack a-type cytochromes but contain approximately three times as much cytochrome c-500, ten times as much cytochrome c-552 and P-450, and two times as much b-type cytochromes as are found in free-living cells. Consistent with the loss of a-type cytochromes, the proportion of b-type cytochromes which may fulfill an oxidase function (i.e., that which complexes carbon monoxide) increases about ten-fold. The cytochrome pattern of anaerobic-nitrate cells of the two rhizobia belonging to the slow-growing group resemble that of bacteroids. However, cultured cells of the fast grower, R. leguminosarum, produce a similar pattern of cytochromes whether grown aerobically or anaerobically with nitrate as the terminal oxidant.

Besides the changes in cytochromes, oxygen also regulates the production of nitrogenase, H_2 uptake system and NO_3^- respiration pathways. These enzyme systems are produced only under low oxygen concentrations and not under air (Keister and Ranga Rao, 1977; Lim and Shanmugam, 1979).

These experiments suggest that when the bacterium differentiates into the bacteroid state, some of the major electron transport components present in aerobically-grown cells are completely repressed and are replaced by alternate electron carriers. The mechanism of control of the synthesis of heme proteins and ultimately of the energy-linked electron transport pathways is an important unanswered question about the nature of symbiotic nitrogen fixation carried out by Rhizobium.

REGULATION OF NITROGENASE SYNTHESIS BY NO_3^-

Addition of nitrate to a nitrogen fixing culture is known to reduce nitrogenase activity and cultures that are growing in a medium containing NO_3^- as the sole source of nitrogen failed to

Table 1. Cytochrome composition of *Rhizobium* species under different growth conditions

Strain	Growth condition	Respiration rate ($\mu l\ O_2$/hr·mg protein)	Cytochrome (μmole heme/g, protein) a	b	c	c_{550}	c_{552}	P_{450}	Reference
R. japonicum strain 505	Aerobic		0.09	0.19	0.25	0.08	0.002	0.006	Daniel & Appleby, 1972
	Anaerobic + NO_3^-		0.00	0.19	0.78	0.26	0.023	0.017	"
	Bacteroids		0.00	0.43	0.87	0.27	0.030	0.050	"
R. lupini strain 371 (effective)	Aerobic		0.11	0.15	0.20	-	-	-	Romanov et al., 1976
	Microaerophilic + NO_3^-		0.04	0.40	0.75	-	-	-	"
	Bacteroids	75-80	0.00	0.36	0.55	0.25	0.060	0.160	Kretovich et al., 1972
R. lupini strain 400 (ineffective)	Bacteroids	23-25	+	0.23	0.32	(0.21)		0.000	Kretovich et al., 1972; Matus et al., 1973
R. leguminosarum strain 96 (effective)	Aerobic		0.13	0.20	0.20	-	-	-	Kretovich et al., 1973
	Semi-anaerobic		0.13	0.38	0.40	-	-	-	Kretovich et al., 1973; Romanov et al., 1974
	Bacteroids	90-100	0.00	0.51	0.56	-	-	-	"
R. leguminosarum strain 87 (ineffective)	Aerobic		0.10	0.17	0.19	-	-	-	Kretovich et al., 1973
	Semi-anaerobic		0.11	0.36	0.36	-	-	-	"
	Bacteroids	20-16	0.004	0.06	0.05	-	-	-	Kretovich et al., 1973; Romanov et al., 1974

produce nitrogenase activity. It is believed that this effect is mediated by NH_4^+, a reduction product of NO_3^- in many nitrogen fixing bacteria. During our initial experiments on the effect of NO_3^- on nitrogenase biosynthesis, we observed that NO_3^- repressed nitrogenase biosynthesis even in mutant strains of *K. pneumoniae* that are fully derepressed for nitrogenase biosynthesis in the presence of NH_4^+ (Hom et al., 1980). This raised the possibility that the observed effect of NO_3^- in these mutant strains is mediated by mechanisms other than the NH_4^+ control. Further experiments demonstrated that this repression of nitrogenase biosynthesis by NO_3^- is not observed in strains that are NO_3^- respiration deficient (isolated as spontaneous chlorate resistant mutants). In ClO_3^- resistant strains, NO_3^- had no detectable effect on nitrogenase synthesis or activity, which is similar to the effect of NO_3^- on hydrogenase production in chlorate resistant mutants. This suggests that the repressive effect of NO_3^- on nitrogenase biosynthesis in *K. pneumoniae* is similar to the repressive effect of O_2 on nitrogenase biosynthesis (i.e., 'redox control').

A UNIFIED CONCEPT FOR THE REGULATION OF NITROGENASE BIOSYNTHESIS (BY O_2 AND OTHER OXIDANTS)

Nitrogenase, like hydrogenase is also capable of functioning as terminal electron transport protein (enzyme). Nitrogenase utilizes H^+ and N_2 as terminal electron acceptors. In the absence of N_2, nitrogenase reduces H^+ and the product is H_2. For this reduction, nitrogenase requires electrons at low potentials for catalysis. At least two genes (*nifF* and *nifJ*) coding for nitrogenase specific electron transport proteins have been identified in the *nif* gene cluster (Roberts et al., 1978; Hill and Kavanagh, 1980). These proteins presumably accept electrons from the fermentative pathway of the cell at some unique but still unclarified location(s) and in turn couple to nitrogenase reductase (Yoch, 1974). If the anaerobic (fermentative) electron transport proteins are absent, no reductant will flow to nitrogenase, even under conditions in which the *nif* gene cluster is fully derepressed. Thus, electron transport, originating from the fermentative metabolism and ending in nitrogenase can be visualized as one pathway. Some of the proteins in this electron transport chain are produced by the *nif* gene cluster (*nifF* and *nifJ*) and coordinately controlled by the presence or absence of O_2 and NH_4^+.

If nitrogenase acts as a terminal electron acceptor, then the availability of electron acceptors that are more positive (E_o' of nitrogenase is ~ -0.69 V) may lead to repression of the electron transport proteins feeding electrons to nitrogenase and also nitrogenase. Repression of the synthesis of proteins in an electron transport pathway that is a net energy consumer (reduction of N_2 and H^+ by nitrogenase) could also occur under conditions in which another oxidant, whose reduction is coupled to production of ATP, is freely available. In other words, nitrogenase synthesis may also be a part of the cellular reactions regulated by 'redox control.'

We have tested such a working hypothesis in K. pneumoniae utilizing NO_3^- as the terminal oxidant since oxygen irreverisbly inactivates nitrogenase. As described before nitrogenase and formate hydrogen-lyase were not produced in the presence of NO_3^-. Nitrogenase was not produced even in the presence of NO_2^-, a product of nitrate metabolism or nitric oxide, a presumed intermediate of NO_2^- metabolism to N_2O (denitrification pathway, Hom and Shanmugam, manuscript in preparation). Nitrate-respiration defective mutants produced both nitrogenase and hydrogenase. However, we do not know at this time about the control of other anaerobic proteins in K. pneumoniae in the presence of nitrite and nitric oxide.

A typical experiment that exemplifies the interaction between e^- transport and nitrogenase is the one involving nitric oxide. K. pneumoniae utilizes NO and reduces it to N_2O. Addition of nitric oxide to a culture of strain SK-512 inhibited nitrogenase activity (Fig. 1). Nitrogenase biosynthesis was elevated immediately after addition of nitric oxide. Repression of nitrogenase synthesis was detectable after 1 hr in strain SK-512. Measurable quantities of N_2O were detected by 2 hrs. Chlorate-resistant strains of K. pneumoniae were found to be defective in the production of N_2O from NO_2^- besides their defects in NO_3^- respiration. However, these strains were found to be normal in their nitric oxide metabolism. The effect of nitric oxide on nitrogenase biosynthesis and activity was found to be similar in both Chl^R strain (SK-1400) and Chl^S strain (Fig. 1). Nitric oxide stimulated nitrogenase synthesis soon after addition, and in this regard nitric oxide acts like other inhibitors of enzyme activity in both trp and his operons of E. coli and S. typhimurium, respectively, in which the inhibited enzyme is actually derepressed (Lester and Yanofsky, 1961; Moyed, 1961). These experimental results suggest that metabolism of NO (N_2O production) is essential for repression of nitrogenase synthesis by nitric oxide. The final detectable product of NO metabolism in K. pneumoniae is N_2O. Nitrous oxide had no effect on the induction or activity of nitrogenase in this organism, although N_2O is known to be an inhibitor of purified nitrogenase activity (Burris, 1979).

Since both the substrate and product (NO and N_2O) failed to repress nitrogenase biosynthesis, one is left with other unknown intermediates in the production of N_2O for explaining the repression. For example, as soon as the metabolism of NO starts nitrogenase synthesis is repressed (Fig. 1) indicating that the repression observed in the presence of NO is not mediated by either

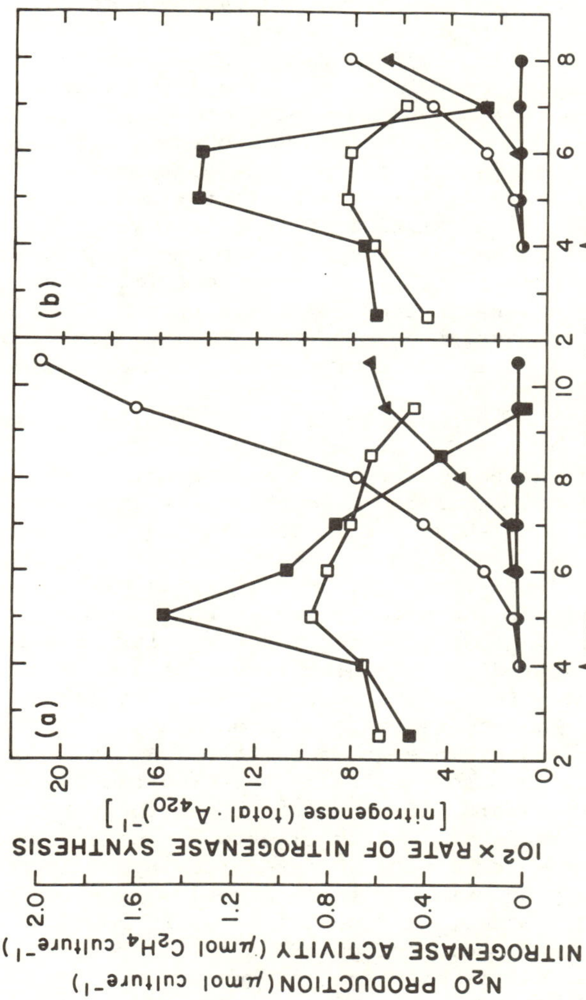

Fig. 1. Effect of nitric oxide (NO) on the rate of nitrogenase synthesis in K. pneumoniae strain SK-512(a) and SK-1400(b). Cultures were grown with 0.028 atm NO (closed symbols) or without NO (open symbols). Whole cell nitrogenase activity (o, ●), rate of nitrogenase synthesis (□, ■), and N_2O production (▲).

NO or N_2O but by compounds in between the two in the metabolism and induction of this "repressor" is essential and is mediated by nitric oxide.

ROLE OF CYCLIC MONONUCLEOTIDES IN 'REDOX CONTROL'

Adenosine-3',5'-cyclic monophosphate (cAMP) is known to play a major role in the regulation of cellular metabolism in several organisms (Pastan and Adhya, 1976). In E. coli cAMP levels are normally low in a glucose containing medium and the intracellular cAMP pool levels increase upon exhaustion of glucose from the medium. Glucose also represses the expression of other catabolite (glucose) sensitive operons (for example, lactose operon) involved in energy production for cellular growth, in the absence of glucose. A direct correlation exists in E. coli between the levels of cAMP and β-galacotsidase in the cell. cAMP in conjunction with catabolite repressor protein (CRP) is known to activate the transcription of catabolite sensitive operons (for example, lac, ara) even in an in vitro transcription system. These results show that cAMP plays a crucial role in the energy metabolism of the cell.

Dobrogosz and his co-workers (Patrick and Dobrogosz, 1973; Broman et al., 1974) observed that cAMP is required in E. coli for optimal production of aerobic electron transport proteins as well as for production of the anaerobic electron transport system, formate hydrogen lyase, suggesting a role for cAMP in 'redox control'. Our experiments with R. japonicum show that cAMP is required for the production of electron transport protein(s) involved in H_2 uptake by the cell (Lim and Shanmugam, 1979). R. japonicum strain 110 produced similar amounts of hydrogenase activity (measured as ^3H uptake activity) in media containing either malate or glutamate as sole carbon source. In the presence of malate as carbon source, the rate at which H_2 was removed and oxidized to H_2O by the culture was less than 40% of a glutamate grown culture. Addition of 1.0 mM cAMP to the malate culture completely restored the hydrogen uptake (Hup) activity. This effect was specific for cAMP. These results suggest that in malate medium electron transport proteins capable of coupling the reductant generated from H_2 by hydrogenase are not produced and production of these proteins requires cAMP. In agreement with this model, cAMP pool sizes were found to be low in cells grown in malate medium compared to cells grown in a glutamate containing medium (as sole carbon source). In R. japonicum both hydrogenase and nitrogenase are produced under microaerophilic conditions ($< 0.1\%$ O_2). This raises the possibility that if electron transport components involved in H_2 uptake are controlled by cAMP, then the electron transport proteins feeding electrons to nitrogenase and nitrogenase itself may also be controlled by cAMP. Although malate decreased the cAMP pool levels of the cell, nitrogenase activity was found to be maximal (Lim and Shanmugam, 1979). Also K. pneumoniae mutant strains that are adenyl cyclase defective produced normal levels of nitrogenase activity without cAMP supplementation (A. Valentine and Shanmugam, unpublished results).

However, addition of cyclic GMP, a known antagonist of cAMP in E. coli, repressed nitrogenase biosynthesis in R. japonicum strain 110 (Lim et al., 1979). This repressive effect is observed in the production of hydrogenase, hydrogen uptake as well as the respiratory nitrate reductase also (Table 2). The pleiotropic nature of this effect indicates that all three microaerophilic enzyme systems are regulated by cGMP concentrations in the cell. In the presence of 0.1 mM cGMP, no nitrogenase activity was detected in other R. japonicum strains (strains 138 and 142) as well as the cowpea Rhizobium strain 32H1 (Lim, unpublished data). At this concentration, cGMP had very little effect on the production of other proteins like malate dehydrogenase (NAD^+) and glutamine synthetase. It is possible that the cAMP and cGMP have opposing effects in the regulation of the microaerophilic electron transport pathways and nitrogenase in R. japonicum. Production of nitrogenase activity in the presence of low intracellular cAMP levels (malate medium) may indicate that the cAMP requirement for production of nitrogenase is low, as against the hydrogen uptake system. The other possibility is that cGMP and not cAMP is the small molecular effector in 'redox control'. Additional experiments involving mutants are necessary to understand the molecular mechanisms of 'redox control' of nitrogen fixation, in both free-living as well as symbiotic nitrogen-fixing organisms.

Table 2. Effect of oxygen and cGMP on the production of nitrogenase, hydrogenase and nitrate reductase in R. japonicum.*

Additions to the medium	Nitrogenase		Hydrogenase		Nitrate reductase
	Activity	Protein	H_2 uptake	^3H-exchange	
	(percent control)				
None (0.1% O_2)	100	100	100	100	100**
+ O_2 (20%)	0	<1	UD	UD	not done
+cGMP (0.1 mM)	<1	<1	30	19	26

*Data from Lim et al. (1979); Lim and Shanmugam (1979) and Scott et al. (1979).

**Both control and cGMP supplemented culture contained 4 mM NO_3^-.

UD = undetectable.

ACKNOWLEDGMENT

This report was prepared with the support of a National Science Foundation Grant PFR 77-07301 and a Solar Energy Research Institute Subcontract XR-9-8036-1. The authors wish to acknowledge the help provided by Dr. R. Bartsch, A. Valentine, and Dr. R. C. Valentine.

REFERENCES

Brill, W. J., 1975, Regulation and genetics of bacterial nitrogen fixation, Ann. Rev. Microbiol., 29:109.

Broman, R. L., Dobrogosz, W. J., and White, D. C., 1974, Stimulation of cytochrome synthesis in Escherichia coli, Arch. Biochem. Biophys., 162:595.

Burris, R. H., 1979, Inhibition, in: "A Treatise on Dinitrogen Fixation," R. W. F. Hardy, F. Bottomley and R. C. Burns, eds., J. Wiley and Sons, New York.

Carithers, R. P., Yoch, D. C., and Arnon, D. I., 1979, Two forms of nitrogenase from the photosynthetic bacterium Rhodospirillum rubrum, J. Bact., 137:779.

Cole, J. A., 1976, Microbial gas metabolism, Adv. Micro. Physiol., 14:1.

Daniel, R. M., and Appleby, C. A., 1972, Anaerobic-nitrate, symbiotic and aerobic growth of Rhizobium japonicum: effect on cytochrome P-450, other haemoproteins, nitrate and nitrite reductases, Biochim. Biophys. Acta, 275:347.

Douglas, M. W., Ward, F. B., and Cole, J. A., 1974, The formate hydrogen lyase activity of cytochrome c_{552}-deficient mutants of Escherichia coli K12, J. Gen. Micro., 80:557.

Drozd, J., and Postgate, J. R., 1970, Effects of oxygen on acetylene reduction, cytochrome content and respiratory activity of Azotobacter chroococcum, J. Gen. Micro., 63:63.

Eady, R. R., Smith, B. E., Cook, K. A., and Postgate, J. R., 1972, Nitrogenase of Klebsiella pneumoniae, Biochem. J., 128:655.

Eady, R. R., Issack, R., Kennedy, C., Postgate, J. R., and Ratcliffe, H. D., 1978, Nitrogenase synthesis in Klebsiella pneumoniae: comparison of ammonium and oxygen regulation, J. Gen. Micro., 104:277.

Giordano, G., Grillet, L., Rosset, R., Dou, J. H., Azoulay, E., and Haddock, B. A., 1978, Characterization of an Escherichia coli K12 mutant that is sensitive to chlorate when grown aerobically, Biochem. J., 176:553.

Haddock, B. A., and Jones, C. W., 1977, Bacterial respiration, Bact. Rev., 41:47.

Harrison, D. E. F., 1972, Physiological effects of dissolved oxygen tension and redox potential on growing populations of microorganisms, J. Appl. Chem. Biotech., 22:417.

Hennecke, H., and Shanmugam, K. T., 1979, Temperature control of nitrogen fixation in Klebsiella pneumoniae, Arch. Microl., 123:259.

Hill, S., and Kavanagh, E. P., 1980, Roles of nifF and nifJ gene products in electron transport to nitrogenase in Klebsiella pneumoniae, J. Bact., 141:470.

Hom, S. S. M., Hennecke, H., and Shanmugam, K. T., 1980, Regulation of nitrogenase biosynthesis in Klebsiella pneumoniae: effect of nitrate, J. Gen. Micro., 117:169.

Keister, D. L., and Ranga Rao, V., 1977, The physiology of acetylene reduction in pure cultures of rhizobia, in: "Recent Developments in Nitrogen Fixation," W. E. Newton, J. R. Postgate and C. Rodriguez-Barrueco, eds., Academic Press, New York.

Kretovich, V. L., Melik-Sarkisyan, S. S., and Matus, V. K., 1972, Cytochrome composition of yellow lupine nodules, Biokhimiya, 37:711.

Kretovich, V. L., Romanov, V. I., and Korolyov, A. V., 1973, Rhizobium leguminosarum cytochromes (Vicia faba), Plant and Soil, 39:619.

Lester, G., and Yanofsky, C., 1961, Influence of 3-methyl anthranilic and anthranilic acids on the formation of tryptophan synthetase in Escherichia coli, J. Bact., 81:81.

Lim, S. T., and Shanmugam, K. T., 1979, Regulation of hydrogen utilization in Rhizobium japonicum by cyclic AMP, Biochim. Biophys. Acta, 584:479.

Lim, S. T., Hennecke, H., and Scott, D. B., 1979, Effect of cyclic guanosine 3',5'-monophosphate on nitrogen fixation in Rhizobium japonicum, J. Bact., 139:256.

Ludden, P. W., and Burris, R. H., 1976, Activating factor for the iron protein of nitrogenase from Rhodospirillum rubrum, Science, 194:424.

Matus, V. K., Melik-Sarkisyan, S. S., and Kretovich, V. L., 1973, Cytochromes and respiration rate of bacteroids from nodules of lupine inoculated with effective and ineffective strains of Rhizobium lupini, Mikrobiologiya, 42:112.

Morris, J. G., 1975, The physiology of obligate anaerobiosis, Adv. Micro. Physiol., 12:169.

Mortenson, L. E., 1978, Regulation of nitrogen fixation, Current Topics in Cell Reg., 13:179.

Moyed, H. S., 1961, Interference with the feedback control of histidine biosynthesis, J. Biol. Chem., 236:2261.

O'Brien, R. W., and Morris, J. G., 1971, Oxygen and the growth and metabolism of Clostridium acetobutylicum, J. Gen. Micro., 68:307.

Pastan, I., and Adhya, S., 1976, Cyclic adenosine 5'-monophosphate in Escherichia coli, Bact. Rev., 40:527.

Patrick, J. M., and Dobrogosz, W. J., 1973, The effect of cyclic AMP on anaerobic growth of Escherichia coli, Biochem. Biophys. Res. Commun., 54:555.

Payne, W. J., 1973, Reduction of nitrogenous oxides by microorganisms, Bact. Rev., 37:409.

Pichinoty, F., 1962, Inhibition par l'oxygene de la biosynthese et de l'activite de l'hydrogenase et de l'hydrogenlyase chez les bacteries anaerobies facultatives, Biochim. Biophys. Acta, 64:111.

Roberts, G. P., MacNeil, T., MacNeil, D., and Brill, W. J., 1978, Regulation and characterization of protein products coded by the nif (nitrogen fixation) genes of Klebsiella pneumoniae, J. Bact., 136:267.

Romanov, V. I., Korolev, A. V., and Kretovich, V. L., 1974, The cytochrome composition of bacteroids of Rhizobium leguminosarum (Vicia faba), Biokhimiya, 39:719.

Romanov, V. I., Fedulova, N. G., Korolev, A. V., and Kretovich, V. L., 1976, Effect of nitrate and nitrite on respiration and the cytochrome system of Rhizobium lupini, Mikrobiologiya, 45:85.

Scott, D. B., Hennecke, H., and Lim, S. T., 1979, The biosynthesis of nitrogenase MoFe protein polypeptides in free-living cultures of Rhizobium japonicum, Biochim. Biophys. Acta, 565:365.

Shanmugam, K. T., and Hennecke, H., 1979, Microbial genetics and nitrogen fixation, in: "Recent Advances in Biological Nitrogen Fixation," N. S. Subba Rao, ed., Oxford International Book House, New Delhi.

Shanmugam, K. T., and Valentine, R. C., 1975a, Microbial production of ammonium ion from nitrogen, Proc. Natl. Acad. Sci. USA, 72:136.

Shanmugam, K. T., and Valentine, R. C., 1975b, Molecular biology of nitrogen fixation, Science, 187:919.

St. John, R. T., Shah, V. K., and Brill, W. J., 1974, Regulation of nitrogenase synthesis by oxygen in Klebsiella pneumoniae, J. Bact., 119:266.

Stouthamer, A. H., 1976, Biochemistry and genetics of nitrate reductase in bacteria, Adv. Micro. Physiol., 14:315.

Watt, G. D., and Bulen, W. A., 1976, Calorimetric and electrochemical studies on nitrogenase, in: "Proceedings of the 1st International Symposium on Nitrogen Fixation," Vol. I, W. E. Newton and C. J. Nyman, eds., Washington State University Press, Pullman, Washington.

Wimpenny, J. W. T., 1969, The effect of E_h on regulatory processes in facultative anaerobes, Biotech. Bioeng., 11:623.

Wimpenny, J. W. T., and Cole, J. A., 1967, The regulation of metabolism in facultative bacteria. III. The effect of nitrate, Biochim. Biophys. Acta, 148:233.

Wimpenny, J. W. T., and Necklen, D. K., 1971, The redox environment and microbial physiology. I. The transition from anaerobiosis to aerobiosis in continuous cultures of facultative anaerobes, Biochim. Biophys. Acta, 253:352.

Yamamoto, I., and Ishimoto, M., 1975, Effect of nitrate reduction on the enzyme levels in carbon metabolism in Escherichia coli, J. Biochem., 78:307.

Yates, M. G., 1977, Physiological aspects of nitrogen fixation, in: "Recent Development in Nitrogen Fixation," W. Newton, J. R. Postgate, and C. Rodriguez-Barrueco, eds., Academic Press, New York.

Yoch, D. C., 1974, Electron transport carriers involved in nitrogen fixation by the coliform, *Klebsiella pneumoniae*, *J. Gen. Micro.*, 83:153.

A MUTANT OF RHIZOBIUM JAPONICUM 110 WITH ELEVATED NIF ACTIVITY IN FREE-LIVING CULTURE

Sui-Sheng T. Hua[1], D. Barry Scott[2], and Soo T. Lim[2]

[1]Western Regional Research Laboratory, USDA/SEA,
Berkeley, CA 94710
[2]Plant Growth Laboratory, University of California,
Davis, CA 95616

The enhancement of biological nitrogen fixation in the legume-*Rhizobium* symbiosis involves selection strategies at both the plant and symbiont level to identify and define traits suitable for improving the symbiosis. One of the major advances at the symbiont level was the identification of the hydrogen uptake (Hup) enzyme system as being important in the overall energy efficiency of nitrogen fixation (Schubert and Evans, 1976; also this volume). Rhizobium strains that possess the uptake hydrogenase are able to reutilize some of the hydrogen evolved from nitrogenase (Schubert and Evans, 1976) and thus regenerate ATP (Emerich et al., 1979). This is important in a system which has such a high energy (ATP) demand for biological nitrogen fixation (Andersen and Shanmugam, 1977).

An alternative approach is to select mutants with enhanced rates of nitrogen fixation, or with improved energy efficiencies, over the wild type strains. This approach has been used successfully by Maier and Brill (1978) to obtain mutants of *R. japonicum* (strain 61A76, Hup^-) with an increased ability to fix nitrogen by forming more nodules. These mutants were selected by a plant effectiveness assay (Maier and Brill, 1978).

The development of the appropriate conditions necessary for the induction of nitrogenase in free living cultures of *Rhizobium* (Kurz and LaRue, 1975; McComb et al., 1975; Pagan et al., 1975) now provides a system suitable for screening the N_2-fixing capacity of different *Rhizobium* mutants, separated from the host plant. This approach has been used successfully to study the hydrogenase enzyme system in different *Rhizobium japonicum* strains (Lim, 1978;

Lim et al., 1980; also this volume). Using this approach we have selected a mutant of Rhizobium japonicum (strain 3Ilb 110, Hup^+) that has increased rates of N_2 fixation (C_2H_2 reduction) in free living culture. A preliminary report of this work has been presented (Hua, 1978).

MATERIALS AND METHODS

Bacterial Strains and Growth Conditions

Rhizobium japonicum strain 3Ib 110 was obtained from D. E. Weber, U. S. Department of Agriculture, Beltsville. Stock cultures of Rhizobium were maintained on agar slants of mannitol-salts-yeast extract (MSY) medium (O'Gara and Shanmugam, 1976). The medium used for the induction of nitrogenase in free-living cultures of R. japonicum was as described previously (O'Gara and Shanmugam, 1976). The carbon source used was malate (4 g l^{-1}) and the nitrogen source was L-glutamate (1.0 g l^{-1}). The medium was buffered with 50 mM morpholinopropane sulphonic acid (MOPS), pH 6.8. Cultures were inoculated with a 2% (v/v) inoculum (A_{420} of final culture was about 0.15; 1 cm light path) and grown under microaerophilic conditions (O'Gara and Shanmugam, 1976) at 28°C. For the determination of nitrogenase activity, the gas phase was adjusted to 0.1% oxygen, 2.9% acetylene and 97% argon. For experiments involving the determination of hydrogenase activity a glutamate minimal medium (1.0 g l^{-1}) was used and the gas phase consisted of 0.1% oxygen, 5% carbon dioxide, 10% hydrogen, and the remainder, argon. Growth was followed by measuring the absorbance at 420 nm in a Gilford Model 300 N spectrophotometer (1 cm light path) and also by the increase in cell protein as previously described (O'Gara and Shanmugam, 1976). In one experiment, where indicated in the text, growth was measured by direct colony counting. Protein was routinely assayed by the procedure of Lowry et al. (1951), using bovine serum albumin as the standard.

Mutagenesis

Cells of R. japonicum (3Ilb 110) were grown in MSY medium to a cell density of 100 Klett units. The cells were harvested by centrifugation (7000 g, 15 min) and then resuspended at a cell density of 50 Klett units in fresh MSY medium, containing 200 μg/ml of N-methyl-N'-nitro-N'nitrosoguanidine (Aldrich Chemical Company). The cells were then incubated at 30°C for 1 hr. This treatment resulted in approximately 99% killing. The mutagenized culture was washed twice with distilled water and then resuspended in fresh MSY medium, and grown for three days at 25°C. Chlorate resistant mutants were then isolated by plating a suitable dilution of the mutagenized culture on minimal agar (1.5% w/v) plates containing potassium chlorate (0.2% w/v), L-glutamate (1 g/l^{-1}) as the nitrogen source, and arabinose (2 g/l^{-1}) and succinate (1 g/l^{-1}) as carbon sources, and then the plates were incubated under a nitrogen atmosphere with 0.1% (v/v) oxygen at 28°C for 10 days. One hundred of these chlorate resistant colonies were then analyzed for whole cell nitrogenase activity and nitrate reductase activity. Five of these colonies were found to have elevated whole cell nitrogenase activity. One of these (strain C33) was chosen for further characterization. We have been unable to get this mutant to revert back to chlorate sensitive phenotype.

Enzyme Assays

Whole cell nitrogenase activity was determined using the acetylene reduction procedure (Hardy et al., 1978) using a Varian Aerograph Model 1400 equipped with a flame ionization detector and a Porapak R column.

Hydrogenase Activity and Nitrate Reductase Activity

Hydrogenase activity, determined by H_2 uptake or $^3H_2$ exchange, and nitrate reductase activity were determined as previously described (Lim, 1978; Scott et al., 1979).

Respiration Rates

Rates of respiration of *R. japonicum* strains were determined as previously described (O'Gara and Shanmugam, 1977).

Radioactive Labeling of *R. japonicum*

Samples (2 ml) of cells grown under aerobic and microaerophilic (0.1% v/v O_2) conditions in minimal medium (see Methods) containing gluconate (4 g l^{-1}) and malate (1.5 g l^{-1}) as the carbon sources, were labeled with L-(^{35}S)methionine (Scott et al., 1979).

Electrophoresis and Autoradiography

The labeled cells were analyzed by two-dimensional polyacrylamide gel electrophoresis, following the method of O'Farrell (1975). The preparation of samples and the conditions of electrophoresis were as previously described (Scott et al., 1979). After electrophoresis, the amount of Component I of nitrogenase synthesized relative to the total SDS soluble proteins was determined by scanning the autoradiograph of the gel (Scott et al., 1979).

Immunization and Preparation of Antiserum

Rhizobium japonicum (3Ilb 110) cells for immunization were grown aerobically for 5 days using a glutamate minimal medium (2.0 g l^{-1}). The cells were harvested and washed with 0.85% (w/v) saline and resuspended to a concentration of approximately 10^9 cells ml^{-1}.

The culture was heated to 100°C and maintained at that temperature for 1 hr to inactivate flagellar antigens. Immunization and preparation of the antiserum was carried out using the procedure of Schmidt et al. (1968).

Ouchterlony Immunodiffusion

Antigen-antibody complex formation was followed in double-diffusion plates (Campbell et al., 1970) which contained 10 mM sodium borate (pH 7.8), 0.85% (w/v) sodium chloride, 1 mM sodium azide, and 0.9% agar. Plates were allowed to develop for 24 to 48 hr at room temperature.

RESULTS

Identification of R. japonicum (Strain C33) as a Derivative of R. japonicum (Strain 3Ilb 110)

In order to confirm that strain C33 is a derivative of strain 110, several tests were carried out comparing this strain with the wild type. Using crude antiserum prepared against strain 110, an Ouchterlony immunodiffusion test was carried out with strain C33 (Fig. 1). An intense precipitin line was obtained with cells of strain C33 and this fused with the precipitin lines formed with strain 110. Several minor precipitin lines also were observed with both strains. These results suggest that strain C33 is of the same sero-group as strain 110. Immunodiffusion tests conducted with twenty different USDA strains of *R. japonicum* showed that strains 3Ilb 117, 3Ilb 129, and CB1809 also cross react with anti-serum to strain 110, indicating that these strains all belong to the 110 sero-group.

Strain C33 is also sensitive to lysis by a bacteriophage (D1) that is specific for strain 110 (Mielenz et al., 1979). No lysis was observed when this phage was tested with twenty other different strains of *R. japonicum*.

The two-dimensional polyacrylamide gel protein patterns (Fig. 3) for cells grown under aerobic and microaerophilic (0.1% O_2) growth conditions are very similar between strains C33 and 110, suggesting that C33 is a derivative of 110. However, some protein differences are observed and these are discussed later.

Nitrogenase Activity in Free-Living Cultures of Rhizobium japonicum

Under microaerophilic growth conditions (0.1% O_2) the rate of acetylene-dependent ethylene production by cultures of strain C33 was significantly greater than strain 110 (Fig. 2b). This increase was not due to differences in growth rates, as the doubling time for strain C33 (24 hr) was similar to that of strain 110 (22 hr)

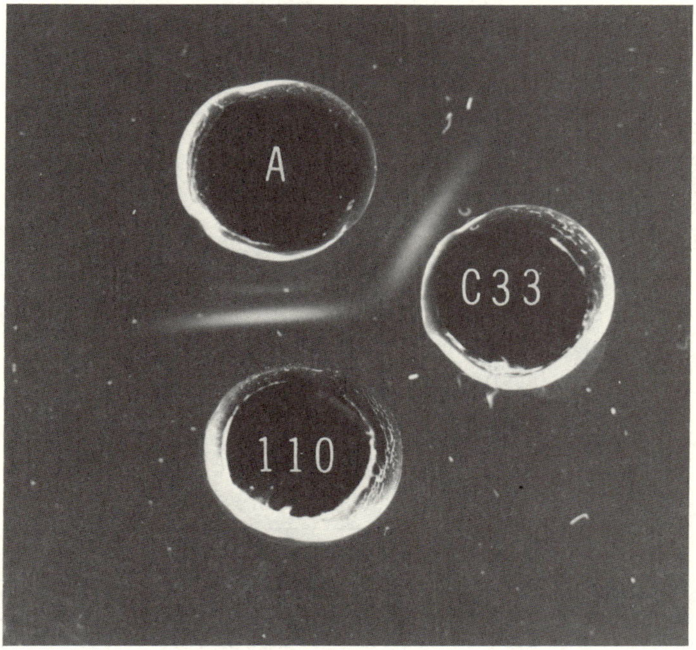

Fig. 1. Antigenic cross-reactions with anti-serum prepared against Rhizobium japonicum (strain 3Ilb 110). The wells contained: antiserum against R. japonicum (strain 3Ilb 110) (A), R. japonicum (strain 110) cells and R. japonicum (strain C33) cells.

(Fig. 2a). This difference in whole cell nitrogenase activity has been observed in a number of experiments. From the original 100 chlorate-resistant mutants isolated, five clones (including C33) were found to have elevated levels of whole cell nitrogenase activity.

Other Properties of R. japonicum Strain C33

In an effort to understand the biochemical differences between strains 110 and C33, other enzymes induced under microaerophilic growth conditions including hydrogenase and nitrate reductase (Lim et al., 1979) were also examined. Hydrogenase activity, as measured by uptake and $^3H_2$ gas exchange, was not significantly different between C33 and 110 (Table 1). High levels of nitrate reductase were found in crude extracts of strain 110 but no activity was detected in crude extracts of strain C33 (Table 1). The high level of nitrogenase activity in strain C33 in the presence of KNO_3 reflects the lack of nitrate reductase in this strain, as

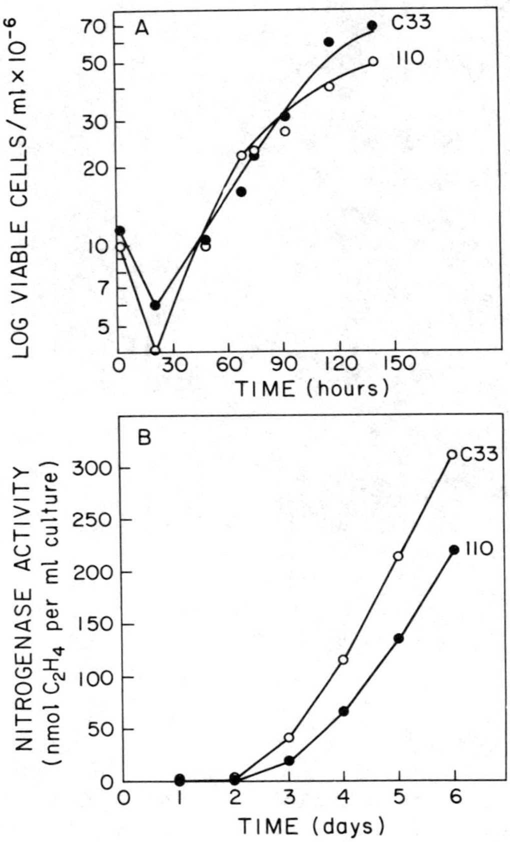

Fig. 2. (A) Time course of growth of R. *japonicum* strain 3Ilb 110 and strain C33 under micoraerophilic growth conditions. Viable cell numbers were determined by plating out a suitable dilution of the cells on MSY-agar plates. (B) Whole-cell nitrogenase activity in cultures of R. *japonicum* grown under microaerophilic growth conditions. o strain C33; ● strain 110.

Table 1. A comparison of the biochemical properties of strains C33 and 110 of *R. japonicum*.

		Rhizobium japonicum strain 110	*Rhizobium japonicum* strain C33
Nitrogenase activity (nmol h^{-1} mg $protein^{-1}$)	Control	25.2	48.8
	$+NO_3^-$ (4 mM)	0	47.3
Nitrogenase polypeptide synthesis (% of total protein)		5.3	5.6
Hydrogenase activity: H_2 uptake (μmol h^{-1} mg $protein^{-1}$)		14.9	13.1
Hydrogenase activity: $^3H_2$ exchange (μmol h^{-1} mg $protein^{-1}$)		2.45	2.47
Nitrate reductase activity (nmol min^{-1} mg $protein^{-1}$)		93.2	0
Respiration rate: aerobically grown cells (nmol min^{-1} mg $protein^{-1}$)		72	69
Respiration rate: cells grown under 0.1% O_2 (nmol min^{-1} mg $protein^{-1}$)		56	66

nitrite inhibits nitrogenase activity (Kennedy et al., 1975). The amount of Component I of nitrogenase synthesized in strain C33 was similar to strain 110 (Table 1 and Scott et al., 1979). The respiration rates of cells grown under both aerobic and microaerophilic conditions were also determined (Table 1), but no significant differences were observed.

When cells of strain C33 and 110 were analyzed by two-dimensional polyacrylamide gel electrophoresis the protein patterns were very similar but some differences were observed between the two strains (Fig. 3). For cells grown under microaerophilic conditions, the most dramatic difference observed was the absence of a 40,000 dalton protein in strain C33 (Fig. 3d), which was a major protein in strain 110 (Fig. 3c). This protein is unique to cells of strain 110 grown under microaerophilic conditions (c.f., Fig. 3a and Fig. 3c) as are the constituent polypeptides of Component I

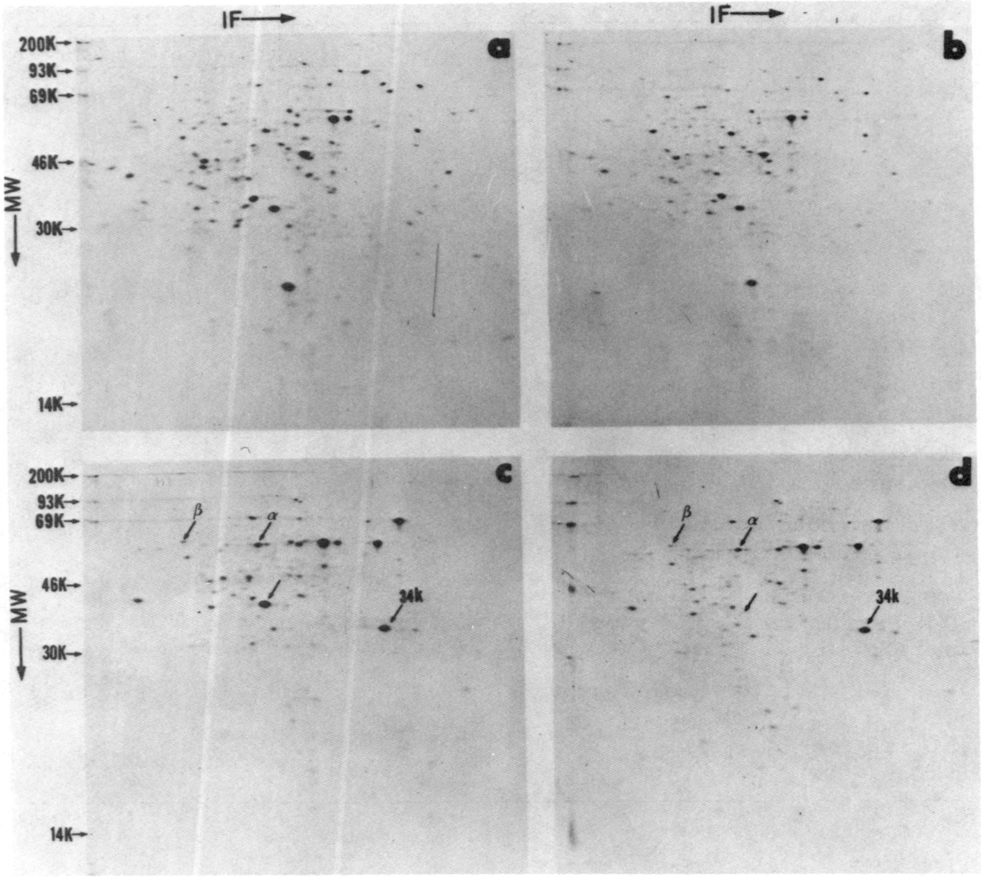

Fig. 3. Autoradiographs of two-dimensional gels of in vivo labeled R. japonicum proteins. The arrows indicate the positions of the α and β polypeptides of Component I of nitrogenase, a 34,000 dalton protein (Component II of nitrogenase) and an unidentified 40,000 dalton protein. (a) Cells of strain 110 grown under aerobic conditions; (b) cells of strain C33 grown under aerobic conditions; (c) cells of strain 110 grown under microaerophilic conditions; (d) cells of strain C33 grown under microaerophilic conditions.

(α and β) and Component II (34K) of nitrogenase (Scott et al., 1979). Minor protein differences can also be observed between aerobic cells of strain 110 (Fig. 3a) and strain C33 (Fig. 3b), but as nitrogenase is synthesized only under microaerophilic growth conditions, protein differences between the two strains in the microaerophilic cultures are probably of greatest importance.

DISCUSSION

A mutant (strain C33) of R. japonicum (strain 3Ilb 110) has been isolated with elevated levels of nitrogenase activity in free-living culture. This difference was not a consequence of an increased synthesis of the constituent polypeptides of nitrogenase as the amount of nitrogenase synthesized in strain C33 was similar to strain 110 (Table 1). The genetic lesion(s) in this mutant have not been determined because of the lack of genetic tools in this organism, but several protein differences were observed on two-dimensional polyacrylamide gels (Fig. 3). Maier and Brill (1978) used a plant effectiveness assay to screen clones from a mutagenized culture. The screening of mutants in a free-living culture system, as described here, has considerable advantages. This approach has also been used successfully to study the hydrogenase enzyme system in different R. japonicum strains (Lim, 1978). Strain C33 has also been shown to lack nitrate reductase activity, but this property does not appear to affect the nitrogenase activity, as we have screened 200 spontaneous chlorate resistant (also nitrate reductase negative) clones and none had higher nitrogenase activity in free-living culture. This would suggest that the elevated nitrogenase activity is a consequence of NTG induced mutation. The lack of nitrate reductase in this mutant provides a useful additional property for studying the interaction of nitrate on nitrogen fixation in the symbiotic situation.

Experiments in controlled environment as well as field experiments are important to evaluate the true symbiotic capability of this Rhizobium strain (Burton, 1976). The rate of nitrogen utilization in the host plant may also play an important role in the overall evaluation of yield (see Williams and Phillips, this volume).

SUMMARY

A culture of Rhizobium japonicum (strain 3Ilb 110) was mutagenized and then screened for chlorate resistance. These clones were isolated and assayed for nitrogenase activity in free-living culture. Five out of one hundred colonies tested were found to have elevated levels of whole cell nitrogenase activity. Several protein differences were observed between one of these mutants (strain C33) and the wild type strain on two-dimensional polyacrylamide gels. Strain C33 was also defective in nitrate reductase activity and had high nitrogenase activity in free-living culture in the presence of nitrate whereas no activity was detected with the wild type strain under similar conditions.

REFERENCES

Andersen, K. and Shanmugam, K. T., 1977, Nitrogen fixation: determination of the ratio of formation of H_2 to NH_4^+ catalyzed by nitrogenase of Klebsiella pneumoniae in vivo, J. Gen. Microbiol., 103:107.

Burton, J., 1976, Pragmatic aspects of the Rhizobium-leguminous plant association, in: "Proceedings of the First International Symposium on Nitrogen Fixation," Vol. 2, W. E. Newton and C. J. Nyman, eds., Washington State University Press, Pullman.

Campbell, D. H., Garvey, J. S., Cremer, N. E., and Sussdorf, D. H., 1970, Methods in Immunology: A laboratory text for instruction and research, 2nd ed., W. A. Benjamin, Inc., Menlo Park, CA.

Emerich, D. W., Ruiz-Argüeso, T., Ching, T.-M., and Evans, H. J., 1979, Hydrogen-dependent nitrogenase activity and ATP formation in Rhizobium japonicum bacteroids, J. Bacteriol., 137: 153.

Hardy, R. W. F., Holsten, D., Jackson, E. K., and Burns, R. C., 1968, The acetylene-ethylene assay for N_2-fixation: laboratory and field evaluation, Plant Physiol., 43:1185.

Hua, S. T., 1978, Energy efficient mutants of Rhizobium japonicum 110, Abstr. Annu. Meet. Am. Soc. Microbiol.

Kennedy, I. R., Rigaud, J., and Trinchant, J. C., 1975, Nitrate reductase from bacteroids of Rhizobium japonicum: enzyme characteristics and possible interaction with nitrogen fixation, Biochim. Biophys. Acta, 397:24.

Kurz, W. G. W., and LaRue, T. A., 1975, Nitrogenase activity in rhizobia in absence of a plant host, Nature, 256:407.

Lim, S. T., 1978, Determination of hydrogenase in free-living cultures of Rhizobium japonicum and energy efficiency of soybean nodules, Plant Physiol., 62:609.

Lim, S. T., Andersen, K., Tait, R., and Valentine, R. C., 1980, Genetic engineering in agriculture: Hydrogen uptake (Hup) genes, Trends in Biochem. Sci., 5:167.

Lowry, O. H., Rosebrough, N. J., Farr, A. L., and Randall, R. J., 1957, Protein measurement with the Folin phenol reagent, J. Biol. Chem., 193:265.

Maier, R. J., and Brill, W. J., 1978, Mutant strains of Rhizobium japonicum with increased ability to fix nitrogen for soybean, Science, 201:448.

McComb, J. A., Elliot, J., and Dilworth, M. J., 1975, Acetylene reduction by Rhizobium in pure culture, Nature, 256:409.

Mielenz, J. R., Jackson, L. E., O'Gara, F., and Shanmugam, K. T., 1979, Fingerprinting bacterial chromosomal DNA with restriction endonuclease EcoRl: comparison of Rhizobium spp. and identification of mutants, J. Microbiol., 25:803.

O'Farrell, P. H., 1975, High resolution two-dimensional electrophoresis of proteins, J. Biol. Chem., 250:4007.

O'Gara, F., and Shanmugam, K. T., 1976, Regulation of nitrogen fixation by rhizobia: export of fixed N_2 as NH_4^+, Biochim. Biophys. Acta, 437:313.

O'Gara, F., and Shanmugam, K. T., 1977, Regulation of nitrogen fixation in Rhizobium spp. Isolation of mutants of Rhizobium trifolii which induce nitrogenase activity, Biochim. Biophys. Acta, 500:277.

Pagan, J. D., Child, J. J., Scowcroft, W. R., and Gibson, A. H., 1975, Nitrogen fixation by Rhizobium cultured on a defined medium, Nature, 256:406.

Schmidt, E. L., Bankole, R. O., and Bohlool, B. B., 1968, Fluorescent-antibody approach to study of rhizobia in soil, J. Bacteriol., 95:1987.

Schubert, K. R. and Evans, H. J., 1976, Hydrogen evolution: a major factor affecting the efficiency of nitrogen fixation in nodulated symbionts, Proc. Natl. Acad. Sci., 73:1207.

Scott, D. B., Hennecke, H., and Lim, S. T., 1980, The biosynthesis of nitrogenase Mo-Fe protein polypeptides in free-living cultures of Rhizobium japonicum, Biochim. Biophys. Acta, in press.

DETRIMENTAL AND BENEFICIAL EFFECTS OF OXYGEN EXERTED ON HYDROGEN-OXIDIZING BACTERIA

H. G. Schlegel, M. E. K. Ibrahim, E. Wilde, K. Schneider, M. Schlesier, B. Friedrich, and K. A. Malik
Institut für Mikrobiologie der Georg-August-Universität Göttingen, Grisebachstr. 8, 3400 Göttingen, FRG and Deutsche Sammlung von Mikroorganismen, GBF, Grisebachstrasse 8, 3400 Göttingen, FGR.

INTRODUCTION

The presence and absence of oxygen divides the biosphere into oxic and anoxic ecosystems, provides aerobic and anaerobic living conditions and enables either aerobiontic or anaerobiontic organisms to grow. As an electron acceptor of a high positive redox-potential oxygen enables aerobiontic cells to channel the substrate-derived electrons through a long respiratory chain and to generate metabolic energy with great efficiency. Oxygen enables cells to utilize aliphatic, aromatic and isoprenoid hydrocarbons as substrates which are not biodegradable under anaerobic conditions. There is no doubt that the beneficial effects of oxygen prevail. Looking closer, however, one realizes that almost all organisms, irrespective of their metabolic type, are exposed to possible oxygen toxicity. Oxygen exerts detrimental effects on anaerobionts such as the strictly anaerobic bacteria like *Succinivibrio*, *Butyrivibrio*, *Clostridium haemolyticum* and the methanogenic bacteria (Loesche, 1969). But even the strictly aerobic bacteria and higher organisms suffer from damage by oxygen (Gottlieb, 1975). This is seen in the defense mechanisms which enable aerobic organisms to cope with oxygen toxicity. At least some of the toxic species of oxygen, such as hydrogen peroxide, superoxide anions, hydroxyl radicals and singlet oxygen, produced by the metabolizing cell can be detoxified or quenched by special enzymes such as superoxide dismutase, catalase, peroxidases and carotenoids and possibly other pigments (Fridovich, 1976; Hassan and Fridovich, 1978; Krinsky, 1978; Morris, 1975, 1978).

There are at least three metabolic processes and key enzyme systems in aerobic organisms which are known to be exceptionally sensitive to oxygen:

1) Nitrogen fixation and nitrogenase are very oxygen-sensitive. This is reflected by the low oxygen tolerance of the majority of aerobic nitrogen-fixing bacteria under conditions of nitrogen fixation *in vivo*. With the exception of *Azotobacter* (Robson, 1979) and the heterocysts of cyanobacteria nitrogen fixation by suspensions of single cells is confined to less than 0.02 bar oxygen.

2) The evolution and utilization of molecular hydrogen is known to be oxygen-sensitive, too. All hydrogenases so far studied are susceptible to inactivation by oxygen, especially the hydrogenases from strict and facultative anaerobes (Schlegel and Schneider, 1978).

3) The function of ribulose bisphosphate carboxylase/oxygenase depends on the partial pressures of carbon dioxide and oxygen, and the oxygenase reaction prevails when the p_{O_2} is high and the p_{CO_2} is low (Lorimer and Andrews, 1980b). This has recently been demonstrated in elegant experiments using a glycolate-negative mutant of *Alcaligenes eutrophus* type strain (King and Andersen, 1980). Although the ribulose bisphosphate carboxylase itself is not extremely oxygen-sensitive, its misfunction is drastically increased by a high oxygen partial pressure resulting in a detrimental loss of CO_2-acceptor molecules and consequently in a futile cycle of organic carbon compounds accompanied by an enormous waste of energy.

These considerations allow the prediction that the aerobic hydrogen-oxidizing and the nitrogen-fixing bacteria will belong to the most oxygen-sensitive bacteria among the strict aerobes. Since an increasing number of species of hydrogen bacteria is found to be able to fix nitrogen (Gogotov and Schlegel, 1974; Wiegel and Schlegel, 1976; Wiegel et al., 1978) and vice versa nitrogen-fixing bacteria are able to grow chemolithotrophically with hydrogen (Simpson et al., 1979; Lepo et al., 1980; J. Döbereiner, personal communication; Malik and Schlegel, 1980), and since all of these bacteria have in common to fix carbon dioxide via the ribulose bisphosphate cycle, one can expect that this physiological group faces more problems of oxygen damage than other bacteria.

The present paper summarizes some recent studies on the tolerance of hydrogen-oxidizing bacteria to 1) hydrogen peroxide and 2) oxygen and presents 3) the characteristics of growth of the hydrogen-oxidizing bacterium *Alcaligenes latus* and its ability to fix nitrogen. Furthermore, 4) the mechanism of oxygen damage to the soluble NAD-reducing hydrogenase of *Alcaligenes eutrophus* H16 will be presented, and finally 5) a short review will be given on the effects of severe oxygen limitation on cells of A. eutrophus. See articles n this volume for further reading on the importance of oxygen (Evans et al., Shanmugam et al., and Tait et al.).

The bacteria used for most experiments to be described were Alcaligenes eutrophus strain H16 (ATCC 17699, DSM 428), the mutant PHB 4 (DSM 541) deficient in the synthesis of poly-β-hydroxybutyric acid (PHB) derived from the wild-type H16, A. eutrophus type strain (ATCC 17697, DMS 531), A. eutrophus strain N9A and the PHB-free mutant and the double mutant N9A-PHB$^-$O3-HB$^-$ which is not able to utilize β-hydroxybutyrate (HB).

OXYGEN SUPPLY WITH HYDROGEN PEROXIDE PLUS CATALASE

Oxygen can be supplied to suspensions of growing cells by continuous addition of hydrogen peroxide provided catalase (bovine liver catalase) is added to the medium. This has been demonstrated for various bacteria and yeasts with fructose or glucose as substrates when growing up to cell densities of 1-2 g cell dry weight/liter (Schlegel, 1977). Under autotrophic conditions it was also possible to provide oxygen to cells of Alcaligenes eutrophus H16 by supplying H_2O_2 directly to the liquid medium under a gas atmosphere containing only H_2 + CO_2. However, the doubling times of the cells were higher than during normal growth under a gas mixture of H_2, O_2 and CO_2. H_2O_2 supported aerobic heterotrophic growth at high cell densities, too (Ibrahim, 1979; Ibrahim and Schlegel, 1980a). However, when with increasing cell densities and consequently increasing oxygen demands of the cells the flow rate and the concentration of the H_2O_2 solution was increased to maintain a concentration of 1.8 mg dissolved oxygen in the medium, the growth rate slowly decreased (Fig. 1). The retardation and finally the suppression of growth could not be overcome by increasing the concentration of catalase in the medium or by adding scavengers of toxic oxygen species. Samples taken at various times after the start of H_2O_2 supply resumed growth after lag phases that were positively correlated to the time the cells were exposed to H_2O_2 aeration (Fig. 2). The reason for the failure to cultivate A. eutrophus at high çell densities, resp. high H_2O_2 flow rates, was finally traced back to the cells' sensitivity to the H_2O_2 steady state concentrations of more than 15 μM H_2O_2 which are reached under such conditions. Three to five μM H_2O_2 were well tolerated. The high steady-state concentrations could not be avoided by adding catalase in excess or continually. The incomplete cleavage of H_2O_2 by catalase is due to the low affinity of catalase toward H_2O_2 and to the rapid inactivation of the enzyme during continuous addition of H_2O_2 (Ibrahim and Schlegel, 1980b). From these experiments we learned that the growth inhibitory effect exerted by low concentrations of H_2O_2 is a slow process.

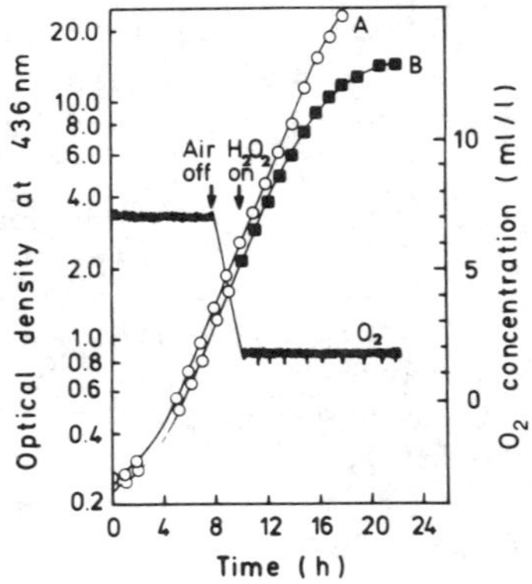

Fig. 1. Growth curves of Alcaligenes eutrophus H16-PHB$^-$4 growing on fructose in 2 l fermentors (800 rpm, 30 C) and aerated either with air (A:300 ml air/l·min) or hydrogen peroxide (B). Suspension B was aerated with air; when an OD (436 nm) of 1.2 was reached the air flux was discontinued. When the oxygen concentration decreased to reach 1.8 mg O_2/liter, 20 mg catalase/liter was added and a solution of 12% (v/v) H_2O_2 was continually pumped to maintain this O_2 concentration.

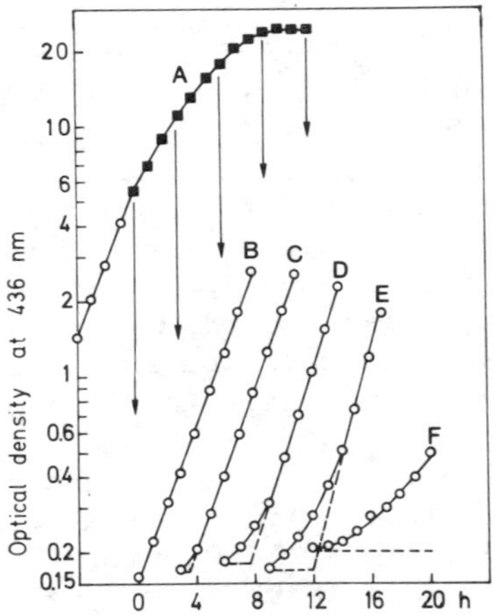

Fig. 2. Influence of the duration of hydrogen peroxide aeration on the growth ability of cells of A. eutrophus PHB$^-$4. A, suspension before and after transition from conventional to H_2O_2 aeration; B, C, D, E, and F, growth curves of cell samples taken from suspension A after a time period of H_2O_2 aeration of 0, 3, 6, 9, and 12 hours and transferred to fresh, warm fructose media. Symbols: O——O growth with air; ■——■ growth with H_2O_2 aeration.

TOLERANCE OF HYDROGEN-OXIDIZING BACTERIA TO VARIED OXYGEN CONCENTRATIONS

When growing heterotrophically, e.g., on fructose, Alcaligenes eutrophus H16-PHB$^-$4 demonstrates a remarkable tolerance to high concentrations of dissolved oxygen in the medium (Fig. 3). A. eutrophus shares this tolerance with many bacteria (Fig. 4) and yeasts (E. Wilde, unpublished).

In contrast, growth under autotrophic conditions under an atmosphere of H_2, O_2 and CO_2 was severely affected by the oxygen concentration. Growth curves of several hydrogen bacteria, measured in shaken Erlenmeyer flasks at low cell densities at varied O_2 concentrations in the gas phase, indicate common features of response as well as marked differences (Fig. 4). At an oxygen concentration higher than 20% (v/v) the growth rates were lower than at 20 or less percent oxygen. Contrary to the effect of hydrogen peroxide, increased oxygen partial pressures did not result in inhibition and arrest of growth but allowed exponential growth at a lower rate. This indicates that the lower growth rate is due either to rate-limiting enzyme synthesis or continuous enzyme destruction.

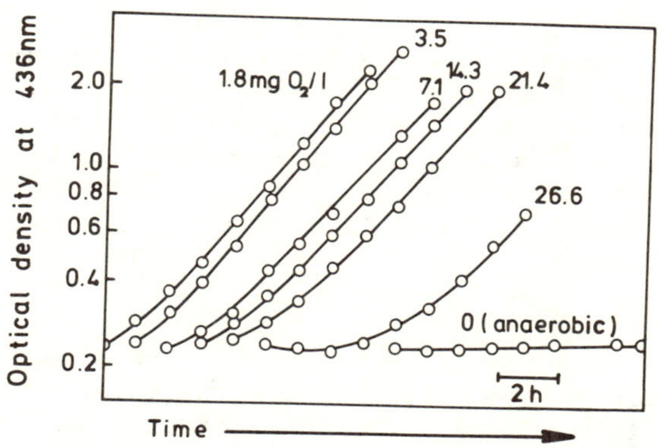

Fig. 3. Growth curves of A. eutrophus H16-PHB$^-$4 at various O_2 concentrations. The O_2 concentrations were measured by means of an O_2-electrode in the suspension and monitored by varying the composition of the N_2-O_2-mixture used for aeration (300 ml/min).

The sensitivity to O_2 concentrations above 20% varied from species to species. While A. eutrophus H16 responded to the increase of the oxygen concentration with a gradual decrease of the growth rate, the type strain and Paracoccus denitrificans reached a threshold value between 40 and 60% O_2 above that the growth rate decreased abruptly to almost zero. The growth inhibiting effect of oxygen on autotrophically growing hydrogen-oxidizing bacteria has been known a long time, however, no experimental data were available to explain this phenomenon.

MECHANISM OF OXYGEN DAMAGE TO THE SOLUBLE, NAD-REDUCING HYDROGENASE OF ALCALIGENES EUTROPHUS H16

The hydrogenases so far isolated are iron-sulfur proteins (Mortenson and Chen, 1974; Schlegel and Schneider, 1978). While some hydrogenases are easily and irreversibly inactivated by oxygen, such as that of Clostridium pasteurianum (Nakos and Mortenson, 1971; Mortenson and Chen, 1976), others have been described to be "oxygen-stable" but to form an inactive, reversible complex with O_2 (see Schneider and Schlegel, 1980, for details).

A. eutrophus H16 contains two completely different hydrogenases which have both been purified and characterized. One enzyme is soluble, contains FMN as chromophor, reduces NAD and is of complex structure (Schneider and Schlegel, 1976, 1978; Schneider et al., 1979; Schneider and Cammack, 1978). The second hydrogenase is membrane-bound, smaller, flavin-free and unable to react with pyridine-nucelotides (Schink and Schlegel, 1978, 1979; Schink, 1978; Weiss et al., 1980).

The purified soluble hydrogenase (hydrogen: NAD^+ oxidoreductase, EC 1.12.1.2; specific activity 57 µmol H_2/min·mg protein) consists of 4 subunits (Fig. 5) three of which are non-identical (molecular weights 68,000, 60,000, 29,000; molar ratio 1:1:2). The enzyme contains 12 atoms each of iron and labile sulfur per molecule localized in two (2Fe-2S) and two (4Fe-4S) centers (Schneider and Cammack, 1978; Schneider et al., 1979). The assignment of the iron-sulfur centers and the FMN molecules to the subunits is still lacking.

The purified hydrogenase turned out to be highly stable under air and could even be stabilized in the bacterial crude extract by flushing with 60% (v/v) oxygen or addition of 1 mM ferricyanide (Table 1). Instability and irreversible inactivation of hydrogenase by oxygen was observed if reducing and electron donating compounds, such as H_2 and NADH, were present in enzyme solutions. Inactivation kinetics of the purified hydrogenase revealed that the rate of inactivation was significantly dependent on the presence and on the concentration of oxygen (Fig. 6).

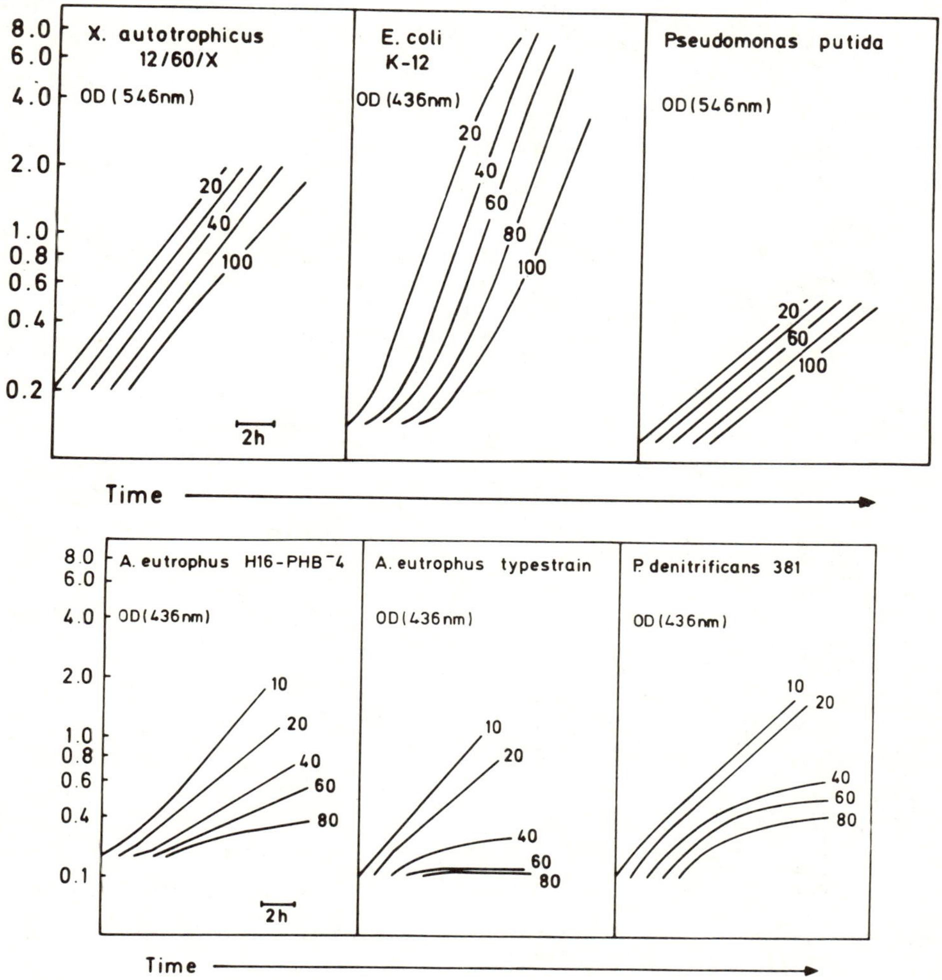

Fig. 4. Growth curves of various bacteria grown at varied O_2 concentrations. Left side: the cells of Xanthobacter autotrophicus, Escherichia coli K-12 and Pseudomonas putida were grown on fructose under gas mixtures containing 20, 40, 60, 80, or 100% (v/v) O_2 (in N_2). Right side: the cells of Alcaligenes eutrophus H16-PHB$^-$4, A. eutrophus type strain and Paracoccus denitrificans 381 were grown autotrophically under gas mixtures containing 10, 20, 40, 60, or 80% (v/v) oxygen, 10% CO_2 and 10% H_2. Suspensions were shaken in 300 ml Erlenmeyer flasks at 30 C.

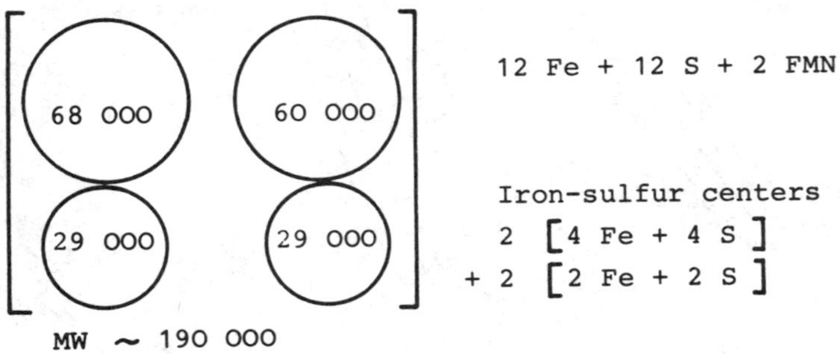

Fig. 5. The components of the soluble, NAD-reducing hydrogenase from *A. eutrophus* H16.

Table 1 Stability of hydrogenase in crude extracts prepared from autotrophically grown cells of *Alcaligenes eutrophus* H16

Conditions of storage (4° C)		Hydrogenase activity after 12 hours storage (in percent)
Atmosphere	Additions	
Air	Ferricyanide (1 mM)	100
Air	-	38
H_2	-	16
H_2	NADH (5 µM)	0
N_2	Mercaptoethanol (1 mM)	3
30 % O_2 + 70 % N_2	-	50
60 % O_2 + 40 % N_2	-	85

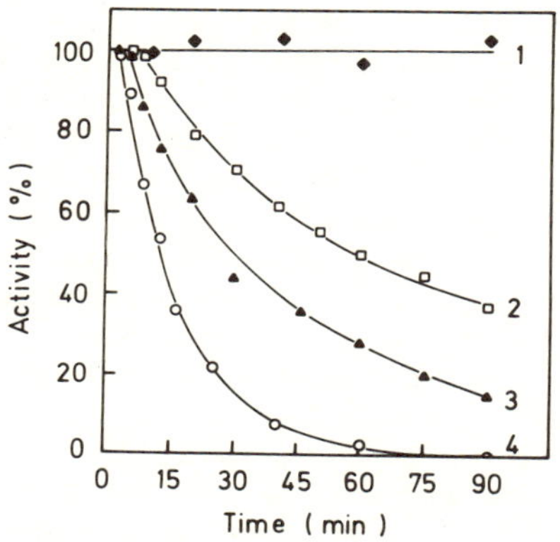

Fig. 6. Kinetics of inactivation of soluble hydrogenase from A. eutrophus H16 in the presence of NADH (25 µM), hydrogen and varied concentrations of oxygen. (1) no NADH under air, (2) with NADH under 5% (v/v) O_2 in H_2, (3) with NADH under 20% O_2 in H_2, (4) with NADH under 50% O_2 in H_2.

The experimental conditions were as follows: The enzyme was kept under optimal conditions in 50 mM Tris-HCl buffer, pH 8.0, at 30 C. Then the solutions were subjected to a gas atmosphere containing a mixture of H_2 and O_2; the latter was present at concentrations from 5 to 50% (v/v). Finally 25 µM NADH was added. At time intervals samples were taken, and the activity of soluble hydrogenase was determined by NADH formation. As shown in Figure 6 a rapid loss of enzyme activity was observed. The starting point of the inactivation by oxygen was dependent on the concentration of oxygen. In the presence of 5% O_2 5 to 8 min elapsed, with 20% O_2 2 to 4 min and with 50% O_2 only 1 to 2 min, before the activity declined. This correlated with half lives of 60, 30 and 12 min, respectively. Under H_2-free air the enzyme activity did not decrease at all within 90 min. Other controls (not shown in Fig. 6), H_2 and H_2 + O_2 mixtures in the absence of NADH or in the presence of only low concentrations of NADH (absence of H_2) caused only insignificant effects. This indicated that each of the assay components, H_2, NADH and O_2 are necessary to effect the complete inactivation of the purified hydrogenase.

These observations and the nature of the enzyme as a conjugated iron-sulfur protein suggested that in the presence of electron donors and oxygen, superoxide radicals are formed which cause the inactivation of the enzyme.

For the detection of superoxide anions hydroxylamine was used; it is oxidized by $O_2^{-}\cdot$ to form nitrite (Elstner and Heupel, 1976) and did not affect the enzyme or its catalytic reaction. Parallel assays both containing enzyme, NADH, buffer, and H_2 + O_2 and one with hydroxylamine for O_2 measurements and the other without hydroxylamine to follow the decrease of enzyme activity were prepared. From the enzyme inactivation kinetics and the kinetics of superoxide anion formation (Fig. 7) as well as from Table 2 the conclusion can be drawn, that the rate of O_2^- production is correlated with the oxygen concentration in the gas mixture. Furthermore, the degree of enzyme inactivation and the concentration of O_2^- anions formed are positively correlated. Under air, pure hydrogen or H_2 + O_2 mixtures in the absence of NADH no radicals were formed.

When superoxide dismutase was added to the reaction mixtures, the formation of nitrite was much lower; only 5 to 10% of the rate observed without superoxide dismutase were detected (Fig. 7). This observation confirmed the assumption that nitrite formation is caused by superoxide anions which were produced by hydrogenase under the reaction conditions.

The possibility that hydrogen peroxide or hydroxyl radicals are involved in the inactivation process, too, was examined by adding catalase and hydroxyl radicals trapping agents (formate, benzoate) to the assay mixture. No stabilizing effect became evident. Furthermore, the hydrogenase activity was not affected by 1 mM H_2O_2. These experiments led to the conclusion that hydrogenase is indeed inactivated merely by self-produced superoxide radicals.

Further experiments aiming at the stabilization of hydrogenase in the presence of H_2, NADH and O_2 resulted in the findings that the addition of cytochrome c and superoxide dismutase provide the most favorable conditions to stabilize the enzyme (Table 3).

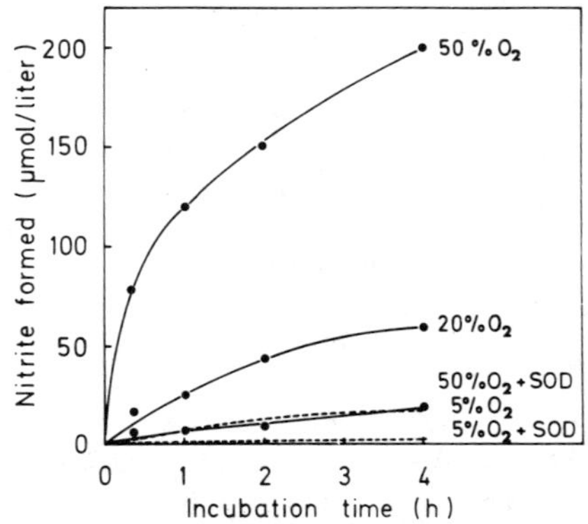

Fig. 7. Production of superoxide radicals in the presence of hydrogenase (0.1 mg/ml), 25 µM NADH and 0.5 mM NH_2OH in 50 mM Tris buffer, pH 8.0, at 30 C. The gas atmosphere contained 5, 20 or 50% (v/v) O_2 in H_2 as indicated; two samples contained 2 mg superoxide dismutase (SOD)/ml.

Table 2 Production of superoxide radicals by purified hydrogenase of A. eutrophus H16 and loss of enzyme activity under various reducing conditions

Conditions of storage: Atmosphere	Conditions of storage: Additions	Nitrite formed (E_{530}) after 12 hours	Hydrogenase activity after 12 hours (in percent)
H_2	-	0	100
Air	-	0	100
80 % H_2 + 20 % O_2	-	0	98
99.9 % H_2 + 0.1 % O_2	NADH, 25 µM	0.137	51
99.0 % H_2 + 1.0 % O_2	NADH, 25 µM	0.502	17.5
95.0 % H_2 + 5.0 % O_2	NADH, 25 µM	1.420	7.9
80 % H_2 + 20 % O_2	NADH, 25 µM	1.610	$<$ 1
Air	NADH, 25 µM	0.022	89.8
Air	NADH, 1 mM	0.420	5.6

Table 3 Stabilization of hydrogenase under reducing conditions in the presence of 25 µM NADH

Conditions of storage (4° C): Atmosphere	Stabilizers added	Hydrogenase activity after 24 h storage (in percent)
Air	none	100
99.9 H_2/0.1 O_2	none	7
99.9 H_2/0.1 O_2	Glucose + glucose oxidase	53
99.9 H_2/0.1 O_2	Dithionite, 1 mM	62
99.9 H_2/0.1 O_2	NH_2OH, 2 mM	48
99.9 H_2/0.1 O_2	Cytochrome c	67
99.9 H_2/0.1 O_2	Superoxide dismutase	72
H_2	none	81
H_2	Cytochrome c	95
H_2	Superoxide dismutase	97

In contrast to many so-called "oxygen-stable" hydrogenases, which are very oxygen-sensitive during the catalytic reaction, the catalytic activity of soluble hydrogenase of A. eutrophus H16 is not or only to a negligible extent inhibited by oxygen (Schneider and Schlegel, 1977; Schneider et al., 1979). This concerns, as documented in Table 4, both hydrogenase-catalyzed reactions, the reduction of NAD as well as the evolution of H_2.

At first sight, the statements that hydrogenase is oxygen-tolerant under reaction conditions and is inactivated by oxygen under the same conditions appear to be contradictory. However, the inactivation process caused by the action of superoxide radicals did not start immediately. It was a relatively slow process and did not affect the enzyme during the short period of the enzyme assay in the almost oxygen-free solution.

On the basis of this study the terms "O_2-insensitivity" and "O_2-stability" should be reconsidered. First, the environment and the substances present during exposure to oxygen should be respected and indicated; second, one should clearly differentiate between storage conditions and the conditions of enzyme function. Many hydrogenases are sensitive to oxygen in both the non-catalytic mode and in the catalytic mode, while others are O_2-sensitive only in the catalytic mode. The present results ask for differentiating three conditions: 1) storage, 2) assay conditions, i.e., short-term catalytic function, and 3) long-term exposure to oxygen under functional coditions.

Table 4 Hydrogenase activity in the presence of oxygen

H_2 atmosphere containing oxygen, in percent (v/v)	Hydrogenase activity, initial rate (in percent) NAD-reduction	H_2-production from NADH
0	100	100
0.1	100	100
5	100	n.d.
10	95	n.d.
20	92	87
40	87	n.d.
60	82	80

The data are yet too preliminary to explain the various interactions of oxygen with hydrogenase. More experiments have to be done to elucidate the action of oxygen on this enzyme. However, as a heuristic principle a model may be presented and briefly discussed.

The oxidized hydrogenase, as it is obtained by enzyme purification, binds oxygen to give a catalytically non-functional but very stable enzyme conformation. This is unable to react with hydrogen or to reduce oxygen with electrons derived from hydrogen. In this state oxygen acts as a stabilizer in the sense that the binding of oxygen prevents the enzyme from being converted into the catalytically active conformation. If, however, low amounts of a reduced electron carrier, such as NADH or its non-reactive analogue NADPH, are present, hydrogenase is converted into the catalytically active conformation and rapidly reduced by H_2. Apparently, by binding the pyridine nucleotide the active center of the enzyme becomes accessible to NAD and other electron acceptors, O_2 included. It is not oxygen itself but rather the product of univalent electron transfer to O_2, i.e., the superoxide radical, which when reaching a certain concentration, affects the enzyme and leads to its irreversible inactivation.

Recent comparison of EPR spectra indicated that some oxygen-stable hydrogenases (from Desulfovibrio desulfuricans, D. gigas, Escherichia coli, Chromatium vinosum, Rhodospirillum rubrum, and also from the aerobic hydrogen-oxidizing bacteria Pseudomonas pseudoflava, Nocardia opaca and Alcaligenes eutrophus) when examined at 25 K in the state as prepared in the presence of oxygen, gave a signal centered about g = 2.02 (Cammack et al., 1980). This signal was, although with less intensity, detectable with the soluble, NAD-reducing hydrogenase of A. eutrophus, too. Upon reduction, the g = 2.02 signals disappeared, and in the case of the hydrogenases of hydrogen bacteria new signals around g = 1.95 appeared. Possibly the Fe-S cluster giving the g = 2.02 signal is that site which binds oxygen and thereupon converts the enzyme into an inactive form in which the hydrogen activating site is protected.

The questions to be studied are obvious. The soluble hydrogenase of A. eutrophus H16 with its (2Fe-2S) centers, (4Fe-4S) centers, and FMN poses the problem which of the three electron transferring components are involved in the reduction of electron acceptors, the production of superoxide radicals and possibly hydrogen peroxide.

IN VIVO INACTIVATION OF SOLUBLE HYDROGENASE

Occasional observations on the high fluctuations of hydrogenase activity in cells of A. eutrophus in the past remained unexplained until kinetic experiments under well-defined conditions were carried out. When hydrogenase-rich cells were transferred to a medium containing e.g. pyruvate, the synthesis of soluble hydrogenase was completely repressed. However, the specific activity of hydrogenase decreased at a much faster rate than that to be expected merely due to dilution of the enzyme during growth. Results of representative experiments shown in Table 5 and Figure 8 were obtained with cells which had been grown heterotrophically with pyruvate under substrate-limiting conditions under air for 48 h. These cells contained a high level of hydrogenase. When kept under air or 100% hydrogen, in the absence of an electron donor or under nitrogen, the hydrogenase was perfectly stable. When the cells were, however, exposed to a gas mixture of H_2 and O_2 the hydrogenase activity decreased rapidly. A less marked but significant decrease occurred also when the cells were exposed to air in the presence of pyruvate. Upon removal of oxygen the decrease instantaneously stopped. The inactivation process proceeded even under autotrophic growth conditions (Fig. 8). When chloramphenicol (120 g/ml) or rifampicin (20 g/ml) were added to autotrophically growing cells growth ceased immediately. Concomitantly an abrupt decrease of hydrogenase activity occurred which followed 1st order reaction kinetics. No decrease, but only a slight increase in hydrogenase activity was found when the cells were kept under pure hydrogen. Under growth conditions with H_2, O_2 and CO_2 hydrogenase activity remained at an almost constant level. On the basis of these results the following statements can be made: 1) during autotrophic growth with H_2, O_2 and CO_2 hydrogenase is inactivated continuously; 2) in order to maintain a constant level of catalytically active hydrogenase the enzyme has to be resynthesized at a high rate; 3) since the enzyme assays were performed with whole cells it is unlikely that there exists a mechanism which protects the hydrogenase from being inactivated *in vivo*; 4) since protein synthesis inhibitors do not arrest the inactivation process and no significant reactivation could ever be observed it is unlikely that hydrogenase is inactivated reversibly by a regulatory protein as described for the enzyme-catalyzed covalent modification; 5) the time course of inactivation and the half life of hydrogenase are similar to those measured for the purified enzyme; 6) the conditions of hydrogenase inactivation are the same *in vivo* and *in vitro*.

In conclusion there is increasing evidence that the *in vivo* inactivation of soluble hydrogenase is also due to superoxide radicals which are formed by the enzyme itself, when oxygen and an electron donor such as hydrogen or organic substrate are present. Thus the destruction of hydrogenase activity is a result of a process we would like to designate as "autogenous inactivation." Under autotrophic conditions the permanent inactivation of hydrogenase does not lead to the cessation of growth because the cell compensates for the loss of enzyme activity by a high differential rate of hydrogenase synthesis.

Table 5 Stability of soluble hydrogenase in whole cells of *A. eutrophus* H16

Conditions of incubation		Hydrogenase activity after 5 h (in percent)	Half life (in h)
Gas atmosphere	Additions		
Air	-	1oo *	stable
H_2	-	1oo	stable
$H_2 + O_2$ (8:2)	-	1o	1.5
Air	Pyruvate (o.2 %)	65	8
N_2	Pyruvate (o.2 %)	1oo	stable

* The hydrogenase activity of the heterotrophically grown cells was about 5oo µmol H_2/min · g cell protein, as measured by a whole cell enzyme assay.

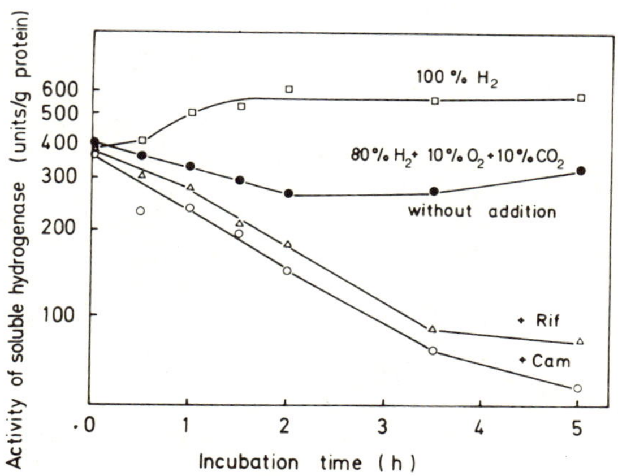

Fig. 8. Inactivation kinetics of soluble hydrogenase in whole cells of *A. eutrophus* H16 under autotrophic growth conditions. Chloramphenicol (120 µg/ml; O) or rifampicin (20 µg/ml; Δ) were added to suspensions of autotrophically growing cells. Samples were taken at intervals and tested for soluble hydrogenase. A control was left without addition (●) and in one flask the gas mixture was replaced by hydrogen (□).

N_2 FIXATION AND AUTOTROPHIC GROWTH OF ALCALIGENES LATUS

Among the hydrogen-oxidizing bacteria the ability to fix nitrogen has been detected in strains 7 C and 14 g (Gogotov and Schlegel, 1974) and further strains (Wiegel and Schlegel, 1976; Berndt et al., 1976) of the new genus and species *Xanthobacter autotrophicus* (Wiegel et al., 1978). *Alcaligenes latus* which was, when first described, regarded to be unable to fix nitrogen (Palleroni and Palleroni, 1978), proved to be a nitrogen-fixing bacterium. It is a comparatively huge bacterium (2 x 5 μm) and is able to grow heterotrophically on glucose with a doubling time of 1.2 h at 37 C. With N_2 as the sole nitrogen source the doubling time is 16 h. The efficiency of nitrogen fixation was 14 mg N/g glucose. Both autotrophic growth and nitrogen fixation exhibit O_2-sensitivity and proceed at a higher rate under 2% (v/v) O_2 than under 20% O_2 (Figs. 9a and 9b). The bacterium tends to accumulate poly-β-hydroxybutyric acid (PHB) when growth is limited. When grown under conditions of O_2-limitation up to 8 g dry weight/liter the cells contained 0.74 g PHB per g dry weight (Ch. Jung, unpublished).

METABOLIC CHANGES DURING EXTREMELY LOW OXYGEN SUPPLY TO VARIOUS ALCALIGENES EUTROPHUS STRAINS

The metabolism of strictly aerobic bacteria such as *Alcaligenes eutrophus*, *Pseudomonas putida* or *Paracoccus denitrificans* is in general considered to be purely respiratory. The growth of these bacteria is known to depend on an external electron acceptor (oxygen, nitrate, nitrite and others) to make energy conversion by electron-transport phosphorylation possible. The failure to grow anaerobically without external electron acceptor was considered to be due to the lack of fermentation enzymes which are necessary for the disposal of excess electrons. Furthermore, on the basis of the present evolutionary philosophy it was assumed that useless genetic information is deleted and eliminated. Contrary to this general belief there are observations on the existence of "cryptic", "silent" or "dormant" genes defined as those which are not expressed under ordinary cultural conditions (Riley and Anilionis, 1978). *Alcaligenes eutrophus* contains such genes which are not expressed under ordinary aerobic conditions with unlimited oxygen supply, however, become recognizable, when the oxygen demand of the cells is not satisfied.

The excretion of significant amounts of ethanol, lactate, succinate, butanediol, acetate, β-hydroxybutyrate and others by *A. eutrophus* (Vollbrecht et al., 1978) prompted us to investigate this process more closely. Studies on the conditions, by which the excretory process is promoted, indicated that the deficiency of oxygen is the decisive factor (Vollbrecht and Schlegel, 1978, 1979). The metabolites were excreted when aerobically grown cells

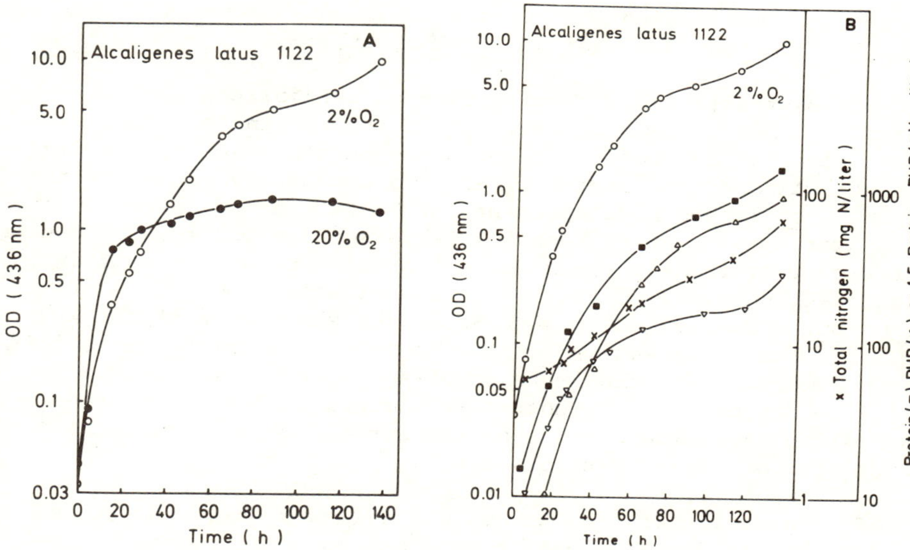

Fig. 9. Heterotrophic growth of Alcaligenes latus (DSM 1122) with dinitrogen as the sole nitrogen source under gas mixtures containing 2 and 20% (v/v) oxygen in nitrogen. A: growth in the presence of 2% (O) and 20% (●) (v/v) O_2. B: growth curve for 2% O_2 (O) and (∇) protein, (Δ) poly-β-hydroxybutyric acid, (X) total nitrogen, and (■) 1.5 protein + PHB in mg/l. The cells were grown in 2 l fluted Erlenmeyer flasks with an initial suspension volume of 600 to 800 ml at 30 C. Mineral medium without combined nitrogen in the presence of 0.5% glucose; magnetic stirring at 500 rpm. The inoculum was grown in a glutamate medium and transferred about 5 mg total (Kjeldahl) nitrogen into 1 l main culture medium.

were subsequently incubated under conditions of restricted oxygen supply. The nature of the metabolite was specifically related to the relative respiration rate (RRR); that is the ratio of the actual respiration rate imposed on the cells by restricting the aeration rate to the maximum respiration rate which the cells expressed under conditions of unrestricted oxygen supply during exponential growth (Vollbrecht et al., 1979). When cells of A. eutrophus strains H16, N9A, B19, G27, or G29 and mutants (PHB^-, HB^-) derived therefrom were grown in 5 l fermentors at 30 C and high aeration rates to reach 2 g dry weight cells/liter and were then exposed to low aeration rates (RRR's of about 2.4 to 80%) in the presence of gluconate (1.5%) as substrate, fermentation products were excreted. Excretion started about 4 h after the shift-down of oxygen supply and continued for about 20 h. A list of the excreted metabolites related to the conditions of excretion (RRR) recently appeared (Schlegel and Vollbrecht, 1980).

The synthesis and excretion of some metabolites such as ethanol, lactate, acetate, butanediol, and succinate indicated the presence of enzymes which are not expected to be synthesized in a strictly aerobic bacterium and had previously not been looked for in A. eutrophus. Measurements of lactate dehydrogenase, alcohol dehydrogenase, and butanediol dehydrogenase confirmed the expectation that these enzymes are not present or are present in insignificant specific activities in aerobically grown cells of A. eutrophus. However, 10 to 20 h after the shift-down of the RRR to about 5%, the three enzymes were present with significant activities. Comparison of the increase of specific enzyme activities in cells that had been incubated at various different relative respiration rates indicated that the enzymes were formed at RRR values of 5 or 14.5%, but not at 40%.

The formation of the enzymes was a slow process and lasted up to 20 h. At 3 and 5% RRR lactate and butanediol dehydrogenases reached specific activities of 100 to 700 $\mu mol \cdot min^{-1}$ g protein (Schlegel and Vollbrecht, 1980). When after aerobic exponential growth the cells were incubated under strictly anoxic conditions butanediol dehydrogenase was the only enzyme which was formed. The synthesis of hydrogenase and acetohydroxyacid synthase was not affected.

The causative regulation mechanisms resulting in metabolite excretion and fermentative enzyme synthesis have not been studied so far. Several successive regulatory events may be involved; their sequence is a matter of speculation. The enforced low respiration rates result in a high NADH/NAD ratio (Harrison, 1973, 1976). This may directly cause inhibition of various enzymes, such as 2-oxoglutarate dehydrogenase, citrate synthesis, pyruvate dehydrogenase and others leading to the excretion of 2-oxoglutarate, β-hydroxybutyrate, acetate and others. The change of the ATP level and energy charge may give signals in addition to the NADH/NAD ratio. Subsequently, the accumulation of intermediary metabolites may function at the genetic level and allow the derepression of the fermentative enzymes required for disposal of excess electrons.

Thus, oxygen exerts a repressive effect on the genes for fermentative enzymes in A. eutrophus and a variety of other strictly aerobic bacteria. Under aerobic conditions these genes are not expressed due to active respiratory metabolism. Only under "semianaerobic" conditions allowing only decreased respiration is the genetic information expressed. This mode of regulation reminds us of the regulation of nitrogenase in nitrogen-fixing bacteria, which is apparently only formed under conditions of very restricted oxygen supply.

SUMMARY AND CONCLUSIONS

Hydrogenases and ribulose bisphosphate carboxylase involved in the metabolism of the hydrogen-oxidizing bacteria so far studied are sensitive to oxygen. Studies on the influence of oxygen concentration on the growth rates of various bacteria indicated an almost insignificant influence of the oxygen concentration up to 100% (v/v) oxygen in the gas phase on heterotrophic growth but a strong growth retarding and inhibitory effect on autotrophic growth. The soluble NAD-reducing hydrogenase of A. eutrophus H16 was recognized as a target of the detrimental effects of oxygen. The purified enzyme was found to produce superoxide radicals under reaction conditions, and the conditions resulting in enzyme inactivation were found to be positively correlated with the conditions promoting superoxide anion formation. The decline of hydrogenase activity under growth conditions in whole cells after arrest of protein synthesis by chloramphenicol indicated that the oxygen-dependent decay of hydrogenase occurs continuously in the growing cells, too.

Nitrogen fixation by the nitrogen-fixing hydrogen bacteria, Xanthobacter autotrophicus and Alcaligenes latus, is even more sensitive to oxygen. In the absence of combined nitrogen these bacteria tolerate less than 2% O_2 (v/v) in the atmosphere.

The strong regulatory effect of oxygen via the respiration rate on suspensions of A. eutrophus and various other strictly aerobic bacteria became evident when the activity of fermentation enzymes were detected in cells kept for some hours under conditions allowing only low respiration rates. The studies indicate that the formation of lactate, alcohol and butanediol dehydrogenase is repressed under ordinary cultural conditions (in the presence of oxygen allowing maximum respiration rates); however, these enzymes are derepressed when respiration is decreased by reduced oxygen supply.

The studies on the effect of oxygen on the growth of aerobic bacteria certainly need more attention and appreciation.

ACKNOWLEDGMENTS

This paper contains results of recent unpublished experimental work of the authors and of E. Heine, and Ch. Jung. The enzyme studies depended on the experience and support of Dr. R. Brinkmann. The technical assistance of M. Meyer, A. Finck and K. Jochim is gratefully acknowledged. The investigations were supported by grants from the Deutsche Forschungsgemeinschaft and by Förderungsmittel des Landes Niedersachsen.

REFERENCES

Berndt, H., Ostwal, K.-P., Lalucat, J., Schumann, Ch., Mayer, F., and Schlegel, H. G., 1976, Identification and physiological characterization of the nitrogen fixing bacterium *Corynebacterium autotrophicum* GZ 29, *Arch. Microbiol.*, 108:17.

Cammack, R., Lalla-Maharajh, W. V., and Schneider, K., 1980, EPR studies of some oxygen-stable hydrogenases, *in*: "Interactions Between Iron and Proteins in Oxygen and Electron Transport," C. Ho, ed., Elsevier-North Holland, New York.

Elstner, E. F., and Heupel, A., 1976, Inhibition of nitrite formation from hydroxylammoniumchloride: A simple assay for superoxide dismutase, *Anal. Biochem.*, 70:616.

Fridovich, I., 1976, Oxygen radicals, hydrogen peroxide, and oxygen toxicity, *in*: "Free Radicals in Biology", Vol. 1, W. A. Pryor, ed., Academic Press, New York.

Gogotov, J. N., and Schlegel, H. G., 1974, N_2-fixation by chemoautotrophic hydrogen bacteria, *Arch. Microbiol.*, 97:359.

Gottlieb, S. F., 1975, Effect of hyperbaric oxygen on microorganisms, *in*: "Annual Review of Microbiology," Vol. 25, C. E. Clifton, S. Raffael, and M. P. Starr, eds., Annual Review Inc., Palo Alto, California.

Harrison, D. E. F., 1973, Growth, oxygen, and respiration, *Crit. Rev. Microbiol.*, 2:185.

Harrison, D. E. F., 1976, The regulation of respiration rate in growing bacteria, *in*: "Advances in Microbial Physiology," Vol. 14, A. H. Rose, and D. W. Tempest, eds., Academic Press, London.

Hassan, H. M., and Fridovich, I., 1978, Superoxide dismutase and its role for survival in the presence of oxygen, *in*: "Life Sciences Research Report 13: Strategies of Microbial Life in Extreme Environments," M. Shilo, ed., Verlag Chemie, Weinheim.

Ibrahim, M. E. K., 1979, Sauerstoffversorgung von Mikroorganismen durch Wasserstoffperoxid, Ph.D. Thesis.

Ibrahim, M. E. K., and Schlegel, H. G., 1980a, Oxygen supply to bacterial suspensions of high cell densities by hydrogen peroxide, *Biotechnol. Bioeng.*, in press.

Ibrahim, M. E. K., and Schlegel, H. G., 1980b, Efficiency of bovine liver catalase as a catalyst to cleave H_2O_2 added continually to buffer solutions, *Biotechnol. Bioeng.*, in press.

King, W. R., and Andersen, K., 1980, Efficiency of CO_2 fixation in the glycollate oxidoreductase mutant of *Alcaligenes eutrophus* which exports fixed carbon as glycollate, *Arch. Microbiol.*, in press.

Krinsky, N. I., 1978, Carotenoid pigments: Multiple mechanisms for coping with the stress of photosensitized oxidations, *in*:

"Life Sciences Research Report 13: Strategies of Microbial Life in Extreme Environments," M. Shilo, ed., Verlag Chemie, Weinheim.

Lepo, J. E., Hanus, F. J., and Evans, H. J., 1980, Chemoautotrophic growth of hydrogen-uptake-positive strains of Rhizobium japonicum, J. Bacteriol., 141:664.

Loesche, W. J., 1969, Oxygen sensitivity of various anaerobic bacteria, Appl. Microbiol., 18:723.

Lorimer, G. H., and Andrews, T. G., 1980, The C_2 photo- and chemo-respiratory carbon oxidation cycle, Plant Biochem., in press.

Malik, K. A., and Schlegel, H. G., 1980, Enrichment and isolation of new nitrogen-fixing hydrogen bacteria, FEMS Microbiol. Lett., 8:101.

Morris, J. G., 1975, The physiology of obligate anaerobiosis, Adv. Microb. Physiol., 12:169.

Morris, J. G., 1978, Nature of oxygen toxicity in anaerobic microorganisms, in: "Life Sciences Research Report 13: Strategies of Microbial Life in Extreme Environments," M. Shilo, ed., Verlag Chemie, Weinheim.

Mortenson, L. E., and Chen, J. S., 1974, Hydrogenase, in: "Microbial Iron Metabolism," J. B. Nielands, ed., Academic Press, New York.

Mortenson, L. E., and Chen, J. S., 1976, Properties of the hydrogenase from Clostridium pasteurianum, in: "Microbial Production and Utilization of Gases (H_2, CH_4, CO_2)," H. G. Schlegel, G. Gottschalk, and N. Pfennig, eds., E. Goltze KG, Göttingen.

Nakos, G., and Mortenson, L. E., 1971, Structural properties of hydrogenase from Clostridium pasteurianum W5, Biochem., 10:2442.

Palleroni, N. J., and Palleroni, A. V., 1978, Alcaligenes latus, a new species of hydrogen-utilizing bacteria, Int. J. Syst. Bacteriol., 28:416.

Riley, M., and Anilionis, A., 1978, Evolution of the bacterial genome, in: "Annual Review of Microbiology," Vol. 32, M. P. Starr, J. L. Ingraham, and S. Raffael, eds., Annual Review Inc., Palo Alto, California.

Robson, R. L., 1979, Characterization of an oxygen-stable nitrogenase complex isolated from Azotobacter chroococcum, Biochem. J., 181:569.

Schink, B., 1978, Membrane-bound hydrogenase from Alcaligenes eutrophus: Biochemical and immunological characterization of the solubilized and purified enzyme, in: "Hydrogenases: Their Catalytic Activity, Structure and Function," H. G. Schlegel, and K. Schneider, eds., Erich Goltze KG, Göttingen.

Schink, B., and Schlegel, H. G., 1978, Hydrogen metabolism in aerobic hydrogen-oxidizing bacteria, Biochimie, 60:297.

Schink, B., and Schlegel, H. G., 1979, The membrane-bound hydrogenase of Alcaligenes eutrophus. I. Solubilization, purifica-

tion, and biochemical properties, Biochim. Biophys. Acta, 567:315.
Schlegel, H. G., 1977, Aeration without air. Oxygen supply by hydrogen peroxide, Biotechnol. Bioeng., 19:413.
Schlegel, H. G., and Schneider, K., 1978, Introductory report: Distribution and physiological role of hydrogenases in microorganisms, in: "Hydrogenases: Their Catalytic Activity, Structure and Function," H. G. Schlegel, and K. Schneider, eds., Erich Goltze KG, Göttingen.
Schlegel, H. G., and Vollbrecht, D., 1980, Formation of the dehydrogenases for lactate, ethanol and butanediol in the strictly aerobic bacterium Alcaligenes eutrophus, J. Gen. Microbiol., 117:475.
Schneider, K., and Cammack, R., 1978, Soluble hydrogenase from Alcaligenes eutrophus, an iron-sulfur flavoprotein, in: "Hydrogenases: Their Catalytic Activity, Structure and Function," H. G. Schlegel, and K. Schneider, eds., Erich Goltze KG, Göttingen.
Schneider, K., Cammack, R., Schlegel, H. G., and Hall, D. O., 1979, The iron-sulfur centers of soluble hydrogenase from Alcaligenes eutrophus, Biochim. Biophys. Acta, 578:445.
Schneider, K., and Schlegel, H. G., 1976, Purification and properties of soluble hydrogenase from Alcaligenes eutrophus H16, Biochim. Biophys. Acta, 452:66.
Schneider, K., and Schlegel, H. G., 1977, The NAD reducing soluble hydrogenase from Alcaligenes eutrophus, in: "Second International Symposium - Microbial Growth on C_1-Compounds," G. K. Skryabin, M. V. Ivanov, E. N. Kondratjeva, G. A. Zavarzin, Yu. A. Trotsenko, and A. I. Nesterov, eds., Scientific Centre for Biological Research USSR, Academy of Sciences, Pushchino.
Schneider, K., and Schlegel, H. G., 1978, Identification and quantitative determination of the flavin component of soluble hydrogenase from Alcaligenes eutrophus, Biochem. Biophys. Res. Commun., 84:564.
Schneider, K., and Schlegel, H. G., 1980, Production of superoxide radicals by soluble hydrogenase from Alcaligenes eutrophus H16, Biochem. J., in press.
Simpson, F. B., Maier, R. J., and Evans, H. J., 1979, Hydrogen-stimulated CO_2 fixation and coordinate induction of hydrogenase and ribulosebisphosphate carboxylase in a H_2-uptake positive strain of Rhizobium japonicum, Arch. Microbiol., 123:1.
Vollbrecht, D., and Schlegel, H. G., 1978, Excretion of metabolites by hydrogen bacteria. II. Influences of aeration, pH, temperature and age of cells, Eur. J. Appl. Microbiol. Biotechnol., 6:157.
Vollbrecht, D., and Schlegel, H. G., 1979, Excretion of metabolites by hydrogen bacteria. III. D(-)-3-hydroxybutanoate, Eur. J.

Appl. Microbiol. Biotechnol., 7:259.
Vollbrecht, D., El Nawawy, M. A., and Schlegel, H. G., 1978, Excretion of metabolites by hydrogen bacteria. I. Autotrophic and heterotrophic fermentations, Eur. J. Appl. Microbiol. Biotechnol., 6:145.
Vollbrecht, D., Schlegel, H. G., Stoschek, G., and Janczikowski, A., 1979, Excretion of metabolites by hydrogen bacteria. IV. Respiration rate-dependent formation of primary metabolites and of poly-3-hydroxybutanoate, Eur. J. Appl. Microbiol. Biotechnol., 7:276.
Weiss, A. R., Schneider, K., and Schlegel, H. G., 1980, Purification and properties of the membrane-bound hydrogenase of Pseudomonas pseudoflava GA3, Curr. Microbiol., 3:317.
Wiegel, J., and Schlegel, H. G., 1976, Enrichment and isolation of nitrogen fixing hydrogen bacteria, Arch. Microbiol., 107:139.
Wiegel, J., Wilke, D., Baumgarten, J., Opitz, R., and Schlegel, H. G., 1978, Transfer of the nitrogen-fixing hydrogen bacterium Corynebacterium autotrophicum Baumgarten et al. to Xanthobacter gen. nov., Int. J. Syst. Bacteriol., 28:573.

HYDROGEN UPTAKE (HUP) PLASMIDS: CHARACTERIZATION OF MUTANTS AND REGULATION OF THE EXPRESSION OF HYDROGENASE

R. C. Tait, K. Andersen, G. Cangelosi, and S. T. Lim

Plant Growth Laboratory
University of California
Davis, CA 95616

The supply of energy may often be a rate-limiting step in symbiotic N_2 fixation and plant productivity (Hardy and Havelka, 1975). The efficient use of metabolic energy during N_2 fixation should involve recovery and utilization of the hydrogen that is generated by nitrogenase. It has been demonstrated that several agronomically important *Rhizobium* spp. do not have an active H_2 uptake system (Schubert and Evans, 1976; Lim, 1978), suggesting that a significant portion of the energy supplied by the host plant may be lost via nitrogenase-catalyzed H_2 evolution. One method of enhancing the N_2 fixation of legumes would be to introduce hydrogen uptake (*hup*) genes into *Rhizobium* strains that lack *hup* activity.

IDENTIFICATION OF HUP PLASMIDS

In the course of our work with the autotrophic gram-negative bacterium *Alcaligenes eutrophus*, we observed that the presence of agents that interfere with plasmid replication, such as mitomycin C, resulted in the loss of the Hup^+ phenotype. When examined by agarose gel electrophoresis, DNA isolated from Hup^+ strains of *A. eutrophus* contained a covalently closed circular DNA of approximately 200 megadaltons (pAE1). Following treatment of cultures with mitomycin C, two types of Hup^- derivatives were obtained: cells which were unable to grow with H_2 as the energy source, and cells which grow with a greatly reduced rate with H_2 as the energy source. Agarose gel electrophoresis revealed that the completely Hup^- derivatives no longer contain the plasmid pAE1, while the partially Hup^- derivatives all contain large plasmids. Co-electrophoresis of plasmids from Hup^+ and Hup^- cells suggested that many of the plasmids present in the Hup^- cells contain deletions of 5-20 megadaltons. When the Hup^+ phenotype was conjugally transferred to a completely Hup^- strain that contained no large plasmid, the Hup^+ derivative acquired the plasmid pAE1. This suggests that pAE1 contains information necessary for hydrogen uptake.

CHARACTERICATION OF HUP^- MUTANTS

A. eutrophus contains two types of hydrogenase, one membrane-bound enzyme coupled to the respiratory chain, and one soluble NAD-reducing enzyme (Rittenberger and Repaske, 1961; Schneider and Schlegel, 1976, 1977; Schink and Schlegel, 1979). In order to determine which of the hydrogenases were affected during the construction of the Hup^- derivatives, six independently derived mutants of A. eutrophus H1 were chosen for further characterization. Total hydrogenase activity was determined using methylene blue as electron acceptor, and activity of the soluble hydrogenase was determined using NAD as electron acceptor in cell-free extracts. The results, shown in Table 1, revealed three classes of mutants. The completely Hup^- derivatives H1-3 and H1-6 contained no large plasmids and were lacking in both methylene blue and NAD reducing activities, suggesting that neither the membrane-bound nor the soluble hydrogenase is present in cells cured of the plasmid pAE1. The mutants H1-1, H1-4 and H1-6 have reduced growth rates under autotrophic conditions and contain undetectable levels of NAD-dependent hydrogenase activity although they contain significant levels of hydrogen uptake with methylene blue as electron acceptor. The deletions present in these plasmids apparently prevent the function of the soluble NAD-reducing hydrogenase but allow the function of the membrane-bound hydrogenase. These two classes of mutants have been previously reported by Schink and Schlegel (1978) with A. eutrophus H16. The mutant H1-2 has normal hydrogen uptake with methylene blue but only 10% of the normal level of NAD-dependent hydrogenase activity. This mutant may synthesize an altered NAD-dependent hydrogenase of reduced activity, may be the result of a reduced level of expression of the hydrogenase, or might be deficient in a component necessary for optimal function of the soluble hydrogenase. The non-reverting nature of these deletion mutants and the inactivation of either the soluble hydrogenase or both hydrogenases makes these strains ideal candidates for the isolation of genes involved in hydrogen metabolism. We have begun construction of recombinant plasmids containing DNA from several autotrophic bacteria, including A. eutrophus H1 and R. japonicum 110 utilizing the cloning vector pRK290 (Ditta and Helinski, this volume) and hope to use these A. eutrophus H1 Hup^- mutants in the isolation and characterization of hydrogenase genes.

Table 1. Properties of A. eutrophus H1 derivatives defective in H_2 utilization.

	Generation time (hr)		Plasmid size	Hup activity (μmol/hr/mg protein) e^- acceptorP	
	$H_2/O_2/CO_2$	Ile	(mdaltons)	methylene blue	NAD
A. eutrophus H1	2.7	6-7	~200	83	114
H1-1	7.5	6-7	~180	30	11
H1-2	8.5	6-7	~200	30	<0.2
H1-3	-	6-7	-	<1	<0.2
H1-4	10.5	6-7	~180	29	<0.2
H1-5	10.5	6-7	~180	29	<0.2
H1-6	-	6-7	-	1	<0.2

Stationary-phase cells grown in isoleucine minimal medium under air were used. Hup activity with 5 mM methylene blue as electron acceptor was determined by following the disappearance of H_2 by gas chromatography (Lim, 1978). Activity using NAD as electron acceptor was determined by following the reduction of NAD spectrophotometrically (Schneider and Schlegel, 1977).

REGULATION OF HYDROGENASE UPTAKE

The expression of hydrogenase in both R. japonicum and A. eutrophus H1 is subject to some form of modulation. O'Gara and Shanmugam (1978) demonstrated that free-living cultures of Rhizobium spp. are capable of inducing hydrogenase activity under low partial pressures of O_2, and Evans and co-workers have reported that cultures of R. japonicum will consume H_2 under certain conditions (Maier et al., 1978; Evans et al., this volume). As shown in Table 2, the nature of the carbon source present during the growth of R. japonicum and A. eutrophus H1 can affect the expression of hydrogenase by as much as 15-fold. Because of the similarity of this modulation of expression of hydrogenase to the

Table 2. Effect of carbon source on hydrogen uptake.

	Carbon source	Hydrogenase (μmol H_2/hr/mg protein)
1. *R. japonicum* 110		
	L-glutamate	2.9
	D-fructose	2.6
	Glycerol	2.0
	D-ribose	1.8
	Sodium malate	0.9
	Sodium malate + 2 mM cyclic AMP	2.3
2. *A. eutrophus*		
	Succinate	2.4
	Sodium malate	9.3
	Fructose	23.8
	Isoleucine	33.6
	Sodium pyruvate	40.6
	*Fructose	12.3
	*Fructose + 2 mM cyclic AMP	25.0

1. Cells were grown with 0.1% glutamate as nitrogen source and 0.4% of the indicated carbon source. Hydrogenase activity was determined 90-100 hr after inoculation. Modified from previously published data (Lim and Shanmugam, 1979).

2. Cells were grown in mineral medium containing 0.25% of the indicated carbon source. Hydrogenase activity was determined during the exponential phase of growtn or (*) during the lag phase prior to exponential growth.

catabolite repression observed in enteric bacteria, we examined the effects of cyclic AMP on hydrogenase expression. As shown in Table 2, the presence of 2 mM cAMP during growth of *R. japonicum* and *A. eutrophus* H1 under conditions that normally repressed hydrogenase function was found to increase hydrogenase activity. The presence of a transcriptional inhibitor (rifampicin) or translational inhibitors (chloramphenicol, tetracycline) prevented the stimulation of hydrogen uptake by cAMP, suggesting that the mechanism involves both transcription and translation of genes involved in hydrogenase function.

Because the cAMP effect in other bacteria has been found to involve a cAMP binding protein (CRP) that functions as a transcriptional modulator, cellular extracts of *R. japonicum* 110 and *A. eutrophus* H1 were examined for the presence of such a receptor protein. CRP activity was detected in both organisms, and the activity from *A. eutrophus* H1 has been purified to apparent homogeneity using the method of Boone and Wilcox (1978). This CRP protein has a subunit molecular weight of 14,000 daltons and appears to function as a dimer with a molecular weight of approximately 30,000 daltons as determined by Sephadex G-100 gel filtration and polyacrylamide gel electrophoresis. Although this protein is approximately one-half of the size of the protein isolated from *E. coli* (Anderson et al., 1971), the affinities of the two proteins for cAMP are similar (apparent dissociation constant of 2.5×10^{-6} vs. 1.3×10^{-6} M (Emmer et al., 1970)). The *A. eutrophus* H1 CRP protein does not stimulate the *in vitro* transcription of the arabinose operon in the presence of cAMP, suggesting that although the proteins from *A. eutrophus* and *E. coli* may be functionally similar, they are of sufficiently different specificity that they are not interchangeable. The isolation of hydrogenase genes from *A. eutrophus* H1 will allow investigation of the *in vitro* interaction of this CRP protein with DNA and may establish a role for cAMP and CRP protein in the regulation of the expression of hydrogenase.

ACKNOWLEDGMENTS

This material is based upon research supported by the National Science Foundation under Grant No. PFR 77-07301. Any opinions, findings, and conclusions or recommendations expressed in this publication are those of the authors and do not necessarily reflect the views of the National Science Foundation.

REFERENCES

Andersen, W. B., Schneider, A. B., Emmer, M., Perlman, R. L., and Pastan, I., 1971, Purification and properties of the cyclic 3',5'-monophosphate receptor protein which mediates cyclic adenosine 3',4'-monophosphate-dependent gene transcription in Escherichia coli, J. Biol. Chem., 246:5929.

Boone, T., and Wilcox, G., 1978, A rapid high-yield purification procedure for the cyclic 3',5'-monophosphate receptor protein from Escherichia coli, Biochim. Biophys. Acta, 541:528.

Emmer, M., deCrombrugghe, B., Pastan, I., and Pertman, R., 1970, Cyclic AMP receptor protein of E. coli: its role in the synthesis of inducible enzymes, Proc. Natl. Acad. Sci. USA, 66: 480.

Hardy, R. W., and Havelka, U. D., 1975, Nitrogen fixation research: a key to world food?, Science, 188:633.

Lim, S. T., 1978, Determination of hydrogenase in free-living cultures of Rhizobium japonicum and energy efficiency of soybean nodules, Plant Physiol., 62:609.

Lim, S. T., and Shanmugam, K. T., 1979, Regulation of hydrogen utilization in Rhizobium japonicum by cyclic AMP, Biochim. Biophys. Acta, 584:479.

Maier, R. J., Campbell, N. E. R., Hanus, F. J., Simpson, F. B., Russell, S. A., and Evans, H. J., 1978, Expression of hydrogenase activity in free-living Rhizobium japonicum, Proc. Natl. Acad. Sci. USA, 75:3258.

O'Gara, F., and Shanmugam, K. T., 1978, Mutant strains of clover rhizobium (Rhizobium trifolii) that form nodules on soybean (Glycine max), Proc. Natl. Acad. Sci. USA, 75:2343.

Schink, B., and Schlegel, H. G., 1978, Mutants of Alcaligenes eutrophus defective in autotrophic metabolism, Arch. Microbiol., 117:123.

Schink, B., and Schlegel, H. G., 1979, The membrane-bound hydrogenase of Alcaligenes eutrophus. I. Solubilization, purification, and biochemical properties, Biochim. Biophys. Acta, 567:315.

Schneider, K., and Schlegel, H. G., 1976, Purification and properties of soluble hydrogenase from Alcaligenes eutrophus H16, Biochim. Biophys. Acta, 452:66.

Schneider, K., and Schlegel, H. G., 1977, Localization and stability of hydrogenases from aerobic hydrogen bacteria, Arch. Microbiol., 112:229.

Schubert, K. R., and Evans, H. J., 1976, Hydrogen evolution: a major factor affecting the efficiency of nitrogen fixation in nodulated symbionts, Proc. Natl. Acad. Sci. USA, 73:1207.

Wittenberger, C. L., and Repaske, R., 1961, Studies on hydrogen oxidation in cell-free extracts of Hydrogenomonas eutropha, Biochim. Biophsy. Acta, 47:542.

DETECTION OF PLASMIDS IN *RHIZOBIUM JAPONICUM*

D. B. Scott* and R. C. Tait**

*Applied Biochemistry Division, D.S.I.R., Private Bag, Palmerston North, New Zealand, and **Plant Growth Laboratory, University of California, Davis, CA 95616

SUMMARY

Plasmids from 20-155 megadaltons have been detected in a variety of strains of *R. japonicum*. No relationship between these plasmids and hydrogen uptake capability was apparent.

Recent results indicate that in various *Rhizobium* species, host-range, bacteriocin production, pigment production, and hydrogenase may be associated with the presence of plasmids (Brewin et al., this volume). Autotrophic hydrogen bacteria, including strains of *Alcaligenes eutrophus*, are able to use hydrogen gas (H_2) as an energy source. Recent observations indicate that the hydrogen uptake (hup) capability of *A. eutrophus* is associated with the presence of a large plasmid (Andersen et al., submitted for publication; Tait et al., this volume). The demonstration by Evans and co-workers (Hanus et al., 1979; Evans et al., this volume) that strains of *Rhizobium japonicum* are capable of autotrophic growth in the presence of H_2 suggested that large plasmids may also be involved in the H_2 uptake phenotype of *R. japonicum*. To investigate this possibility we have examined a variety of *R. japonicum* strains for the presence of plasmid DNA.

A variety of *R. japonicum* strains were examined, including 110 (hup^+), 110/RP4 (hup^+), 31 (hup^-), 76 (hup^-), 83 (hup^-), and soy 440 (hup^-). All strains were grown in 20 ml of MY medium without mannitol (Lim, 1978) to minimize production of extracellular polysaccharides, which interfere with DNA isolation. Cultures were harvested by centrifugation and DNA extracted by the SDS rapid screen procedure for large plasmids described by Andersen et al. (submitted for publication). DNA was subjected to electrophoresis

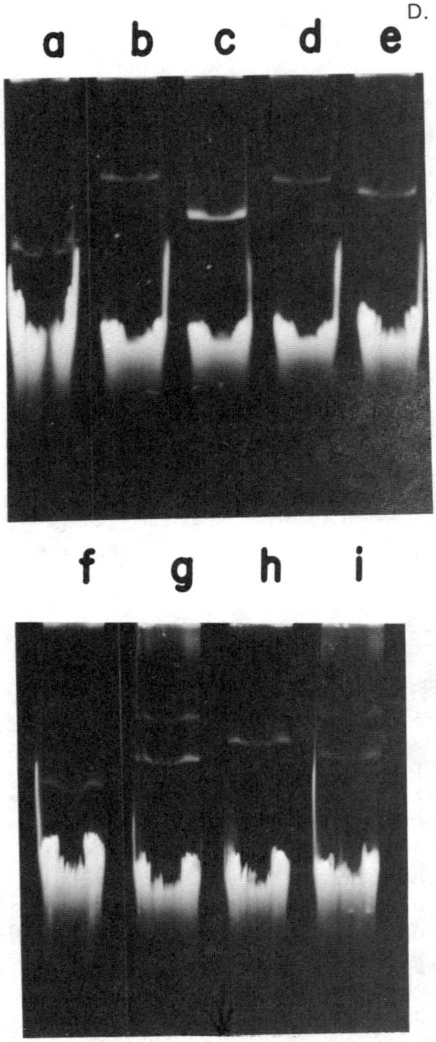

Fig. 1. Plasmids detected in *R. japonicum* strains. DNA was isolated and examined by electrophoresis on an 0.8% agarose gel. Illustrated is the DNA isolated from: a) *R. japonicum* 110/RP4 (hup^{+}); b) *R. japonicum* 76 (hup^{-}); c) *R. japonicum* 83 (hup^{-}); d) *R. japonicum* 31 (hup^{-}); e) *R. meliloti* U45; f) *R. japonicum* 110/RP (hup^{+}); g) *R. japonicum* soy 440 (hup^{-}); h) *R. meliloti* U45; i) *R. japonicum* 110 + soy 440. *R. meliloti* U45 was included as a standard, as the plasmid in this strain has previously been determined to have a mass of 101 megadaltons (Casse et al., 1979).

on an 0.8% agarose gel and visualized by ethidium bromide staining with UV-fluorescence photography. Several of the strains examined, including those shown in Figure 1, were found to contain large plasmids, ranging in size from 20 to 155 megadaltons. No plasmids could be detected in the hup^+ strain 110. To verify that plasmids can be purified from strain 110, a derivative of 110 containing the plasmid RP4 (36 megadaltons) was examined. As shown in Figure 1a and 1f, RP4 was easily detected in extracts of R. japonicum 110/RP4. To determine whether extracts of strain 110 have a tendency to degrade large plasmids during the isolation procedure, 10 ml of a culture of strain 110 was mixed with 10 ml of a culture of soy 440 and DNA was prepared from the mixed cells. As can be seen by comparing Figure 1g with 1i, the same three plasmid DNA bands were detected in both the mixed culture and the pure culture of soy 440. The results suggest that the failure to detect plasmids in strain 110 is not the result of a property of the strain itself, such as nucleolytic activity, for both the plasmid RP4 and the plasmids present in soy 440 can be detected in the presence of extracts of 110.

Because no correlation could be demonstrated between hup phenotype and plasmid presence, one might be tempted to conclude that plasmids are not involved in hydrogen uptake in R. japonicum. However, it should be emphasized that the failure to detect a plasmid with a particular purification protocol does not indicate that the plasmid does not exist, but only that it has not been detected. As an illustration of this point, we have observed that the large plasmid present in A. eutrophus (type strain) could not be detected when cellular lysis was performed with the detergents Brij, deoxycholate, sodium lauryl sarcosinate, or Triton-X100, but could be detected following lysis in the presence of SDS (R. Tait, unpublished observation).

Although we have detected a variety of plasmids in strains of R. japonicum, there is no obvious correlation between the presence of these plasmids and hydrogen uptake. Further genetic and biochemical work will be necessary to determine whether plasmids are involved in hydrogen uptake in R. japonicum.

ACKNOWLEDGMENTS

This material is based upon research supported by the National Science Foundation under Grant No. PFR 77-07301. Any opinions, findings, and conclusions or recommendations expressed in this publication are those of the authors and do not necessarily reflect the views of the National Science Foundation.

REFERENCES

Casse, F., Boucher, C., Julliot, J. S., Michel, M., and Denarie, J., 1979, Identification and characterization of large plasmids in *Rhizobium meliloti* using agarose gel electrophoresis, *J. Gen. Microbiol.*, 113:229.

Hanus, F. J., Maier, R. J., and Evans, H. J., 1979, Autotrophic growth of H_2-uptake-positive strains of *Rhizobium japonicum* in an atmosphere supplied with hydrogen gas, *Proc. Natl. Acad. Sci. USA*, 76:1788.

Lim, S. T., 1978, Determination of hydrogenase in free-living cultures of *Rhizobium japonicum* and energy efficiency of soybean nodules, *Plant Physiol.*, 62:609.

CHEMOLITHOTROPHY IN *RHIZOBIUM*

H. J. Evans, J. E. Lepo, F. J. Hanus, K. Purohit, and S. A. Russell

Laboratory for Nitrogen Fixation Research
Oregon State University
Corvallis, OR 97331

INTRODUCTION

The first pure cultures of *Rhizobium* species were described by Beijerinck in 1888 who surface disinfected nodules from *Vicia*, *Lathrus* and *Trifolium* and cultured the nodule endophytes on gelatin plates containing plant extracts, sucrose, and asparagine. Since this pioneering discovery, *Rhizobium* species have been considered chemoorganotrophs and always have been supplied with carbon substrates such as hexoses, pentoses, or complex carbohydrates (Vincent, 1978).

A series of reviews (Evans et al., 1978, 1980a, 1980b; Dixon, 1978) have summarized recent research showing that a relatively small proportion of strains of most of the *Rhizobium* species possess the capacity for synthesis of a H_2-oxidizing hydrogenase in nodules. This system catalyzed the oxidation of the H_2 that is evolved as a by-product of the N_2-fixation reaction. Further investigations have defined the conditions necessary for the expression of activity of the hydrogenase system in pure cultures of H_2-uptake-positive (Hup^+) *R. japonicum* (Maier et al., 1978a; Lim, 1978). These studies have led to the discovery in our laboratory that Hup^+ strains of *R. japonicum* are capable of growing on a medium containing a trace of vitamins, inorganic salts, and a supply of H_2 and CO_2 in the gas phase (Hanus et al., 1979). The Hup^+ *R. japonicum* strains derive their metabolic energy from the oxidation of H_2 and thus may be categorized as chemolithotrophs (Rittenberg, 1969). It is the purpose of this paper to summarize some of the background information that led to these findings and to discuss their possible significance.

CARBON DIOXIDE UTILIZATION

Most of the bacteria are known to require CO_2 for growth. One hundred different bacterial isolates representing a variety of families and species of heterotrophic bacteria were examined by Valley and Rettger (1927) who observed that all exhibited a requirement for CO_2. A suggestion of an involvement of CO_2 in the N_2-fixing process by legume root nodules was reported by Mulder and Veen (1960) who grew nodulated peas, beans, and clover in nitrogen-free nutrient solutions (pH 4.8) that were deprived of exogenous CO_2 in the aeration stream. Lack of added CO_2 caused marked decreases in growth, N_2 fixation, and extent of nodulation. No response to the removal of CO_2 from the aeration stream was observed when plants were cultured in solution at pH 6.8. Although no effects of the removal of effects of added CO_2 on photosynthesis were ruled out, Mulder and Veen (1960) suggested that CO_2 was needed for the growth of the host legume rather than the nodule bacteria.

Lowe and Evans (1962) cultured five different species of *Rhizobium* in a medium containing mineral salts, L arabinose, and vitamins and found that all of them exhibited an absolute requirement for CO_2 in the aeration stream. The addition to the medium of yeast extract or any one of a series of metabolites suspected of being direct or indirect products of carboxylation reactions failed to substitute for CO_2. From these results it was concluded that CO_2 is required for heterotrophic growth of at least five *Rhizobium* species.

Phosphoenol Pyruvate Carboxylase

Phosphoenol pyruvate (PEP) carboxylase has been identified in the cytosol fraction of nodules from *Vicia faba* (Lawrie and Wheeler, 1975), soybeans (Peterson and Evans, 1979), and lupins (Christeller et al., 1977), but this enzyme apparently is not present in either bacteroids or free-living rhizobia. Five isoenzymes of PEP carboxylase have been detected in soybean nodule extracts and the properties of the major ones have been described (Peterson and Evans, 1979). A positive correlation has been observed between the activities of nitrogenase and PEP carboxylase of nodule cytosol during the growth cycle of lupins (Christeller et al., 1977). It was proposed (Laing et al., 1979) that oxalacetate, formed via PEP carboxylase in lupin nodules serves as a substrate for the synthesis of aspartate and α-ketoglutarate through a transamination reaction. Furthermore, Laing et al. (1979) have pointed out that aspartate is required for the synthesis of asparagine, a known major nitrogenous constituent exported from lupin nodules. The scheme proposed by Scott et al. (1976) illustrating a relationship between N_2 fixation and oxalacetate synthesis via PEP carboxylase requires 1 mole of CO_2

fixed per mole of N_2 reduced, a stoichiometry that is consistent with the observations of Christeller et al., 1977.

Propionyl CoA Carboxylase

Carbon dioxide also is fixed in nodule bacteroids and free-living Rhizobium species via the propionyl CoA carboxylase reaction:

$$\text{Propionyl-CoA} + \text{ATP} + HCO_3^- \rightleftharpoons \text{Methylmalonyl-CoA} + \text{ADP} + \text{Pi}$$

Catalysis of this reaction has been demonstrated by extracts from R. japonicum bacteroids (Lowe and Evans, 1962), free-living cultures of R. japonicum (Simpson et al., 1979), and R. meliloti (De Hertogh et al., 1964). In R. meliloti and probably other Rhizobium species, this reaction is part of a pathway involving propionate activation to form propionyl CoA, carboxylation to yield methylmalonyl CoA and then conversion of methylmalonyl CoA to succinyl CoA via the methylmalonyl CoA racemase and mutase reactions. This series of reactions constitutes a route whereby propionate and CO_2 may be incorporated into the heme moiety of leghemoglobin or metabolized via the citric acid cycle (Jackson and Evans, 1966).

Ribulose Bisphosphate Carboxylase in Hydrogenase Induced Cells

Carbon dioxide fixation in the chemolithotrophic bacteria takes place via the Calvin cycle and the two enzymes that are unique in this cycle are ribulosebisphosphate (RuBP) carboxylase and phosphoribulokinase (Schlegel and Eberhardt, 1972). An indication that the Calvin cycle might be operative in the Hup^+ R. japonicum strains was provided by Simpson et al. (1979) who added 0.1 atm of H_2 over cultures of hydrogenase-induced and non-induced R. japonicum 122 DES and observed a low but reproducible activity of RuBP carboxylase in the induced cells. No activity was detected in those cells that had not been induced for hydrogenase. If succinate at, 10 mM, were added either with or without H_2 during the 24-hour preincubation period necessary for hydrogenase induction, RuBP carboxylase activity was not expressed. Further studies by Simpson et al. (1979) demonstrated that the time courses of induction of the hydrogenase and RuBP carboxylase in R. japonicum 122 DES roughly paralleled each other. In contrast, the activity of propionyl CoA carboxylase did not change appreciably during the hydrogenase and RuBP carboxylase induction period. Additional evidence of a regulatory relationship between hydrogenase and RuBP carboxylase in free-living R. japonicum was indicated by the observation that Hup^- mutants derived from the Hup^+ SR strain expressed neither hydrogenase nor RuBP carboxylase activities during conditions that led to the expression of both activities in the Hup^+ parent SR. Nodule extracts from bacteroids formed by the Hup+ SR strain, however, consistently contained relatively high hydrogenase activities but RuBP carboxylase activity was never detected (Simpson et al., 1979). These observations are consistent with a recent report by Maier (1980) describing mutants derived from *R. japonicum* SR, some of which lacked the capacity for expression of hydrogenase and others that expressed hydrogenase activity but failed to express activity of the carboxylase. Mutants of *R. japonicum* SR lacking hydrogenase activity and expressing low activity of RuBP carboxylase also have been isolated by J. E. Lepo in our laboratory (unpublished results, 1980). Obviously the regulation of hydrogenase and RuBP carboxylase in *R. japonicum* is complex and requires further study.

HYDROGENASE IN *RHIZOBIUM*

The hydrogenase system in *R. leguminosarum* bacteroids was discovered by Phelps and Wilson (1941) and was more intensively investigated by Dixon (1968-1972) who proposed that the enzyme system might benefit the N_2-fixing process by: (a) removal of O_2 from the immediate environment of the O_2-sensitive nitrogenase, (b) preventing the inhibition of nitrogenase by the H_2 produced as a by-product of the N_2-fixation reaction, and (c) conserving some of the energy expended through nitrogenase-catalyzed H_2 evolution by providing a means of H_2 supported oxidative phosphosphorylation. Dixon found two Hup+ strains of *R. leguminosarum* and no Hup+ strains of other species. For some unknown reason, *R. leguminosarum* is subject to loss of the capability for synthesis of the enzyme under laboratory conditions and the Hup+ strains that Dixon investigated now have lost their capability for hydrogenase expression (Dixon, personal communication, 1980).

The magnitude of H_2 loss from nodulated legumes was initially pointed out by Schubert and Evans (1976), and the occurrence of the hydrogenase activity in cowpea *Rhizobium* and *R. japonicum* was observed by Schubert and Evans (1977) and Schubert et al. (1977, 1978). Prior to this, Marauyama et al. (1967) and Marauyama (1975) reported that O_2 was required for tritium (T_2) uptake by soybean and lupin nodules and that O_2 inactivated the enzyme during attempts at purification.

Since the possession of the hydrogenase characteristic in legume inoculants has the potential of significant economic importance in the production of legumes, our laboratory has intensively investigated the properties, regulation, and possible transfer of this enzyme complex. Progress in these areas has been discussed in reviews by Evans et al. (1978, 1980a, 1980b), Dixon, 1978), and Anderson et al. (1979).

Table 1. Some properties of the hydrogenase system in Hup$^+$ strains of Rhizobium japonicum.

Properties	Type of preparation	Observations	References
Location in cells	Bacteroids (122 DES and 110)	Membranes	Ruiz-Argueso et al., 1979 Arp and Burris, 1979
Reaction	Bacteroids (110)	$H_2 + \frac{1}{2}O_2 \rightarrow H_2O$	McCrae et al., 1978
Apparent K_m for O_2	Bacteroids (122 DES) Bacteroids (122 DES) with LbO_2 Free-living (110)	1 μM 10.6 ηM 1.05 μM	Ruiz-Argueso et al., 1979 Emerich et al., 1980a Lim and Shanmugan, 1979
Apparent K_m for H_2	Purified extract (110 Bacteroids (122 DES) Free-living cells (110)	1.4 μM 0.05 μM 1.8 μM	Arp and Burris, 1979 Emerich et al., 1980b Lim and Shanmugan, 1979
Catalysis of exchange	Free-living Hup$^+$ strains Bacteroids and free-living cells (122 DES) Purified enzyme (110)	Rapid T_2 exchange D_2-H_2 exchange not detected Weak H_2-O_2 exchange	Lim, 1978 Emerich et al., 1980b Arp and Burris (unpublished, 1980)
Molecular weight	Purified enzyme (110)	63,300	Arp and Burris, 1979
Acceptors	Purified enzyme (110)	MB, $K_3(FeCN)_6$, DCIP Cyto.c, inactive with O_2	Arp and Burris, 1979
	Bacteroids (122 DES) Bacteroid membranes (122 DES)	O_2, MB, DCIP, PMS, TPTC $K_3(FeCN)_6$, DCIP, PMS, TPTC, O_2 weak	Ruiz-Argueso et al., 1979 Ruiz-Argueso et al., 1979 and Emerich et al., 1980a
Inhibitors	Bacteroids (122 DES)	KCN, NaN_3, DNP, CCCP, TTFA	Emerich et al., 1979-1980b
Regulation	Free-living cells (122 DES)	Expression requires H_2, low O_2, low carbon substrates	Maier et al., 1978a, 1979
	Free-living cells (110)	Low O_2, low carboxylic acids. Cyclic AMP prevents malate inhibition of expression. Expression inhibited by cyclic GMP.	Lim, 1978; Lim and Shanmugam, 1979; Lim et al., 1979

[1]Strains of R. japonicum are in parentheses.

[2]Abbreviations: 2,6, Dichlorophenolindophenol, DCIP; methlene blue, MB; phenazine methosulfate, PMS; potassium ferricyanide, $K_3(FeCN)_6$; cytochrome c, Cyto.c; 2,3,5,triphenyltetrazolium chloride, TPTC; theonyltrifluoroacetone, TTFA; carbonyl cyanide m-chlorophenyl-hydrazone, CCCP; oxyleghemoglobin, LbO_2; adenosine monophosphate, AMP; guanosine monophosphate, GMP.

Properties of the Hydrogenase in R. japonicum

Most of the recent investigations on the H_2 oxidizing hydrogenase in Rhizobium have been carried out with R. japonicum. Table 1 summarizes some of the important observations.

A tentative scheme suggesting possible relationships of the hydrogenase to components of an electron transport chain are outlined in Figure 1. Insufficient evidence is available to accurately describe the hydrogenase complex in R. japonicum; however, it seems apparent that the hydrogenase system is membrane-bound and that it transfers electrons to O_2 via cytochrome and unidentified components in a manner similar in some respects to that described for Alcaligenes eutrophus H.16 (Probst and Schlegel, 1976).

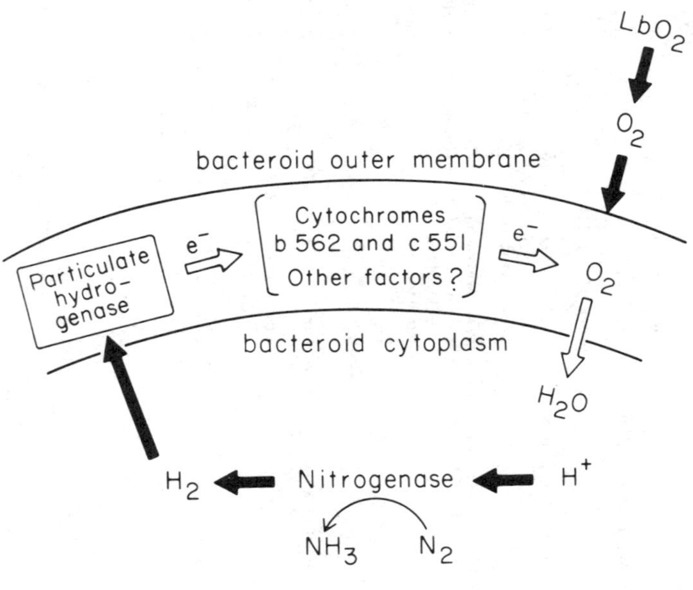

Fig. 1. A tentative scheme showing possible relationships of the membrane bound hydrogenase in R. japonicum 122 DES bacteroids to electron transport components and to nitrogenase. A supply of O_2 from dissociation of oxyleghemoglobin (LbO_2) is shown. The diagram is based upon unpublished results of D. W. Emerich and H. J. Evans (1980) and Emerich et al. (1980a, 1980b).

On the basis of difference spectrophotometry, cytochromes of the b and c types may be involved, however, further study is essential. The immediate electron acceptor for the hydrogenase and all of the components involved in the electron transport chain from H_2 to O_2 have not been identified. So far, no evidence has been obtained indicating that R. japonicum SR possesses a soluble hydrogenase of the type that occurs in A. eutrophus and catalyzes the reduction of pyridine nucleotides by H_2 (Probst and Schlegel, 1976). From Table 1 it is clear that the solubilized purified hydrogenase from the membranes of R. japonicum bacteroids no longer has a capacity for transferring electrons to O_2. Separation of the hydrogenase from its native membrane complex during purification produces an enzyme that no longer is coupled to the electron transport chain and as a consequence is not inhibited by KCN, NaN_3, and other compounds that are known to interact with electron transport components (Ruiz-Argüeso et al., 1979; Emerich et al., 1980b).

Hydrogenase Dependent Exchange

The information in Table 1 reveals inconsistent observations concerning the capacity of the hydrogenase system in intact R. japonicum cells to catalyze the exchange reaction. In our laboratory, exposure of either Hup^+ bacteroids or cultures of Hup^+ R. japonicum to (D_2) or to a combination of D_2 and 50% D_2O in the liquid phase has consistently indicated no detectable rate of exchange (Emerich et al., 1980). In contrast, Lim (1978) claims that Hup^+ cultured R. japonicum cells catalyze a rapid exchange reaction in which T_2 is incorporated into HTO. Obviously an assay method involving the radioactive T_2 isotope is far more sensitive than a method based upon mass differences in isotopes that must be measured by mass spectrometry. Despite this, Emerich et al. (1980), using the mass spectrometric method, easily detected H_2-D_2 exchange catalyzed by crude extracts of Clostridium pasteruianum which were used in control reactions that were conducted at the same time as those reactions that included R. japonicum cells. To measure hydrogenase exchange, Anand and Krasna (1965) recommended the use of the following reaction:

$$H_2 + HTO \rightleftharpoons HT + H_2O$$

They monitored exchange by incorporation of T_2 into HT in the gas phase. In a strictly anaerobic environment where the oxyhydrogen reaction is non-functional, it is possible to monitor exchange by use of the following reaction:

$$T_2 + HOH \rightleftharpoons HT + HTO$$

Under strictly anaerobic conditions, exchange may be determined by measurement of HTO in the liquid phase (Anand and Krasna, 1965). In the paper by Lim (1978) no control reaction lacking O_2 was included and the lowest level of O_2 used in his experiment where the effect of O_2 on exchange was tested was 0.05% in the gas phase over cultures. This is equivalent to about 0.67 μM O_2 in solution. Since the K_m of the hydrogenase system is extremely low (i.e., 10.6 nM for bacteroid hydrogenase) (Table 1), it seems clear that the great majority of incorporation of T_2 into HTO by *R. japonicum* cells, as reported by Lim (1978), took place via the oxyhydrogen reaction rather than exchange. Furthermore, use of the T_2 exchange reaction for measurement of hydrogenase activity in intact nodules of legumes is not reliable because the nitrogenase system *per se* catalyzes D_2-H_2 exchange (Turner and Bergersen, 1969). Bethlenfalvay and Phillips (1979) measured the incorporation of T_2 into HTO in the liquid phase of nodules as a monitor of exchange; however, this method is not recommended because it is not practical to eliminate sufficient free O_2 or leghemoglobin bound O_2 from nodules to prevent the oxyhydrogen reaction from being operative. Great care must be used in choosing the proper method if one wishes to measure hydrogenase dependent exchange between H_2 and T_2 or H_2 and D_2 and distinguish this from the oxhydrogen reaction.

Benefits to the N_2-Fixation Process

Convincing evidence now has been obtained that the oxidation of H_2 via the hydrogenase system in Hup$^+$ *R. japonicum* results in protection of the nitrogenase system from O_2 damage (Emerich et al., 1979) and provides a mechanism whereby H_2 may be utilized for support of nitrogenase activity. Dixon (1972) demonstrated H_2 dependent ATP formation in membrane preparations of a Hup$^+$ *R. leguminosarum* and Emerich et al. (1979) has shown that the oxidation of H_2 by Hup$^+$ 122 DES bacteroids led to a marked increase in the steady-state level of cellular ATP. In a series of greenhouse and field experiments where groups of Hup$^-$ strains were compared with groups of Hup$^+$ strains as inocula for soybeans statistically significant increases in yield and N_2 fixation in greenhouse trials and increases in total nitrogen contents of grain in field experiments, in favor of the Hup$^+$ strains, were consistently obtained (Albrecht et al., 1979; Hanus et al., unpublished results, 1980; Evans et al., 1980b). Lepo (unpublished results, 1980) recently has developed Hup$^-$ mutant strains derived from Hup$^+$ *R. japonicum* SR and demonstrated that these will revert to the Hup$^+$ type. These isogenic strains are being used in more critical evaluations of the benefits of the hydrogenase system to the N_2-fixing process. The Hup$^-$ mutants developed by Maier et al. (1978b) lack a capability to revert to the Hup$^+$ phenotype and thus cannot be claimed to be isogenic with their parent SR (Evans et al., 1980a).

AUTOTROPHIC GROWTH IN THE PRESENCE OF H_2 AND CO_2

Since *R. japonicum* has a CO_2 requirement for growth and the Hup^+ strains show a capability for synthesis of a membrane bound hydrogenase with properties similar to those that have been described for the particulate hydrogenase in *A. eutrophus* and some other hydrogen bacteria (Schink and Schlegel, 1978), a series of experiments were initiated to determine whether the Hup^+ strains of *R. japonicum* might utilize H_2 as an energy source for chemolithotrophic growth (Hanus et al., 1979). During the period while these studies were underway, Simpson et al. (1979) examined hydrogenase induced *R. japonicum* SR for carboxylases and observed coordinate induction of RuBP carboxylase and hydrogenase. This evidence provided further support for the view that Hup^+ strains of *R. japonicum* might grow chemolithotrophically.

Initial evidence of chemolithotrophic growth was obtained in experiments where cultures of the Hup^+ SR and Hup^- SR3 *R. japonicum* strains were utilized (Hanus et al., 1979). These mutants were described by Maier et al. (1978b). Both organisms grew vigorously on a medium supplied with yeast extract and mannitol; however, when Noble agar plates containing mineral salts (NH_4Cl as the source of N) and a trace of vitamins (0.27 μg of organic carbon per ml) were inoculated with these strains and cultured in an atmosphere containing 1% O_2, 5% CO_2, 10% H_2, and 84% N_2, the Hup^+ SR strain grew well, but the Hup^- SR3 exhibited only a trace of growth (Hanus et al., 1979). In the absence of CO_2 or H_2 over cultures neither strain SR nor SR3 grew appreciably on the mineral salts-vitamins medium. An examination of a series of wild-type Hup^+ and Hup^- *R. japonicum* strains confirmed the observation that Hup^+ strains were capable of chemolithotrophic growth whereas Hup^- strains were not (Figure 2).

Cultures of the Hup^+ SR in liquid media also showed a capacity for chemolithotrophic growth. Under these conditions a stimulatory effect, but not an absolute requirement, for the vitamins addition was observed (Hanus et al., 1979). In the initial experiments growth rates were extremely slow in liquid cultures as indicated by optical densities of no more than about 0.12 after a 16-day period. Lepo et al. (1980) now have show that the slow growth rates were due to O_2 limitation and have described procedures in which the O_2 supply is progressively increased from an initial 1% to 8% over cultures after a growth period of 270 hours (Fig. 3). Cultures having low cell densities (2×10^7 per ml or less) will tolerate no more than 1% O_2, but as cell densities increase an increasing O_2 partial pressure in the atmosphere is needed. Under optimum conditions, cultures may continue to grow, reaching optical densities of 1.3 or more after a period of 14 days (Fig. 3). Without added H_2 essentially no growth is observed.

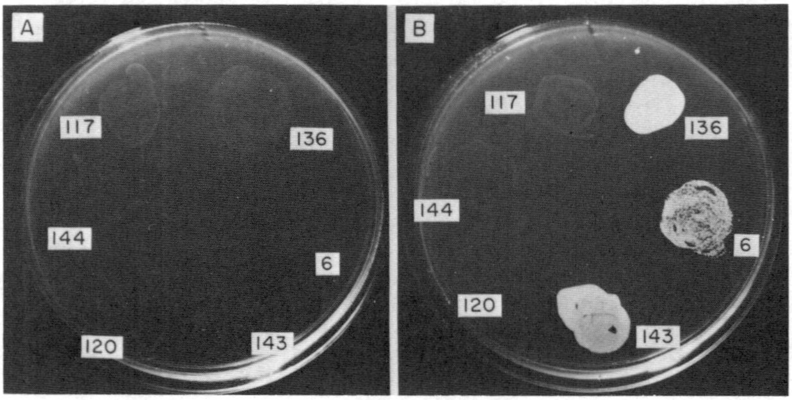

Fig. 2. (A) Growth of Hup *R. japonicum* strains 117, 144, and 120 and Hup$^+$ strains 136, 6, and 143 on Noble agar containing inorganic salts and a trace of vitamins in an atmosphere of 1% O_2, 5% CO_2, 10% He, and 84% N_2. (B) Growth of the same *R. japonicum* strains on an inorganic salts-vitamins medium in an atmosphere as described in (A) above with the exception that 10% H_2 was substituted for 10% He. After Hanus et al. (1979).

From carbon analyses of samples of cells from cultures taken at several points during the growth period of R. japonicum SR (Fig. 3) and calculated quantities of carbon accumulated in cells as determined from a series of $^{14}CO_2$ fixation rates, Lepo et al. (1980) have concluded that at least 89% of the carbon in 11-day-old chemolithotrophically cultured cells was derived from CO_2. Furthermore, Lepo et al. (1980) have show that the RuBP carboxylase activity of *R. japonicum* cultured chemolithotrophically is sufficient to account for the CO_2 uptake rates. The product of the RuBP carboxylase reaction was identified as 3-phosphoglycerate and the expected stoichiometry of about 1 mole of CO_2 fixed for every two moles of 3-phosphoglycerate formed was observed.

The conditions where Hup$^+$ strains of *R. japonicum* are grown have a marked effect on both RuBP carboxylase and hydrogenase activities of cells. No activity of either enzyme was detected in cells cultured heterotrophically in air (Table 2).

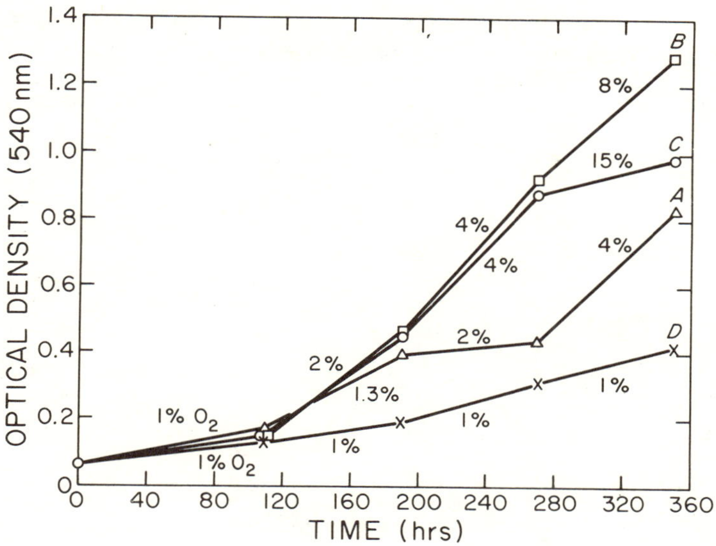

Fig. 3. The effect of progressively increasing the percentage of O_2 in the atmosphere over cultures on chemolithotrophic growth of R. japonicum strain SR. Four groups of cultures were grown with shaking in 125-ml bottles in atmospheres containing 10% H_2, 5% CO_2, and O_2 percentages at different periods during growth as indicated. The balance of the mixtures were adjusted to 1 atm with N_2. This experiment is described in detail by Lepo et al. (1980).

Easily measured activities of both enzymes (57 and 6.3 moles H_2 oxidized or CO_2 fixed per minute per mg protein for the hydrogenase and RuBP carboxylase respectively) were observed in cells cultured under conditions that allowed the hydrogenase system to be induced. R. japonicum 122 cultured under chemolithotrophic conditions, however, showed a specific activity of the hydrogenase and RuBP carboxylase of 1030 and 80 nmoles/min/mg of protein, respectively (Table 2). The hydrogenase activities of mixotrophically cultured cells and nodule bacteroids are comparable but for reasons that remain to be clarified the RuBP carboxylase activity apparently is not expressed in nodule bacteroids.

Purification of RuBP Carboxylase From Chemolithotrophically Cultured Cells

RuBP carboxylase has been purified from R. japonicum cells grown in the presence of H_2, CO_2 and O_2 under conditions similar

Table 2. Effect of growth conditions on the expression of hydrogenase and RuBP carboxylase in *R. japonicum*.[1]

Conditions of growth	Hydrogenase (nmol · min^{-1}	RuBP carboxylase mg · protein^{-1})[2]	Reference
Heterotrophic	0.05	0.2	Simpson et al. (1979)
Mixotrophic[3] (limited O_2 and C, + H_2)	57	6.3	Simpson et al. (1979)
Chemolithotrophic	1030	80	Lepo et al. (1980)
Symbiotic (bacteroids)	62	0.2	McCrae et al. (1978) Simpson et al. (1979)

[1] Strains 122 or 122 SR

[2] Expressed as H_2 oxidized or RuBP-dependent CO_2 fixed

[3] Grown in air on H_2 uptake medium of Maier et al. (1978).

to those used for culturing the hydrogen bacteria. Phosphoribulokinase activity has also been detected in extracts of such cells. The purification procedure involved pelleting the enzyme from the high-speed bacterial extract, $(NH_4)_2SO_4$ fractionation, and chromatography on DEAE-cellulose followed by sucrose gradient centrifugation. The specific activity of the enzyme was 1.6 μmoles RuBP dependent CO_2 fixed per min per mg of protein. The enzyme was homogeneous by the criterion of electrophoresis on gels polymerized from 4 to 7.5% acrylamide. Electrophoresis of the SDS-β-mercaptoethanol-dissociated *R. japonicum* RuBP carboxylase on polyacrylamide gels in the presence of SDS showed the presence of two types of subunits having approximate molecular weights of 55,000 and 15,000. *R. japonicum* enzyme seems to be similar to the ubiquitous dual subunit type RuBP carboxylase (Purohit and Evans, 1980).

Significance of Chemolithotrophy in Rhizobium

We now have conclusive evidence that the Hup^+ strains of *R. japonicum* have a capability for utilizing H_2 as a source of energy for use in the fixation of CO_2, and are capable of deriving the great majority of their required organic carbon through RuBP-catalyzed CO_2 fixation. From this information, Hup^+ *R. japonicum* strains may be categorized as aerobic hydrogen bacteria (Schrink and Schlegel, 1978). Although relatively high activities of the H_2 oxidizing hydrogenase has been observed in several strains of cowpea Rhizobium (Schubert et al., 1977) and two strains of R. leguminosarum (Ruiz-Argueso et al., 1978), attempts to grow these species under chemolithotrophic conditions have not been successful. It seems probable that species other than R. japonicum, that contain the Hup^+ characteristic, may also have a capacity for chemolithotrophic growth, but these may not have been observed because appropriate nutritional and environmental conditions have not been discovered.

It would seem reasonable to expect that the capacity of the Hup^+ strains of R. japonicum to grow as chemolithotrophs would provide an advantage in the competition and survival of these bacteria in the soil. When rhizobia are released from senescing nodules they must exist as free-living bacteria in the soil until an opportunity for infection and nodulation of another legume root is available. Also, when commercial inocula are applied to surfaces of seeds prior to planting the rhizobial cells must exist in the soil as free-living bacteria until actual infection takes place. As free-living microorganisms, the Hup^- heterotrophic rhizobia are dependent upon organic carbon in the soil and must compete with the multitude of other soil heterotrophs for usable carbon compounds. In contrast, the Hup^+ chemolithotrophic strains have the advantage of an alternate possibility of supporting their life processes through the utilization of H_2 that may be produced by fermentative bacteria in anaerobic niches in the soil. Under aerobic conditions, H_2 also may be produced by Hup^- rhizobial strains or by poorly coupled free-living N_2-fixing bacteria in the soil. H_2 produced from these sources would be available for use by aerobic hydrogen bacteria which now include the Hup^+ strains of R. japonicum. The possible advantages of chemolithotrophy in Hup^+ rhizobia are illustrated in the diagram of Figure 4.

Experiments are now in progress to evaluate the importance of chemolithotrophic capability in R. japonicum as a factor in their survival and competitiveness in the soil. The recognition of chemolithotrophy in Rhizobium japonicum raises interesting questions about the taxonomic relationships between Rhizobium and the various species of hydrogen bacteria. More research is necessary for the provision of further insight into these problems. Discovery of chemolithotrophic capability of Hup^+ R. japonicum strains also has provided us with a new and powerful tool for use in the isolation of Hup^- mutants from Hup^+ parent strains. Hup^- mutants isogenic with their parent strains, with the exception of the hydrogenase phenotype, will be most important in future evaluations of the significance of H_2 recycling in the growth of N_2-fixing legumes.

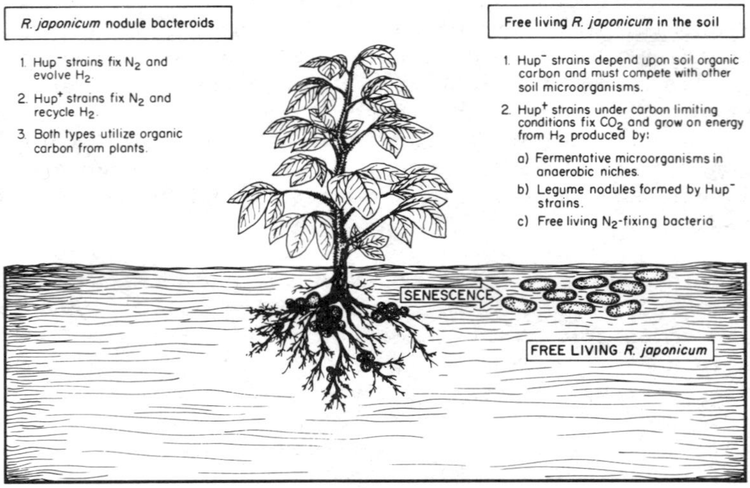

Fig. 4. An illustration of the possible significance of H_2 recycling in N_2 fixation and H_2 utilization during chemolithotrophic growth of Hup$^+$ rhizobia.

SUMMARY

This paper summarizes research that has led to the discovery that the H_2-uptake-positive strains of Rhizobium japonicum are capable of using H_2 gas as an energy source for the fixation of CO_2 and for growth in a medium composed of inorganic salts and a trace of vitamins. For chemolithotrophic culture of Hup$^+$, R. japonicum nitrogen may be supplied either as the ammonium or nitrate forms, but no evidence has been obtained that the chemolithotrophically cultured Hup$^+$ strains have a capability of utilizing N_2. Chemolithotrophically cultured R. japonicum cells contain a membrane-bound hydrogenase system and a soluble ribulose bisphosphate carboxylase. The activity of the RuBP carboxylase is sufficient to account for CO_2 fixation by intact cells. In nodule bacteroids, activity of the H_2 oxidizing hydrogenase is expressed and the enzyme is involved in the utilization of the H_2 that is produced as a by-product of N_2 fixation. The beneficial effects of the hydrogenase system to the N_2-fixing process are summarized. No activity of the RuBP carboxylase has been detected in bacteroids formed by Hup$^+$ rhizobia. Activities of both enzymes, however, are coordinately induced in free-living R. japonicum under conditions where carbon substrates and O_2 are limited and H_2 is provided. When added organic carbon substrates are removed from media and cells are cultured chemolithotrophically, the specific activities of the hydrogenase and RuBP carboxylase increase 18-fold and 13-fold, respectively, over activities observed under carbon limited conditions. The properties of the hydrogenase system are summarized and the significance of chemolithotrophy in *Rhizobium* is briefly discussed.

ACKNOWLEDGMENTS

The authors express their appreciation to Dr. Michael Cantrell and Miss Ruth Hickok for their suggestions and to Mrs. Joyce Cooper and Mrs. Donna Norvell for typing the manuscript. This research was supported by NSF Grant PCM 08784 and The Oregon Agricultural Experiment Station from which this is Technical Paper No. 5558.

REFERENCES

Albrecht, S. L., Maier, R. J., Hanus, F. J., Russell, S. A., Emerich, D. W., and Evans, H. J., 1979, Hydrogenase in *R. japonicum* increases nitrogen fixation by nodulated soybeans, *Science*, 203:1255.

Anand, S. R., and Krasna, A. I., 1965, Catalysis of the H_2-HTO exchange by hydrogenase: A new assay for hydrogenase, *Biochemistry*, 4:2747.

Anderson, K., Shanmugam, K. T., Lim, S. T., Csonka, L. N., Tait, R., Hennecke, H., Scott, D. B., Hom, S. S. M., Haury, J. F., Valentine, A., and Valentine, R. C., 1980, Genetic engineering in agriculture with emphasis on nitrogen fixation, *Trends in Biochem. Sci.*, 5:35.

Arp, D. J., and Burris, R. H., 1979, Purification and properties of the particulate hydrogenase from the bacteroids of soybean root nodules, *Biochim. Biophys. Acta*, 570:221.

Beijerink, M. W., 1888, Die Bacterien der Papilionaceenknöllchen, *Botanische Zeitung*, 45:725.

Bethlenfalvay, G. J., and Phillips, D. A., 1979, Variation in nitrogenase and hydrogenase activity of Alaska pea root nodules, *Plant Physiol.*, 63:816.

Christeller, J. T., Laing, W. A., and Sutton, W. D., 1977, Carbon dioxide fixation by lupin root nodules. I. Characterization, association with phosphoenolpyruvate carboxylase, and correlation with nitrogen fixation during nodule development, *Plant Physiol.*, 60:47.

DeHertogh, A. A., Mayeux, P., and Evans, H. J., 1964, The relationship of cobalt requirement to propionate metabolism in *Rhizobium*, *J. Biol. Chem.*, 239:2446.

Dixon, R. O. D., 1968, Hydrogenase in pea root nodule bacteroids, Arch. Microbiol., 62:272.

Dixon, R. O. D., 1972, Hydrogenase in legume root nodule bacteroids: Occurrence and properties, *Arch. Microbiol.*, 85:193.

Dixon, R. O. D., 1978, Nitrogenase-hydrogenase interrelationships in rhizobia, Biochimie., 60:233.

Emerich, D. W., Albrecht, S. L., Russell, C. A., Ching, T. M., and Evans, H. J., 1980a, Oxyleghemoglobin-mediated H_2 oxidation by R. japonicum. USDA 122 DES bacteroids, Plant Physiol., 65:605.

Emerich, D. W., Ruiz-Argüeso, T., Ching, T. M., and Evans, H. J., 1979, Hydrogen-dependent nitrogenase activity and ATP formation in R. japonicum bacteroids, J. Bacteriol., 137:153.

Emerich, D. W., Ruiz-Argüeso, T., Russell, S. A., and Evans, H. J., 1980b, Investigations of the H_2 oxidation system in Rhizobium japonicum (122 DES) nodule bacteroids, Plant Physiol., accepted.

Evans, H. J., Emerich, D. W., Lepo, J. E., Maier, R. J., Carter, K. R., Hanus, F. J., and Russell, S. A., 1980a, The role of hydrogenase in nodule bacteroids and free-living rhizobia, in: "International Symposium on Nitrogen Fixation," W. D. P. Stewart and J. R. Gallon, eds., Academic Press, London, in press.

Evans, H. J., Emerich, D. W., Ruiz-Argüeso, T., Albrecht, S. L., Maier, R. J., Simpson, F., and Russell, S. A., 1978, Hydrogen metabolism in legume nodules and rhizobia: Some recent developments, in: "Hydrogenases: Their Catalytic Activity, Structure, and Function," H. G. Schlegel and K. Schneider, eds., Erich Goltze KG, Gottingen.

Evans, H. J., Emerich, D. W., Ruiz-Argüeso, T., Maier, R. J., and Albrecht, S. L., 1980b, Hydrogen metabolism in the legume rhizobium symbiosis, in: "Nitrogen Fixation: Symbiotic Associations and Cyanobacteria," Vol. II, W. H. Orne-Johnson and W. E. Newton, eds., University Park Press, Baltimore.

Hanus, F. J., Maier, R. J., and Evans, H. J., 1979, Autotrophic growth of H_2-uptake positive strains of R. japonicum in an atmosphere supplied with hydrogen gas, Proc. Natl. Acad. Sci., 76:1788.

Jackson, E. K., and Evans, H. J., 1966, Propionate in heme biosynthesis in soybean nodules, Plant Physiol., 41:1673.

Laing, W. A., Christeller, J. T., and Sutton, W. D., 1979, Carbon dioxide fixation by lupin root nodules. II. Studies with ^{14}C-labeled glucose, the pathway of glucose catabolism and the effects of some treatments that inhibit nitrogen fixation, Plant Physiol., 63:450.

Lawrie, A. C., and Wheeler, C. T., 1975, Nitrogen fixation in the root nodules of Vicia faba L. in relation to the assimilation of carbon. I. Plant growth and metabolism of photosynthetic assimilates, New Phytologist, 74:429.

Lepo, J. E., Hanus, F. J., and Evans, H. J., 1980, Further studies on the chemautotrophic growth of hydrogen uptake positive strains of R. japonicum, J. Bacteriol., 141:664.

Lim, S. T., 1978, Determination of hydrogenase in free-living cultures of Rhizobium japonicum and energy efficiency of soybean nodules, Plant Physiol., 62:609.

Lim, S. T., Hennecke, H., and Scott, D. B., 1969, Effect of cyclic guanosine 3',5'-monophosphate on nitrogen fixation in R. japonicum, J. Bacteriol., 139:256.

Lim, S. T., and Shanmugam, K. T., 1969, Regulation of hydrogen utilization in Rhizobium japonicum by cyclic AMP, Biochim. Biophys. Acta, 584:479.

Lowe, R. H., and Evans, H. J., 1962, Carbon dioxide requirement for growth of legume nodule bacteria, Soil Sci., 94:351.

Maier, R. J., 1980, Rhizobium japonicum mutant strains unable to grow chemoautotrophically with H_2, in: "Abstracts of the Annual Meeting of the American Society for Microbiology," R. A. Finkelstein, ed., American Society of Microbiology, Washington, DC.

Maier, R. J., Campbell, N. E. R., Hanus, F. J., Simpson, F. B., Russell, S. A., and Evans, H. J., 1978a, Expression of hydrogenase activity in free-living R. japonicum, Proc. Natl. Acad. Sci., 75:3258.

Maier, R. J., Hanus, F. J., and Evans, H. J., 1969, Regulation of hydrogenase in R. japonicum, J. Bacteriol., 137:824.

Maier, R. J., Postgate, J. R., and Evans, H. J., 1978b, Mutants of R. japonicum unable to utilize hydrogen, Nature, 276:494.

Maruyama, Y., 1975, Studies on symbiotic nitrogen fixation: Nitrogen-fixing and hydrogenase activities in extracts of legume root nodules, in: "Nitrogen Fixation and Nitrogen Cycle," H. Takahashi, ed., University of Tokyo Press, Tokyo.

Maruyama, Y., Onodera, K., and Funahashi, S., 1967, Nitrogen and hydrogen fixation in extracts of legume root nodules, in: "Abstracts of VII Intern. Congr. Biochem.," J52, Tokyo.

McCrae, R. E., Hanus, F. J., and Evans, H. J., 1978, Properties of the hydrogenase system in Rhizobium japonicum bacteroids, Biochem. Biophys. Res. Comm., 80:384.

Mulder, E. G., and Veen, W. L. van, 1960, The influence of carbon dioxide on symbiotic nitrogen fixation, Plant and Soil, 13: 265.

Peterson, J. B., and Evans, H. J., 1979, Phosphoenolpyruvate carboxylase from soybean nodule cytosol: Evidence for isoenzymes and kinetics of the most active component, Biochim. Biophys. Acta, 567:445.

Phelps, A. S., and Wilson, P. W., 1941, Occurrence of hydrogenase in nitrogen fixing organisms, Proc. Soc. Exp. Biol., 47:473.

Probst, I., and Schlegel, H. G., 1976, Respiratory components and oxidase activities in Alcaligenes eutrophus, Biochim. Biophys. Acta, 440:412.

Purohit, K., and Evans, H. J., 1980, D-Ribulose 1,5-bisphosphate carboxylase/oxygenase from autotrophically grown Rhizobium japonicum, in: "Abstracts of American Society of Plant Physiologists Meeting at Washington State University,

Pullman, Washington," American Society of Plant Physiologists, Bethesda.
Rittenberg, S. C., 1969, The roles of exogenous organic matter in the physiology of chemolithotrophic bacteria, in: "Advances in Microbial Physiology," Vol. 3, A. H. Rose and J. F. Wilkinson, eds., Academic Press, London.
Ruiz-Argüeso, T., Emerich, D. W., and Evans, H. J., 1969, Characteristics of the hydrogen oxidizing system in soybean nodule bacteroids, Arch. Microbiol., 121:199.
Ruiz-Argüeso, T., Hanus, F. J., and Evans, H. J., 1978, Hydrogen production and uptake by pea nodules as affected by strains of Rhizobium leguminosarum, Arch. Microbiol., 116:113.
Schink, B., and Schlegel, H. G., 1978, Hydrogen metabolism in aerobic hydrogen-oxidizing bacteria, Biochimie, 60:297.
Schlegel, H. G., and Eberhardt, U., 1972, Regulatory phenomena in the metabolism of Knallgasbacteria, in: "Advances in Microbial Physiology," Vol. 7, A. H. Rose and D. W. Tempest, eds., Academic Press, London.
Schubert, K. R., Engelke, J. A., Russell, S. A., and Evans, H. J., 1977, Hydrogen reactions of nodulated leguminous plants. I. Effect of rhizobial strain and plant age, Plant Physiol., 60:651.
Schubert, K. R., and Evans, H. J., 1976, Hydrogen evolution: A major factor affecting the efficiency of nitrogen fixation in nodulated symbionts, Proc. Natl. Acad. Sci., 73:1207.
Schubert, K. R., and Evans, H. J., 1977, The relation of hydrogen reactions to nitrogen fixation in nodulated symbionts, in: "Proceedings of the II International Symposium on N_2 Fixation: Recent Developments in N_2 Fixation," W. E. Newton, J. R. Postgate, and C. Rodriguez-Barrueco, eds., Academic Press, London.
Schubert, K. R., Jennings, N. T., and Evans, H. J., 1978, Hydrogen reactions of nodulated leguminous plants. II. Effects on dry matter accumulation and nitrogen fixation, Plant Physiol., 61:398.
Scott, D. B., Farnden, K. J. F., and Robertson, J. G., 1976, Ammonia assimilation in lupin nodules, Nature, 263:703.
Simpson, F. B., Maier, R. J., and Evans, H. J., 1979, Hydrogen-stimulated CO_2 fixation and coordinate induction of hydrogenase and ribulosebisphosphate carboxylase in an H_2-uptake positive strain of R. japonicum, Arch. Microbiol., 123:1.
Turner, G. L., and Bergersen, F. J., 1969, The relationship between nitrogen fixation and the production of HD from D_2 by cell-free extracts of soya-bean nodule bacteroids, Biochem. J., 115:529.
Valley, G., and Rettger, L. F., 1927, The influence of carbon dioxide on bacteria, J. Bacteriol., 14:101.
Vincent, J. M., 1977, Rhizobium: General microbiology, in: "A Treatise on Dinitrogen Fixation, Section III," R. W. F. Hardy and W. S. Silver, eds., John Wiley and Sons, New York.

HYDROGEN UPTAKE (HYDROGENASE) ACTIVITY OF *RHIZOBIUM JAPONICUM* STRAINS FORMING NODULES IN SOYBEAN PRODUCTION AREAS OF THE U.S.A.[1]

S. T. Lim[2], S. L. Uratsu[2], D. F. Weber[3], and H. H. Keyser[3]

[2]Plant Growth Laboratory/Department of Agronomy & Range Science, University of California, Davis, CA 95616 and [3]USDA-ARC-West, Cell Culture and Nitrogen Fixation Laboratory, Plant Physiology Institute, Beltsville, MD 20705

INTRODUCTION

Root nodules of soybeans (*Glycine max*) inoculated with most strains of *Rhizobium japonicum* release considerable amounts of hydrogen whereas nodules from a minority of strains evolve little if any H_2 (Lim, 1978; Schubert and Evans, 1977; Evans et al., this volume, and the extensive reference list of this paper). A number of studies have pointed toward a significant benefit of the H_2 uptake system to overall plant productivity (see above references) since energy limitation is one of the major factors in nitrogen fixation in soybeans (Hardy and Havelka, 1975). A comparison of soybean productivity in experiments conducted with Hup^+ (strains with an active H_2 uptake system) and Hup^- (strains which lack an active H_2 uptake system) strains of *Rhizobium japonicum* have shown that the Hup^+ strains are more efficient symbionts (see Evans et al., this volume). In addition, a comparison of plants (grown in the greenhouse and growth chamber) inoculated with isogenic Hup^- mutants (derived from Hup^+ parents) fixed significantly less nitrogen and had reduced yields than the parent Hup^+ strain which synthesized the hydrogenase system (Albrecht et al., 1979; and

[1]This report was prepared with the support of the National Science Foundation Grant No. PFR 77-07301. Any opinions, findings, and conclusions or recommendations expressed in this publication are those of the authors and do not necessarily reflect the views of the National Science Foundation.

discussion by Evans et al., this volume, on the isogenic nature of these strains). Thus, the capacity to synthesize an active H_2 uptake system may be a desirable characteristic of Rhizobium spp.

These considerations, along with the development of the sensitive tritium assay (Lim, 1978) for evaluating hydrogenase activity in Rhizobium japonicum stimulated us to make a more comprehensive survey of the H_2 uptake activity in commercial inoculants as well as strains isolated from nodules obtained from plants grown in the major soybean production areas.

MATERIALS AND METHODS

Nodule Collection and Isolation of Bacterial Strains

Random samples of soybean (Glycine max) nodules from 70 locations in 28 different states representing the major soybean production areas in the country were collected during the 1979 growing season and shipped to Davis in glass vials containing silica gel dessicant (Vincent, 1970). The nodules were rehydrated immediately upon receiving the samples and six representative nodules from three plants in each location were selected at random and surface sterilized with 10% (v/v) sodium hypochlorite for 10 min and washed four times with sterile water. The nodules were crushed and plated on mannitol-salts-yeast extract (MSY) medium as previously reported (Lim, 1978) and twenty single colonies were isolated at random and used as inocula for the assay of hydrogenase activity.

Commercial Inoculants

Commercial soybean inoculants were supplied by the Nitragin Company, Inc., Milwaukee, WI 53209; Urbana Laboratories, Urbana, IL 61801; Research Inoculants, St. Joseph, MO 64501; and Land 'O Lakes, Inc., Fort Dodge, IA 50501.

Rhizobium japonicum strain USDA 110 was from the USDA Rhizobium culture collection at Beltsville, Maryland.

Hydrogenase Induction and Assay

Hydrogenase activity was induced in free-living cultures of Rhizobium japonicum as previously described (Lim, 1978; Lim and Shanmugam, 1979) using a glutamate-gluconate medium with a gas phase consisting of 10% H_2, 5% CO_2, and 84% argon (v/v). After approximately 6 days incubation at 25 C on a rotary shaker (150 rpm), hydrogenase activity was assayed using the tritium assay (Lim, 1978) and H_2 uptake activity monitored by following the uptake of H_2 using gas chromatography (Lim, 1978; Lim and Shanmugam, 1979). Note that in previous publications from this laboratory that the term "tritium exchance" has been coined for this sensitive measure of H_2 uptake activity. Evans et al. in this volume have criticized the use of the word "exchange." Until this issue is settled the term "tritium assay" may be more appropriate. It should also be stressed that this assay, regardless of semantics, is a highly valid measure of the Hup phenotype (see Lim, 1978, and below for comparison with other assays).

RESULTS

Properties of Hydrogenase (H_2 Uptake) System in *Rhizobium japonicum*

The hydrogen uptake system has been shown to be inducible in free-living cultures of *Rhizobium japonicum* under conditions of low oxygen tension ($\leq$ 1% O_2) in the gas phase and with a poor carbon source in the medium (Lim and Shanmugam, 1979; Maier et al., 1979). Later experiments by other workers showed that the H_2 uptake system is stimulated by H_2 and CO_2 (Maier et al., 1979). Several of the parameters which influence hydrogenase activity in free-living cultures are summarized in Table 1. In the absence of H_2, only low levels of hydrogenase activity are detected. In the presence of 10% H_2 (v/v) in the gase phase a ten-fold increase in activity is obtained. Five percent CO_2 by itself caused only a slight stimulation of activity over the endogenous levels (see Table 1). Maximum hydrogenase activity was obtained in the presence of 5% CO_2 and 10% H_2 in the gas phase. The results also show that both the H_2 uptake and tritium assays are valid for the measurement of hydrogenase activity.

Commercial Soybean Inoculants

In this section, eight different inoculants (consisting of both single and multiple strains) from four inoculant companies were used to inoculate *Glycine max* var. Evans and analyzed for acetylene reduction and H_2 evolution (Table 2). All the inoculants produced effective nodules on this variety of soybean. A majority of the nodules were found to evolve significant amounts of H_2. At best, three (XC, XE, and CG) of the eight inoculants were found to contain both Hup^+ and Hup^- strains since two distinct kinds of nodules were obtained (Table 2). From the known number of strains (and assuming all strains are equally effective on this variety of soybean) present in each inoculant it can be calculated that the majority of the strains (68.8%) present lack a H_2 uptake system.

To confirm the correlation between H_2 evolution and the lack of a H_2 uptake system, isolated single colonies from the nodules of three different inoculants (Table 3) were assayed for hydrogenase activity and then compared with the H_2 evolution data from Table 2. Several colonies isolated from each nodule were tested to exclude the possibility of more than one strain in a single

Table 1. Effect of hydrogen and CO_2 on the induction and hydrogenase in free-living culture of Rhizobium japonicum 3Ilb 110 (see Evans et al., this volume, and references therein for further discussion of assay conditions).

Conditions	Hydrogenase Activity	
	H_2 uptake[a] $\mu mol\ H_2\ hr^{-1} mg^{-1}$ protein	Tritium assay[b] $nmol\ ^3H_2$ mg^{-1} protein
1. $-H_2$	1.19	116.5
2. + 10% H_2	11.15	1428.6
3. + 5% CO_2	2.97	607.3
4. + 5% CO_2 + 10% H_2	14.94	2450.0

Cells were grown in a medium containing glutamate (0.1%) under a constant gas phase consisting of argon and O_2 (0.1%), with H_2 (10%) and CO_2 (5%) as indicated.

[a]Assayed in the presence of 1% O_2.

[b]Assayed "anaerobically" with $^3H_2$ (35.87 mCi $mmol^{-1}$); following discussions with H. J. Evans several cycles of evacuation and flushing with argon were utilized to remove traces of O_2 (as discussed by Evans et al. (this volume), traces of O_2 may still be present).

nodule (Lindemann et al., 1974). Inoculant CG which is comprised of multiple strains produced two kinds of nodules in the evolution assay (Table 3). When the colony isolates were assayed for hydrogenase activity using the tritium method, two kinds of strains were obtained, indicating that this particular inoculant contains both Hup^+ and Hup^- strains. It is interesting to note that no single nodule isolated contains both Hup^+ and Hup^- strains -- these strains were always on different nodules but on the same plant. However, the presence of more than one

Table 2. Analysis of commercial inoculants (Rhizobium japonicum) for C_2H_2 reduction, H_2 evolution and nodule weight. Values were obtained from approximately 5-week soybeans (Glycine max var. Evans) grown in the greenhouse.

Inoculants	C_2H_4 formed $\mu mol\ hr^{-1}\ mg^{-1}$ Nodule wt	H_2 evolution $\mu mol\ hr^{-1} g^{-1}$ Nodule wt	Nodule wt $mg\ plant^{-1}$
SA	14.3 ± 2.5[a]	<0.05	168.8
CB	10.5 ± 3.3	5.4 ± 0.6	82.7
XC	19.1 ± 1.4	<0.05	98.5
		5.1 ± 0.9	
CD	31.1 ± 4.7	11.9 ± 0.1	109.0
XE	19.9 ± 3.2	<0.05	94.9
		5.7 ± 0.1	
XF	23.0 ± 2.1	5.5 ± 0.5	68.2
CG	18.4 ± 7.5	<0.05	112.3
		5.7 ± 0.2	
SH	26.7 ± 2.7	<0.05	N.D.[b]

[a]Results are means ± s.e.m.

[b]Not determined.

Hup^+ or Hup^- strain in a single nodule cannot be ruled out. Inoculant CD, although also consisting of multiple strains, produced nodules which all evolved H_2 while the isolated clones also lacked hydrogenase activity (Table 3). In a single strain inoculant (SH) no H_2 evolution was detected in the nodules and the isolated clones all had high hydrogenase activity when assayed in free-living cultures. In summary, the results in Table 3 show that all strains

Table 3. Hydrogen uptake activity of free-living cultures of Rhizobium japonicum isolated from nodules inoculated with commercial inoculants. Single colonies were isolated from the nodules induced and assayed for hydrogenase activity as previously described in Table 1.

Commercial inoculants	H_2 evolution from nodule $\mu mol\ hr^{-1} g^{-1}$ Nodule wt	Tritium assay nmol $^3H_2$ taken up $hr^{-1} mg^{-1}$ Cell protein
CG (Composite strains)	5.7 ± 0.2	<0.1
	<0.05	2548.5
CD (Composite strains)	11.9 ± 0.1	<0.1
SH (Single strain)	<0.05	2397.2

of R. japonicum isolated from nodules evolving H_2 possess the Hup negative phenotype as expected.

In order to minimize the possibility that the large number of H_2 evolving strains reported in Table 2 is biased because of strain-cultivar interaction (Caldwell and Vest, 1968), inoculants CB, CD, and XF (which contained only Hup^- strains, see Table 2) were inoculated on four different soybean cultivars. Table 4 shows that although the degree of H_2 evolution from the nodules might vary from cultivar to cultivar, a Hup^- strain (scored using the tritium assay) always evolves H_2 during symbiosis regardless of the cultivar tested (see Schubert and Evans, 1977, for further discussion of this point). Similarly, when Rhizobium japonicum USDA 110 (a Hup^+ phenotype) was used to inoculate these four cultivars, little or no H_2 evolution was obtained (data not shown).

Analysis of Strains From Major Soybean Regions

Soybeans grown in the U.S. are usually inoculated with one or more of the commercially available soybean inoculants. In view of the results presented above, it is important to sample nodules from the soybean producing areas to ascertain the nitrogen fixing

Table 4. Effect of different soybean (Glycine max) cultivars on H_2 evolution.

Commercial inoculum	Soybean cultivar	H_2 evolution from nodule $\mu mol\ H_2\ hr^{-1} g^{-1}$ Nodule wt
CB	var Beeson	5.9 ± 2.7
	var Dixon	6.0 ± 2.0
	var Evans	5.4 ± 0.6
	var Harosoy	6.0 ± 3.2
CD	var Beeson	5.3 ± 2.6
	var Dixon	7.4 ± 1.6
	var Evans	11.9 ± 0.1
	var Harosoy	5.7 ± 1.8
XF	var Beeson	2.7 ± 1.8
	var Dixon	4.4 ± 1.0
	var Evans	5.7 ± 0.1
	var Harosoy	3.4 ± 2.5

efficiency of these nodules as measured by their H_2 uptake activities. Table 5 presents the results of one such survey. Note that strains isolated from nodules obtained from the southeastern states, especially Alabama, Florida, Mississippi, North and South Carolina, and Louisiana contained mainly Hup$^-$ strains. (A second survey to confirm this finding is underway.) In most other states (where more than one location was tested) both Hup$^+$ and Hup$^-$ strains were found except Delaware, Kansas, South Dakota, New Jersey, and Pennsylvania where mainly Hup$^-$ strains were isolated. Certain regions, especially locations in New York and Maryland, had high percentages of Hup$^+$ strains. The distribution of Hup$^+$ strains in different locations within a state (where several locations were tested) was found to vary greatly. In Illinois, for example, one location in Beltsville had over 90% Hup$^+$ strains (Table 5).

Table 5. A survey of H_2 uptake activity (Hup) in *Rhizobium japonicum* strains isolated from nodules from the major soybean production areas of the U.S.A.

State and location	Soybean cultivar	Soil type	Percentage of Hup^+ strains
Alabama:			
Belle Mina	Braxton	Decatur clay loam	0
Fairhope	Braxton	Malbis fine sandy loam	0
Marion Junction	Braxton	Sumpter clay	0
Arkansas:			
Pine tree	Dare	Calloway silt loam	0
Delaware:			
Two locations at Georgetown	Essex	Sassafras sandy loam	0
Florida:			
Gainesville	Braxton	Arredondo fine sand	0
Gainesville	Cobb	Wakeland fine sand	0
Georgia:			
Blairsville	Lee 74	Dyke loam	0
Experiment	Wright	Cecil sandy loam	0
Plains	Bragg	N.D.	11.1
Midville	Ransom	Marlboro loamy sand	0
Illinois:			
Belleville	Williams	Weir silt loam	90.0
Dekalb	X1878	N.D.	0
Urbana	Wayne	Drummer	0
Indiana:			
N.D.	Union	Warsaw	47.4
Bluffton	N.D.	N.D.	0
Greenfield	N.D.	N.D.	0
Iowa:			
Ames	A2440	Clarion-Nicollet-Webster	7.1
Keota	A3860	Nicollete-Webster	0
Rudd	A1564	Kenyon-Floyd-Clyde	0

Table 5 (continued)

Kansas:			
Three locations not determined	N.D.	N.D.	0
Kentucky:			
Lexington	Williams	Maury silt loam	33.3
Louisiana:			
LeCompte	Davis	Moreland clay	0
N.D.	Dare	Olivier silt loam	0
Winnsboro	Bragg	Grenada silt	0
Maryland			
Four locations at Beltsville	Kent	Silt loam	31.7
Minnesota:			
Blooming Prairie	A2858	N.D.	0
Mississippi:			
Stoneville	Tracy-M	Boshet fine sandy loam	0
Stoneville	Forrest	Sharkey clay	0
Verona	Tracy-M	N.D.	0
Missouri:			
N.D.	Ransom	N.D.	16.7
N.D.	Williams	N.D.	0
N.D.	Essex	N.D.	0
Nebraska:			
Lincoln	Woodworth	Wabash silty clay loam	5.6
Lincoln	Nebsoy	Sharbsburg silty clay loam	0
Lincoln	Williams	Wabash silty clay loam	0
New York:			
Volatie	Coles	Hoosic gravely silt loam	50.0
Chazy	SRF 150P	Sandy silt loam	38.9
North Carolina:			
Clinton	Tracy	Norfolk sandy loam	0
Kingston	Tracy	Norfolk sandy loam	0
Clayton	Lee 74	Norfolk sandy loam	0
Plymouth	Tracy	Portsmouth	0
North Dakota:			
Fargo	Evans	Clay	22.2
Fargo	McCall	Clay	0

Table 5 (continued)

Ohio:			
Darby	Williams	Crosby silty clay loam	0
S. Charleston	Elf	Brookston silt loam	33.3
New Jersey:			
Adelphia	Adelphia	Freehold sandy loam	0
Adelphia	N.D.	Freehold sandy loam	0
Wooster	Elf	Wooster silt loam	0
Henry Ct	Williams	Locustrene lake plain	0
Pennsylvania:			
Landisville	Williams	Hagerstown silt loam	0
Manheim	Cutler 71	Hagerstown silt loam	0
South Carolina:			
Darlington County	Bragg	Norfolk	0
Darlington County	Bragg	Corville	0
South Dakota:			
Brookings	Swift	Lismore Pachic Udie Haploboroll (fine silty clay loam)	0
Brookings	Swift	Egeland Udie Haploboroll (mixed sandy loam)	0
Tennessee:			
Jackson	Bedford	Sandy loam	5.3
Jackson	Forrest	Sandy loam	20.0
Jackson	Bedford	Sandy loam	0
Texas:			
Beaumont	Dowling	Beaumont clay	5.6
Beaumont	Bragg	Morey silt loam	0
Eagle Lake	N.D.	Very fine sandy loam	0
Virginia:			
Blacksburg	Essex	Droseclose silt loam	0
Warsaw	Essex	Sassafras sandy loam	0
Suffolk	Essex	Othello fine silt loam	44.4
Wisconsin:			
Arlington	Hodgson	Plano silt loam	0
Arlington	Hodgson	Plant silt loam	37.5

N.D. = Not Determined

DISCUSSION

The H_2 uptake system (hydrogenase system) can be monitored qualitatively and quantitatively by a variety of assays using whole plants, detached nodules (Bethlenfalvay and Phillips, 1979; Lim, 1978; Schubert and Evans, 1977), or free-living cultures induced for hydrogenase activity (Lim, 1978; Lim and Shanmugam, 1979). Hydrogen uptake can be measured chromatographically (Lim, 1978; Schubert and Evans, 1977), amperometrically (Schubert and Evans, 1977), or by the tritium assay (Bethlenfalvay and Phillips, 1979; Lim, 1978; Suzuki and Maruyama, 1979). The strong correlation between hydrogenase activity in free-living cultures and in root nodules when measured by the tritium assay (Bethlenfalvay and Phillips, 1979; Lim, 1978; Suzuki and Maruyama, 1979; Table 3) was the basis for using this procedure in this report (see Materials and Methods).

A recent survey using the tritium assay of a variety of different strains of *Rhizobium japonicum* for hydrogenase activity has shown that less than 30% of the strains surveyed are Hup^+ (Lim, 1978). Interestingly, a variety of different strains recently located from China, effective on several different soybean cultivars were found to evolve large amounts of H_2 indicating a lack of an active hydrogenase system (data not shown). It is of considerable interest to analyze the presence of the Hup^+ phenotype in commercial inoculants since such strains are used on millions of acres of soybeans. Analysis of several such inoculants showed that the majority (68.8%) of the commercial strains tested are Hup^- (Table 2). In Table 5, results from a survey of some 1400 isolates of *Rhizobium japonicum* isolated from nodules from seventy different locations in the major soybean production areas are presented. Although the percentage of Hup^+ strains varies from location to location, approximately 75% of the strains tested were missing the H_2 uptake activity. Since many commercial soybean inoculants do not contain strains that are Hup^+, it is not surprising that this trait is missing from most of the production areas surveyed. Caldwell and Vest (1970) reported that as much as 90% of the nodules formed in the fields were produced by indigenous strains despite prior inoculation of the seeds with agronomically desirable strains. This could indicate the dominance of different strains in different locations or be attributed to strain-cultivar interactions (Caldwell and Vest, 1968; Hardy and Havelka, 1975). Soybean cultivars grown in past years could also affect the relative abundance of *Rhizobium japonicum* strains in field soils (Caldwell and Vest, 1968; Hardy and Havelka, 1975) as can soil properties (Weber and Miller, 1972). There appears to be no simple correlation between the variety of soybean cultivar or soil type versus the percentage of Hup^+ strains.

Although the use of soybean inoculants containing Hup^+ strains would be preferable this may not be feasible in some cases where local soil and other environmental factors preclude the introduction of new strains or where local strains have evolved to be more competitive (Caldwell and Vest, 1970; Hardy and Havelka, 1975; Smith and Miller, 1974). Introduction of the *hup* genes into Hup^- strains via a small plasmid (Hup^+) might improve the efficiency of N_2 fixation leading to higher soybean yields (see Lim et al., 1980, for discussion of Hup plasmids). One strategy would involve first the isolation from nodules from specific locations of predominate Hup^- strains followed by introduction into these established strains of *hup* genes using vectors such as broad host-range R factors. The major point of such an approach would be the use of strains already established (or adapted to) a given production area as recipients for small and defined segments of Hup^+ DNA with the hoped for results being hybrid cells retaining their original properties while acquiring the Hup trait.

SUMMARY

The hydrogen uptake (Hup) activity of *Rhizobium japonicum* strains isolated from nodules from seventy different locations in 28 states of the major soybean production areas were analyzed using the tritium assay as described previously (Lim, 1978). In addition, soybean inoculants from four different commercial sources were tested for H_2 uptake activity. The major conclusion is that a majority (> 75%) of the *Rhizobium japonicum* strains isolated from major production areas as well as commercial inoculants are missing H_2 uptake activity.

REFERENCES

Albrecht, S. L., Maier, R. J., Hanus, F. J., Russell, S. A., Emerich, D. W., and Evans, H. J., 1979, Hydrogenase in *Rhizobium japonicum* increases nitrogen fixation by nodulated soybeans, *Nature*, 203:1255.

Bethlenfalvay, G. H., and Phillips, D. A., 1979, Variation in nitrogenase and hydrogenase activity of Aladka pea root nodules, *Plant Physiol.*, 63:816.

Caldwell, B. E., and Vest, G., 1968, Nodulation interactions between soybean genotypes and serogroups of *Rhizobium japonicum*, *Crop Sci.*, 8:680.

Caldwell, B. E., and Vest, G., 1970, Effects of *Rhizobium japonicum* strains on soybean yields, *Crop Sci.*, 10:19.

Hardy, R. W. F., and Havelka, U. D., 1975, Nitrogen fixation research: a key to world food?, *Science*, 188:633.

Johnson, H. W., Means, U. M., and Weber, C. R., 1965, Competition for nodule sites between strains of *Rhizobium japonicum* applied as inoculum and strains in the soil, *Agron. J.*, 57:179.

Lim, S. T., 1978, Determination of hydrogenase in free-living cultures of Rhizobium japonicum and energy efficiency of soybean nodules, Plant Physiol., 62:609.

Lim, S. T., and Shanmugam, K. T., 1979, Regulation of hydrogen utilization in Rhizobium japonicum by cyclic AMP, Biochim. Biophys. Acta, 684:479.

Lim, S. T., Andersen, K., Tait, R. C., and Valentine, R. C., 1980, Genetic engineering in agriculture: hydrogen uptake (hup) genes, Trends in Biochem. Sci., 5:167.

Lindemann, W. C., Schmidt, E. L., and Ham, G. E., 1974, Evidence for double infection within soybean nodules, Soil Sci., 18: 274.

Maier, R. J., Hanus, F. J., and Evans, H. J., 1979, Regulation of hydrogenase in Rhizobium japonicum, J. Bacteriol., 137:824.

Schubert, K. R., and Evans, H. J., 1977, The relation of hydrogen reactions to nitrogen fixation in nodulated symbionts, in: "Recent Developments in Nitrogen Fixation," W. Newton, J. R. Postgate, and C. Rodriguez-Barrueco, eds., Academic Press, New York.

Smith, R. S., and Miller, R. H., 1974, Interaction between Rhizobium japonicum and soybean rhizosphere bacteria, Agron. J., 66:564.

Suzuki, T., and Maruyama, Y., 1979, Studies on hydrogenases in legume root nodule bacteroids: effect of ATP, Agric. Biol. Chem., 43:1833.

Vincent, J. M., 1970, A Manual for the Practical Study of Root-Nodule Bacteria, Blackwell Scientific Publications, Oxford.

Weber, D. F., and Miller, V. L., 1972, Effect of soil temperature on Rhizobium japonicum serogroup distribution in soybean nodules, Agron. J., 64:796.

PROLINE OVER-PRODUCTION ENHANCES NITROGENASE ACTIVITY UNDER OSMOTIC STRESS IN KLEBSIELLA PNEUMONIAE

Daniel Le Rudulier, Su Sheng Yang, and Lászlo N. Csonka

Plant Growth Laboratory
University of California
Davis, CA 95616

INTRODUCTION

In 1955, J. H. B. Christian made the interesting observation that proline, added exogenously at low concentrations, specifically stimulated the growth rate (Christian, 1955a,b) and respiration rate (Christian and Waltho, 1966) of some bacteria in media of inhibitory osmolarity. In order to test whether, in general, the accumulation of proline at high intracellular concentrations would directly result in enhanced osmotolerance, we isolated mutants of Salmonella typhimurium which over-produce proline (L. N. Csonka, manuscript in preparation). Some of the proline over-producing strains were indeed found to grow faster than the wild type organism in media of elevated osmolarity. These mutants were obtained as strains which were resistant to the toxic proline analogue: L-azetidine-2-carboxylic acid. The mutations in these strains were shown to be closely linked to $proB^{+}A^{+}$, the genes for the enzymes that catalyze the first two reactions of proline biosynthesis. The strain used for the isolation of the mutants was one in which the $proB^{+}A^{+}$ region (of E. coli K12) was carried on an autonomously replicating, self-transmissible plasmid F'_{128}. Therefore, the mutations conferring enhanced osmotolerance could be transferred easily to other strains of Salmonella typhimurium and related bacteria.

One of our goals is to develop a procedure to obtain osmotolerant mutants of other bacteria, including Rhizobia. As a first step in testing whether the selection of proline over-producing mutants might be a pratical approach, we transferred into Klebsiella pneumoniae one of the mutant F's which conferred enhanced osmotolerance on S. typhimurium and determined the effect of proline overproduction on nitrogenase activity under osmotic stress.

MATERIALS AND METHODS

Bacterial Strains

Salmonella typhimurium strain JL2468 (del(proBA) leuD$^-$/F'$_{128}$ proB$^+$A$^+$ argF$^+$) is the parent of the osmotolerant mutant, TL88 (del(proB$^+$A$^+$) leuD$^-$/F'$_{128}$ pro-74 argF$^+$) (L. N. Csonka, manuscript in preparation). Strains SY1(del(proBA) leuD$^-$/F'$_{128}$ pro-74 argF::Tn10) and SY2(del(proBA) leuD$^-$/F'$_{128}$ proB$^+$A$^+$ argF::Tn10) were derived from TL88 and JL2468, respectively, by transduction to tetracylcine resistance by phage P22, grown on TT670 (F'$_{128}$ proB$^+$A$^+$ argF::Tn10). Klebsiella pneumoniae strain Pro3 (pro-3) is a proline auxotrophic (proA$^-$ or ProB$^-$) derivative of wild type strain M5A1. Strains KY1 and KY2 were the progeny of mating Pro3 with SY1 and SY2, respectively, selecting tetracycline resistant, leucine independent exconjugants. The construction of these strains is outlined in Figure 1.

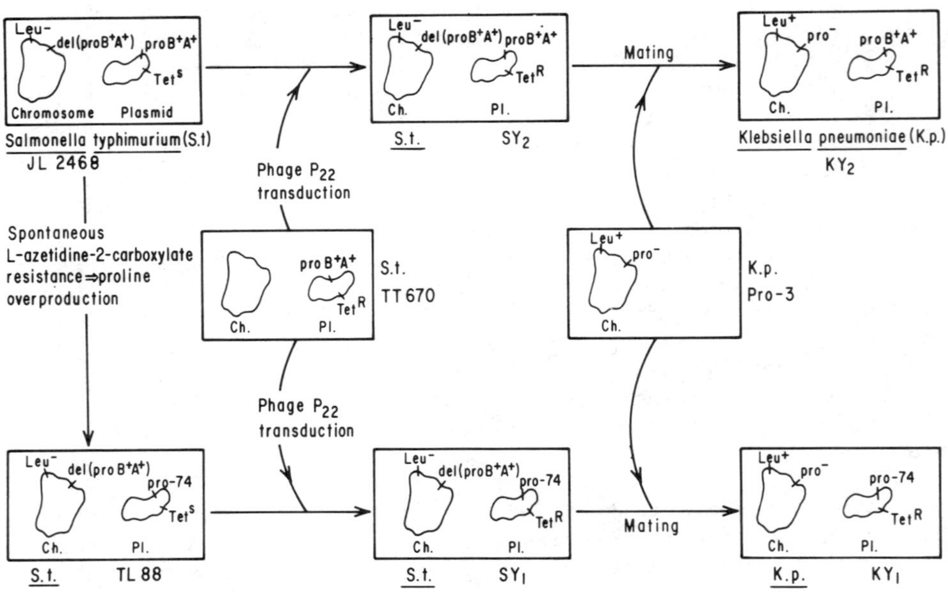

Fig. 1. Construction of the Klebsiella pneumoniae strains KY1 and KY2.

Analytical Methods

For the determination of nitrogenase activity, cultures were prepared as described (Shanmugam et al., 1974) using sucrose minimal medium (Streicher et al., 1971), supplemented with 0.68 mM L-glutamine as sole nitrogen source. The osmolarity of the medium was increased by the addition of NaCl as indicated. When NaCl was added, it was necessary to readjust the pH to that of the minimal medium (pH 7.4) which was done by the addition of NaOH. The cultures were evacuated, sparged four times with argon, and incubated at room temperature. When the OD_{420} of the cultures reached 1.0 (0.14 mg protein/ml) they were sparged once more with argon and acetylene was added to a final concentration of 10% (v/v). Nitrogenase was measured as acetylene reduction activity, as described by Shanmugam et al. (1974). Whole cell protein was determined by the method of Drews (1965), using bovine serum albumin as the standard. The growth rates were determined by monitoring optical density (420 nm) of cultures growing at room temperature under the conditions that were used for the nitrogenase assay.

RESULTS

We have examined the effect of osmotic inhibition, achieved by the addition of NaCl, on the growth rate (Table 1) and nitrogenase activity (Table 2) of *Klebsiella pneumoniae*.

Nitrogenase activity was much more sensitive to osmotic inhibition than the overall growth rate of the cells. For instance, in the absence of proline, 0.4 M NaCl caused a ten-fold decrease in the nitrogenase activity of strain M5A1 (Table 2, column 1), whereas it caused only a two-fold decrease in the growth rate (Table 1, column 1). Similarly, in the absence of proline, 0.6 M NaCl decreased the nitrogenase activity of strain M5A1 over a hundred-fold, while it caused only about a six-fold reduction in growth rate. Analogous results were obtained with strain KY2 (*pro-3*/F'_{128} *proB*$^{+}$*A*$^{+}$).

The stimulatory effect of proline on the growth rate is manifested only under conditions of extreme osmotic inhibition ($\geq$ 0.6 M NaCl with *K. pneumoniae*, unpublished results). However, proline exerted a much greater stimulatory effect on nitrogenase activity at lower osmolarities. Thus, in the presence of 0.4 M NaCl, 0.5 mM proline caused approximately a four-fold enhancement of the nitrogenase activity of strain M5A1, and approximately fifty-fold enhancement in the presence of 0.5 M NaCl (Table 2, columns 1 and 2).

The effect of the mutation resulting in proline over-production in strain KY1 (F'_{128} *pro-74*) was similar to the effect seen when proline was supplied exogenously in that the growth rate was

Table 1. The effect of exogenously added proline and of proline over-production on the growth rate of *K. pneumoniae* under conditions of osmotic inhibition.*

NaCl (M)	Growth rate (generation/h) M5A1 (wild type) -proline	M5A1 (wild type) +0.5 mM proline	KY1 (F'pro-74) -proline	KY2 (F'proA$^+$B$^+$) -proline	KY2 (F'proA$^+$B$^+$) +0.5 mM proline
0.0	0.33	0.33	0.31	0.33	0.33
0.3	0.25	0.26	0.24	0.25	0.26
0.4	0.17	0.24	0.17	0.17	0.24
0.5	0.15	0.16	0.14	0.14	0.17
0.6	0.054	0.096	0.079	0.033	0.15

*Growth was under anaerobic conditions at room temperature (cf. Materials and Methods).

Table 2. The effect of exogenously added proline and of proline over-production on nitrogenase activity under conditions of osmotic inhibition.

NaCl (M)	Nitrogenase activity (μmoles ethylene produced per hour per mg protein) M5A1 (wild type) -proline	M5A1 (wild type) +0.5 mM proline	KY1 (F'pro-74) -proline	KY2 (F'proA$^+$B$^+$) -proline	KY2 (F'proA$^+$B$^+$) +0.5 mM proline
0.0	2.55	2.66	2.64	2.71	2.59
0.3	0.42	1.26	1.55	0.60	1.31
0.4	0.25	0.96	1.53	0.14	0.74
0.5	0.02	0.94	0.42	0.04	0.73
0.6	0.02	0.12	0.26	0.01	0.19

stimulated only under extreme osmotic inhibition (compare strains KY1 and KY2, without proline, with 0.6 M NaCl; Table 1, columns 3 and 4). However, the mutation had a much more pronounced stimulatory effect on nitrogenase activity: in the absence of proline, at 0.4 and 0.5 M NaCl, the nitrogenase activity of strain KY1 was at least ten times greater than that of strain KY2. At 0.6 M NaCl, the stimulatory effect of the mutation was greater than twenty-five-fold.

DISCUSSION

Previously we have isolated mutations which resulted in proline over-production and enhanced osmotolerance in *Salmonella typhimurium* (L. N. Csonka, manuscript in preparation). In the experiments presented in this manuscript, we transferred one such mutation into *Klebsiella pneumoniae*. We found (Table 2) that the mutation stimulated nitrogenase activity ten- to twenty-five-fold in media of elevated osmolarity (0.4 to 0.6 M NaCl).

It might be argued that proline over-production might result in enhanced nitrogenase activity because it leads to NH_4^+ (or other nitrogen compounds) limitation and, hence, might cause derepression of the *nif* genes. This possibility can be discounted because in the absence of osmotic stress the nitrogenase activity of the proline over-producing strain KY1 was similar to that of the control strain KY2. Also, exogenously added proline had stimulatory effects similar to those of the mutation causing proline over-production.

It is important to note that nitrogen fixation activity summarized in Table 2 is expressed in terms of "acetylene reduction", an assay which reflects the overall nitrogenase activity of the cell, but does not take into account the relative efficiency of nitrogenase expressed as the ratio of fixed N (as NH_4^+) to H_2 evolved. The nitrogenase mediated H_2 evolution and consequently the overall energy efficiency of the enzyme might vary under conditions of osmotic stress. Experiments to determine whether this might be the case are now underway.

Why does proline alleviate osmotic inhibition? There have been two, not necessarily incompatible, explanations suggested. First, proline might be an osmotic balancer which, when present at high intracellular concentrations, might act to prevent osmotic dehydration of the cytoplasm (Measures, 1975). According to the second hypothesis, formulated by Schobert (1977) (Schobert and Tschesche, 1978), proline is stimulatory in media of high osmolarity because it has special interactions with proteins to stabilize them and keep them in solution under conditions of low water activity. At present, there is not sufficient evidence to decide which explanation is correct. Nevertheless, whatever the mechanism, proline over-production led to faster growth in Salmonella typhimurium and enhanced nitrogen fixation activity in Klebsiella pneumonaie under conditions of osmotic stress. Thus, it seems plausible that selection of proline over-producing mutants of other organisms, including Rhizobia could yield derivatives with enhanced osmotolerance.

SUMMARY

A mutation, which was isolated in Salmonella typhimurium and which resulted in proline over-production and enhanced osmotolerance in that organism, was transferred via F' plasmid mediated conjugation into the free-living nitrogen fixing bacterium Klebsiella pneumoniae. The mutation caused a ten-to twenty-five-fold enhancement of nitrogenase specific activity under conditions of osmotic stress.

ACKNOWLEDGMENTS

We thank Raymond C. Valentine for encouragement and advice, and Sydney G. Kustu for helpful comments on the manuscript. This work was supported by the National Science Foundation under Grant PFR 77-07301. D. Le Rudulier was the recipient of a NATO Fellowship. S. S. Yang was supported by the Ministry of Education of the People's Republic of China. Any opinions, findings, and conclusions or recommendations expressed in this publication are those of the authors and do not necessarily reflect the views of the National Science Foundation.

REFERENCES

Christian, J. H. B., 1955a, The influence of nutrition on the water relations of Salmonella orianenburg, Aust. J. Biol. Sci., 8:75.

Christian, J. H. B., 1955b, The water relations of growth and respiration of Salmonella orianenburg at 30°, Aust. J. Biol. Sci., 8:490.

Christian, J. H. B., and Waltho, J. A., 1966, Water relations of Salmonella orianenburg: stimulation by amino acids, J. Gen. Microbiol., 43:345.

Drews, G., 1965, Untersuchungen zur Regulation der Bacteriochlorophyll-Synthese bei Rhodospirillum rubrum, Archiv. für Mikrobiologie, 51:186.

Measures, J. C., 1975, Role of amino acids in osmoregulation of nonhalophilic bacteria, Nature, 257:398.

Schobert, B., 1977, Is there an osmotic regulatory mechanism in algae and higher plants?, J. Theor. Biol., 68:17.

Schobert, B., and Tschesche, H., 1978, Unusual properties of proline and its interaction with proteins, Biochim. Biophys. Acta, 541:270.

Shanmugam, K. T., Loo, A. S., and Valentine, R. C., 1974, Deletion mutants of nitrogen fixation in *Klebsiella pneumoniae*: mapping of a cluster of *nif* genes essential for nitrogenase activity, *Biochim. Biophys. Acta*, 338:545.

Streicher, S., Gurney, E., and Valentine, R. C., 1971, Transduction of the nitrogen-fixation genes in *Klebsiella pneumoniae*, *Proc. Natl. Acad. Sci. USA*, 68:1174.

SELECTION OF NATURALLY OCCURRING STRESS TOLERANT *RHIZOBIUM*

D. N. Munns, K. G. Cassman[1], and H. H. Keyser[2]

Department of Land, Air and Water Resources, University of California, Davis, CA 95616; IRI, Jari Project, Belem, Brazil[1]; USDA, CCNF, Beltsville, MD[2]

INTRODUCTION

Symbiotic nitrogen fixation is commonly limited by soil infertility conditions, especially phosphate deficiency, acidity and salinity. Phosphate sorption buffers phosphate at low concentrations in soil solutions, the resulting immobility causing deficiency to plants in most soils with moderate or high degrees of mineral weathering. Acidity, the consequence of prolonged leaching with inputs of carbonic, nitric, and sulphuric acid, is attended by calcium deficiency and toxicities of aluminum or manganese, and is common where the rainfall is high, the landsurface old, or the agriculture old and productive. Salinity, the accumulation of leachable products, attended by alkalinity and sodicity, is common in irrigated agriculture of arid regions. Some legumes become specially sensitive to these disorders when they are dependent on symbiotic nitrogen fixation.

The sensitivity of the symbiosis is partly a consequence of complexity. Effective symbiosis requires other things as well as vigorous plant growth. It requires colonization of soil by rhizobia as saprophytes in competition with endogenous microbes; and there is evidence that this phase is limited by soil infertility (Parker et al., 1977). Another requirement is rapid rhizobial growth during invasion of the rhizosphere; and this region may rapidly become acid, saline, or phosphate-depleted even in soils of normal fertility (Helyar and Munns, 1975; Nye, 1975). Finally, infection and nodule development can suffer from soil disorders even if other phases have been completed (Munns, 1977). Ability of rhizobia to tolerate adverse soil conditions might be compromised by adaptation to nutritionally luxurious conditions routinely supplied in conventional media used for culture and testing (Parker et al., 1977).

Little research has been done on mineral nutrition of rhizobia (Robson, 1978). We and our colleagues at Davis have initiated research to add information in this area. The particular aims of the work described in this paper have been:

1) To identify and measure requirements for phosphate and tolerances of the compounds of soil acidity stress; and to determine whether strains of rhizobia differ in these requirements and tolerances.

2) To develop laboratory screening procedures for distinguishing tolerant from sensitive strains.

3) To test the laboratory screening procedures against symbiotic performance under stress in soil.

LABORATORY SCREENING FOR GROWTH UNDER STRESS

Calcium Deficiency and Manganese Toxicity

Strains have been screened for ability to grow in defined acid (pH 4.5) liquified medium at 50 μM Ca, which represents the extreme low range of concentrations found in soil solutions, or at 200 μM Mn, which is toxic to most legumes (Keyser and Munns, 1979). These concentrations can be maintained during rhizobial growth without special precautions. Of 65 strains of cowpeas and soybean rhizobia tested, almost all grew as well as in control medium (200 μM Mn, 50 μM Ca), implying that neither Mn toxicity nor Ca deficiency is likely to limit slow-growing rhizobia in the field. This is in contrast with the other important soil-acidity components, viz, Al and H.

Acidity and Aluminum Toxicity

In acid soil solutions, Al^{+3} concentrations rise sharply as pH decreases. The system behaves like a saturated solution in equilibrium with solid $Al(OH)_3$ (Lindsay, 1980). Similarly, in artificial media Al solubility is limited by the dissolution of either $Al(OH)_3$ or $Al(OH)_2H_2PO_4$. Aluminum toxicity occurs only at low pH, low P, and in organisms that tolerate acidity. Our screening media have been low in P (below 10 μM), acid (pH 4.4 to 4.8), and free of organic buffers whose anions could complex and detoxify Al^{+3}. Control of pH is essential.

Control of pH by chemostat or flowing culture is too unwieldy for the purpose. Supplying galactose or arabinose instead of mannitol stops some slow-growing rhizobia from shifting the pH

(Date and Halliday, 1978), but we found it ineffective for most strains. Fortunately, sufficient pH control can result simply from operating at low population density. No significant pH-shift was found to occur until populations exceeded 10^6 cells/ml. It is easy to study growth at lower densities by starting with small inocula, and either terminating the experiment early (Keyser and Munns, 1979; Keyser et al., 1979) or imposing a growth limitation such as low phosphate supply (Munns and Keyser, 1980).

Some 65 strains of rhizobia have been screened using this approach (Table 1). The principal conclusions are as follows:

1) Strains vary in tolerance of low pH, confirming previous evidence (e.g., Graham and Parker, 1964).

2) Strains that grow at pH 4.5 grow more slowly if the medium contains 50 or 60 µM Al, concentrations that would normally occur in soil at pH 4.5 (Fig. 1).

3) Strains vary in aluminum tolerance: concentrations that merely slow the growth of some strains stop growth of others.

4) Increased Ca concentration does not moderate toxic effects of aluminum in rhizobia (unlike plants).

Table 1. Tolerance of acid and Al stress among 65 strains of rhizobia

Category	Cowpea rhizobia		Soybean rhizobia	
	number of strains	% of total	number of strains	% of total
Sensitive to pH 4.5	14	27	5	38
Tolerate pH 4.5, but sensitive to 50 µM Al	13	25	5	38
Tolerate pH 4.5 with 50 µM Al	25	48	3	23

(From Keyser and Munns, 1979).

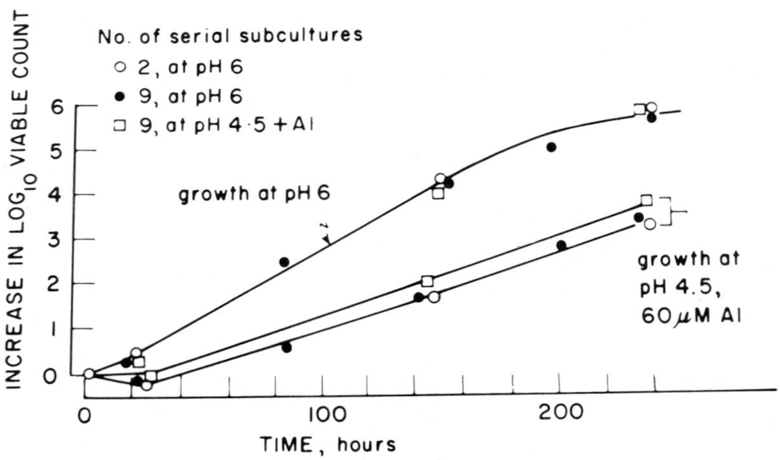

Fig. 1. Effect of low pH and aluminum toxicity on growth of Rhizobium sp. (strain TAL189) in defined liquid medium. Effect on growth rate is unaffected by serial subculture 9 times (72 generations) under stress (from Munns and Keyser, 1980).

Combined Acid and Aluminum Stress in Solid Medium

In solid media, difficulties of control of pH and other ionic activities are accentuated by diffusional limitations: gradients should develop rapidly around growing colonies. A solid medium was needed, however, to test whether putative variants tolerant of combined acidity and Al could be separated from single-strain populations (Munns and Keyser, 1980). A partially satisfactory medium was devised, citrate-glutamate-aluminum hydroxide agar. Aluminum is supplied as solid $Al(OH)_3$, mobilized by chelation with citrate at 10 μM concentration. Phosphate is low (5 μM). These features mimic the system that probably operates in acid soils (Lindsay, 1980). The $Al(OH)_3$ provides a reservoir of Al buffering the Al^{+3} activity according to the pH and the solubility product $(Al)(OH)^3 = 10^{-32}$. A little pH-buffering is provided by high (0.2%) concentration of glutamate/glutamic acid, pK = 4.6. Some strains can locally neutralize this medium; once colonies have grown to visible size they develop halos of bromcresol purple included as indicator. But control is probably adequate during early growth. Only a fraction of inoculant cells develop colonies, the fraction ranging from 1:10 for strains tested as tolerant by independent tests down to less than 1:10^6 for strains rated highly sensitive.

Isolates from colonies developed on this medium are no more tolerant than isolates of the same strain from non-stress media. This corroborates other evidence that tolerance of acidity and Al is a stable characteristic of strains of *Rhizobium* (see Fig. 1).

Tolerance of Phosphate Deficiency: Storage and Reutilization

Rhizobia accumulate P from high-P media, sufficient to support considerable growth after transfer to low-P media. This storage and reutilization could be an important aspect of tolerance of P stress in nature, where rhizobia are probably confronted by large variations in ambient P concentrations (e.g., 10^{-6} M in bulk soil, less than 10^{-7} M in the rhizosphere, perhaps 10^{-3} M in nodules (and microbiologists' media)).

To evaluate P storage, Cassman et al. (in press a.) grew several strains of *Rhizobium japonicum* in a conventional high P medium (2×10^{-3} M P) and then determined their internal P concentration and their ability to grow upon transfer to medium with no added P (about 1×10^{-7} M by analysis). All the strains stored P to high internal concentrations (1.5-2.5%) and made 2 to 5 generations' growth on the stored P as sole source. There were large differences among strains, and the number of generations depended on the concentration of P stored (Fig. 2a). Some of the storage was in the form of electron-dense granules, possibly polyphosphate. The relationship between cell phosphate and granule volume (Fig. 2b) suggests also that there is considerable stored P apart from the granules. Cells grown in media buffered at 6×10^{-6} M P, as in a fertile soil, contained 1.3-1.8% P, with few granules. This storage is still large, and could be useful for cells exposed to the extreme P deficient conditions associated with invasion of the rhizosphere; but calculations based on reasonable assumptions suggest that storage must be supplemented by efficient uptake from low concentrations.

Tolerance of Phosphate Deficiency: External Requirement for Growth

To measure growth response to P at low concentrations, Cassman et al. (in press b.) inoculated P-depleted cells into "oxide dialysis cultures", in which P concentration is buffered by sorption of P on goethite (αFeOOH). The oxide is separated from the liquid medium by a dialysis membrane for convenience of analysis and rhizobial enumeration. The reserve of sorbed P maintains a concentration of P in solution according to the desorption isotherm (Fig. 3a). Phosphate desorbs and diffuses into solution fast enough to counter the removal by rhizobia, until the population reaches a critical density dependent on the total P in the system

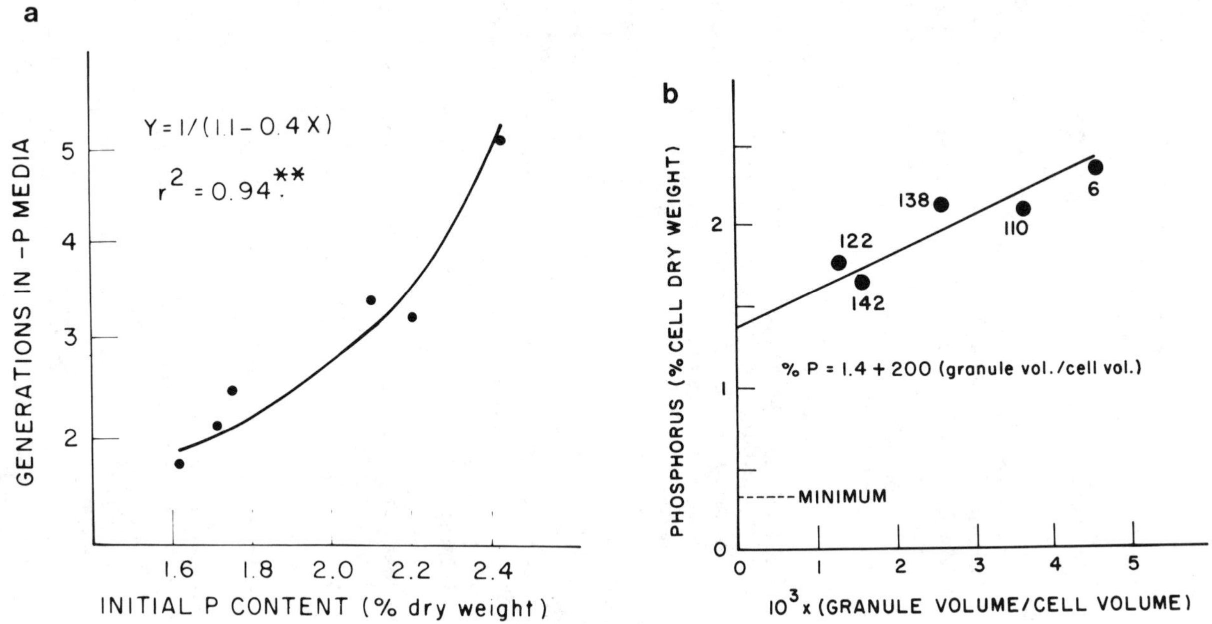

Fig. 2. Strain variation in phosphate storage. (a) Relationship between ability to grow in P-deficient medium and internal P previously stored during growth in high-P medium (points are data for 6 strains of *R. japonicum*). (b) Relationship between internal P concentration and volume of presumed polyphosphate granules (points identified by USDA strain number).

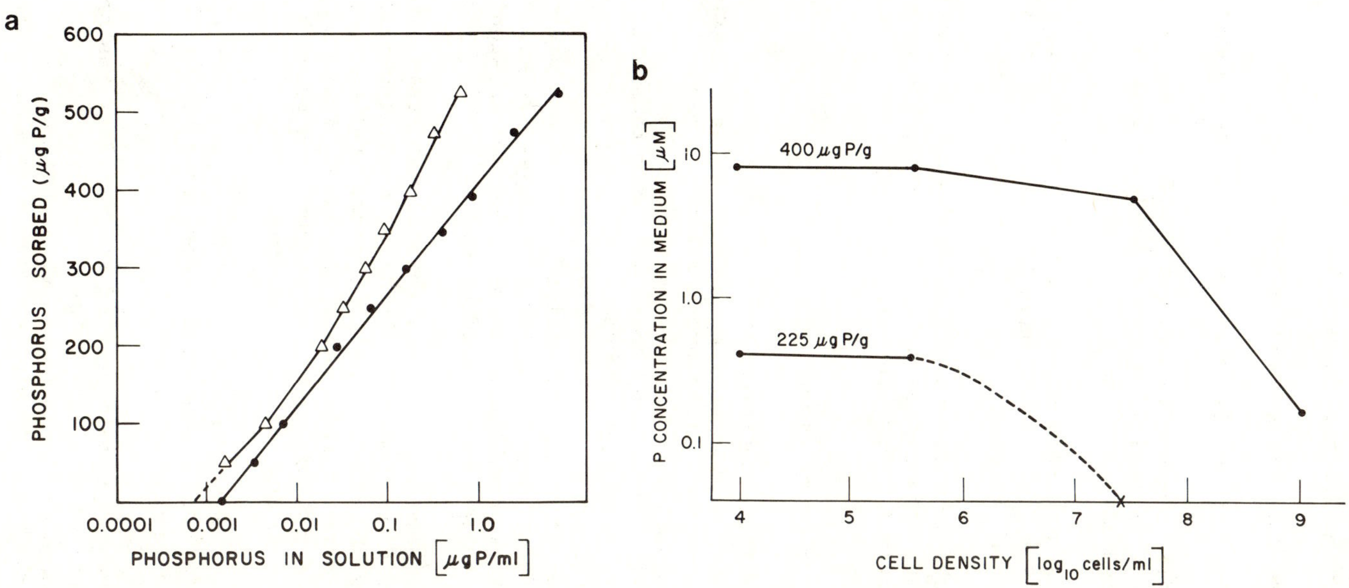

Fig. 3. Behavior of iron oxide dialysis culture system for study of growth response to external P concentration. (a) Sorption isotherm (black) and desorption isotherm after autoclaving (open) of typical batch of limonite. (b) Dependence of buffering of solution P on cell density.

(Fig. 3b). Below this critical density, growth response can be assessed at constant solution P concentrations.

Using the oxide dialysis culture system we have found that:

1) Strains differ in ability to grow at low P concentrations.

2) Agriculturally successful strains (USDA 110 and CB756, for example) grow unimpaired at the lowest concentration we have provided (5×10^{-8} M).

3) Some strains grow very slowly even at concentrations as high as 10^{-6} M, and should have difficulty coping with normal soil environments.

4) Exopolysaccharide production (as indicated by culture viscosity) is enhanced at solution P concentrations which approximate those in soil, and suppressed in high P medium.

SYMBIOTIC PERFORMANCE UNDER STRESS IN SOIL

The laboratory screening procedures we have been describing require validation. This can only be done by comparing strain ratings from the laboratory procedures with performance in symbiosis with host legumes under stress in soils. These symbiotic tests are essentially simple greenhouse (or preferably field) trials. All that is needed is to choose soil, treatments, host, and test conditions so that N is deficient, interfering rhizobia are absent, the subject stress is defined in kind and severity, the host cultivar is tolerant, and other stresses are minimized so that differences in nodulation can be expressed as significant differences in plant growth and yield of N.

Our test soils have been Ultisol subsoil materials from California's Coast Range and Sierra foothills. Mineralogically and chemically they resemble many soils of agricultural regions of the eastern U.S. and the humid tropics and subtropics. Our host cultivars are selected on the basis of a companion project dealing with variation of tolerance among legume varieties.

Testing at this level for P responsiveness has only begun, but preliminary data with soybean and cowpea rhizobia are at least consistent with the laboratory tests.

Testing for symbiotic tolerance of soil acidity has been much more extensive (Munns et al., 1978; Keyser et al., 1979; Munns et al., 1978). An example of data from such a test are given in Figure 4. In general, with strains of cowpea rhizobia, laboratory tests for sensitivity to acid or Al identify a usefully high

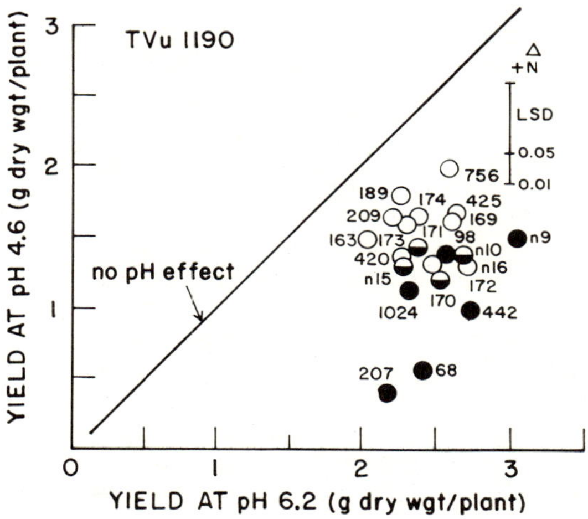

Fig. 4. Results of a greenhouse screening of cowpea rhizobia for symbiotic tolerance of soil acidity; and validation of laboratory screening procedure for Al and H tolerance. Strains were inoculated at 10^4 cells/seed onto TVu1190 cowpea in acid soil with and without calcium carbonate treatment. Graph compares plant yield at acid pH with yield at optimum pH. Tolerant strains approach line of unit slope; sensitive strains drop well below the line. White circles indicate strains found tolerant of Al and H in the laboratory, half-black indicates tolerant of H, black sensitive of Al and H.

proportion of strains that will fail in acid soil with the tolerant cowpea cultivars Blackeye 5, TVu1190, or TVu4557. Some strains that could grow under acid/Al stress in the laboratory failed symbiotically in soil; perhaps because they failed to infect or to function in nodules under stress (Munns et al., 1978). These represent about 35% of the total sensitive strains in the greenhouse tests, not predicted by laboratory test. We regard this as a reasonable failure rate. More important, the laboratory tests did not misidentify as sensitive any strain that proved highly tolerant and effective in soil.

WHEN IS IT JUSTIFIABLE TO SCREEN RHIZOBIUM FOR TOLERANCE?

This is the first question to ask, despite our leaving it to last in this paper. The simple answer is that screening is justifiable if it leads to better performance of the symbiosis; and essentially this answer comes from simple trials like those described in the previous section.

The answer may depend on the kind of stress. Salinity, for instance, affects either nodule function or growth of the legume host much more readily and severely than it inhibits growth of most strains of *Rhizobium*. By and large, rhizobia can grow at NaCl concentrations of the order 100 to 300 mM (Graham and Parker, 1964), well above the levels that kill legumes. Attempting to improve the ability of rhizobia to grow in saline soil is probably pointless.

The answer also depends on the legume involved. With cowpea, for example, it is clear from Figure 4 that choice of *Rhizobium* makes a large difference to symbiotic performance in acid soil, and that selection for stress tolerance need not entail loss of potential performance. This conclusion does not hold for all legumes however. In our experience, it holds for mund and peanut, but not for soybean or certain aluminum-sensitive cultivars of cowpea. In the latter cases, host-plant sensitivity limits the performance of the symbiosis; and host-plant selection then has priority over selection of *Rhizobium*.

ACKNOWLEDGMENT

This work was supported in part by a grant from the National Science Foundation (NSF PFR 77-07301).

REFERENCES

Cassman, K. G., Munns, D. N., and Beck, D. P., 1981, Phosphate nutrition of *Rhizobium japonicum*: strain differences in storage and utilization, *Soil Sci. Soc. Amer. J.*, in press a.

Cassman, K. G., Munns, D. N., and Beck, D. P., 1981, Growth of Rhizobium strains at low concentrations of phosphate, *Soil Sci. Soc. Amer. J.*, in press b.

Date, R. A., and Halliday, J., 1978, Selecting *Rhizobium* for acid infertile soils of the tropics, *Nature*, 277:62.

Graham, P. H., and Parker, C. A., 1964, Diagnostic features in the characterization of the root nodule bacteria of legumes, *Plant and Soil*, 20:383.

Helyar, K. R., and Munns, D. N., 1975, Phosphate fluxes in the soil-plant system: a computer simulation, *Hilgardia*, 43:103.

Keyser, H. H., and Munns, D. N., 1979, Tolerance of rhizobia to acidity, aluminum, and phosphate, *Soil Sci. Soc. Amer. J.*, 43:519.

Keyser, H. H., and Munns, D. N., 1979, Effects of calcium, manganese, and aluminum on growth of rhizobia in acid media, Soil Sci. Soc. Amer. J., 43:500.

Keyser, H. H., Munns, D. N., and Hohenberg, J. S., 1979, Acid tolerance of rhizobia in culture and in symbiosis with cowpea, Soil Sci. Soc. Amer. J., 43:719.

Lindsay, W. L., 1980, "Chemical Equilibria in Soils," Wiley Interscience, New York.

Munns, D. N., 1977, Mineral nutrition and the legume symbiosis, in: "A Treatise on Dinitrogen Fixation. IV. Agronomy and Ecology," R. W. F. Hardy and A. H. Gibson, eds., Wiley Interscience, New York.

Munns, D. N., 1978, Soil acidity and nodulation, in: "Mineral Nutrition of Legumes in Tropical and Subtropical Soils," C. S. Andrew and E. J. Kamprath, eds., C.S.I.R.O., Melbourne.

Munns, D. N., Keyser, H. H., Fogle, V. W., Hohenberg, J. S., Righetti, T. L., Lauter, D. L., Zaroug, M. G., Clarkin, K. L., and Whitacre, K. A., 1979, Tolerance of soil acidity in symbioses of mung bean with rhizobia, Agron. J., 71:256.

Munns. D. N., and Keyser, H. H., 1980, Response of Rhizobium to acid and aluminum stress, Soil Biol. Biochem., in press.

Nye, P. H., 1978, Soil properties controlling the supply of nutrients to root surfaces, in: "The Soil-Root Interface," J. L. Harley and R. Scott Russell, eds., Academic Press, London.

Parker, C. A., Trinick, M. J., and Chatel, D. L., L977, Rhizobia as soil and rhizosphere inhabitants, in: "A Treatise on Dinitrogen Fixation. IV. Agronomy and Ecology," R. W. F. Hardy and A. H. Gibson, eds., Wiley Interscience, New York.

Robson, A. D., 1978, Mineral nutrients limiting nitrogen fixation in legumes, in: "Mineral Nutrition of Legumes in Tropical and Subtropical Soils," C. S. Andrew and E. J. Kamprath, edsl, C.S.I.R.O., Melbourne.

III. PLANT FACTORS IMPACTING NITROGEN ASSIMILATION

EVALUATING POTENTIALLY SUPERIOR *RHIZOBIUM* STRAINS IN SOYBEANS

L. E. Williams, S. S. Jue, K. G. Cassman, and
D. A. Phillips

Department of Agronomy & Range Science
University of California
Davis, CA 95616

INTRODUCTION

Legumes infected with the symbiotic bacterium *Rhizobium* obtain nitrogen from both soil N and N_2. It would be economically advantageous to conserve existing soil N by increasing the proportion of N derived from symbiotic N_2 fixation in legumes such as soybeans, if agronomic yields could be maintained. One method to achieve that goal is the identification of superior *Rhizobium* strains.

Recent results have reemphasized the classical concept (Allen and Allen, 1939) that effective *Rhizobium* strains differ in their efficiency of N_2 fixation. Schubert et al. (1978) reported that *Rhizobium* strains with H_2-oxidation capacity promoted dry matter yield in soybean and cowpea. Bethlenfalvay et al. (1978) found greater rates of apparent photosynthesis on both a whole-plant and leaf-area basis in peas inoculated with *Rhizobium* strains that reduced more N_2. In the latter study enhanced N_2 fixation, measured by Kjeldahl N accumulation in the absence of combined N, was associated with several different factors including decreased H_2 evolution in some strains and increased acetylene reduction activity in others (Bethlenfalvay et al., 1978). Maier and Brill (1978) reported the selection of mutant *R. japonicum* strains that promoted soybean seedling growth and N_2 fixation under controlled environmental conditions. Such reports of the physiological superiority of natural variants (Bethlenfalvay et al., 1978; Schubert et al., 1978) and selected mutants (Maier and Brill, 1978) in *Rhizobium* suggest that the identification of strains with increased capacity to reduce N_2 may be an important strategy for the conservation of soil N.

The single most important criterion that must be met to identify *Rhizobium* strains with enhanced N_2 fixation capacity and agronomic potential is effective nodule formation under field conditions with a resulting increase in total N_2 reduced. All workers recognize that replicated field tests of separate *Rhizobium* strains are difficult and should be performed only when it is clear that a particular strain shows potential superiority. The problem, therefore, is to find a reliable test procedure that eliminates inferior strains before field trials are run. Many investigators have reported data from *Rhizobium* strains tested on plants grown under controlled environmental conditions (Bethlenfalvay et al., 1978; Maier and Brill, 1978; Schubert et al., 1978). The time and space requirements of such trials could be reduced significantly if *Rhizobium* strains could be screened in free-living culture for triats associated with superior N_2 fixation in legumes.

Relationships between physiological traits of *Rhizobium* in free-living culture and in root nodules are poorly understood. Differences in colony morphology and carbohydrate utilization of free-living *R. japonicum* genotypes were associated with differences in acetylene reduction by the same strains on 30-day-old soybean plants (Kuykendall and Elkan, 1976). Uptake of H_2 and $^3H_2$-exchange activity of free-living cultures of *R. japonicum* 3Ilb31 and 3Ilb110 (Lim, 1978) correlated well with H_2-uptake values reported for the same strains in symbiosis with Anoka soybeans (Schubert et al., 1978). Nitrate-reductase-deficient strains of *Rhizobium* have been identified in culture and tested on legumes (Gibson and Pagan, 1977; DeVasconcelos et al., 1980; Manhart and Wong, 1980).

Scott et al. (1979) characterized an interesting mutant of *R. japonicum* 3Ilb110 that showed a 100% increase in specific activity of C_2H_2 reduction in free-living cultures. That mutant, *R. japonicum* C33, which was selected for chlorate resistance and was shown to be nitrate-reductase-deficient, maintained similar C_2H_2-reduction activity in pure culture with 4 mM KNO_3, while the wild type strain 3Ilb110 showed no activity. The purpose of the present study was to evaluate whether a mutant of *R. japonicum* identified as having high C_2H_2-reduction activity and nitrate-reductase deficiency in free-living culture would have enhanced N_2 fixation in soybean root nodules.

METHODS

Growth Chamber Studies

Soybeans (Glycine max (L.) Merr.) cv. Clark were germinated in the dark at 28 C. Day 0 was determined as the day of seed imbibition. On day 3 seedlings were planted in 15 cm diameter plastic pots that permitted repeated C_2H_2 reduction and apparent photosynthesis measurements of intact plants (Williams and Phillips, 1980). The plants were grown with 2 mM KNO_3 complete nutrient solution under controlled environmental conditions (Williams and Phillips, 1980). Plants were inoculated with either R. japonicum 3Ilb110 or strain C33, a nitrate-reductase-deficient mutant of 3Ilb110 (obtained from D. B. Scott). Apparent N_2 fixation was measured with the C_2H_2-reduction assay on both intact plants and detached root systems (Williams and Phillips, 1980). Plants were harvested, dried at 70 C for 48 h, and weighed. Total N was determined by Kjeldahl analysis (Burris and Wilson, 1957).

Field Studies

Clark soybeans and a non-nodulating, near isogenic line were grown in the field at Davis in 1979. Seeds were planted 5 cm apart with a single row on 76 cm beds. Normal irrigation practices were followed throughout the season. A randomized, complete plot design with six replicate blocks was used. Four treatments consisted of 1) uninoculated Clark seeds, 2) Clark seeds + C33, 3) Clark seeds + 3Ilb110, 4) non-nodulating Clark seeds. Each plot consisted of four 8 m rows; data were collected only from the two center rows. Two replicate blocks were sacrificed during the growing season to collect physiological and microbiological data; four replicate blocks were analyzed for yield parameters.

Microbiological Studies

Rhizobium strains were inoculated onto growth-chamber-grown seedlings on day 3 and 10 as a slurry of cells removed from agar slants of mannitol/salts/yeast extract (MSY) medium (Lim and Shanmugam, 1979). Inoculants for field studies were grown in liquid MSY medium which was mixed with sterile peat and pelleted onto seeds before planting. Single-cell isolates of bacteria were recovered from root nodules periodically and were tested for the following phenotypes: 1) sensitivity to a bacteriophage which lyses Rhizobium in the 110 serogroup (Vincent, 1970), and 2) resistance to 2 mg $KClO_3$/ml in MSY medium when tested microaerophilically on Petri plates. Strain 3Ilb110 was known to be positive only for test 1, while strain C33 was positive for both tests 1 and 2.

RESULTS

Growth Chamber Studies

Mutant Rhizobium strain C33 enhanced whole-plant apparent photosynthesis and whole-plant apparent N_2 fixation in Clark soybeans grown with 2 mM KNO_3 at 1500 $\mu E \cdot m^{-2} \cdot sec^{-1}$ irradiance, compared with the wild type parent strain 3Ilb110. The advantage measured for plants inoculated with strain C33 was uniformly significant for apparent photosynthesis from day 17 to 27, at which time plants became too large to measure. Acetylene reduction by plants inoculated with strain C33 was greater in 9 of 10 assays between day 17 and 42. Actual plant dry weight and N content reflected the short-term measures of CO_2 and N_2 reduction in plants grown at the higher irradiance and harvested on days 31, 45 and 52 (Table 1). Plants grown under similar conditions at 700 $\mu E \cdot m^{-2} \cdot sec^{-1}$ irradiance, however, showed a significant effect of Rhizobium strain only on dry weight of 45-day-old plants (Table 1).

Field Studies

Field studies of mutant C33 conducted in 1979 at Davis showed no significant effect of Rhizobium strain on the seasonal pattern of C_2H_2 reduction by detached root systems of Clark soybeans (Fig. 1). Uninoculated plants became nodulated later and showed an altered C_2H_2-reduction profile (Fig. 1). No root nodules were observed on the non-nodulating isoline of Clark. Total seed yield and seed N content of that material were not affected by Rhizobium strain, and no increase in those parameters over uninoculated, nodulating plants were recorded (Table 2). Plants inoculated with Rhizobium lodged 115 days after planting, but both the uninoculated and the non-nodulating plants were erect at harvest. Total vegetative dry matter and N content at the R5 stage of development were not significantly greater in plants inoculated with C33, but there were trends in that direction (Table 3). Symbiotic N_2 fixation calculated by the depletion of ^{15}N content of nodulated relative to natural ^{15}N abundance in non-nodulating plants suggested that plants inoculated with either Rhizobium strain obtained significantly ($P \leq .05$) more N from N_2 than those that were uninoculated (Table 4). The fraction of seed N derived from N_2 was estimated as approximately 48% for inoculated plants on the basis of ^{15}N depletion (Table 2). Acetylene reduction data and total N differences between nodulated and non-nodulated plants supported the conclusion that inoculated plants fixed more N_2 than uninoculated plants that were nodulated by field contaminants (Table 4).

Table 1. The effect of mutant Rhizobium japonicum C33 on dry weight and N content of soybeans grown at two irradiances. Data represent the percent change in parameters of five replicate plants relative to plants inoculated with the parent wild-type R. japonicum 3Ilb110.

Plant age (days)	Irradiance			
	$700\ \mu E \cdot m^{-2} \cdot sec^{-1}$		$1500\ \mu E \cdot m^{-2} \cdot sec^{-1}$	
	Dry weight (%)	N content (%)	Dry Weight (%)	N content (%)
24	+29	+28	+10	+2
31	+15	+15	+40**	+29*
45	+25*	+14	+43*	+38*
52	+1	-5	+16*	+6

*,** Strain effect significant at $p \leq 0.05$ or 0.01, respectively.

Table 2. Seed yield, total N, and ^{15}N content of Clark soybeans grown during 1979 in the field with a wild-type Rhizobium or a potentially superior mutant strain C33. Data were collected from four, 4-row replicate plots of each treatment. The fraction of seed N derived from N_2 was calculated by dividing the ^{15}N values of the nodulating treatments by the value of the non-nodulating isoline and subtracting that ratio from 1.

Plant nodulation genotype	Rhizobium strain	Seed yield	Seed N content	Seed ^{15}N content	Fraction of seed N from N_2
		(kg/ha)		(atom % excess)	(%)
Nodulating	3Ilb110	2420	153	.00058	47.7
Nodulating	C33	2630	163	.00058	47.7
Nodulating	None	2640	160	.00082	26.1
Non-nodulating	None	1820	82.4	.00111	-
LSD (0.05)		396	24.0	.00026	-

Table 3. Vegetative yield and N content of Clark soybeans inoculated with *R. japonicum* 3Ilb110 or a potentially superior mutant strain C33. Data were collected in 1979 field trails at the R5 stage (Fehr et al., 1971) from four, 4-row replicate plots of each treatment.

Plant nodulation genotype	Rhizobium strain	Dry Matter		N Content	
		Shoot	Root	Shoot	Root
		(kg/ha)			
Nodulating	3Ilb110	7710	485	220	8
Nodulating	C33	8500	535	241	8
Nodulating	None	8000	645	182	7
Non-nodulating	None	7090	604	101	4
LSD (0.05)		1220	106	60	2

Table 4. Estimates of symbiotic N_2 in Clark soybeans grown in the field with *R. japonicum* 3Ilb110 or a potentially superior mutant strain C33. Fixation estimated by N difference is based on total plant N of nodulated plants relative to non-nodulating controls. The ^{15}N dilution estimate was derived from dilution of natural abundance of ^{15}N in seeds of non-nodulating controls relative to nodulated plants, multiplied by total plant N content at the R5 stage. Seasonal N_2 fixation from C_2H_2 reduction was determined from data in Figure 3.

Rhizobium strain	N_2 Fixed		
	N Difference	^{15}N Dilution	C_2H_2 Reduction
	(kg/ha)		
3Ilb110	123	109	74
C33	144	119	64
Field contaminants	84	49	44

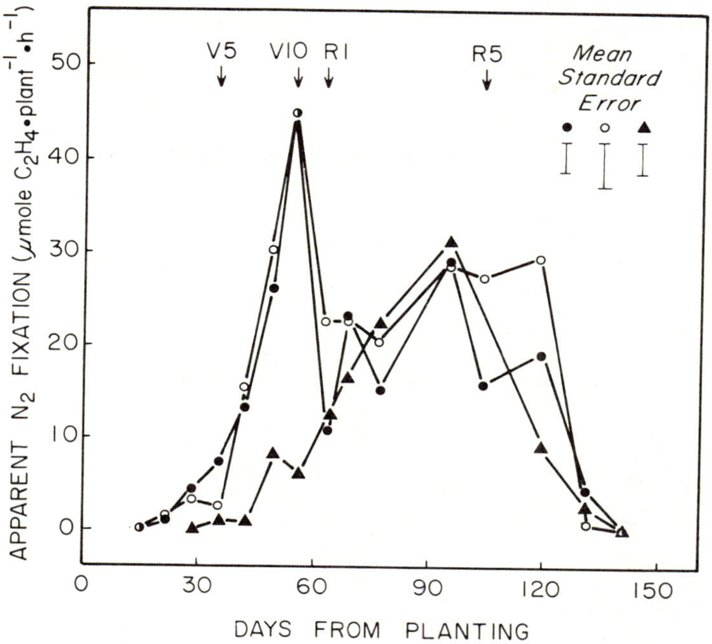

Fig. 1. Apparent N_2 fixation of field-grown Clark soybeans. Plants were inoculated and nodulated primarily by a wild-type R. japonicum 3Ilb110 (O) or mutant C33 (●). Uninoculated plants (▲) were nodulated by both indigenous bacteria and the two introduced strains. Data are mean values of detached root systems harvested from six replicate plants. Symbols at the top of the figure represent stages of soybean development (Fehr et al., 1971).

Microbiological Studies

Field trials with specific Rhizobium strains, require verification that nodules were formed by bacteria in the inoculant. On three occasions during the 1979 growing season bacteria were isolated from root nodules of the Clark soybeans. Three to five nodules were used from each of the five replicate plants harvested. Two to five individual bacterial clones were obtained from each nodule. All Rhizobium clones isolated from plants inoculated with 3Ilb110 or C33 were lysed by a bacteriophage specific for the 110 serogroup; 33% of isolates from the uninoculated plants were lysed under the same conditions. None of the clones derived from uninoculated plants or those inodulated with strain 3Ilb110 were resistant to 2 mg/ml chlorate, but 84% of the isolates from plants inoculated with C33 were resistant to chlorate.

DISCUSSION

In spite of a 100% increase in specific activity of C_2H_2-reduction by strain C33 relative to strain 3Ilb110 in free-living cultures (Scott et al., 1979) and a 38% increase in plant N content of 45-day-old soybeans (Table 1), only suggestive and not statistically significant increases were recorded for seed yield, seed N content, and dry matter production in the field (Tables 2 and 3). Such results do not necessarily invalidate testing of potentially superior Rhizobium strains under controlled conditions, but they do emphasize the complexity of natural environments where any genetically-altered Rhizobium strain must function eventually.

The C33-induced promotion of plant dry weight and N content observed with 2 mM KNO_3 under 1500 $\mu E \cdot m^{-2} \cdot sec^{-1}$ (Table 1) deserves further comment. The 30-40% increase in N content and dry weight found in plants harvested on days 31 and 45 were similar in magnitude to promotive effects reported for other Rhizobium mutants analyzed at one stage of plant growth (Maier and Brill, 1978). Data in Table 1, however, suggest that plants inoculated with strain 3Ilb110 assimilated C and N more rapidly between day 31 and 52 than those inoculated with strain C33. One might suggest, therefore, that the promotive effect of C33 on dry weight and N content was transitory. Unfortunately that possibility could not be verified with measures of whole-plant apparent photosynthesis because the plants become too large for the assay chambers. Such results do not justify a recommendation that plants should be sampled repeatedly when evaluating Rhizobium strains, but they do suggest that data from controlled environments be interpredted with caution. Likewise, the importance of environmental parameters is emphasized by the fact that under 700 $\mu E \cdot m^{-2} \cdot sec^{-1}$ irradiance strain C33 produced fewer significant differences in dry weight and N content but still showed a general superiority to strain 3Ilb110 (Table 1). Additional evidence that supports the importance of environment comes from the observation that when Clark soybeans were grown in Leonard jar assemblies, rather than the well-drained containers of the present study, plants nodulated by strain 3Ilb110 or C33 were indistinguishable in all parameters measured at five nitrate concentrations between 0 and 16 mM (Williams, unpublished data).

Table 5. Hypothetical relationship between soil nitrogen availability and symbiotic performance of *Rhizobium* strains with varying effectivity.

		Relative N yield		
Rhizobium strain	Comparative effectivity+	N derived from soil	N derived from symbiosis++	Relative total N yield
		75	25	100
		65	35	100
C33	100	55	45	100
		45	55	100
		35	65	100
		25	75	100
		75	18	93
		65	25	90
3Ilb110	72.5	55	33	88
		45	40	85
		35	47	82*
		25	54	79**

+ Relative symbiotic effectiveness expressed as a percentage of the most effective strain (strain C33) which is given a 100% rating, 38% greater than strain 3Ilb110.

++ Comparative effectivity multiplied by the N derived from symbiosis by the most effective strain (strain C33).

*,** Indicates a total N yield significantly lower ($p \leq 0.05, 0.01$) than that of the most effective strain (strain C33) assuming a coefficient of variation of 10% in a field experiment with two rhizobial treatments and four replications.

One may conclude from data in this study that mutant Rhizobium strain C33 has not yet been proved to be agronomically superior to strain 3Ilb110. It may be possible with additional field tests on various sites to prove that C33 is slightly better, but the concept of a simple free-living or growth-chamber assay that predicts field performance of marginally superior Rhizobium strains was weakened by this study. Hypothetically, it can be calculated (Table 5) that a 38% increase in symbiotic N_2 fixation (Table 1) could be detected in a field environment when less than 43% of the total plant N (35 of 82 relative N yield units, Table 5) is derived from the soil. Whether large increases in efficiency of N_2 fixation can be accomplished by improving a single physiological trait such as C_2H_2 reduction or H_2 uptake remains to be determined. It seems reasonable that several potentially advantageous traits will have to be combined to show positive benefits under field conditions.

ACKNOWLEDGMENT

This research was supported in part by National Science Foundation Grant PFR 77-07301.

REFERENCES

Allen, O. N., and Allen, E. K., 1939, Root nodule bacteria of some tropical leguminous plants: II. Cross-inoculation tests within the cowpea group, Soil Sci., 47:63.

Bethlenfalvay, G. J., Abu-Shakra, S. S., and Phillips, D. A., 1978, Interdependence of nitrogen nutrition and photosynthesis in Pisum sativum L. II. Host plant response to nitrogen fixation by Rhizobium strains, Plant Physiol., 62:131.

Burris, R. H., and Wilson, P. W., 1957, Methods for measurement of nitrogen fixation, in: "Methods in Enzymology," S. P. Colowick and N. O. Kaplan, eds., Academic Press, New York.

DeVasconcelos, L., Miller, L., and Neyra, C. A., 1980, Free-living and symbiotic characteristics of chlorate resistant mutants of Rhizobium japonicum, Can. J. Microbiol., 26:338.

Fehr, W. R., Canviness, C. E., Burmood, D. T., and Pennington, J. S., 1971, Stage of development descriptions for soybeans, Glycine max (L.) Merrill, Crop Sci., 11:929.

Gibson, A. H., and Pagan, J. D., 1977, Nitrate effects on the nodulation of legumes inoculated with nitrate-reductase-deficient mutants of Rhizobium, Planta, 134:17.

Kuykendall, L. D., and Elkan, G. H., 1976, Rhizobium japonicum derivatives differing in nitrogen-fixing efficiency and carbohydrate utilization, Appl. Environ. Microbiol., 32:511.

Lim, S. T., 1978, Determination of hydrogenase in free-living cultures of Rhizobium japonicum and energy efficiency of soybean nodules, Plant Physiol., 62:609.

Lim, S. T., and Shanmugam, K. T., 1979, Regulation of hydrogen utilization in Rhizobium japonicum by cyclic AMP, Biochim. Biophys. Acta, 584:479.

Maier, R. J., and Brill, W. J., 1978, Mutant strains of Rhizobium japonicum with increased ability to fix nitrogen for soybean, Science, 201:448.

Manhart, J. R., and Wong, P. P., 1980, Nitrate effect on nitrogen fixation (acetylene reduction). Activities of legume root nodules induced by rhizobia with varied nitrate reductase activities, Plant Physiol., 65:502.

Schubert, K. R., Jennings, N. T., and Evans, H. J., 1978, Hydrogen reactions of nodulated leguminous plants. II. Effects on dry matter accumulation and nitrogen fixation, Plant Physiol., 61:398.

Scott, D. B., Hennecke, H., and Lim, S. T., 1979, The biosynthesis of nitrogenase MoFe protein polypeptides in free-living cultures of Rhizobium japonicum, Biochim. Biophys. Acta, 565:365.

Vincent, J. M., 1970, A Manual for the Practical Study of Root-Nodule Bacteria, Blackwell, Oxford.

Williams, L. E., and Phillips, D. A., 1980, The effect of irradiance on development of apparent nitrogen fixation and photosynthesis in soybean, Plant Physiol., in press.

PHYSIOLOGICAL INTERACTIONS BETWEEN ALASKA PEAS AND STRAINS OF RHIZOBIUM LEGUMINOSARUM THAT DIFFER IN PLASMID-LINKED GENES

T. M. DeJong*, N. J. Brewin†, and D. A. Phillips*

*Dept. of Agronomy & Range Science
University of California, Davis, CA

†Dept. of Genetics
John Innes Institute
Colney Lane, Nowwich, NR4 7UH, UK

INTRODUCTION

Recent advances in *Rhizobium* genetics have led to increased optimism regarding the possibility of enhancing symbiotic N_2 fixation by generating superior strains of *Rhizobium* that combine the best attributes from different parental strains. The procedures for developing such strains involve the identification of desired traits, the genetic marking and mapping of such traits, and the transfer of the traits with their markers to another strain which already has desirable traits of a different nature. Since at least some of the genetic information that codes for both root infectivity and nodule effectiveness appears to be located on plasmids in *Rhizobium* species (Higashi, 1967; Duncan et al., 1967; Nuti et al., 1979; Johnston et al., 1978; Brewin et al., 1980a,b;) it may be possible to develop "superior" *Rhizobium* strains by plasmid transfers between strains.

A non-nodulating mutant of *R. leguminosarum* strain 300, strain 16015, carries a substantial deletion in nodulation functions on one of its plasmids (Hirsch et al., 1980; Buchanan-Wollaston et al., 1980). Two transmissible plasmids that are capable of suppressing the mutant phenotype of strain 16015 have been identified in other field isolates of *R. leguminosarum*. One of those transmissible nodulation plasmids pRL1JI (obtained from field isolate 248) also specifies the production of a non-dialysable, medium-sized bacterocin (Hirsch, 1979). A derivative of pRL1JI, pJB5JI (Johnston et al., 1978) carries the transposon Tn5 (specifying kanamycin resistance, Kan-r) as an insertion into the determinants for medium bacteriocin production (Med). The other transmissible nodulation plasmid, pRL5JI, was derived from the field isolate TOM (strain 3622). It confers the ability to nodulate the primitive pea cultivar 'Afghanistan' in addition to western pea varieties such as 'Alaska' (Brewin et al., 1980b). When transferred to derivatives of R. leguminosarum strain 300, neither pRL1JI nor pRL5JI eliminated any of the resident plasmids of that strain (Hirsch et al., 1980; Brewin et al., 1980b), hence it is likely that more than one copy of certain plasmid-linked symbiotic genes are present in derivatives of R. leguminosarum strain 300 that contain one of these supplementary nodulation plasmids.

Two other transmissible bacteriocinogenic plasmids (pRL3JI and pRL4JI) also have been identified (Hirsch, 1979). Although apparently related to pRL1JI, pRL3JI and pRL4JI lack many, if not all, of the symbiotic determinants carried by pRL1JI, and they fail to suppress the nod fix deletion of strain 16015 (Brewin et al., 1980a). Kanamycin resistant, Med$^-$ derivatives of pRL3JI and pRL4JI (termed pVW3JI and pVW5JI, respectively) have been created by inter-plasmid transduction from pJB5JI (Brewin et al., 1980a). Using pRL3JI, pRL4JI and the Kan-r derivatives, it has been shown that these plasmids can mobilize determinants for nodulation ability from strains such as 300 in which the nodulation plasmid is not normally self-transmissible (Brewin et al., 1980a). Apparently that mobilization results from inter-plasmid recombination. Thus by suitable genetic manipulation of natural plasmids from strains of R. leguminosarum, it is possible to vary the strain of origin for groups of symbiotic genes, the number of copies of symbiotic genes and the pattern of bacteriocin production.

Little is known about the quantitative aspects of legume-Rhizobium symbioses when an association involves strains of Rhizobium altered by plasmid transfer. This paper reports the nodulation characteristics and host-plant responses of Alaska peas (Pisum sativum L.) inoculated with genetically-altered strains of R. leguminosarum. The goals of this research were: 1) to provide a functional analysis of a series of R. leguminosarum strains that differed only in their plasmid profiles; 2) to determine if the insertion of genetic markers for medium bacteriocin production or Kanamycin resistance into the Rhizobium genome necessarily is associated with changes in symbiotic effectiveness; 3) to compare various measurements of Rhizobium effectiveness so that future screening of Rhizobium mutants for symbiotic effectiveness can be simplified.

METHODS

Alaska peas (Pisum sativum L.) plants were grown in a controlled environment chamber under a 14/10-hr light/dark cycle at 21/20 C, 50% relative humidity and a photosynthetic photon flux density of 650 $\mu E \cdot m^{-2} \cdot sec^{-1}$. Plants were grown individually in modified "Leonard jar" (Leonard, 1943) assemblies consisting of a bottomless, 750 ml wine bottle that was inserted neck-first into a 1-liter glass jar. The neck of the wine bottle was plugged with a two-holed rubber stopper and filled with vermiculite and a 1-cm thick surface layer of perlite. The jar acted as a nutrient solution reservoir and initially contained 750 ml of N-free nutrient solution. The nutrient solution contained 2 mM $CaSO_4$, 1 mM K_2SO_4, 1 mM K_2HPO_4, 2 mM $MgSO_4$, 4 μM $CoCl_2$, 1 ml micronutrient solution/liter after Johnson et al. (1957) and 18.7 mg/liter of Sequestrene 138Fe iron chelate (Ciba-Geigy). The acidity of the solution was adjusted to pH 7.0 with HCl and changed less than 0.5 pH units during the course of each experiment. The complete Leonard jar assemblies were autoclaved prior to planting. Sides of the assemblies were covered with aluminum foil to prevent light absorption. Pea seeds weighing 0.21-0.23 grams were surface-sterilized with 70% ethanol, rinsed and germinated on sterile paper towels with distilled water. After 3 days of imbibition seedlings were selected for uniformity, planted in the sterile Leonard jar assemblies, and inoculated with specified Rhizobium strains (Table 1)

Ineffective R. leguminosarum strains used in this experiment can be divided into four separate groups. The first three groups consisted of a field isolate and one to three genetically-altered strains for which the field isolate served as a control. The altered strains in each of those groups were constructed by transferring plasmids with symbiotic genes derived from the common field isolate into the symbiotically defective (nod^- fix^-) mutant strain 16015 (Tables 1 and 2). Furthermore the first two groups contained pairs of strains that had symbiotic genes from common backgrounds but differed in their capacity to produce medium bacteriocin (Med) or to show Kanamycin resistance (Kan-r) Table 2). The fourth group consisted of three strains in which plasmids were transferred to a field isolate (strain 300) that already had the capacity to form an effective symbiosis (Table 1).

Data were collected from 25-day-old plants. Each strain was used to inoculate six plants. At the end of each experiment bacteria from four randomly selected nodules on each plant were reisolated and identified (Josey et al., 1979) to confirm that root nodules were formed by the appropriate inoculant. Acetylene reduction and H_2 evolution were measured (Bethlenfalvay and Phillips, 1977). Leaf area was measured using an area meter (Li-Cor LI-3000). Dry weights of plant parts were measured after

Table 1. Characteristics of Rhizobium leguminosarum strains studied

Rhizobium strain	Symbiotic phenotype	Characters	Reference
16015	Nod^-	300 str-37,spc-54, nod 6007	Brewin et al. (1980a)
248	Nod^+Fix^+	Field isolate	Hirsch (1979)
3747	Nod^+Fix^+	16015 pRL1JI	Brewin et al. (1980a)
3738	Nod^+Fix^+	16014 pVW1JI	Brewin et al. (1980a)
300	Nod^+Fix^+	Field isolate	Johnston & Beringer (1975)
3745	Nod^+Fix^+	16015 pRL4JI[1]	Brewin et al. (1980a)
3740	Nod^+Fix^+	16015 pVW5JI[1]	Brewin et al. (1980a)
3622	Nod^+Fix^+	Field isolate	Brewin et al. (1980b)
3737	Nod^+Fix^+	16015 pRL5JI	Brewin et al. (1980b)
3711	Nod^+Fix^+	300 pRL3JI	This study
3713	Nod^+Fix^+	300 pRL1JI	This study
3718	Nod^+Fix^+	300 pJB5JI	This study

[1] These strains acquired the Nod^+ determinants from strain 300, presumably as a result of recombination with pRL4JI or pVW5JI prior to transfer to strain 16015.

Table 2. Characteristics of plasmids studied in Rhizobium leguminosarum

Plasmid	Source	Characters[1]	Reference
pRL1JI	Strain 248	$Med^+Nod^+Fix^+Tra^+$	Hirsch (1979)
pRL3JI	Strain 306	Med^+Tra^+	Hirsch (1979)
pRL4JI	Strain 309	Med^+Tra^+	Hirsch (1979)
pRL5JI	Strain 3622	$Nod^+Fix^+Tra^+$	Brewin et al. (1980b)
pJB5JI	pRL1JI::Tn5[2]	Med^- Kan-r $Nod^+Fix^+Tra^+$	Johnston et al. (1978)
pVW1JI	pRL1JI::Tn5	Med^- Kan-r $Nod^+Fix^+Tra^+$	Brewin et al. (1980a)
pVW5JI	pRL4JI::Tn5	Med^- Kan-r Tra^+	Brewin et al. (1980a)

[1] Med^+ means that the plasmid specifies medium bacteriocin; Nod^+ and Fix^+ refer to nodulation and N_2-fixation functions that are absent from R. leguminosarum strain 16015; Tra^+ means transmissible by conjugation; Kan-r indicates Kanamycin resistance.

[2] pRL1JI::Tn5 means that transposon Tn5 was inserted into the region determining production of medium bacteriocin (Med) of pRL1JI.

48 hr at 75 C. Plant N content was determined by Kjeldahl analysis (Burris and Wilson, 1957).

RESULTS

Functional Analysis of Gene Transfer

The three field isolates (strains 248, 300, and 3622) showed distinct strain effects on two short-term measures of root nodule function. Differences in rates of C_2H_2 reduction and H_2 evolution between plants nodulated by those three strains were statistically significant (data not shown). However, those effects were not observed in the genetically altered strains which should have contained some genetic information for symbiosis from the field isolate within their respective groups. Rates of C_2H_2 reduction and H_2 evolution of all plants inoculated with genetically altered strains were more similar to each other than to the field isolate from which the nodulation determinants were derived.

Change in Kjeldahl N contant, a more reliable measure of N_2 fixation than C_2H_2 reduction for plants grown without combined N, supported the conclusions drawn from C_2H_2 reduction and H_2 evolution data in this study (Table 3). Symbiotic N_2 fixation differed significantly among field isolates (strains 248, 300, and 3622) but not among effective strains produced by transferring plasmids from those field isolates into the non-nodulating mutant 16015 (strains 3747, 3738, 3745, 3740, and 3737). Plants nodulated by the three field isolates showed statistically significant differences in plant dry weight, plant nitrogen content, nodule dry weight, and leaf area; plants nodulated by the genetically-altered strains, with the exception of strain 3713, were more similar to each other than to the plants nodulated by the field isolates.

Number of nodules per plant, one measure of root nodule development, was altered by plasmid transfer. Field isolates 248, 300, and 3622 formed different numbers of nodules on Alaska peas, but the number of nodules per plant produced by the genetically altered strains was more similar to the number formed by the field isolate within each group than to strains in other groups (Table 3). The only notable exception to that trend was the set of plants inoculated with strain 3745. Those plants had more nodules than plants inoculated by any other strain.

Brewin et al. (1980a) reported the transfer of symbiotic genes with bacteriocinogenic plasmids in *R. leguminosarum*. The quantitative data presented in this paper provide information on the symbiotic functionality of *R. leguminosarum* strains altered by plasmid transfers. The data on nodule number per plant indicate that the genetic information on plasmids introduced from the field isolates into strain 16015 may code for effective nodulation

Table 3. Effects of *R. leguminosarum* strains on root nodule development and symbiotic N_2 fixation. Fixation activity was calculated as the difference between total Kjeldahl N at harvest and initial seed N content for plants grown without combined N. Values represent the mean of six replicate plants.

Strain	Root nodule development	Symbiotic N_2 fixation
	(nodules/plant)	(mg/plant)
16015	0.0	0.0
248	134	17.3
3747	145	15.4
3738	175	15.2
300	266	24.7
3745	392	14.9
3740	275	17.0
3622	314	4.4
3737	322	14.6
3711	268	15.5
3713	225	24.2
3718	209	20.9
LSD(0.05)	54	4.7

and control the number of nodules formed while the level of overall symbiotic effectiveness may be controlled by genetic information inherent (but previously unexpressed) in strain 16015. For a single *Rhizobium* strain, nodule number generally was correlated with plant dry weight or some other measure of nodulation effectiveness (Phillips et al., 1976), but comparisons among different strains in this study indicate that nodule number may be controlled genetically by the bacteria. Furthermore differences in nodule numbers are unrelated to strain effectiveness. It is difficult to imagine how nodule number is controlled independently from

other factors that control the effectiveness of the symbiosis, and this may indicate another unforeseen level of complexity in the Rhizobium-legume symbiosis.
Although the total dry weight and leaf area of plants inoculated with genetically-altered strains (3740, 3745, and 3711) often were significantly ($p \leq .05$) different from those inoculated with field isolate controls, in other cases the altered strains produced plant growth similar to the controls. The one case in which a genetically-altered strain produced superior plant N content and leaf area and enhanced short-term measures of N_2 fixation relative to a control strain (strain 3737 vs. 3622) demonstrates the positive potential for the use of genetic manipulation in developing Rhizobium strains with a set of desired characteristics.

One might question whether plasmid transfer per se affects symbiotic efficiency of Rhizobium, but insertion of plasmids carrying symbiotic genes from one strain into another strain that already had symbiotic capabilities had little effect on the symbiotic properties. Three genetically-altered strains 3711, 3713, and 3718 were formed by transferring plasmids into the effective field isolate strain 300 (Table 1). Strains 3713 and 3718 were constructed by insertion of plasmids pRL1JI and pJB5JI which are thought to carry symbiotic genes from strain 248. Therefore, those strains may have duplicates of symbiotic genes. Both of those strains produced plants with significantly greater N content than all other altered strains except 3740 (Table 3). Their performance, however, was only equal and not superior to the field isolate strain 300. Strain 3711 was constructed by the insertion of plasmid pRL3JI which lacks at least some of the symbiotic genes determined by pRL1JI (Brewin et al., 1980a). Its effectiveness, measured as N content, was decreased significantly compared to strain 300 but was not worse than many other strains in the study.

The Effect of Antibiotic-Resistance Markers

Some controversy exists over the relationship between antibiotic resistance and Rhizobium effectiveness (Schwinghamer, 1964, 1967; Pankhurst, 1977; Hagedorn, 1979; Bromfield and Jones, 1979). Schwinghamer (1967) reported no correlation between Kanamycin resistance and symbiotic effectiveness for R. leguminosarum. A similar lack of correlation between those traits was observed in the limited number of comparisons that could be made in the present study. Neither strains 3747 and 3738 nor 3740 and 3745 showed significant differences in symbiotic N_2 fixation (Table 3). The apparent lack of a negative association between medium bacteriocin production and symbiotic effectiveness is important because it indicates a possible method for enhancing the competitive advantage of "superior" rhizobia over indigenous Rhizobium strains when they are used as an inoculum for agronomic crops.

Measurements of Nitrogen Fixation Activity

Actual N_2 fixation measured by Kjeldahl N accumulation in Alaska peas grown under uniform conditions in the absence of combined N was correlated most highly with total leaf area ($r^2 = .963$). The high correlation reflects the fact that leaves are a large sink for N in young plants (Pate et al., 1979). Acetylene reduction and apparent N_2 fixation calculated from C_2H_2 reduction and H_2 evolution data were less well correlated with total Kjeldahl N data than were the plant growth measurements. That is not surprising because such short-term rates only reflect nodule activity at a single point in time. Plants nodulated by two different bacterial strains with the same rate of N_2 fixation at day 25 could have different amounts of total Kjeldahl N because the time of initiation and rate of increase in N_2 fixation may differ for the two bacterial strains. Alternatively, plants nodulated by two different strains could have different rates of N_2 fixation but identical amounts of Kjeldahl N for the same reasons.

CONCLUSIONS

Plasmid transfers produced symbiotically effective bacteria from a strain previously unable to form root nodules on Alaska peas. The effective, genetically-altered strains formed did not differ significantly in N_2-reduction capacity, but differences in number of root nodules formed were associated with the source of the plasmid. A strong positive correlation between plant leaf area and total N accumulation in the absence of combined N showed that leaf area can be used as a rapid screen for symbiotic effectiveness of genetically-altered *R. leguminosarum* strains in Alaska peas grown under uniform environmental conditions.

ACKNOWLEDGMENT

We acknowledge the technical assistance of S. S. Jue. This research was supported in part by NATO Research Grant 1533 and National Science Foundation Grants PFR 77-07301 and PCM 78-01146.

REFERENCES

Bethlenfalvay, G. J., and Phillips, D. A., 1977, Ontogenetic interactions between photosynthesis and symbiotic nitrogen fixation in legumes, *Plant Physiol.*, 60:419.

Brewin, N. J., Beringer, J. E., Buchanan-Wollaston, A. F., Johnston, A. W. B., and Hirsch, P. R., 1980a, Transfer of symbiotic genes with bacteriocinogenic plasmids of *Rhizobium leguminosarum*, *J. Gen. Microbiol.*, 116:261.

Brewin, N. J., Beringer, J. E., and Johnston, A. W. B., 1980b, Plasmid-mediated transfer of host range specificity between two strains of *Rhizobium leguminosarum*, *J. Gen. Microbiol.*, in press.

Bromfield, E. S. P., and Jones, D. G., 1979, The competitive ability and symbiotic effectiveness of doubly labeled antibiotic resistant mutants of Rhizobium trifolii, Ann. Appl. Biol., 91:211.

Buchanan-Wollaston, A. V., Beringer, J. E., Brewin, N. J., Hirsch, P. R., and Johnston, A. W. B., 1980, Identification of symbiotically defective mutants by insertion of the transposon Tn5 into a transmissible plasmid, Mol. Gen. Genet., 178:185.

Burris, R. H., and Wilson, P. W., 1957, Methods for measurement of nitrogen fixation, in: "Methods in Enzymology," S. P. Colowick and N. O. Kaplan, eds., Academic Press, New York.

Dunican, L. K., O'Gara, F., and Tierney, A. B., 1976, Plasmid control of effectiveness in Rhizobium: transfer of nitrogen-fixing genes on a plasmid from Rhizobium trifolii to Klebsiella aerogenes, in: "Symbiotic Nitrogen Fixation in Plants," P. S. Nutman, ed., Cambridge University Press, Cambridge.

Hagedorn, C., 1979, Relationship of antibiotic resistant to effectiveness in Rhizobium trifolii populations, Soil Sci. Soc. Amer. J., 43:921.

Higashi, S., 1967, Transfer of clover infectivity of Rhizobium trifolii to Rhizobium phaseoli as mediated by an episomic factor, J. Gen. Appl. Microbiol., 13:391.

Hirsch, P. R., 1979, Plasmid-determined bacteriocin production by Rhizobium leguminosarum, J. Gen. Microbiol., 113:219.

Hirsch, P. R., Johnston, A. W. B., Brewin, N. J., Van Montagu, M., and Schell, J., 1980, Physical identification of bacteriocinogenic, nodulation and other plasmids in strains of Rhizobium leguminosarum, J. Gen. Microbiol., in press.

Johnson, C. M., Stout, P. R., Broyer, T. C., and Carlton, A. B., 1957, Comparative chlorine requirements of different plant species, Plant Soil, 8:337.

Johnston, A. W. B., and Beringer, J. E., 1975, Identification of the Rhizobium strains in pea root nodules using genetic markers, J. Gen. Microbiol., 87:343.

Johnston, A. W. B., Beynon, J. L., Buchanan-Wollaston, A. V., Setchell, S. M., Hirsch, P. R., and Beringer, J. E., 1978, High frequency transfer of nodulating ability between strains and species of Rhizobium, Nature, 276:635.

Josey, D. P., Beynon, J. L., Johnston, A. W. B., and Beringer, J. E., 1979, Strain identification in Rhizobium using intrinsic antibiotic resistance, J. Appl. Bact., 46:343.

Leonard, L. T., 1943, A simple assembly for use in testing of cultures of rhizobia, J. Bact., 45:523.

Nuti, M. P., Lepidi, A. A., Prakash, R. K., Schilperoort, R. A., and Cannon, F. C., 1979, Evidence for nitrogen fixation (nif) genes on indigenous Rhizobium plasmids, Nature, 282:533.

Pankhurst, C. E., 1977, Symbiotic effectiveness of antibiotic-resistant mutants of fast- and slow-growing strains of Rhizobium nodulating Lotus species, Can. J. Microbiol., 23:1026.

Pate, J. S., Layzell, D. B., and Atkins, C. A., 1979, Economy of

carbon and nitrogen in a nodulated and non-nodulated (NO_3-grown) legume, Plant Physiol., 64:1083.
Phillips, D. A., Newell, K. D., Hassell, S. A., and Felling, C. E., 1976, The effect of CO_2 enrichment on root nodule development and symbiotic N_2 reduction in *Pisum sativum* L., Amer. J. Bot., 63:356.
Schwinghamer, E. A., 1964, Association between antiboitic resistance and ineffectiveness in mutant strains of *Rhizobium* spp., Can. J. Microbiol., 10:221.
Schwinghamer, E. A., 1967, Effectiveness of *Rhizobium* as modified by mutation for resistance to antibiotics, Antonie van Leeuwenhoek, 33:122.

EVALUATING ELITE ALFALFA LINES FOR N_2-FIXATION UNDER FIELD CONDITIONS

G. H. Heichel, C. P. Vance, and D. K. Barnes

Agricultural Research, Science and Education Administration, U. S. Department of Agriculture, and the University of Minnesota
St. Paul, MN 55108

INTRODUCTION

The nitrogen (N_2) fixation capacity of a legume-*Rhizobium* symbiosis is associated with several metabolic processes and morphological characteristics of the host and of the bacterium. These processes and characteristics are quantitatively regulated with gene expression often dependent upon the environment in which the symbiosis develops. Because of the close association between gene expression and environment, we must understand how metabolic processes investigated in the artificial and relatively uniform environments of laboratories and glasshouses respond to the uncontrolled environments of the field. This understanding is essential if research to improve N_2-fixation of legumes is to impact beneficially on agricultural cropping systems. For this reason, a major thrust of our research program is to conduct field experiments to determine the N_2-fixation capabilities of alfalfa (*Medicago sativa* L.) previously selected in artificial environments for traits associated with N_2-fixation. The purpose of this communication is to report three investigations of alfalfa N_2-fixation and *Rhizobium* strain preference using germplasm from our breeding program, and to interpret the results in the context of plant improvement.

NITROGEN FIXATION OF SELECTED MNNC AND MNPL POPULATIONS

Methods

Experiments were conducted on Waukegan silt loam that previously was in 4 years of alfalfa culture. They were located at the Rosemount Experiment Station, University of Minnesota. The experiments utilized two alfalfa populations that had been subjected to one cycle of recurrent phenotypic selection in the glasshouse for five physiological and morphological characteristics associated with whole-plant nitrogenase activity. A commercially-available mixture of strains of R. meliloti was used to inoculate all plants during the selection program. Other details of the selection procedure for $N_2(C_2H_2)$ reduction activity, fibrous root score, secondary root score, nodule mass score, and herbage mass were reported by Viands et al. (1981). The population designated MnNC-5 was moderately winterhardy with a fall dormancy response similar to 'Ranger'. The population designated MnPL-8 was similar in parentage, winterhardiness, and fall dormancy characteristics to 'Vernal'.

Symbiotic N_2-fixation was measured as nitrogen fixed on a land area or on a plant basis (N_f), and as the proportion of total sample nitrogen derived from symbiosis (N_{sy}) using the isotope dilution technique (McAuliffe et al., 1958) as described by Heichel et al. (manuscript in preparation). Isotope was incorporated into soil in 1 x 1 m plots as $(^{15}NH_4)_2SO_4$ in aqueous solution. Plots were established in five replicates with seed inoculated with a commercial preparation of R. meliloti (Nitragin, Co.)[1] containing five strains. Reed canarygrass (Phalaris arundinacea L.) was used as the perennial, non-nitrogen-fixing control species. Plants were sampled on four occasions throughout the growing season, subjected to Kjeldahl analysis, and the distillates were analyzed for nitrogen isotope composition by mass spectrometry (Heichel et al., unpublished data).

Results and Discussion

Nitrogen fixation of alfalfa communities measured either as N_f or N_{sy} increased in the seeding year from the first to the second or third harvest, and then declined in the fourth harvest (Table 1). Although a significant ($P < 0.05$) population x harvest interaction for N_f occurred at harvest 3, the mean values of N_f or N_{sy} for the two populations were not significantly different on a seasonal basis. The two populations averaged about 44% of their total nitrogen needs from symbiosis during the seeding year. During the second and third harvest intervals, MnNC-5-HB-A derived nearly two-thirds of its total nitrogen from symbiosis. In comparison, soybean (Glycine max L. Merr) obtained 39 to 66% of its seasonal nitrogen needs from symbiosis (Ham, 1978; Deibert et al., 1979).

---

[1] Mention of a trademark, vendor, or proprietary product does not constitute a guarantee or warranty of the product by the USDA and does not imply its approval to the exclusion of other trademarks, vendors, or proprietary products that may also be suitable.

Table 1. Nitrogen fixation of two alfalfa populations measured by the isotope dilution technique at Rosemount, MN.

	MnNC-5-HB-A		MnPL-8-HB-A	
	N_f †	N_{sy} ‡	N_f †	N_{sy} ‡
	gm^{-2}	%	gm^{-2}	%
Harvest 1	2.0 ± 0.6	29.4 ± 8.2	1.3 ± 0.3	25.6 ± 6.1
Harvest 2	4.8 ± 0.5	63.3 ± 7.1	5.7 ± 0.3	64.8 ± 3.2
Harvest 3	8.7 ± 0.4	66.8 ± 3.1	4.7 ± 0.4	58.5 ± 5.2
Harvest 4	1.5 ± 0.2	18.8 ± 11.7	0.8 ± 0.4	20.1 ± 10.9
Total N_f (gm^{-2}) or mean N_{sy} (%)	17.0	44.6 12.0	12.5	42.3 11.3

† Symbiotic nitrogen fixation expressed on a land area basis.

‡ Proportion of nitrogen in the total plant derived from symbiosis.

White clover (Trifolium repens L.) and subterranean clover (T. subterraneum L.) derived more than 80% of their nitrogen budget from atmospheric sources (Edmeades and Goh, 1978; Phillips and Bennett, 1978). On a land area basis, the two alfalfa populations averaged about 14.8 g/m^2 of fixed nitrogen over the growing season of the establishment year (Table 1).

Although there was no evidence that N_2-fixation of field communities of MnNC-5-HB-A and MnPL-8-HB-A differed in the establishment year, statistical analysis showed significant population x harvest interactions (Table 1).

Because of a 14% difference in stand between the two populations, we made further comparisons on an individual plant basis. Nitrogen fixation varied with N_{sy} (Table 1) and with the total nitrogen yield of the plant. The latter in turn varied with the average nitrogen concentration in the plant, and the total dry matter yield. Over four harvests, the total dry matter yield of MnPL-8-HB-A was significantly less ($P < 0.05$) than that of MnNC-5-HB-A, and a significant population x harvest interaction for yield of the two populations was evident (Fig. 1). The herbage yields per plant were similar at the first three harvests for both populations, which is consistent with glasshouse evaluations of these populations (Viands et al., manuscript in preparation). Thus, the difference in dry matter yield was attributable to contrasting root growth of the two populations.

Combination of total dry matter yield (Fig. 1), nitrogen concentration, and N_{sy} data for whole plants of seeded material allowed comparisons among average rates of growth (GR), symbiotic nitrogen fixation (SNF), and soil nitrogen uptake (SNU) for each of the four seeding year harvests (Figs. 2 and 3).

For both populations, patterns of GR and SNF were very similar throughout the season. GR and SNF were slow in the first harvest interval and increased several-fold through either the second or third harvest. The onset of dormancy occurred more quickly in MnPL-8-HB-A than in MnNC-5-HB-A, so that the latter population continued to grow and to fix nitrogen until later in the season than did MnPL-8-HB-A. For this reason, GR and SNF were significantly ($P < 0.05$) greater for MnNC-5-HB-A than for MnPL-8-HB-A for all harvest intervals except the last.

Rates of SNU, calculated by assuming that the nitrate reductase system fulfilled plant nitrogen needs not met by symbiotic nitrogen fixation, were similar between populations and relatively invariant during the first three harvest intervals (Figs. 2 and 3). With the onset of dormancy, SNU mirrored the precipitous decline of SNF and GR. In vivo nitrate reductase activity reflected the rate of SNU for both populations (Vance and Heichel, unpublished).

Although there were no differences between the original, unselected populations and between the subsequent cycles of these populations selected in the glasshouse for traits associated with N_2-fixation (Viands et al., manuscript in preparation), significant differences in N_2-fixation per plant between the populations were observed in the field (Figs. 2 and 3). The contrast between glasshouse and field performance occurred because the glasshouse evaluations were made under a long photoperiod in a relatively narrow range of temperatures and evaluated at one stage of growth. The field-grown plants responded to the continuously changing natural environment for 5 months and were sampled on several occasions. The differences in fall dormancy between the populations clearly influenced the N_2-fixation of field-grown material, but not of material grown in a noninductive environment in the glasshouse. Thus, considerable caution should be exercised in attempts to anticipate the field performance for N_2-fixation of alfalfas with varying dormancy responses that are initially evaluated in the glasshouse under noninductive conditions.

NITROGEN FIXATION OF MNNC AND MNPL SUBPOPULATIONS DEVELOPED BY BIDIRECTIONAL SELECTION

Methods

Field plots for this experiment were like those described above. The investigation utilized the MnNC and MnPL gene pools that had been subjected to two cycles of bidirectional recurrent selection for both high (H) and low (L) acetylene reduction rate (A), nodule mass score (N), fibrous root mass score (F), and top dry weight (T). The selection procedures, which resulted in eight subpopulations per cycle from each gene pool, were previously reported (Viands et al., manuscript in preparation). This experiment compared N_2-fixation of the original, unselected MnNC and MnPL gene pools with that of the eight subpopulations after the first cycle of glasshouse selection.

Seed of the original gene pools and eight subpopulations were sown in glasshouse sandbenches in late March, inoculated with commercial inoculum (Nitragin Co.)[1] containing five *Rhizobium* strains, and transplanted as bare-root seedlings with clipped tops to field plots at 7 weeks of age. Seedlings were established at 200 plants/m^2 in five replicates of a randomized complete block design. Herbage of 10 plants per replicate was sampled on three occasions throughout the growing season, and N_2-fixation was measured by the isotope dilution technique.

Results and Discussion

Glasshouse experiments (Viands et al., 1981) showed substantial selection progress for A, N, F, and T in both H and L

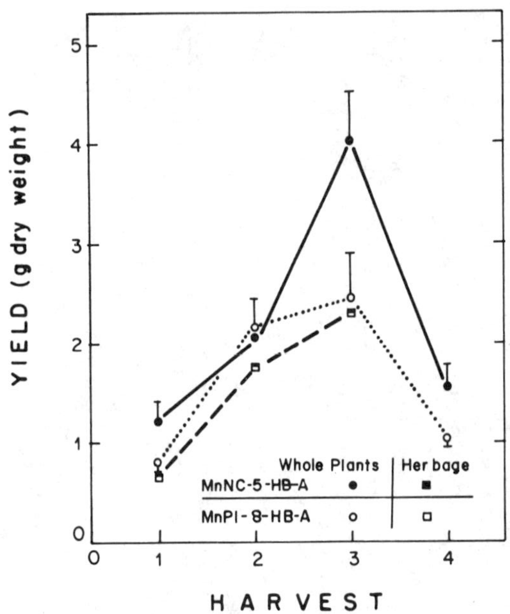

Fig. 1. Total dry matter yield and herbage yield per plant of two alfalfa populations established from seed and harvested at four intervals in the seeding year. Values are mean + or - SE.

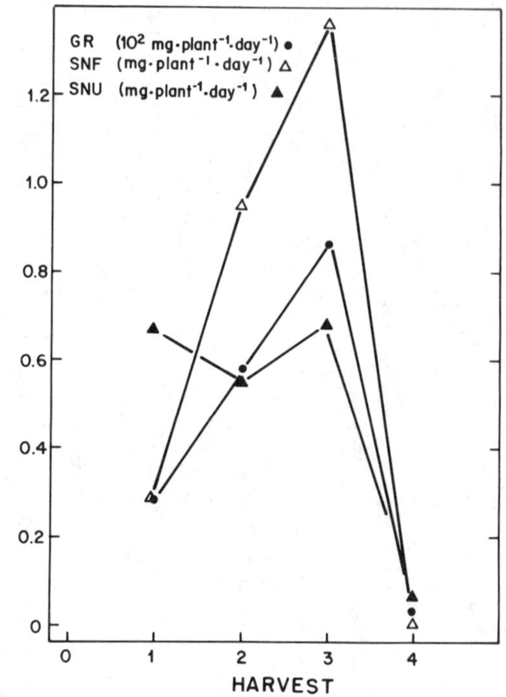

Fig. 2. Average values of growth rate (GR), rate symbiotic nitrogen fixation (SNF), and rate of soil nitrogen uptake (SNU) over each of four harvest intervals in the seeding year for MnNC-5-HB-A.

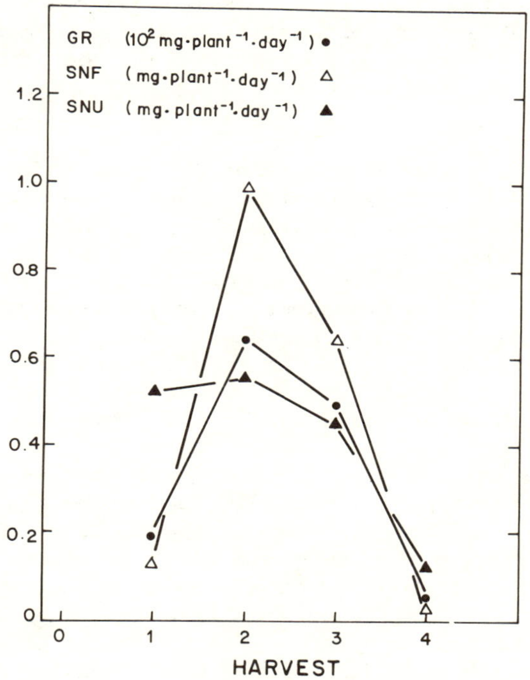

Fig. 3. Average values of growth rate (GR), rate of symbiotic nitrogen fixation (SNF) and rate of soil nitrogen uptake (SNU) over each of four harvest intervals in the seeding year for MnPL-8-HB-A.

directions. When the MnNC-4 entries were grown in the field, significant ($P < 0.05$) differences of N_{sy} among entries over the three harvests were observed (Table 2). Differences in herbage mass among entries were also apparent in the field.

Despite the significant improvement in A observed in glasshouse selection, the HA subpopulation derived less nitrogen from symbiosis and fixed a similar amount of nitrogen per plant compared with the original population (Table 2). The LA subpopulation derived less nitrogen from symbiosis in the field, and fixed nitrogen equally well on a per plant basis compared with the original gene pool. The HA and LA subpopulations differed appreciably in N_{sy} and nitrogen fixed per plant under field conditions, which mirrored the per plant nitrogenase activities observed in the glasshouse.

Although bidirectional selection progress for F and N was effected in the glasshouse, no significant differences between the HF and LF or HN and LN subpopulations for N_{sy} or for nitrogen fixed per plant were observed in the field (Table 2). Interestingly, selection in the glasshouse for increased herbage mass decreased N_{sy} and nitrogen fixed per plant compared with the original gene pool, but selection for decreased herbage mass increased N_{sy} and nitrogen fixed per plant. The results suggest that selection for herbage yield is probably not an effective procedure to increase N_2-fixation on an individual plant basis.

Many of the results from the MnNC gene pool were also exemplified by the MnPL gene pool (Table 3). Selection for increased T in the glasshouse did not improve herbage yield in the field. In fact, the HT subpopulation (selected for high herbage yield) actually performed the poorest of all entries in the field. The subpopulations of MnPL differed significantly ($P < 0.05$) for N_{sy} and for nitrogen fixed per plant. The HA and LA subpopulations differed significantly in N_{sy} in a pattern consistent with the A results from the glasshouse (Table 3). However, the improvement in N_{sy} or nitrogen fixed per plant anticipated for the HA subpopulation because of glasshouse performance was not apparent in the field. Both gene pools were consistent in this result.

Despite the selection progress for F, N, and T in the glasshouse, improvement of these traits in the glasshouse always resulted in both decreased N_{sy} and nitrogen fixed per plant in the field for the H subpopulations relative to the L subpopulations. Furthermore, none of the selected subpopulations were better for N_2-fixation in the field than the unselected gene pools.

The two gene pools were consistent in showing that one cycle of selection for physiological and morphological traits in the glasshouse did not significantly improve seasonal N_2-fixation in

Table 2. Yield, proportion of nitrogen from symbiosis, and nitrogen fixed per plant averaged over three harvests in MnNC-4 and subpopulations from one cycle of bidirectional recurrent selection.

	Herbage Mass	N_{sy}	N Fixed/Plant
	(g/plant)	(%)	(mg)
Original population			
MnNC-4	0.20	72.0	5.6
1st cycle subpopulations			
MnNC-6-HA	0.26	68.0	6.8
-LA	0.24	62.2	5.6
-HF	0.27	51.1	4.7
-LF	0.26	51.5	5.5
-HN	0.32	51.4	6.1
-LN	0.31	49.2	6.9
-HT	0.22	63.9	5.4
-LT	0.23	77.7	6.5
LSD (0.05)	0.07	10.8	1.4

Table 3. Yield, proportion of nitrogen from symbiosis, and nitrogen fixed per plant averaged over three harvests in MnPL-6 and subpopulations from one cycle of bidirectional recurrent selection.

Entry	Herbage Mass	N_{sy}	N Fixed/Plant
	(g/plant)	(%)	(mg)
Original population			
MnPL-6	0.46	78.9	13.4
1st cycle subpopulations			
MnPL-8-HA	0.32	69.4	7.7
-LA	0.40	59.8	8.9
-HF	0.32	49.4	5.9
-LF	0.36	58.0	8.0
-HN	0.43	58.1	9.3
-LN	0.43	70.9	11.5
-HT	0.27	77.2	8.0
-LT	0.36	81.4	11.4
LSD (0.05)	0.07	9.9	1.2

field evaluations of selected materials. Some of the lower than anticipated performance may be attributable to inbreeding depression for N_{sy} and nitrogen fixed per plant compared with the original gene pool. This is plausible considering that each of the subpopulations was derived by intercrossing only 14% of the best or poorest individuals for each trait from the original population of 800 plants of each gene pool in the glasshouse.

The suggestion of inbreeding and the consistent differences between HA and LA subpopulations in the field support the hypothesis that N_2-fixation capacity of alfalfa can be altered by manipulating the host genome. It is clearly evident, however, that the field environment drastically reduced differences in the N_2-fixation performance expected from glasshouse selection procedures. The full implications of these results await the evaluation of subpopulations from the second selection cycle.

RHIZOBIUM STRAIN PREFERENCE OF SELECTED MNNC AND MNPL POPULATIONS AND SUBPOPULATIONS

Methods

Experiment I. The first set of material was the unselected MnNC and MnPL gene pools, the second cycle selections (MnNC-7-HB-A and MnPL-10-HB-A) developed by Viands et al. (manuscript in preparation), and the reciprocal F_1 between the second cycle selections. Seeds of all entries were scarified, sown into sterile sand benches, and inoculated with a mixture containing 2 x 10^8 cells/ml of R. meliloti strain 102F51 str^ra, and 6.1 x 10^8 cells/ml of 102F77 spc^r. The strains were streptomycin (str) or spectinomycin (spc) mutants, respectively, of wild type R. meliloti prepared and characterized as previously described (Hardardson et al., unpublished data).

Six-week-old seedlings of each population were transplanted to 1 x 1 m plots as described above and reinoculated with the mutant Rhizobium strains. The indigenous Rhizobium in the plots averaged 10^5 viable cells/g dry soil. The experiment was established in a randomized complete block design in five replications with each of the five alfalfa populations as treatments. Ten plants per plot were removed on July 3, August 1, September 18, and October 29. Fifteen nodules per plant were randomly chosen, crushed, and antibiotic mutant Rhizobium strains identified as described above.

Experiment II. The second set of plant material was the unselected MnNC gene pool, and the second cycle selections for high (H) and low (L) acetylene reduction rate (A), nodule mass score (N), fibrous root score (F) and top dry weight (T) described by Viands et al. (manuscript in preparation). Sterilized seeds were sown in 12-cm pots filled with nonsterile Waukegan silt loam obtained from an alfalfa field at the St. Paul Experiment Station, University of Minnesota. The experiment was established in 14 replicates of a randomized complete block design with the nine alfalfa entries as treatments. Five days after sowing, each pot was inoculated with a surface application of inoculum containing equal proportions of R. meliloti strain 102F51 str^ra (effective in nitrogen fixation) and 102F62 spc^rb (ineffective in nitrogen fixation) prepared and characterized as described previously (Hardarson et al., manuscript in preparation).

The plants were grown in the glasshouse at 22 ± 5 C day temperature and 17 ± 3 C night temperature for a 16-hour photoperiod. The experiment was terminated after 9 weeks when nodules were harvested for identification.

Results and Discussion

Experiment I. The unselected and selected alfalfas differed significantly (χ^2_{12} = 38.0; $P < 0.001$) in preference for indigenous and the two effective mutant bacteria (Table 4). Compared with the unselected original gene pool, nodules of MnNC-7-HB-A showed a reduced occupancy by indigenous strains and an increased occupancy by each of the mutant strains. The proportion of nodules containing the mutant 102F77 spc^r was significantly less, and that by the mutant 102F51 str^ra significantly greater, in MnPL-10-HB-A than in the unselected MnPL-6. Selection did not alter the preference for indigenous strains in the MnPL gene pool.

In the F_1, nodule occupancy by indigenous and mutant strains was similar to that of MnNC-7-HB-A (Table 4). The F_1 also responded similarly to the selected populations in showing reduced preference by indigenous bacteria, and increased preference for the mutant bacteria, compared with the unselected MnNC-4 and MnPL-6.

Strain preference averaged across all populations also varied significantly (χ^2_9 = 25.5; $P < 0.01$) with time of harvest. Except for the second harvest, nodule occupancy by indigenous strains decreased, and that by mutant strains increased throughout the season.

Experiment II. The unselected MnNC gene pool and the selected subpopulations differed significantly (χ^2_{27} = 47.9; $P < 0.01$) in their preference for indigenous and mutant Rhizobium (Table 5). Most of the subpopulations had significantly fewer ineffective bacteria (102F62 spc^rb) in their nodules compared with the original population. The high A subpopulation had significantly fewer ineffective nodules than did the low A subpopulation. As a result of the decrease in ineffectively nodulated plants, the

Table 4. Nodule occupancy by indigenous and antibiotic-mutant *Rhizobium* strains in alfalfa populations selected for enhanced nitrogenase activity, and the F_1.

Entry	Rhizobium strains			
	Indigenous %	102F51 str^ra %	102F77 spc^r %	Double infection %
MnNC-4 (original)	46	12	41	1
MnNC-7-HB-A (2nd cycle)	26	23	49	2
MnPL-6 (original)	34	12	54	0
MnPL-10-HB-A (2nd cycle)	39	31	28	2
MnNC-7-HB-A X MnPL-10-HB-A (F_1)	29	15	55	1

(Hardarson et al., manuscript submitted)

Table 5. Nodule occupancy by indigenous and antibiotic-mutant *Rhizobium* strains in MnNC-4 and subpopulations derived by bidirectional selection.

Entry	Rhizobium strains			
	Indigenous %	102F51 str^ra %	102F62 spc^rb %	Double infection %
Original population				
MnNC-4	77	2	20	1
High subpopulations				
MnNC-6-HA	90	5	4	1
-HN	88	6	6	0
-HT	89	4	7	0
-HF	81	7	12	0
Mean	87.0	5.5	7.2	0.2
Low subpopulations				
MnNC-6-LA	82	1	17	0
-LN	91	1	8	0
-LT	87	7	6	0
-LF	86	2	12	0
Mean	86.5	2.7	10.7	0

(Hardarson et al., manuscript submitted)

selected subpopulations showed an increase in nodule occupancy by effective indigenous strains compared with the control. The high subpopulations showed significantly greater nodule occupancy by the effective (102F51 str^{r}a) mutant compared with the low subpopulations and the control. Both experiments are consistent in showing that host selection for physiological and morphological traits associated with N_2-fixation capacity altered the preference of the selected alfalfa for Rhizobium.

CONCLUDING DISCUSSION

Several tentative conclusions seem justified from the foregoing results. First, it is clear that plants selected for improved N_2-fixation capacity in artificial environments may not reliably express this feature throughout a growing season in the field. This was amply illustrated in the first experiments, where experimental materials performing similarly at one stage of growth in the glasshouse behaved quite differently as individual plants in the field. Furthermore, even small changes in crop community characteristics, such as plant population density, may obviate differences in N_2-fixation that are evident in individual plants.

The second experiment reinforced these conclusions and further illustrated that subpopulations that were substantially improved over the unselected gene pool for several traits including nitrogenase activity measured in the glasshouse were no better than the unselected gene pools when all materials were grown in the field. Significant environmental control of N_2-fixation is evident from the observation that subpopulations selected for improved nodule score, root score, or herbage yield in the glasshouse consistently performed poorest in the field in both gene pools. Nevertheless, control of N_2-fixation by the host genome was illustrated by the consistent differences in field performance of subpopulations divergently selected for nitrogenase activity, and by the apparent inbreeding depression for all traits during one cycle of recurrent selection.

Finally, it is clear that host selection for physiological and morphological traits associated with N_2-fixation may inadvertently modify the preference of the plant for Rhizobium. In one alfalfa gene pool, plants subjected to two recurrent selection cycles for N_2-fixation during exposure to a mixture of effective Rhizobium strains exhibited increased preference for effective strains, and decreased preference for other indigenous strains, compared with the original plant material. This characteristic was also expressed by the F_1 between the two second cycle selections. Additionally, selected subpopulations had reduced preference for ineffective Rhizobium compared with the unselected gene pools.

Our most inescapable conclusion is that concurrent field and laboratory experiments are indispensable at several steps of N_2-fixation research involving plant breeding and physiology to realistically determine the potential of metabolic modifications of the host and *Rhizobium*.

REFERENCES

Diebert, E. J., Bijeriego, M., and Olson, R. A., 1979, Utilization of ^{15}N fertilizer by nodulating and non-nodulating soybean isolines, *Agron. J.*, 71:717.

Edmeades, D. C., and Goh, K. M., 1978, Symbiotic nitrogen fixation in a sequence of pastures of increasing age measured by a ^{15}N dilution technique, *N. Z. J. Agric. Res.*, 21:623.

Ham, G. E., 1978, Use of ^{15}N in evaluating symbiotic N_2 fixation of field-grown soybeans, *in*: "Isotopes in Biological Dinitrogen Fixation," International Atomic Energy Agency, Vienna, Austria.

McAuliffe, C., Chamblee, D. S., Uribe-Arango, H., and Woodhouse, W. W., 1958, Influence of inorganic nitrogen on nitrogen fixation of legumes as revealed by ^{15}N, *Agron. J.*, 50:334.

Phillips, D. A., and Bennett, J. P., 1978, Measuring symbiotic nitrogen fixation in rangeland plots of *Trifolium subterraneum* L. and *Bromus mollis* L., *Agron. J.*, 70:671.

Viands, D. R., Barnes, D. K., and Heichel, G. H., 1981, Responses to bidirectional selection for characteristics associated with nitrogen fixation in alfalfa, *USDA Technical Bulletin*, in press.

SUCCESSES AND PROBLEMS ENCOUNTERED WHILE BREEDING FOR ENHANCED N_2 FIXATION IN ALFALFA

D. K. Barnes, G. H. Heichel, C. P. Vance, D. R. Viands*, and G. Hardarson

Agricultural Research, Science and Education Administration, U. S. Department of Agriculture, and University of Minnesota, St. Paul, MN 55108
*Presently at Cornell University, Ithaca, NY 14801

INTRODUCTION

We initiated a modest research effort on breeding alfalfa (Medicago sativa L.) for increased N_2 fixation in 1974. This project became the first concentrated effort in the U. S. to breed a forage legume for increased N_2 fixation. Now six years later we are taking this opportunity to look at the project in retrospect and to examine both the successes and the problems that we have encountered. Hopefully our experiences will be useful in projecting future advances and in suggesting methodologies and philosophies that can be utilized in other breeding programs for N_2 fixation.

Our decision to initiate a research program for increasing the N_2 fixation potential in alfalfa was the direct result of the energy situation in 1973. The forage research group at St. Paul discussed ways that agronomic research could help reduce the energy requirements of U. S. agriculture. Nitrogen is the major limiting nutrient for production of corn, small grains, and for more than 81 million hectares of grassland in humid areas of the U.S. In 1973, information on N_2 fixation research methodology for forage legumes was lacking, there was skepticism about the probability of successfully breeding alfalfa for improved N_2 fixation potential, and new research funds were not readily available. Nevertheless, the potential benefits which could add hundreds of millions of dollars to U. S. agriculture in terms of improved pastures, improved soil tilth, reduced costs of crop production and energy saved seemed great enough for us to reorder our research priorities.

PROGRAM DEVELOPMENT

Our project goals have been: 1) to develop an in-depth understanding of N_2 fixation in alfalfa, 2) to breed alfalfa that is more effective in N_2 fixation, and 3) to develop strategies to apply N_2 fixation information and improved germplasm to agricultural production systems.

Our approach to breeding for increased N_2 fixation was based on the philosophy that the symbiotic association between alfalfa and *Rhizobium meliloti* Dang. is essentially a host:pathogen association. In this association the plant provides an energy source and a suitable environment for the bacteria, while the bacteria provide a source of usable nitrogen for the plant. Vance (1978) described the many steps that must be successfully completed before the nodule is developed and N_2 fixation can occur. The sequence of events by which the plant-bacterial association is established includes bacterial attraction in the vicinity of root hairs, curling of the root hairs, and bacterial penetration. After penetration, infection threads develop through which the bacteria migrate to the root cortex. Nodules are formed as the result of plant cell division and cell enlargement stimulated by the bacteria. The bacteria multiply in the nodule tissue and finally develop the capacity to fix nitrogen.

Symbiotic N_2 fixation will be maximized when the most effective bacterium infects the most productive alfalfa cultivar. However, there are many indigenous strains of *R. meliloti* in U. S. soils and not all alfalfa cultivars are nodulated by all strains. According to Burton (1972) rhizobia by host cultivar interactions have been observed for nodulation and for N_2 fixation after nodulation. Even though host by *Rhizobium* strain interactions exist, we thought that it was unrealistic to breed alfalfas that reacted favorably with only one strain. Under field conditions, a plant will be exposed to many indigenous strains. Weaver and Frederick (1974) showed with soybeans (*Glycine max* (L.) Merr.) that it was difficult to introduce and establish new *Rhizobium* strains into fields containing adapted strains. Because alfalfa is a perennial species exposed to a wide range of environmental conditions, we suspected that different strains of *Rhizobium* may be differentially effective during various periods of plant growth. For these reasons we chose to conduct our breeding program with a mixture of *Rhizobium* strains rather than to work with a single strain. We postulated that plants selected after exposure to a mixture of strains would have a greater potential for N_2 fixation with existing strains under field conditions than plants selected with one strain. This would be similar to the selection of resistant plants after exposing them to a mixture of strains of a pathogen. The success of this procedure is well-documented in disease resistance research.

The effectiveness of the N_2 fixation capacity of a plant can be limited by any factor in the plant-bacteria association. Therefore, improvement of the N_2 fixation potential by plant breeding requires that the efficiency of many processes be simultaneously improved. It is critical for plant breeders to have a rapid, inexpensive method that can be used to measure the N_2 fixation on thousands of plants each year. The acetylene-reduction (AR) technique described by Hardy et al. (1968) was adapted for use with alfalfa by Seetin and Barnes (1977). Twelve-week-old plants (first regrowth) grown in nil-nitrate sand have been used in most of our research.

GREENHOUSE SELECTION EXPERIMENTS

The first breeding objective with alfalfa was to determine if genetic differences existed for N_2 fixation. Using the AR technique, Seetin and Barnes (1977) demonstrated a four-fold difference in AR values among plants. They also established that crosses among parents with high AR values produced progenies with AR rates more than twice those of progenies from low X low crosses. Low X high crosses were intermediate for AR. These data demonstrated that it should be possible to increase N_2 fixation by breeding. Several plant traits were correlated with AR. These included nodule mass ($r = 0.68^{**}$), fibrous root score ($r = 0.63^{**}$), root fresh weight ($r = 0.38^{*}$), and top dry weight ($r = 0.34^{*}$). Duhigg et al. (1978) also demonstrated genetic differences for AR among alfalfa genotypes and the association between AR and the same group of plant traits.

Although the available data suggested that it should be possible to increase the N_2 fixation potential of alfalfa, it was important that the various interrelationships were understood before a large scale breeding program was initiated. Viands (1979) conducted an extensive study to determine responses to bidirectional selection individually for AR per plant, nodule mass, top dry weight, and fibrous root mass in two alfalfa gene pools. It was intended that this research would determine the associations among AR and the various morphological characteristics, and then provide information on the best morphological characteristics or combination of characteristics, for selection to improve the genetic potential of N_2 fixation in alfalfa.

The two broad-based gene pools used in Viands' selection experiments were chosen to represent winterhardy (MnPL) germplasms and moderately winterhardy (MnNC) germplasms. All screening and evaluation studies were conducted on plants inoculated with commercial inoculum containing five *R. meliloti* strains and grown in greenhouse sand benches with low nitrate levels (12-14 ppm). Each gene pool was divided into subpopulations developed from two cycles of recurrent phenotypic selection for both high and low levels of the four characteristics listed in Table 1. Over 800 plants of each gene pool were screened and approximately 110 plants were selected as parents for each subpopulation in the first cycle of selection. Because of the size of the experiment only 250 plants in each subpopulation were screened in the second cycle of selection and 50 plants were selected as parents. Responses to selection were evaluated by simultaneously growing plants from the original gene pool and plants from the two cycles of selection for each characteristic. The MnNC and MnPL gene pools were evaluated on different dates. Therefore, data could only be compared within gene pools and not between gene pools.

Both the MnNC and MnPL gene pools responded to selection for each of the four plant characteristics (Fig. 1). Highly significant ($P < 0.01$) differences were observed between the high and low subpopulations developed for each characteristic in both gene pools. It should be noted that the AR data for the MnPL gene pool were obtained during several days with very high temperatures. The high temperatures reduced the relative levels of AR activity. When these same populations were reevaluated at cooler temperatures the AR activity of the MnPL gene pool was similar to the MnNC gene pool (Viands et al., manuscript in preparation). The influence of high temperature on N_2 fixation of alfalfa has been described by Munns et al. (1977).

The AR rate per plant was significantly ($P < 0.01$) correlated with top dry weight, nodule mass score, nodule number score, fibrous root score and secondary root score for both populations. This agreed with earlier reports by Seetin and Barnes (1977) and Duhigg et al. (1978). In Viands' study the selection for nodule mass score, fibrous root score, and top dry weight all changed AR rate significantly (Table 2). Because responses to selection did not indicate a good substitute for AR selection, coefficients of determination were computed by stepwise regression analysis to determine how much additional variation due to AR was accounted for by adding each of the morphological characteristics measured to the regression equation. In both gene pools, nodule mass score explained much more of the variation due to AR than any other characteristic (Table 3). Those results suggested to us that nodule mass would be the best morphological characteristic to select for in a breeding program.

Because AR measures nitrogenase activity of the nodules, we concluded that a breeding program should include first selection for high nodule mass followed by evaluation for AR rate. This type of breeding program is presently being conducted successfully at several locations.

Table 1. Scheme of two cycles of bidirectional recurrent phenotypic selection for four plant characteristics and a base index comprised of the four characteristics in two alfalfa gene pools (adapted from Viands, 1979 and Viands et al., manuscript in preparation).

Gene pool		Cycle of selection	Selection intensity	Selection scheme[1]				
MnNC	MnPL							
MnNC-6	MnPL-9	2	20%	HT	HN	HF	HA	HB-B (1%)
MnNC-5	MnPL-8	1	14%	HT	HN	HF	HA	HB-B (20%)
MnNC-4	MnPL-6	0		ORIGINAL GENE POOL				
MnNC-5	MnPL-8	1	14%	LT	LN	LF	LA	
MnNC-6	MnPL-9	2	20%	LT	LN	LF	LA	

[1]H and L = selection for high and low levels, respectively. T = top dry weight, N = nodule mass score, F = fibrous root mass score, A = nitrogenase activity per plant. B=B (1%) and B-B (20%) = Base Index B at the 1% and 20% selection intensities, respectively.

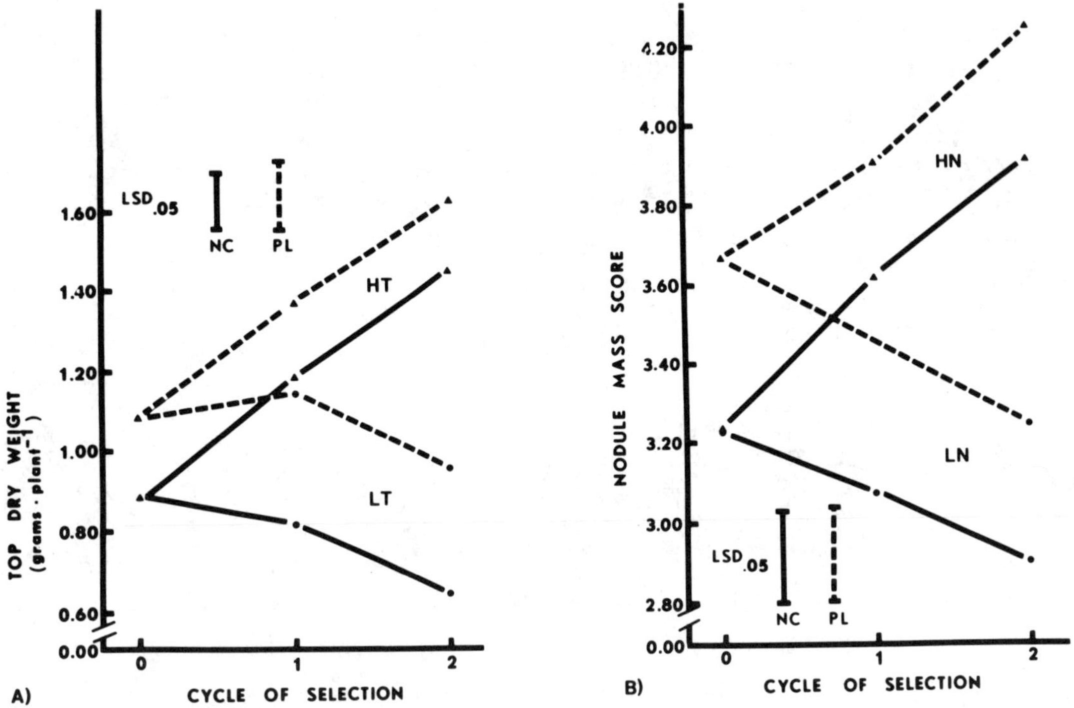

Fig. 1. Responses to two cycles of recurrent phenotypic selection for high and low levels of A) top dry weight, B) nodule mass score (1 = no fibrous roots, 5 = many), C) fibrous root score (1 = no fibrous roots, 5 = many), and D) acetylene reduction rate per plant in the MnNC and MnPL alfalfa gene pools (adapted from Viands, 1979, and Viands et al., manuscript in preparation).

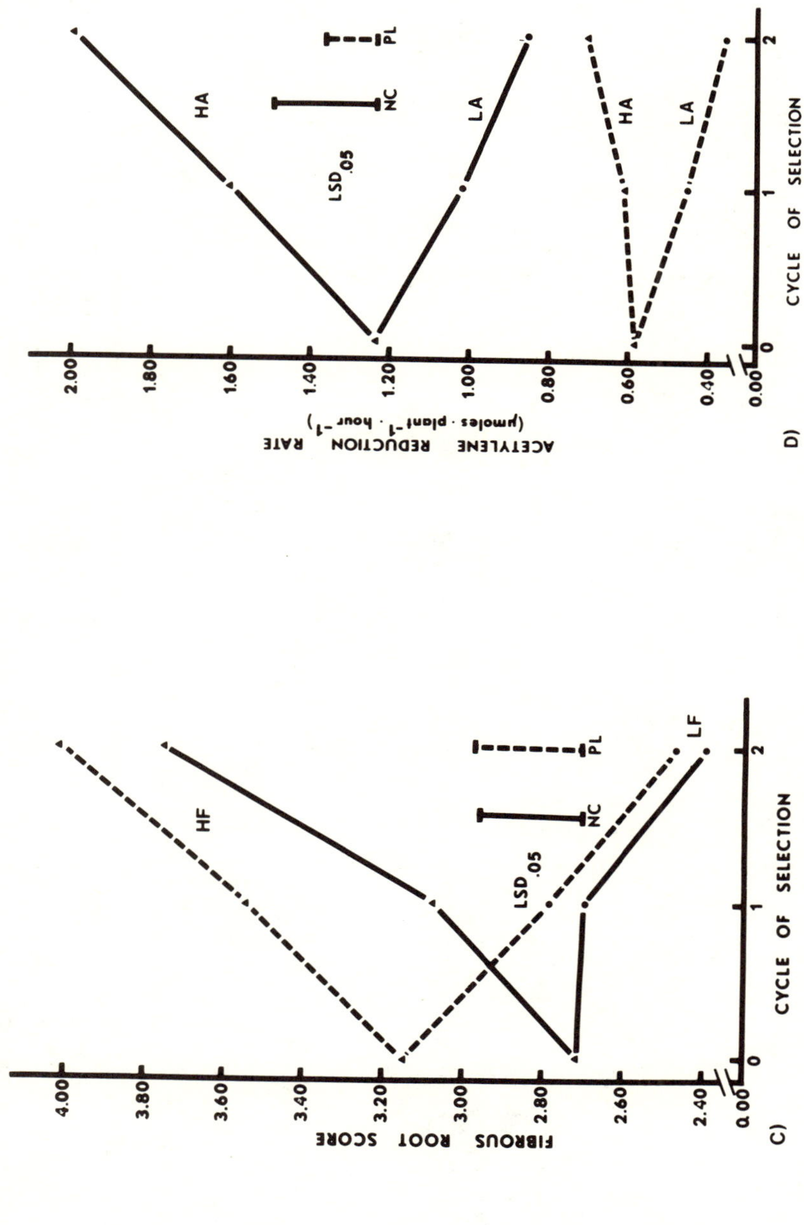

Figure 1. (continued)

Table 2. Acetylene reduction rates for subpopulations resulting from two cycles of bidirectional selection in the MnNC and MnPL alfalfa gene pools (adapted from Viands 1979 and Viands et al., manuscript in preparation.)[†]

Subpopulation/[‡] gene pool	Cycle of selection	Acetylene reduction rate (μmoles/plant/hr) for subpopulations selected for:			
		Acetylene reduction rate	Nodule mass score	Fibrous root score	Top dry weight
MnNC-6-H	2	2.06	1.86	1.97	1.94
MnNH-5-H	1	1.60	1.49	1.62	1.55
MnNC-4 (original)	0	1.24	1.24	1.24	1.24
MnNC-5-L	1	1.02	1.13	1.17	1.18
MnNC-6-L	2	0.85	0.96	0.94	0.75
LSD 0.05		0.27			
MnPL-9-H	2	0.70	0.85	0.84	0.64
MnPL-8-H	1	0.61	0.78	0.69	0.55
MnPL-6 (original)	0	0.58	0.58	0.58	0.58
MnPL-8-L	1	0.45	0.62	0.51	0.51
MnPL-9-L	2	0.36	0.37	0.41	0.50
LSD 0.05		0.14			

[†] The MnNC populations and the MnPL populations were evaluated in the greenhouse during May and September, 1978, respectively.

[‡] H and L = subpopulations selected for high and low levels, respectively.

Table 3. Percentage of variation in acetylene reduction rates accounted for by each of five morphological characteristics in the MnNC and MnPL alfalfa gene pools according to a stepwise regression analysis (Viands, 1979, and Viands et al., manuscript in preparation).

Characteristic	% Variation of nitrogenase activity/gene pool	
	MnNC	MnPL
Nodule mass score	42	31
Nodule mass squared	1	0
Nodule number score	3	0
Top dry weight	3	1
Fibrous root score	1	2
Secondary root score	0	1
Proportion unexplained	50	65

PLANT AND ENVIRONMENTAL FACTORS AFFECTING NODULE ACTIVITY

Viands (1979) and Viands et al. (manuscript in preparation) indicated that nodule mass was the most critical plant characteristic in N_2 fixation of alfalfa. However, more than 50% of the variation in AR was unexplained. We hypothesized that much of the unexplained variation in AR was due to host by rhizobia strain interactions. Therefore, we conducted a series of studies to better elucidate some of the factors affecting nodule activity. Vance et al. (1979) determined that forage removal caused an 88% decline in AR capacity of alfalfa root systems within 24 hours. The AR activity in harvested plants remained low for 13 days until forage regrowth resumed. Following forage harvest, protease and nitrate reductase activity of nodules increased while soluble protein and leghemoglobin levels decreased. All of these activities were reversed with the onset of regrowth. No massive loss in either numbers or mass of nodules occurred after harvest. There was localized senescence at the base of nodules in harvested plants (Fig. 2). However, the meristem and vascular bundles of nodules

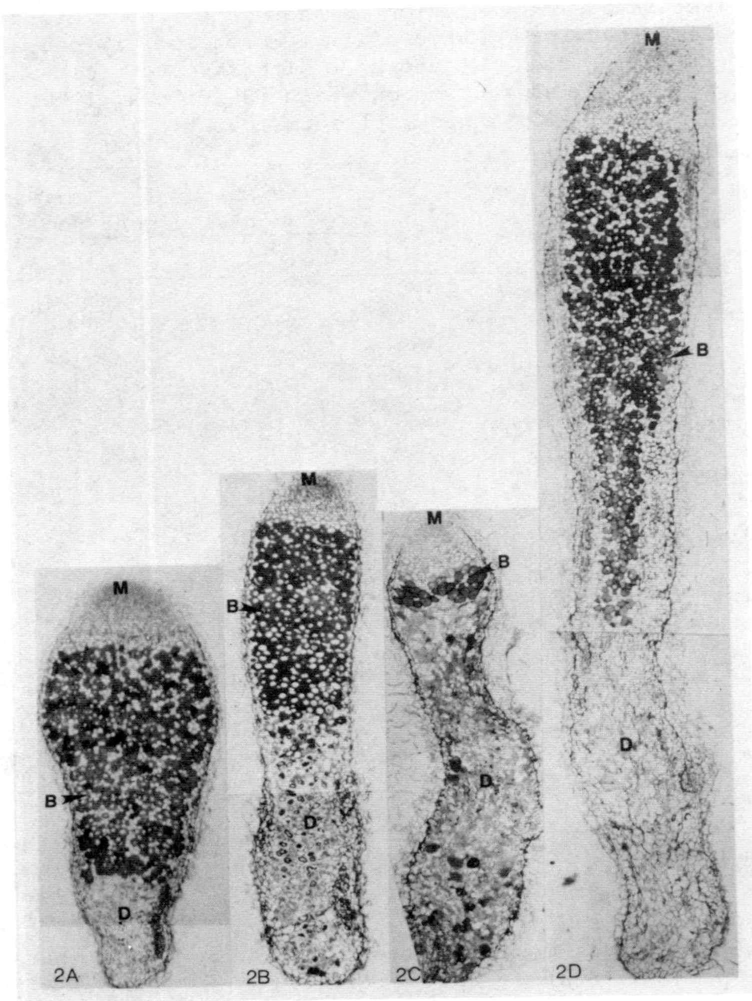

Fig. 2. Alfalfa nodule deterioration and regrowth after harvest (B = bacteroid-containing cells, D = disorganized cells, M = meristem region). (2A) A nodule 4 days after harvest with the loss of bacteroids from the disorganized cells at the base of the nodule; (B) A nodule 7 days after harvest with loss of bacteroids in nearly one-half of nodule; (2C) A nodule 10 days after harvest with only a few bacteroid-containing cells remaining adjacent to the meristematic region; and (2D) A nodule 26 days after harvest showing renewed growth. The disorganized cells did not regenerate after harvest, however, the number of bacteroid containing cells increased (adapted from Vance et al., 1979).

from harvested plants remained intact. The partly senescent nodules began to regrow and fix nitrogen after shoot growth resumed. The processes involved with nodule senescence and repair following defoliation appeared to be very complex (Vance et al., 1980). Nevertheless, as we learn more about these systems we should find selection criteria that will be useful in further increasing the N_2 fixation of alfalfa.

Our greenhouse selection studies have been conducted under essentially nil-nitrate conditions. However, in many mid-western soils the nitrate level is sufficient to delay alfalfa nodulation for weeks or months. Heichel and Vance (1979) investigated growth and nodulation of alfalfa seedlings, inoculated with five strains of R. meliloti individually and in a mixture, grown at 0, 10, 25 and 50 μg N/ml of nutrient solution. With a mixed Rhizobium culture, the proportion of seedlings that nodulated declined from 84% at zero N to 33% at 50 μg/N ml. Significant ($P = 0.05$) differences in frequency of seedling nodulation and amount of growth occurred among individual Rhizobium strains at each N level, and significant ($P = 0.05$) strain X N interactions were observed. Heichel and Vance suggested that it should be possible to improve alfalfa nodulation by either selecting for increased nodule numbers in the presence of nitrate N, or by the identification of Rhizobium strains with maximum effectiveness at ambient N levels. We have not yet evaluated the breeding populations that we have selected for high nodulation and AR activity under nil-nitrate conditions, under high nitrate levels. However, Bray (1977) reported that alfalfa plants selected for good symbiotic characteristics in the presence of Rhizobium were the highest yielding plants when evaluated with applied nitrogen.

We tested our hypothesis that a major portion of the unexplained variation in AR was due to host by Rhizobium strain interactions. A series of Rhizobium strains were isolated from field-grown plants of subpopulations that had been selected for high and low levels of AR and nodule mass (Hardarson et al., unpublished data). Generally, the Rhizobium strains isolated from the high trait subpopulations were more effective when tested on an unrelated cultivar than were the Rhizobium strains isolated from the corresponding low trait subpopulations. Although this research is at a preliminary stage, it appears that we have selected plant genotypes that can select the more effective indigenous strains of Rhizobium. We have concluded that procedures need to be developed so that breeding programs can identify strains of Rhizobium with high competitive ability and then select plants with increased preferences for those strains. Hardarson et al. (unpublished data) has developed a test that should expedite the selection of elite Rhizobium-alfalfa combinations with enhanced N_2 fixation capabity. However, the test has not been evaluated in our plant breeding program.

While conducting our plant breeding program for improved N_2 fixation we learned to recognize alfalfa seedlings that were incapable of fixing nitrogen. They were very small, chlorotic and would normally be lethal. The first ineffective plant we found was described by Viands et al. (1979). Since then we have found many additional non-fixers in a broad range of germplasm. Peterson (1980) determined the inheritance of ineffective nodules from five different clones and one non-nodulating clone. It was surprising to learn that lack of nitrogenase activity in four of the five ineffective parent clones was controlled by different genes. Three of the ineffective traits were each controlled by one recessive gene with tetrasomic inheritance. The fourth ineffective trait and the non-nod trait were each controlled by two recessive genes. These traits should be useful in studying the morphogenesis of alfalfa nodules. The relatively large number of different ineffective traits provide an indication that a great many genes are involved with nodule formation and function.

IMPORTANCE OF FIELD EVALUATIONS

The majority of N_2 fixation research has been conducted under controlled conditions in greenhouse and laboratories. However, before it can be utilized to increase agricultural production it is necessary to evaluate selected germplasm and scientific hypotheses under field conditions. Measurement of N_2 fixation is much more difficult and expensive in the field than in the greenhouse. We have used a ^{15}N dilution technique in the field (Heichel et al., this volume; Heichel et al., manuscript in preparation; and Viands et al., manuscript in preparation). The ^{15}N method measures how much total plant nitrogen comes from the atmosphere through fixation compared to that coming from the soil. It measures N_2 fixation over time rather than at only one point in time like the AR procedure.

Unfortunately, ^{15}N data are not available for our most advanced breeding materials. However, comparisons made during the seeding year from two experimental populations after one cycle of selection in the greenhouse for AR averaged about 43% of their nitrogen needs from symbiosis compared with 36% for the standard cultivar 'Saranac' (Viands et al., manuscript in preparation). The experimental populations fixed an average of about 148 kg/ha of nitrogen during the growing season, compared with 109 kg/ha for 'Saranac'. These data indicated that our greenhouse selection procedures were successful. However, there have been other situations where it has been difficult to relate greenhouse performance to field performance. This is especially true when comparing relative yield levels among populations (Heichel et al., this volume). Factors such as dormancy responses caused by temperature and daylength fluctuations, root and foliar diseases, soil nitrogen levels, soil moisture levels, and indigenous _Rhizobium_ strains can significantly influence N_2 fixation and plant performance.

An example of how unrelated factors can influence N_2 fixation evaluation was observed when we tested the field performance of the single crosses described in the studies by Seetin and Barnes (1977). By chance the single cross with the highest AR values in the greenhouse was completely susceptible to downy mildew caused by *Peronospora trifoliorum* de By. The disease susceptibility prevented the possibility of measuring N_2 fixation in the field.

The subpopulations produced by Viands (1979) were evaluated in field studies for fall dormancy response, resistance to bacterial wilt caused by *Corynebacterium insidiosum* (McCull.) H. L. Jens, Phytophthora root rot caused by *Phytophthora megasperma* Drechs, and Fusarium wilt caused by *Fusarium oxysporum* Schlecht. f. sp. *medicaginis* (Weimer) Synd. & Hans. Viands et al. (1980) reported that lines selected for high levels of N_2 fixation generally were more susceptible to bacterial wilt than unselected populations and lines selected for low levels of N_2 fixation (Table 4). There were no apparent associations between the N_2 fixation selections and the other evaluated traits. The apparent association between genes conditioning one type of resistance to bacterial wilt and those influencing nodule development and nodule activity illustrate how breeding for one characteristic may affect the expression of other traits. Because bacterial wilt is a serious disease of alfalfa it is important that alfalfa breeders be aware of the negative associations between N_2 fixation and bacterial wilt resistance. Fortunately, the association was not with the dominant $\underline{BW_1}$ gene which conditions bacterial wilt resistance in most cultivars (Viands and Barnes, 1980). It should be possible to maintain adequate levels of resistance while improving the level of N_2 fixation. However, it will be necessary to periodically test for level of bacterial wilt resistance in any N_2 fixation selection program.

The few examples of field results that we have reported here are not intended to discourage the development of breeding programs to improve N_2 fixation potential of legume species. It was our intent to describe the complexities of conducting field N_2 fixation evaluations, and to illustrate the importance of coordinating N_2 fixation research efforts with established plant breeding and plant production research programs. Adapted and pest-resistant germplasm with good agronomic characteristics must be the basis for any N_2 fixation improvement program.

PROJECTIONS FOR THE FUTURE

After 6 years of N_2 fixation research on alfalfa we have concluded that the problems are much more complex than we had originally assumed. It appears that nearly all plant systems are influenced or somehow involved with the N_2 fixation process.

Table 4. Mean disease scores and percentages of plants resistant to bacterial wilt in the original MnNC and MnPL alfalfa gene pools and in two subpopulations after one cycle of mass selection for high levels of a base index comprised of characteristics associated with N_2-fixation in each gene pool (adapted from Viands et al., 1980).

Subpopulation/gene pool[†]	Mean disease score[‡]	% Resistant plants[‡]
MnNC-4 Original (Syn. 2)	1.16	70.4
MnNC-5 HB-B (20%) (Syn. 2)	1.37	64.3
MnNC-5 HB-B (1%) (Syn. 2)	1.57	60.7
MnPL-6 Original (Syn. 2)	1.34	58.3
MnPL-8 HB-B (20%) (Syn. 2)	1.63	46.2
MnPL-8 HB-B (1%) (Syn. 2)	2.59	20.7
$LSD_{0.05}$	0.39	11.8

[†] HB-B (20%) subpopulation represents intercross of 160 best plants (20%). HB-B (1%) subpopulation represents intercross of 10 best plants (1%).

[‡] Scored on a 0-5 basis, where 0 = no disease symptoms and 5 = dead plant. Classes 0 and 1 considered resistant.

Nevertheless, our results have rather conclusively demonstrated that it is possible to improve the N_2 fixation potential of alfalfa.

When we initiated our breeding programs we assumed that improved N_2 fixation potential would automatically increase yield and increase nitrogen concentration in tops and roots. None of these attributes have been observed under field conditions. We have since concluded that it will be necessary to first improve the N_2 fixation in germplasm gene pools, then further select for useful traits that can capitalize on the increased levels of symbiotically produced nitrogen. Presently, we have conducted two cycles of selection in several non-dormant germplasm sources to increase the amount of nitrogen stored in the tap root during late fall. Our intent is to breed an annual alfalfa for use as a source of residual nitrogen in crop rotations. We are searching for alfalfa plants that can exude nitrogen to be used by co-habitating crop species. We are also searching for high-yielding high-protein hay types of alfalfa.

The information we have obtained about N_2 fixation and the alfalfa nodule has greatly increased our knowledge about how many production practices are influenced by N_2 fixation processes. We anticipate that nodule formation, activity, and senescence patterns will become an important consideration in alfalfa management research. We must develop a rapid, inexpensive method to compare N_2 fixation rates among cultivars in the field. We need to increase our knowledge about the nodule's role in the production and regulation of growth regulators. We also need a better understanding of the host by Rhizobium strain interactions. Our list of types of information still needed is very long. However, we are convinced that the challenges of the future can be met and that N_2 fixation by alfalfa can be managed to improve agricultural production systems. It will require the continued efforts of scientists from many disciplines cooperating to achieve a common goal. This philosophy is similar to the recommendations suggested by a recent conference on "Selecting and breeding legumes for enhanced nitrogen fixation" (LaRue, 1978). They recommended the establishment of cooperative multidisciplinary N_2 fixation research programs for each major legume species.

REFERENCES

Bray, R. A., 1977, The effect of Rhizobium on the response to recurrent phenotypic selection for seedling weight in Lucerne (Medicago sativa), in: "Plant Breeding Papers," 3rd Intern. Cong. Soc. for Advancement of Breeding Research in Asia and Oceania 30 (V. 2), 14(a):14.

Burton, J. C., 1972, Nodulation and symbiotic nitrogen fixation, in: "Alfalfa Science and Technology," C. H. Hanson, ed., American Soc. of Agron., Madison, WI.

Duhigg, P., Melton, B., and Baltensperger, A., 1978, Selection for acetylene reduction rates in 'Mesilla' alfalfa, Crop Sci., 18:813.

Hardy, R. W. F., Holsten, R. D., Jackson, E. K., and Burns, R. C., 1968, The acetylene-ethylene assay for N_2 fixation: laboratory and field evaluation, Plant Physiol., 53:1185.

Heichel, G. H., and Vance, C. P., 1979, Nitrate-N and Rhizobium strain roles in alfalfa seedling nodulation and growth, Crop Sci., 19:512.

LaRue, T. A., 1978, Selecting and breeding legumes for enhanced nitrogen fixation (recommendations for research and proceedings of a workshop), Boyce Thompson Institute at Cornell, Ithaca, New York.

Munns, D. N., Fogle, V. W., and Hallock, B. G., 1977, Alfalfa root nodule distribution and inhibition of nitrogen fixation by heat, Agron. J., 69:377.

Peterson, M. A., 1980, Inheritance of five ineffective nitrogen fixation traits in alfalfa, M.S. Thesis, Department of Agronomy and Plant Genetics, University of Minnesota.

Seetin, M. W., and Barnes, D. K., 1977, Variation among alfalfa genotypes for rate of acetylene reduction, Crop. Sci., 17:783.

Vance, C. P., 1978, Nitrogen fixation in alfalfa: an overview, Proceedings, Eighth Annual Alfalfa Symposium, Bloomington, Minnesota.

Vance, C. P., Heichel, G. H., Barnes, D. K., Bryan, J. W., and Johnson, L. E. B., 1979, Nitrogen fixation, nodule development, and vegetative regrowth of alfalfa (Medicago sativa L.) following harvest, Plant Physiol., 64:1.

Vance, C. P., Johnson, L. E. B., Halvorsen, A. M., Heichel, G. H., and Barnes, D. K., 1980, Histological and ultrastructural observations of Medicago sativa root nodule senescence after foliage removal, Can. J. Bot., 58:245.

Viands, D. R., 1979, Selection experiments on nitrogen fixation in alfalfa, Ph.D. Thesis, Department of Agronomy and Plant Genetics, University of Minnesota.

Viands, D. R., and Barnes, D. K., 1980, Inheritance of resistance to bacterial wilt in two alfalfa gene pools: Qualitative analysis, Crop Sci., 20:48.

Viands, D. R., Barnes, D. K., and Frosheiser, F. I., 1980, An association between resistance to bacterial wilt and nitrogen fixation in alfalfa, Crop Sci., in press.

Viands, D. R., Vance, C. P., Heichel, G. H., and Barnes, D. K., 1979, An ineffective nitrogen fixation trait in alfalfa, Crop Sci., 19:905.

Weaver, R. W., and Frederick, L. R., 1974, Effect of inoculum rate on competitive nodulation of Glycine max L. Merrill. II. Field studies, Agron. J., 66:233.

BIOCHEMICAL GENETICS OF RIBULOSE BISPHOSPHATE CARBOXYLASE/OXYGENASE: HIGH FREQUENCY TRANSFER AND CHROMOSOME MOBILIZATION BY BROAD HOST-RANGE R-FACTORS IN CO_2-FIXING BACTERIA

Kjell Andersen

Plant Growth Laboratory
University of California
Davis, CA 95616

INTRODUCTION

Much attention has recently been focused on CO_2 fixation, and especially on ribulose bisphosphate (RuBP) carboxylase/oxygenase which may represent a rate-limiting step in crop productivity (see Siegelman and Hind, 1978, for recent symposium). For example, symbiotic N_2 fixation is probably often limited by the availability of energy in the form of photosynthate supplied by the host plant (Hardy and Havelka, 1975).

We think that studies of simple bacterial systems can significantly increase our knowledge of the molecular mechanism of CO_2 fixation and its regulation. Our choice of organism for such studies was the "hydrogen" bacterium (Schlegel, 1976; and Schlegel et al., this volume) *Alcaligenes eutrophus*, which possesses a RuBP carboxylase/oxygenase almost identical in structure and function to the enzyme in higher plants (Andersen et al., 1978). We have isolated mutant strains with RuBP carboxylase/oxygenase having altered catalytic activity (Andersen, 1979), and have also studied the relationship between CO_2/O_2 concentration, efficiency of CO_2 fixation and glycollate production in *A. eutrophus* (King and Andersen, 1980). Identification and properties of plasmids that determine the utilization of H_2 gas in *A. eutrophus*, and the relevance of this work to symbiotic N_2 fixation is discussed elsewhere in this volume by Tait et al.

Chromosome transfer mediated by a natural conjugation system has recently been reported for the nitrogen-fixing hydrogen bacterium *Xanthobacter autotrophicus* (Wilke, 1980). This communication describes progress in establishing systems to allow chromosomal gene transfer for mapping of mutants in *A. eutrophus*, an important tool for further studies of the biochemical genetics of CO_2 fixation in this organism.

INTRODUCTION OF R-FACTORS INTO HYDROGEN BACTERIA

R-Factor RP4

The class of conjugative plasmids known as broad host-range R-factors have been found to be very versatile in development of systems for gene-transfer in a variety of gram-negative bacteria (Holloway, 1979). The P1 incompatibility group R-factors RP4, RK2 and R68.45 were transferred into *A. eutrophus* ATCC 17697 (type strain) from *Escherichia coli* at frequencies of $4\text{-}6 \cdot 10^{-6}$ (Table 1). They were stably maintained in *A. eutrophus* and expressed high levels of resistance to tetracycline (Tc, 25-50 μg/ml) as well as kanamycin (Km, 1-2 mg/ml). Transfer between derivatives of *A. eutrophus* (type strain) occurred at a frequency of about $5 \cdot 10^{-4}$ (Table 1). However, conjugation with *Escherichia coli* as a recipient resulted in very high frequency of R-factor transfer. Thus, *A. eutrophus* type strain behaved as an efficient donor but was a relatively poor recipient of these R-factors. No significant difference in transfer frequencies was observed for the related R-factors RP4, RK2 and R68.45.

Other *A. eutrophus* wild-type strains were also screened for their ability to serve as RP4 recipients. All were about equal to the type strain, with the exception of ATCC 17707 which was a very good recipient (Table 1). RP4 was stably maintained in this strain, conferring high resistance to kanamycin (> 1 mg/ml), moderate resistance to tetracycline (∿ 10 μg/ml), and low resistance to ampicillin (∿ 10 μg/ml).

Other R-Factors

The P1 group R-factors R751 and R772 have been reported to mediate chromosome transfer in *Acinetobacter calcoaceticus* (Towner and Vivian, 1977) and *Proteus mirabilis* (Coetzee, 1978), respectively. These R-factors, and also R906 (Hedges et al., 1974), were all transferred into *A. eutrophus* ATCC 17707 at high frequency (Table 2). They were stably maintained, expressing high levels of resistance to trimethoprim (Tp, > 100 μg/ml), kanamycin (> 200 μg/ml) and sulfathiazole (Su, > 200 μg/ml), respectively.

Transposons, especially translocatable drug-resistance elements, can be valuable tools for *in vivo* genetic engineering, serving as convenient genetic markers, mutagens, as well as facilitating a number of genetic manipulations (Kleckner et al., 1977).

Table 1. Introduction of the broad host-range conjugative plasmid RP4 into hydrogen bacteria. Donor and recipient strains were grown in L-broth to $5 \times 10^8 - 1 \times 10^9$ cells/ml, mixed in a ratio of 5 donor cells per recipient cell, and a total of about 5×10^9 cells collected on a 25 mm diameter membrane filter. Conjugation was allowed to proceed overnight by incubating the filter on L-broth plates at 30 C. The cells were then resuspended and plated on the appropriate media, in this case on autotrophic medium containing 25 μg/ml tetracycline (Tc) when hydrogen bacteria were the recipients. Autotrophic growth was under 85% H_2, 10% CO_2 and 5% O_2 (for composition of growth media, see Andersen, 1979). The methionine-requiring strain AE101 was isolated after NTG mutagenesis of *A. eutrophus* type strain (Andersen, 1979).

Donor	Recipient	Frequency of transfer (Tc^R per recipient)
E. coli C600/RP4	*A. eutrophus* ATCC17697 (type strain)	6×10^{-6}
A. eutrophus (type strain) AE101/RP4 (Met^-)	*E. coli* C600	4×10^{-1}
	A. eutrophus ATCC17697	5×10^{-4}
	A. eutrophus ATCC 17698, 17699, 17702, 17704-17706, 17708, or 17709	$\sim 10^{-4}$
	A. eutrophus ATCC17707	5×10^{-1}
	A. paradoxus ATCC17713	$\sim 10^{-2}$
	A. ruhlandii ATCC15749	5×10^{-1}
	P. facilis ATCC15376	$\sim 10^{-5}$
	P. palleronii ATCC17724	$\sim 10^{-7}$

Table 2. Introduction of R-factors of the Pl incompatibility group into *A. eutrophus*. Crosses were performed as described for Table 1. R-factor containing exconjugants were selected on fructose minimal medium containing the drugs indicated (25 μg/ml Tc, 100 μg/ml Tp, 200 μg/ml Km (R772) or 200 μg/ml Su). Spontaneous rifampicin resistant (Rif^R) derivatives of *A. eutrophus* were used in the cross with *E. coli* 1830, and selection was on L-broth plates with 0.4 mM EDTA, 100 μg/ml Rif and 1 mg/ml Km. R-factor containing donors were kindly supplied by L. N. Csonka (R751, R906, pJB4JI), J. N. Coetzee (R772) and G. Ditta (pRK290 and pRK2013).

R-factor	Donor	Drug resistance marker selected	Frequency of transfer (per recipient) to A. eutrophus	
			ATCC17697	ATCC17707
RP4	*E. coli* C600	Tc	6×10^{-6}	5×10^{-1}
R68.45	*E. coli* C600	Tc	5×10^{-6}	3×10^{-1}
R751	*E. coli* JC32721	Tp	ND	1×10^{-2}
R772	*P. mirabilis* PM5006	Km	$<10^{-7}$	1×10^{-1}
R906	*E. coli* JC32721	Su	ND	2×10^{-1}
pJB4JI	*E. coli* 1830	Km	$<10^{-9}$	1×10^{-6}
pRK290 & pRK2013	*E. coli* HB101	Tc	6×10^{-3}	8×10^{-1}

ND = Not determined

The composite R-factor pJB4JI, carrying Mu and the kanamycin resistance transposon Tn5, has been used to introduce Tn5 into Rhizobium (Beringer et al., 1978). This R-factor has a very low ability to become established in Rhizobium, leaving Tn5 inserted into the chromosome. Attempts to introduce Tn5 into A. eutrophus (type strain) failed. Kanamycin resistance was transferred to A. eutrophus ATCC 17707 at low frequency Table 2). However, examination of the A. eutrophus exconjugants revealed that they, in contrast to the A. eutrophus parent, were also resistant to gentamicin, spectinomycin and streptomycin, resistance markers carried by pJB4JI. The whole plasmid therefore apparently became established at the same frequency as Tn5, making it unsuitable as a vehicle for Tn5 mutagenesis of this strain of A. eutrophus.

The binary R-factor system pRK290 plus pRK2013 was transferred to A. eutrophus at high frequency (Table 2). pRK290 is a cloning vehicle derived from RK2, and has been used to derive a gene bank of Rhizobium meliloti (Ditta et al., this volume). This is also an interesting vehicle for cloning of A. eutrophus DNA. The high transfer frequency should allow identification of cloned genes by their ability to complement genetic lesions in specific A. eutrophus mutants.

R-FACTOR MEDIATED PLASMID AND CHROMOSOME TRANSFER IN A. EUTROPHUS

There is generally a close correlation between the transfer frequency of a specific plasmid and its ability to promote chromosome transfer. We have recently isolated a Tc^S derivative of R68.45 which when present in recipient strains increases the R-factor transfer to 1-2% (Table 1). Such improved recipients allowed R-factor mediated transfer of the plasmid pAE1 which determines the utilization of H_2 for growth (Table 3; Andersen et al., manuscript in preparation; Tait et al., this volume).

The RuBP carboxylase genes in A. eutrophus (type strain) probably reside on the chromosome, since "cured" strains lacking pAE1 still synthesized RuBP carboxylase activity (Andersen et al., manuscript in preparation). However, no transfer of RuBP carboxylase genes was observed even with the improved recipients (Table 3).

A. eutrophus ATCC 17707, which is a much better R-factor recipient than the type strain (Table 2) was also investigated. Here both RP4 and R68.45 promoted transfer of streptomycin resistance (str^R 1) at a frequency of about 5 x 10^{-7}. The auxotrophic markers his1, cys1, phe1, and ade1 were also mobilized at approximately the same frequency. All these auxotrophic mutants were very stable. Controls where donors and recipients were plated separately gave 10-100 times fewer prototrophs than the conjugation experiments. No transfer of streptomycin resistance mediated by R751 or R772 was detected.

Table 3. R-factor mediated transfer of the H_2 utilization plasmid pAE1 in *A. eutrophus* (type strain). Crosses were performed as described for Table 1. The "cured" Hup^- strain AE131 was isolated after exposure to mitomycin C and had lost plasmid pAE1 (Andersen et al., manuscript in preparation). The Tc^S derivative of R68.45 was isolated after NTG mutagenesis of AE113/R68.45 and transferred into a spontaneous Nal^R Rif^R derivative of AE131 to make the second recipient. Strain AE370 is blocked in CO_2 fixation, having an inactive RuBP carboxylase protein (Andersen, 1979). Nalidixic acid (Nal, 500 µg/ml) was included in the media to select against the donors where Nal^R recipients were used. Hup^+ was scored as Aut^+.

Donor	Recipient	Frequency of transfer of: RP4 (Tc^R per recipient)	pAE1 (Hup^+ per RP4 transfer)	Cfx^+ (Aut^+ per RP4 transfer)
AE101/RP4 (Met^-)	AE131 (Hup^-)	6×10^{-4}	$<10^{-6}$	--
AE101/RP4	AE131 Nal^R Rif^R/R68.45 Tc^S	2×10^{-2}	3×10^{-4}	--
AE101/RP4	AE370 Nal^R/ R68.45 Tc^S (Cfx^-)	1×10^{-2}	--	$<2 \times 10^{-7}$

Aut = Autotrophic
Hup = H_2 uptake
Cfx = CO_2 fixation

Table 4. R-factor mediated chromosome transfer in A. eutrophus ATCC 17707. Crosses were performed as described for Table 1, except for using a donor/recipient ratio of 2/1. Mutant strains were isolated after ethyl methane sulfonate (EMS) mutagenesis, except drug resistance which was spontaneous. Prototrophs were selected on fructose minimal medium plates.

Donor	Recipient	Frequency of R-factor transfer	Selected phenotype	Recombination frequency (per recipient)
AE7-20/ R68.45 (Str^R)	AE7-22 ($Rif^R Nal^R$)	8×10^{-1}	$Str^R Rif^R$	6×10^{-7}
AE7-20/ RP4	AE7-22	7×10^{-1}	$Str^R Rif^R$	5×10^{-7}
AE7-66/ RP4 Leu$^-$)	AE7-68 (His$^-$)	8×10^{-1}	$Leu^+ His^+$	4×10^{-7}
AE7-66 RP4	AE7-72 (Cys$^-$)	6×10^{-1}	$Leu^+ Cys^+$	3×10^{-7}
AE7-66/ RP4	AE7-80 (Phe$^-$)	5×10^{-1}	$Leu^+ Phe^+$	3×10^{-7}
AE7-52/ RP4 (Met$^-$)	AE7-74 (Ade$^-$)	7×10^{-1}	$Met^+ Ade^+$	7×10^{-7}

It is concluded that RP4 mediates chromosome transfer in A. eutrophus, mobilizing several different markers at similar frequencies. This system may in the future allow mapping of chromosomal mutations in A. eutrophus, which would be an important tool in the study of the biochemical genetics of CO_2 fixation in this organism.

ACKNOWLEDGMENT

I am indebted to Dr. R. C. Valentine for encouragement, advice and support, and to W. King for assistance with the initial experiments in A. eutrophus (type strain). This work was supported by the National Science Foundation under Grant No. PFR 77-07301 and by the Science and Education Administration of the U. S. Department of Agriculture under Grant No. 5901-0410-8-0142-0 from the Competitive Research Grants Office. Any opinions, findings, and conclusions or recommendations expressed in this publication are those of the authors and do not necessarily reflect the views of the granting agencies.

REFERENCES

Andersen, K., 1979, Mutations altering the catalytic activity of a plant-type ribulose bisphosphate carboxylase/oxygenase in Alcaligenes eutrophus, Biochim. Biophys. Acta, 585:1.

Andersen, K., King, W. R., and Valentine, R. C., 1978, Catalytic mutants of ribulose bisphosphate carboxylase/oxygenase, in: "Photosynthetic Carbon Assimilation," H. W. Siegelman and G. Hind, eds., Plenum Press, New York.

Beringer, J. E., Beynon, J. L., Buchanan-Wollaston, A. V., and Johnston, A. W. B., 1978, Transfer of the drug-resistance transposon Tn5 to Rhizobium, Nature, 276:633.

Coetzee, J. N., 1978, Mobilization of the Proteus mirabilis chromosome by R-plasmid R772, J. Gen. Microbiol., 108:103.

Hardy, R. W. F., and Havelka, U. D., 1975, Nitrogen fixation research: a key to world food?, Science, 188:633.

Hedges, R. W., Jacob, A. E., and Smith, J. T., 1974, Properties of an R-factor from Bordetella bronchiseptica, J. Gen. Microbiol., 84:199.

Holloway, B. W., 1979, Plasmids that mobilize bacterial chromosome, Plasmid, 2:1.

King, W. R., and Andersen, K., 1980, Efficiency of CO_2 fixation in a glycollate oxidoreductase mutant of Alcaligenes eutrophus which exports fixed carbon as glycollate, Arch. Microbiol., in press.

Kleckner, N., Roth, J., and Botstein, D., 1977, Genetic engineering in vivo using translocatable drug-resistance elements, J. Mol. Biol., 116:125.

Schlegel, H. G., 1976, Physiology of hydrogen bacteria, Ant. van Leeuwenhoek, 42:181.

Siegelman, H. W., and Hind, G., 1978, Photosynthetic Carbon Assimilation, Plenum Press, New York.

Towner, K. J., and Vivian, A., 1977, Plasmids capable of transfer and chromosome mobilization in Acinetobacter calcoaceticus, J. Gen. Microbiol., 101:167.

Wilke, D., 1980, Conjugational gene transfer in Xanthobacter autotrophicus GZ 29, J. Gen. Microbiol., 117:431.

PHOTOSYNTHETIC ENZYME REGULATION BY THE FERREDOXIN/THIOREDOXIN AND THE FERRALTERIN MECHANISMS

Bob B. Buchanan

Section of Cell Physiology, Department of Plant and Soil Biology, University of California
Berkeley, CA 94720

INTRODUCTION

One type of covalent enzyme modification that has been used in studies on the structure and mechanism of action of enzymes is a change in the oxidation state of a "nonprosthetic" part of the protein, such as a sulfhydryl group. Limited attention has been given to the idea that oxidation-reduction changes may be used to regulate the activation of enzymes in a manner analogous to other types of reversible covalent modification, such as phosphorylation-dephosphorylation and adenylation-deadenylation. We summarize in this report evidence that enzyme regulation due to a reversible oxidation-reduction change is a process fundamental to photosynthetic and perhaps to other types of living cells. In this article we will first describe work from our own laboratory that led to the finding of a redox-based regulatory mechanism in photosynthesis; we will then relate these findings to other mechanisms of light-dependent enzyme regulation in chloroplasts, including the recently found ferralterin mechanism.

RESULTS

Our first indication that the activity of an enzyme may be altered by redox change came more than 10 years ago when studies with isolated spinach chloroplasts revealed that the enzyme fructose 1, 6-bisphosphatase (Fru-P_2ase) -- a key enzyme of the reductive pentose phosphate cycle of CO_2 assimilation -- is activated by reduced soluble ferredoxin (Buchanan et al., 1967). (Soluble ferredoxin is an iron-sulfur protein that functions as an electron carrier for the light-dependent reduction of chloroplast metabolites such as NADP in the presence of specific ferredoxin-linked reductase enzymes.) In our experiments, ferredoxin, which was reduced photochemically with chloroplast membranes, activated the Fru-P_2ase enzyme component of the stroma or soluble phase of chloroplasts ('chloroplast extract'). Fru-P_2ase was also found, by Bassham and his colleagues (Petersen et al., 1966) to be light-activated in whole cells. Somewhat earlier, the Zieglers showed that another enzyme of the reductive pentose phosphate cycle, NADP-glyceraldehyde-3-phosphate dehydrogenase, is light-activated (Ziegler and Zeigler, 1965). We had at the time no premonition of findings to be made years later that the NADP-glyceraldehyde 3-phosphate dehydrogenase, as well as other light-activated enzymes that had been studied by Bassham's group, share the capacity for activation by reduced ferredoxin.

Following the demonstration of the ferredoxin-linked Fru-P_2ase activation, we began attempts to purify Fru-P_2ase and to study its mechanism of activation. This study revealed that reduced ferredoxin does not interact directly with the enzyme but that another soluble chloroplast component is required (Buchanan et al., 1971). We separated this component (which, because of its properties, was designated the protein factor) from both ferredoxin and Fru-P_2ase (Table 1). Somewhat later, we showed that sedoheptulose 1,7-bisphosphatase (Sed-P_2ase), another enzyme of the reductive pentose phosphate cycle of chloroplasts, resembles Fru-P_2ase in its capacity for activation by ferredoxin and the protein factor (Schürmann and Buchanan, 1975).

Table 1. Requirements for activation of homogeneous chloroplast Fru-P_2ase by ferredoxin.

Treatment	P_i released (μmol)
Complete	4.0
Minus ferredoxin	0.1
Minus Fru-P_2ase	0.1
Minus $MgCl_2$	0.1
Minus protein factor	0.1
Minus fructose bisphosphate	0.0
Complete, ferredoxin not reduced (dark)	0.1

Identification of the Chloroplast Protein Factor

Once the protein factor was isolated, we began an investigation aimed at its purification and identification. These studies led to the finding that the protein factor consists not of one but of two components that we ultimately isolated and gave the names "assimilation regulatory protein a" (ARP_a) and "assimilation regulatory protein b" (ARP_b) (Schürmann et al., 1976) (Table 2).

Table 2. Requirement for ARP_a and ARP_b for activation of chloroplast Fru-P_2ase by reduced ferredoxin.

Treatment	P_i released (nmol/min)
Control	0
+ ARP_a	14
+ ARP_b	0
+ ARP_a and ARP_b	73

To gain insight into the nature of the newly separated regulatory proteins, we exploited our earlier finding (Buchanan et al., 1971) that the sulfhydryl reagent dithiothreitol (DTT) can replace reduced ferredoxin in the activation of Fru-P_2ase, provided the then-called protein factor is present. Our experiments with the resolved protein factor components showed that ARP_b is the active component and that ARP_a has no effect on activation in the presence of DTT (Schürmann et al., 1976). This finding not only provided new information on ARP_b but also gave us a convenient assay for ARP_b that is independent of ferredoxin and ARP_a. We have taken advantage of this assay in much of our later work.

The new ARP_b assay was immediately applied in our investigation of the occurrence of ARP_b in different types of organisms. This study revealed that ARP_b is not confined to photosynthetic cells but occurs in other types of cells and is wide-spread, if not ubiquitous, in nature (Buchanan and Wolosiuk, 1976). We found ARP_b not only in chloroplasts, algae, and photosynthetic bacteria, but also in roots, seeds, etiolated plants shoots, fermentative bacteria, aerobic bacteria, and even in animal cells. In studying mammalian ARP_b we were able to obtain a highly purified ARP_b preparation from rabbit liver.

Just after we observed the wide natural distribution of ARP_b, an article on the protein thioredoxin caught our attention. This article initiated a search that led ultimately to the identification of ARP_b as chloroplast thioredoxin (Table 3) and of ARP_a as an enzyme that catalyzes the reduction of thioredoxin by reduced ferredoxin, i.e., ferredoxin-thioredoxin reductase (Table 4) (Wolosiuk and Buchanan, 1977). The ARP_b protein isolated from liver was identified as authentic liver thioredoxin (Holmgren et al., 1977).

Table 3. Identification of ARP_b as chloroplast thioredoxin

Treatment	P_i released (nmol/min)	
	Chloroplast ARP_b	*E. coli* Thioredoxin
Light, complete	12	10
Light-ARP_b or thioredoxin	2	2
Light-ferredoxin	4	2
Light-ARP_a	1	1
Light-fructose 1,b-bisphosphatase	0	1
Dark, complete	2	1

Table 4. Identification of chloroplast ARP_a as ferredoxin-thioredoxin reductase.

Component added to oxidize photoreduced ferredoxin	Ferredoxin oxidized (nmol/min)
None	0.0
ARP_a	0.6
Chloroplast thioredoxin	0.9
ARP_a + chloroplast thioredoxin	12.0

Thioredoxin is a low-molecular-weight, hydrogen-carrier protein that was first isolated (under different names) in studies on the reduction of sulfoxide and sulfate compounds by enzyme systems isolated from yeast (Black et al., 1960; Wilson et al., 1961). Thioredoxin was later independently isolated, named, and extensively characterized in the pioneering work of Reichard and his colleagues on DNA synthesis (Laurent et al., 1964). Experiments by those investigators revealed that thioredoxin functions as a hydrogen carrier in the reduction of ribonucleoside diphosphates to their deoxyribose derivatives in DNA synthesis by enzyme systems isolated first from bacteria and later from animal tissues (Reichard, 1968). Work by Reichard's group also revealed that the low-molecular-weight protein isolated earlier from yeast (Black et al., 1960; Wilson et al., 1961) is the same as thioredoxin (Porque et al., 1970). At the time of our entry into the field, thioredoxin had not been reported to occur in plants.

Just prior to or parallel with our studies on the identification of chloroplast thioredoxin, new findings were made on the thioredoxin from *Escherichia coli*. Mark and Richardson showed thioredoxin to be a subunit of phage-induced DNA polymerase in that organism (Mark and Richardson, 1976), and Pigiet reported the isolation of a phosphorylated form of *E. coli* thioredoxin (Pigiet and Conley, 1920).

Enzyme Activation by the Ferredoxin/Thioredoxin System

Thioredoxin and ferredoxin-thioredoxin reductase have emerged as components of a ferredoxin-linked regulatory mechanism (designated the ferredoxin/thioredoxin system) by which light regulates selected enzymes during photosynthesis. In the light, electrons from chlorophyll are transferred to ferredoxin and then via the enzyme ferredoxin-thioredoxin reductase to thioredoxin (Wolosiuk and Buchanan, 1977). Reduced thioredoxin, in turn, reduces and thereby activates a number of regulatory enzymes of chloroplasts (Fig. 1). There is now evidence that thioredoxin activates four enzymes of the reductive pentose phosphate cycle (Fru-P_2ase (Wolosiuk and Buchanan, 1977), Sed-P_2ase (Breazeale et al., 1978), NADP-glyceraldehyde 3-phosphate dehydrogenase (Wolosiuk and Buchanan, 1978), phosphoribulokinase (Wolosiuk and Buchanan, 1978)), an enzyme of CO_2 assimilation that is not a part of the carbon cycle (NADP-malate dehydrogenase (Wolosiuk et al., 1977)), and an enzyme of secondary plant metabolism (phenylalanine ammonia lyase (Nishizawa et al., 1979)). The regulatory role of thioredoxin has been extended to the blue-green algae, where it was shown that an enzyme of sulfate reduction (2'-phosphoadenosine 5'-phosphosulfate (PAPS) sulfotransferase) is activated by reduced thioredoxin (Wagner et al., 1978). As discussed below, reduced thioredoxin also activates the ATPase activity associated with chloroplast coupling factor (CF_1) (McKinney et al., 1978).

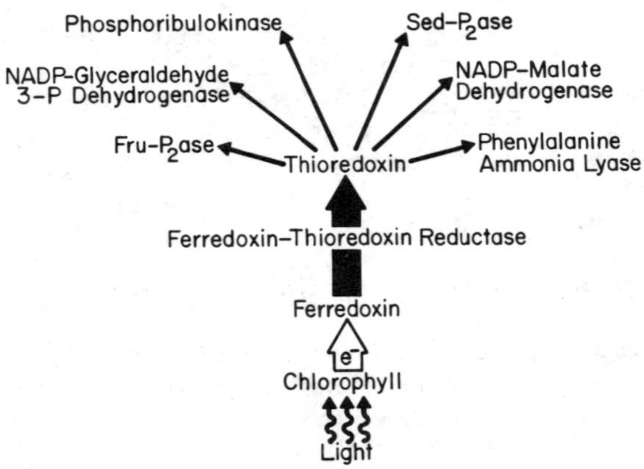

Fig. 1. Role of thioredoxin in the light-dependent activation of chloroplast enzymes.

In the light-dependent activation of enzymes by the ferredoxin/thioredoxin system, the light signal is converted via chlorophyll to a reductant signal (reduced ferredoxin), which is relayed via ferredoxin-thioredoxin reductase to thioredoxin. Thioredoxin thus may be visualized as a regulatory messenger between light and the enzymes of processes that utilize the products formed by light, i.e., ATP and NADPH. In this capacity, thioredoxin "alerts" key enzymes of diverse biosynthetic processes that the light is turned on and that biosynthesis is to proceed.

Deactivation of Thioredoxin-Activated Enzymes

An important aspect of the ferredoxin/thioredoxin system that has not yet been fully elucidated is the way by which the activated enzymes are converted to a less active (deactivated) state. The unique feature of the enzymes of this, and of other, light-mediated regulatory mechanisms of chloroplasts is that deactivation must occur in the dark. Current evidence suggests that, unlike activation, the mechanism for the dark deactivation of thioredoxin-linked enzymes can differ as to the particular enzyme involved. Based on their deactivation properties, thioredoxin-linked enzymes are of three types.

Enzymes of the first type require for deactivation a soluble oxidant, such as the oxidized form of glutathione (GSSG) (Fig. 2) or ascorbate (dehydroascorbate), both of which may be formed by chloroplasts in the dark (Fig. 3) Wolosiuk and Buchanan, 1977; Groden and Beck, 1977). This group of enzymes includes Fru-P_2ase (Wolosiuk and Buchanan), phosphoribulokinase (Wolosiuk and Buchanan, 1978), and phenylalanine ammonia lyase (Nishizawa et al., 1979). The second type of enzymes, of which NADP-malate dehydrogenase is the sole representative, appear to be deactivated by an unidentified membrane-bound oxidant in the absence of soluble oxidants (Table 5) (Wolosiuk et al., 1977). Evidence indicates this oxidant is available for deactivation only when the membranes are maintained in the dark. Enzymes of the third type, typified by NADP-glyceraldehyde 3-phosphate dehydrogenase, have no known mechanism of deactivation (Wolosiuk and Buchanan, 1978).

Despite differences in deactivation properties, the thioredoxin linked regulatory enzymes of chloroplasts share one common feature: in each case, the rate of deactivation (in which the enzyme is converted from an active to an inactive form) and the rate of activation (in which the enzyme is converted from an inactive to an active form) are slow relative to the rate of catalysis. Friden (1971) has designated enzymes of this type as hysteretic enzymes. Ribulose 1,5-bisphosphate carboxylase, the enzyme catalyzing the sole carboxylation reaction of the reductive pentose phosphate cycle, also shows hysteretic behavior (Lorimer

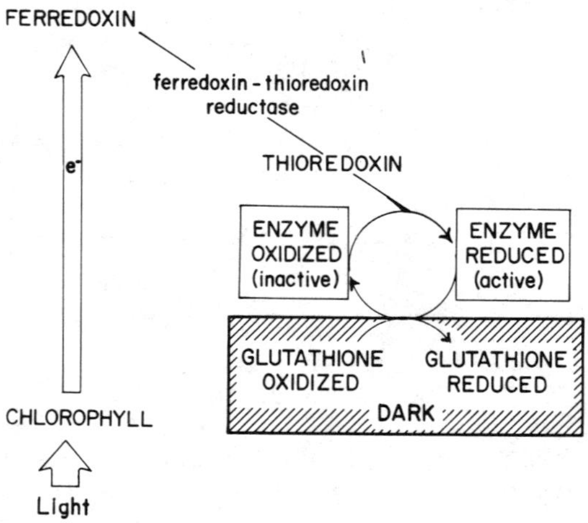

Fig. 2. A mechanism for light-mediated enzyme regulation in chloroplasts.

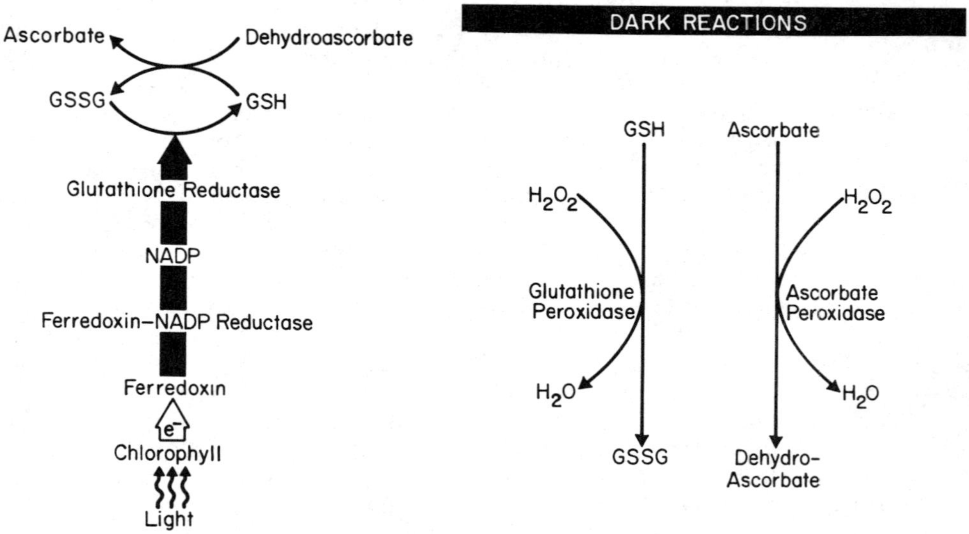

Fig. 3. Possible mechanisms for formation of soluble oxidants functional in the deactivation of thioredoxin-activated enzymes.

Table 5. Evidence for a chloroplast membrane-bound component functional in the dark deactivation of NADP-malate dehydrogenase activated by dithiothreitol-reduced thioredoxin.

Deactivation factor added	Enzyme activity (NADPH oxidized, nmol/min)
None	34
Oxidized glutathione	16
Chloroplast membranes (dark)	16
Chloroplast membranes (light)	33

and Badger, 1976). There is no indication that the activity of this enzyme is modulated by thioredoxin.

Multiple Forms of Thioredoxin in Leaves

The importance of thioredoxin in enzyme regulation in plants is emphasized by the recent finding that algae and leaves contain multiple forms of thioredoxin (Wagner et al., 1978; Buchanan et al., 1978; Jacquot et al., 1978; Wolosiuk et al., 1979; Jacquot et al., 1976). Two different thioredoxins have been found in spinach chloroplasts (thioredoxins f and m) and a third thioredoxin (thioredoxin c) was found outside chloroplasts, possibly in the cytoplasm (Buchanan et al., 1978; Jacquot et al., 1976). (Later work revealed that thioredoxin c consists of two different thioredoxins, designated c_f and c_m (Crawford et al., 1979)). The thioredoxins are distinguished by their stability to heat and their molecular weight (Tables 6 and 7) and by their enzyme specificity described below. A protein that appears to be the equivalent of thioredoxin m in maize has also been isolated in the laboratory of Hatch (Kagawa and Hatch, 1977). It remains to be seen which of the thioredoxin(s) function *in vivo* in the algal ribonucleotide reductase system recently described by Follman and his colleagues (Wagner and Follman, 1977). It is interesting to note that a thioredoxin that may have such a function was recently isolated from wheat embryos (Suske et al., 1979).

Thioredoxins f and m show specificity in their ability to activate enzymes of chloroplasts. Thioredoxin m is more effective in the activation of NADP-malate dehydrogenase, and thioredoxin f is more effective in the activation of the other enzymes tested, i.e., Fru-P_2 qse, phosphoribulokinase, NADP-glyceraldehyde

Table 6. Effect of heat on barley leaf thioredoxins.

Thioredoxin	% Activity retained in samples heated to 80 C for 4 min
Chloroplast	
Thioredoxin *f*	26
Thioredoxin *m*	113
Cytoplasmic	
Thioredoxin c_f	39
Thioredoxin c_m	100

Table 7. Molecular weight of leaf thioredoxins as determined by exclusion chromatography on Sephadex G-75.

	Molecular weight
Thioredoxin *f*	16,000
Thioredoxin *m*	9,000
Thioredoxin c_f	20,000
Thioredoxin c_m	15,000

3-phosphate dehydrogenase, and phenylalanine ammonia lyase (Table 8) (Nishizawa et al., 1979; Buchanan et al., 1978; Jacquot et al., 1978; Wolosiuk et al., 1979). Thioredoxins c_f and c_m selectively activate Fru-P_2ase and NADP-malate dehydrogenase, respectively, but neither their function nor their mechanism of reduction *in vivo* is known. It is noteworthy that an enzyme that reduces thioredoxin with NADPH as donor (NADP-thioredoxin reductase) was recently found in wheat embryo extracts (Suske et al., 1979). As this enzyme has not been found in photosynthetic cells it seems possible that the mechanism of thioredoxin reduction may differ in photosynthetic and nonphotosynthetic tissues: nonphotosynthetic systems would utilize NADPH and photosynthetic systems would utilize reduced ferredoxin.

Table 8. Effectiveness of dithiothreitol-reduced leaf thioredoxins in the activation of regulatory enzymes from chloroplasts.

	Relative activity					
	Fru-P_2ase	Sed-P_2ase	NADP-malate dehydrogenase	Phosphoribulokinase	NADP-glyceraldehyde-3-P dehydrogenase	Phenylalanine ammonia-lyase
Thioredoxin *f*	100	100	51	100	100	100
Thioredoxin *m*	16	18	100	43	40	25
-Thioredoxin	3	18	28	36	42	22

Thioredoxin and CF_1

As indicated above, there is evidence that thioredoxin has the capability of interacting not only with soluble enzymes but also with membranous proteins. Thioredoxin was shown to stimulate the dithiothreitol-dependent ATPase activity that is associated with chloroplast coupling factor (CF_1). Of the different thioredoxins tested, thioredoxim *m* was the most effective in promoting the dithiothreitol-linked activation of CF_1 ATPase preparations.

In an extension of this work, we addressed ourselves to the question of whether CF_1 itself contains a thioredoxin-like component that acts in the basal dithiothreitol-dependent ATPase activity that is observed in the absence of added thioredoxin (McKinney et al., 1979). We observed that when preparations of CF_1 were fractionated by defined procedures, thioredoxin activity was recovered in the fraction enriched in the δ subunit, the 20,000-dalton component that is envisioned to be essential for the attachment of CF_1 to the chloroplast membrane (Nelson, 1976; Binder te al., 1978). The isolated δ subunit could partially replace authentic thioredoxin in the dithiothreitol-linked activation of both chloroplast Fru-P_2 ase and CF_1 ATPase. The isolated ε subunit, by contrast, inhibited both of these reactions (cf. Nelson et al., 1973), whereas other CF_1 subunits had no effect. These experiments raise the interesting question of whether thioredoxin (or its phosphorylated derivative (Pigiet and Conley, 1978)) functions in energy transduction in chloroplasts or other membrane systems. Evidence that this might be the case was recently provided by experiments in which light, ferredoxin and ferredoxin-thioredoxin reductase were shown to increase the ATPase activity of CF_1 *in situ* (Mitchell and Schürmann, 1980). The evidence was consistent with the view that activation was effected by a membrane-bound thioredoxin or thioredoxin-like component.

Discovery of Ferralterin

Earlier this year, we reported the isolation of a soluble chloroplast protein that functions in enzyme regulation in the presence of illuminated chlorophyll-containing membranes (Lara et al., 1980). This protein (designated the "new protein factor") was found to activate chloroplast fructose 1,6-bisphosphatase ($Fru\text{-}P_2ase$) -- a regulatory enzyme of photosynthetic CO_2 assimilation -- in the absence of the ferredoxin/thioredoxin system, i.e., ferredoxin-thioredoxin reductase and a thioredoxin. We recently were able to improve the earlier-described purification procedure and to obtain homogeneous preparations of the new protein factor both from spinach leaves and from the cyanobacterium (blue-green alga) *Nostoc muscorum* (Lara et al., 1980). The new protein factor isolated from both sources is an iron-sulfur protein that, as far as we are aware, has not been described previously. Because of its chemical and functional resemblance to ferredoxin, we gave the protein the name "ferralterin" (alternate iron protein).

Activation of Chloroplast $Fru\text{-}P_2ase$ by Spinach and Nostoc Ferralterins

A characteristic of ferralterins from both spinach chloroplasts and *Nostoc* is the capability to activate photosynthetic $Fru\text{-}P_2ase$ in a reaction dependent on illuminated chlorophyll-containing membranes but not on soluble proteins such as ferredoxin or other components of the ferredoxin/thioredoxin system. Table 9 shows that ferralterin isolated from *Nostoc* and spinach chloroplasts effected the activation of homogeneous chloroplast $Fru\text{-}P_2ase$ in a reaction that required only chloroplast membranes, light and factors needed by the $Fru\text{-}P_2ase$ enzyme itself. Other experiments show that *Nostoc* membrane fragments can replace chloroplast membranes in the light activation of $Fru\text{-}P_2ase$ with each of the two ferralterins.

Some Chemical Properties of Ferralterin

The molecular weight of ferralterin has been determined by chromatography on calibrated Sephadex G_{100} columns and by electrophoresis in native polyacrylamide gels. The two methods gave similar values (average = $\sim$ 30,000) for the ferralterin preparations from both *Nostoc* and chloroplasts. It thus appears that ferralterin, like soluble ferredoxin, shows similar structure and activity in different photosynthetic organisms.

A feature consistently noted throughout the purification of ferralterin from both chloroplasts and _Nostoc_ is the association of activity with a brownish-yellow color. Absorption spectra measured with homogeneous ferralterin preparations revealed that

Table 9. Requirements for the light activation of chloroplast Fru-P_2ase mediated by homogeneous ferralterins.

Treatment	P_i released, nmol/min	
	Chloroplast ferralterin	_Nostoc muscorum_ ferralterin
Light		
Complete	136	86
Minus ferralterin	0	0
Minus Mg^{++}	0	0
Minus fructose 1,6-bisphosphate	0	0
Minus Fru-P_2ase	0	0
Minus chloroplast membranes	0	0
Dark		
Complete	0	0

the visible absorption is due to a broad peak at 410 nm that is accompanied by a weak shoulder in the near-ultraviolet region at 320 nm and by a peak in the ultraviolet region at 280 nm (Fig. 4). The ratios obtained for the most highly purified preparations were 0.34 and 0.32 for _Nostoc_ and chloroplast ferralterin, respectively.

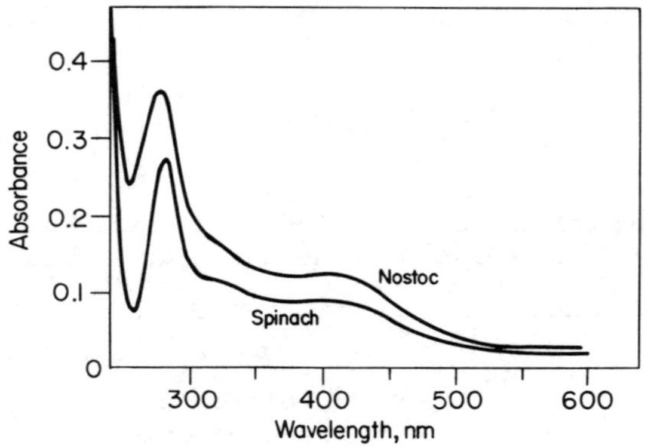

Fig. 4.

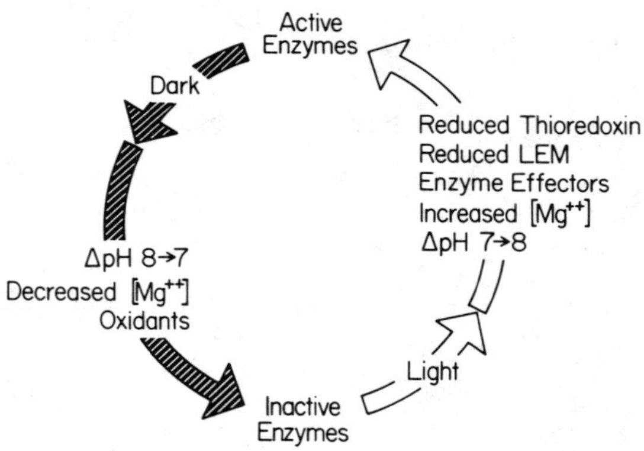

Fig. 5. Light-dependent mechanisms of enzyme regulation in chloroplasts.

The spectral characteristics displayed by ferralterin are similar to those of certain iron-sulfur proteins, e.g., nitrogenase proteins (Hardy and Burns, 1973) and bacterial ferredoxin (Malkin, 1973). Its spectral similarity to these proteins prompted us to analyze ferralterin for iron and acid labile sulfide (henceforth called sulfide). The results of the analyses were positive: ferralterin from chloroplasts and Nostoc contained both iron (measured by atomic absorption) and sulfide (measured colorimetrically). Furthermore, the iron and sulfide were present in equivalent amounts (respective maximum values of 120 and 116 nmmole of iron and sulfide per mg protein). On the basis of a molecular weight of 30,000, this value corresponds to a nearest integer number of 4 g-atoms each of iron and sulfide per mole of ferralterin. Analyses carried out with α,α-dipyridyl, a reagent specific for nonheme iron, have demonstrated that the iron is of nonheme type (Brill et al., 1974).

DISCUSSION

The ferredoxin/thioredoxin system appears to constitute a general mechanism of enzyme regulation that operates in conjunction with other light-actuated regulatory systems in chloroplasts (Fig. 5). One of the mechanisms by which light governs enzyme activity independently of thioredoxin is via specific enzyme effectors whose synthesis is increased in the light (e.g., NADPH, ATP (Miller et al., Pupillo and Giuliani-Piccari, 1975)). It was proposed that, by analogy to energy charge, a reductant charge (ratio of NADPH to NADP) may be operative in the light/dark control of the activity of certain chloroplast enzymes (Lendzian and Bassham, 1975; Wildner, 1975).

Light-induced ion shifts also appear to be important factors in insuring that regulatory enzymes of chloroplasts are fully active only in the light (Heldt et al., 1978). In particular, the light-induced shift in stromal concentrations of H^+ (pH $7_{dark} \rightarrow$ pH 8_{light}) and Mg^{++} (increase of 1 to 3 mM in the light) seem to be important in providing the environment necessary for the optimal activity of certain enzymes. Two of the thioredoxin-linked enzymes (Fru-P_2ase and Sed-P_2ase) seem to be especially sensitive to light-induced changes in H^+ and Mg^{++} concentration as well as to changes in redox state (Heldt et al., 1978).

Finally, there is evidence that membrane-bound reductants may function in enzyme regulation in chloroplasts. It is envisaged that these reductants (designated light-effect mediators, or LEMs) are disulfide components that interact with the electron transport chain or chloroplasts (Anderson and Avron, 1976). The LEM components would occur mainly in the oxidized (disulfide) state in the dark and in the reduced (sulfhydryl) state in the light. The relation of the LEM system to the ferredoxin-thioredoxin system remains to be established. It might be pointed out, nevertheless, that the above-mentioned membrane-bound component functional in the deactivation of thioredoxin-activated NADP-malate dehydrogenase appears to be analogous to LEM, with the exception that it acts in the dark rather than in the light.

It is also noteworthy that recent results suggest that a soluble protein is required for light modulation with the LEM system that is similar, if not identical, to ferralterin (Ashton and Anderson, 1979; Ashton and Anderson, 1980; Anderson, 1980).

The finding that ferralterin can function as a regulatory protein independently of the components of the ferredoxin/thioredoxin system adds a new dimension to our understanding of the way light governs enzyme activity during photosynthesis. The activity of chloroplast Fru-P_2ase appears to be controlled photochemically by ferralterin as well as by the ferredoxin/thioredoxin system. Significantly, the changes induced in Fru-P_2ase by the two regulatory mechanisms seem to be similar in that the activated enzyme in both cases depends on an oxidant, e.g., dehydroascorbate, for deactivation. It remains to be seen whether the activity of photosynthetic enzymes other than Fru-P_2ase is regulated by ferralterin.

It should be noted that, despite their overall functional similarities, the ferralterin and ferredoxin/thioredoxin mechanisms effect enzyme activation via different routs. The ferredoxin/thioredoxin system requires three proteins to achieve its regulatory effects (soluble ferredoxin to receive electrons from the photosynthetic apparatus and ferredoxin-thioredoxin reductase to transfer electrons from ferredoxin to a thioredoxin, in this case thioredoxin *f*), whereas ferralterin can apparently interact with chlorophyllous membranes directly and activate an enzyme independently of other soluble proteins. The mechanism by which ferralterin interacts with photosynthetic membranes and thereupon receives and transmits the membrane-based regulatory signal remains to be determined.

A question that is also raised by recent studies on light modulation of chloroplast enzymes is whether certain enzymes might behave in a manner opposite to that described above and undergo an oxidant-induced activation. Evidence for this possibility is provided by our recent finding that nonspecific acid phosphatases from spinach leaves and potato tubers were activated by oxidized glutathione or dehydroasorbate (Buchanan et al., 1979). Activation was accompanied by a change in the pH optimum so that the enzymes became active in the neutral as well as in the acid region. At neutral pH, the activation induced by oxidized glutathione was reversed by reduced gluthione. The results suggest that oxidized glutathione and dehydroascorbate serve not only in the previously demonstrated deactivation of thioredoxin-linked enzymes but also in the activation of other enzymes from plants (cf. Morens et al., 1972; Haddox et al., 1978). Recent results by Scheibe in Anderson's laboratory indicate that oxidized thioredoxin itself also functions in enzyme activation, i.e., in the activation of chloroplast glucose 6-phosphate dehydrogenase that was previously deactivated by light (Scheibe, 1980).

In view of the central regulatory role that thioredoxin seems to play in chloroplasts, the question arises as to whether thioredoxin might have an as yet unknown regulatory function in heterotrophic cells. The presence of ferredoxin is not prerequisite to such a function because, as noted above, cells of both animal and bacterial origin are enzymatically fitted to reduce thioredoxin independently of ferredoxin, via an NADP-linked thioredoxin reductase (Reichard, 1968). It is possible that thioredoxin-dependent enzyme regulation in heterotrophic cells could be executed, for example, either by an NADP-linked mechanism, for which the thioredoxin-linked reduction of insulin provides a model (Holmgren, 1977) or by the formation of a stable thioredoxin enzyme complex, as found for the *E. coli* DNA polymerase described above. Whether these or other thioredoxin-linked mechanisms act in the regulation of enzymes of heterotrophic cells is one of the exciting problems for the future. An equally exciting problem concerns the distribution and possible function of ferralterin in nonphotosynthetic cells.

SUMMARY

Thioredoxin, a hydrogen carrier protein that functions in the synthesis and replication of DNA and in the transformation of sulfur metabolites, has been found to serve as a regulatory protein in linking light to the activation of enzymes during photosynthesis. In this system, thioredoxin is reduced photochemically via ferredoxin and the enzyme ferredoxin-thioredoxin reductase. The enzymes activated by this mechanism (designated the ferredoxin/thioredoxin system) include four enzymes of the reductive pentose phosphate cycle of CO_2 assimilation (fructose 1,6-bisphosphatase, sedoheptulose 1,7-bisphosphatase, NADP-glyceraldehyde 3-phosphate dehydrogenase, and phosphoribulokinase) as well as three enzymes not associated with the cycle (NADP-malate dehydrogenase, phenylalanine ammonia lyase, and PAPS sulfotransferase). Two different enzyme-specific thioredoxins (thioredoxins *f* and *m*) function in chloroplasts as part of the ferredoxin/thioredoxin system. Two other thioredoxins of unknown function (thioredoxins c_f and c_m) reside outside of the chloroplasts, possibly in the cytoplasm.

The enzymes photochemically activated by thioredoxin appear to be deactivated in the dark by a mechanism that differs as to the enzyme. The mechanisms include deactivation by 1) a soluble oxidant such as oxidized glutathione or dehydroascorbate; 2) an unidentified membrane-bound oxidant; and 3) an unknown chloroplast reaction. Each of the enzymes regulated by the ferredoxin/thioredoxin system shows hysteretic properties, i.e., its rate of activation or deactivation is slow relative to the rate of catalysis.

In addition to its role in light-mediated enzyme regulation, thioredoxin increases the dithiothreitol-dependent ATPase activity of heated chloroplast coupling factor (CF_1). The δ subunit fraction isolated from purified CF_1, which shows thioredoxin f activity, may replace thioredoxin in this activation. Recent experiments show that the ATPase activity of CF_1 *in situ* is also activated by light via ferredoxin-thioredoxin reductase and, presumably, a bound thioredoxin or thioredoxin-like component.

ACKNOWLEDGMENTS

The work from my own laboratory described herein was aided by grants from the National Science Foundation.

REFERENCES

Anderson, L. E., 1980, Light modulation of enzyme activity in green plants: the LEM system, *Federation Proc.*, 39:1931.

Anderson, L. E., and Avron, M., 1976, Light modulation of enzyme activity in chloroplasts, *Plant Physiol.*, 57:209.

Ashton, A. R., and Anderson, L. E., 1979, Reconstitution of the light activation system for pea leaf chloroplast malate dehydrogenase, *Plant Physiol.*, 63, suppl. 24.

Ashton, A. R., and Anderson, L. E., 1980, Resolution and reconstitution of the light-dependent modulation system of chloroplast enzymes, Abstracts, 5th Intl. Congr. Photosynthesis, Halkidiki, Greece.

Binder, A., Jagendorf, A., and Ngo, E., 1978, Isolation and composition of the subunits of spinach chloroplast coupling factor protein, *J. Biol. Chem.*, 253:3094.

Black, S., Harte, E. M., Hudson, B., and Wartofsky, L., 1960, A specific enzymatic reduction of L(-) methionine sulfoxide and a related nonspecific reduction of disulfides, *J. Biol. Chem.*, 235:2910.

Breazeale, V. D., Buchanan, B. B., and Wolosiuk, R. A., 1978, Chloroplast sedoheptulase 1,7-biphosphatase: Evidence for regulation by the ferredoxin/thioredoxin system, *Zeit. Naturforsch.* 33c:521.

Brill, W. J., Westphal, J., Stieghorst, M., Davis, L. C., and Shah, V. K., 1974, Detection of nitrogenase components and other nonheme iron proteins in polyacrylamide gels, Anal. Biochem., 60:237.

Buchanan, B. B., Kalberer, P. P., and Arnon, D. E., 1967, Ferredoxin-activated fructose diphosphatase in isolated chloroplasts, Biochem. Biophys. Res. Commun., 29:74.

Buchanan, B. B., Schürmann, P., and Kalberer, P. P., 1971, Ferredoxin-activated fructose diphosphatase of spinach chloroplasts, J. Biol. Chem., 246:5952.

Buchanan, B. B., and Wolosiuk, R. A., 1976, Photosynthetic regulatory protein found in animal and bacterial cells, Nature, 264:669.

Buchanan, B. B., Wolosiuk, R. A., Crawford, N. A., and Yee, B. C., 1978, Evidence for three thioredoxins in leaves, Plant Physiol., 61:38S.

Buchanan, B. B., Crawford, N. A., and Wolosiuk, R. A., 1979, Activation of plant acid phosphatases by oxidized glutathionine and dehydroascorbate, Plant Sci. Lett., 14:245.

Crawford, N. A., Yee, B. C., Nishizawa, A. N., and Buchanan, B. B., 1979, Occurrence of cytoplasmic f- and m-type thioredoxins in leaves, FEBS Lett., 104:141.

Frieden, C., 1971, Protein-protein interaction and enzymatic activity, Ann. Rev. Biochem., 40:653.

Groden, D., and Beck, E., 1977, Characterization of a membrane bound, ascorbate specific peroxidase from spinach chloroplasts, Abstracts, 4th Int. Cong. Photosynthesis, Reading, U.K.

Haddox, M. K., Stephenson, J. H., Moser, M. E., and Goldberg, N. D., 1978, Oxidative-reductive modulation of Guinea pig splenic cell cyclase activity, J. Biol. Chem., 253:3143.

Hardy, R. W. F., and Burns, C., 1973, Comparative biochemistry of iron-sulfur proteins and dinitrogen fixation, in: "Iron-Sulfur Proteins," W. Lovenberg, ed., Vol. 1, Academic Press, NY.

Heldt, H. W., Chon, C. J., Lilley, R. McC., and Portis, A., 1978, The role of fructose- and sedoheptulose bisphosphatase in the control of CO_2 fixation. Evidence from the effects of Mg^{++} concentration, pH and H_2O_2, Proc. 4th Int. Congr. Photosynthesis, The Biochem. Soc., London.

Holmgren, A., Buchanan, B. B., and Wolosuik, R. A., 1977, Different receptors for somatostatin and opioids in neuroblastoma x glioma hybrid cells, FEBS Lett., 82:351.

Holmgren, A., 1977, Bovine thioredoxin system, J. Biol. Chem., 252:4600.

Jacquot, J. P., Vidal, J., and Gadal, P., 1976, Identification of a protein factor involved in dithiothreitol activation of NADP malate dehydrogenase from French bean leaves, FEBS Lett., 71:223.

Jacquot, J. P., Vidal, J., Gadal, P., and Schürmann, P., 1978, Evidence for the existence of several enzyme specific thioredoxins in plants, FEBS Lett., 96:243.

Kagawa, T., and Hatch, M. D., 1977, Regulation of C_4 photosynthesis: characterization of a protein factor mediating the activation

and inactivation of NADP-malate dehydrogenase, Arch. Biochem. Biophys., 184:290.
Lara, C., de la Torre, A., and Buchanan, B. B., 1980, A new protein factor functional in the ferredoxin-independent light activation of chloroplast fructose 1,6-biphosphatase, Biochem. Biophys. Res. Commun., 93:544.
Lara, C., de la Torre, A., and Buchanan, B. B., 1980, Ferraltrin: an iron-sulfur protein functional in enzyme regulation in photosynthesis, Biochem. Biophys. Res. Commun., 94:1337.
Laurent, T. C., Moore, E. C., and Reichard, P., 1964, Enzymatic synthesis of deoxyribonucleotides, J. Biol. Chem., 239:3436.
Lendzian, K., and Bassham, J. A., 1975, Regulation of glucose-6-phosphate dehydrogenase in spinach chloroplasts by ribulose 1,5-diphosphate NADPH/NADP$^+$ ratios, Biochem. Biophys. Acta, 396:260.
Lorimer, G. H., Badger, M. R., and Andrews, T. J., 1976, The activation of ribulose-1,5-bisphosphate carboxylase by carbon dioxide and magnesium ions. Equilibria, kinetics, a suggested mechanism and physiological implications, Biochemistry, 15:529.
Malkin, R., 1973, The chemical properties of ferredoxins, in: "Iron-Sulfur Proteins," W. Lovenberg, ed., Vol. 2, Academic Press, New York.
Mark, D. F., and Richardson, C. C., 1976, Escherichia coli thioredoxin: a subunit of bacteriophage T7 DNA polymerase, Proc. Natl. Acad. Sci. USA, 73:780.
McKinney, D. W., Buchanan, B. B., and Wolosiuk, R. A., 1978, Activation of chloroplast ATPase by reduced thioredoxin, Phytochemistry, 17:794.
McKinney, D. W., Buchanan, B. B., and Wolosiuk, R. A., 1979, Association of a thioredoxin-like protein with chloroplast coupling factor (CF_1), Biochem. Biophys. Res. Commun., 86:1178.
Mills, J. D., Mitchell, P., and Schürmann, P., 1980, Modulation of coupling factor ATPase activity in intact chloroplasts, FEBS Lett., 112:173.
Moreno, C. G., Aparacio, P. J., Palacian, E., and Losada, M., 1972, Association of a Thioredoxin-like protein with chloroplast coupling factor (CF_1), FEBS Lett., 26:11.
Müller, B., Ziegler, I., and Ziegler, H., 1969, Lichtinduzierté, reversible aktivitatssteigerung der NADP-abhangigen glycerin-aldehyde-3-phosphat-dehydrogenase in chloroplasta, Eur. J. Biochem., 9:101.
Nelson, N., Deters, H., Nelson, D. W., and Racker, E., 1973, Partial resolution of the enzymes catalyzing photophosphorylation, J. Biol. Chem., 248:2049.
Nelson, N., 1976, Structure and function of chloroplast ATPase, Biochem. Biophys. Acta, 456:314.
Nishizawa, A. N., Wolosiuk, R. A., and Buchanan, B. B., 1979, Chloroplast phenylalanine ammonia-lyase from spinach leaves, Planta, 145:7.
Petersen, T. A., Kirk, M., and Bassham, J. A., 1966, Light-dark

transients in levels of intermediate compounds during photosynthesis in air-adapted Chlorella, Physiol. Plant., 19:219.
Pigiet, V., and Conley, R. R., 1978, Isolation and characterization of phosphothioredoxin from Escherichia coli, J. Biol. Chem., 253:1910.
Porque, P. G., Baldesten, A., and Reichard, J. P., 1970, Purification of a thioredoxin system from yeast, J. Biol. Chem., 245:2363.
Pupillo, P., and Giuliani-Piccari, G. G., 1975, The reversible depolymerization of spinach chloroplast glyceraldehyde-phosphate dehydrogenase, Eur. J. Biochem., 51:475.
Reichard, P., 1968, The biosynthesis of deoxyribonucleotides, Eur. J. Biochem., 3:259.
Scheibe, R., 1980, On the dark modulation of chloroplast enzymes, Federation Proc., 39:1771.
Schürmann, P., and Buchanan, B. B., 1975, Role of ferredoxin in the activation of sedoheptulose diphosphatase in isolated chloroplasts, Biochem. Biophys. Acta, 376:189.
Schürmann, P., Wolosiuk, R. A., Breazeale, V. D., and Buchanan, B. B., 1976, Two proteins function in the regulation of photosynthetic CO_2 assimilation in chloroplasts, Nature, 263:257.
Suske, G., Wagner, W., and Follmann, H., NADPH-dependent thioredoxin reductase and a new thioredoxin from wheat, Zeit. Naturforsch., 34:214.
Wagner, W., and Follmann, H., 1977, A thioredoxin from green algae, Biochem. Biophys. Res. Commun., 77:1044.
Wagner, W., Follmann, H., and Schmidt, A., 1978, Multiple functions of thioredoxins, Zeit. Naturforsch., 33c:517.
Wilson, L. G., Asahi, T., and Bandurski, R. S., 1971, Yeast sulfate-reducing system, J. Biol. Chem., 236:1822.
Wildner, G. F., 1975, The regulation of glucose-6-phosphate dehydrogenase in chloroplasts, Zeit. Naturforsch., 30c:756.
Wolosiuk, R. A., Crawford, N. A., Yee, B. C., and Buchanan, B. B., 1979, Isolation of three thioredoxins from spinach leaves, J. Biol. Chem., 254:1627.
Wolosiuk, R. A., and Buchanan, B. B., 1977, Thioredoxin and glutathione regulate photosynthesis in chloroplasts, Nature, 266:565.
Wolosiuk, R. A., and Buchanan, B. B., 1978, Regulation of chloroplast phosphoribulokinase by the ferredoxin/thioredoxin system, Arch. Biochem. Biophys., 189:97.
Wolosiuk, R. A., Buchanan, B. B., and Crawford, N. A., 1977, Regulation of NADP-malate dehydrogenase by the light-activated ferredoxin/thioredoxin system of chloroplasts, FEBS Lett., 81:253.
Wolosiuk, R. A., and Buchanan, B. B., 1978, Activation of chloroplast NADP-linked glyceraldehyde-3-phosphate dehydrogenase by the ferredoxin/thioredoxin system, Plant Physiol., 61:669.
Ziegler, H., and Ziegler, I., 1965, Der einfluss der belichtung auf die $NADP^+$-abhängige glycerinaldehyd-3-phosphat-dehydrogenase, Planta (Berlin), 65:369.

REGULATION OF PHOTOSYNTHETIC CO_2 FIXATION

James T. Bahr and James J. Steffens

Crop Chemicals R&D
Mobil Chemical Company
P. O. Box 240
Edison, NJ 08817

INTRODUCTION

In any consideration of the rate and efficiency of nitrogen utilization by plants, it is obvious that attention must be paid to photosynthetic CO_2 fixation. Photosynthetic CO_2 fixation is both complementary to and competitive with nitrogen metabolism. Photosynthetic electron transport provides the energy for nitrogen metabolism, directly or indirectly; and CO_2 fixation provides the carbon skeletons for amino acid biosynthesis. On the other hand, CO_2 fixation is directly competitive with nitrogen metabolism, especially nitrite reduction and amino acid synthesis, for photosynthetic reducing equivalents and ATP.

In a biochemical system as complex as photosynthesis, the possibilities for regulatory interactions are numerous and complex. A variety of regulatory processes and schemes have been described in the photosynthesis literature over the last ten years. These can be grouped together as follows:

1) Interactions between CO_2 fixation and energy supply (ATP:ADP ratios, stromal pH, ion transport)

2) Changes in enzyme activation (RuBP carboxylase, fructose bisphosphatase and others)

3) Effects of changes in O_2:CO_2 ratios (RuBP carboxylase-oxygenase, O_2 reduction by PS I)

4) Phosphate - 3-P-glycerate effects on starch metabolism

5) Chloroplast stroma - cytoplasmic interactions (phosphate translocator, dicarboxylic acid shuttles, ion movements)

The first two groups might be expected to primarily regulate the overall rate of CO_2 fixation. The last three groups also regulate the distribution of photosynthetic carbon into end products, starch and sucrose. In this paper we wish to focus on regulatory processes which influence primarily the rate of photosynthetic CO_2 fixation.

We will briefly examine the role of changes in the degree of activation of RuBP carboxylase and fructose bisphosphatase in regulating CO_2 fixation. RuBP carboxylase can exist in several forms, only one of which is able to catalyze the carboxylation and oxygenation of RuBP. Fructose bisphosphatase also exists in two forms, governed by an interactions with thioredoxin. These forms differ in their pH dependence, and in other ways, such that at expected stomal pH values one form may be considered active and the other inactive.

CONTROL OF CO_2 FIXATION

Under steady state conditions the rate of photosynthetic CO_2 fixation is the rate of RuBP carboxylation, the rate of catalysis by RuBP carboxylase. At fixed O_2 and CO_2 levels, that rate should be determined by the availability of RuBP, the amount of active RuBP carboxylase and the concentration of compounds which influence carboxylase catalysis (H^+, Mg^{++}, competitive inhibitors). Three regulatory models can therefore be developed.

In the first model, the steady state RuBP level is such that CO_2 fixation is rate-saturated with respect to RuBP. Since carboxylase catalytic sites are present at several millimolar in the stroma, the _in vitro_ Km (RuBP) is irrelevant for estimating whether RuBP is saturating. In this model the CO_2 fixation rate will be determined by the amount of RuBP carboxylase in the catalytically active form.

In the second model, RuBP is not at a rate-saturating concentration and regulation of the overall rate is primarily a function of other enzymes of the cycle via their effect of the RuBP steady state level. The other enzymes most likely to be involved would be those regulated by the thioredoxin system, especially fructose bisphosphatase and phosphoribulokinase.

In the third model, RuBP is again at a rate-saturating concentration. Rather than the amount of active RuBP carboxylase being rate-determining, it is one or more factors affecting carboxylase catalysis which controls the rate; e.g., pH in the stroma.

In a biochemical system such as the Calvin cycle there need not be a single rate-determining step or process. Most studies of regulation in isolated chloroplasts are conducted for convenience at saturating CO_2, yet _in vivo_ in C-3 plants it is clearly CO_2 which limits the rate. Both CO_2 and other processes can limit the rate simultaneously. As will be discussed below, intact chloroplasts doing CO_2 fixation in a medium without added catalase or dithioerythritol do so at a reduced rate due to inhibition of FBPase activation. Photosynthesis can be further reduced by decreasing the CO_2 concentration (Table 1). The rate of photosynthesis by isolated chloroplasts incubated at low CO_2 and without catalase can be stimulated by either increasing CO_2 or by adding catalase. Each rate limitation in this case can be overcome separately, as might be expected from the kinetics of RuBP carboxylase, since RuPB and CO_2 do not influence each others' interaction with the catalytic site (Jensen and Bahr, 1977).

Table 1. Multiple rate-determining steps in photosynthetic CO_2 fixation. Chloroplasts (7.8 μg Chl) were incubated in 500 μl of medium containing 0.3 M soribtol, 0.05 M Hepes, 2 mM EDTA, 1 mM $MgCl_2$, 1 mM $MnCl_2$, 0.5 mM K_2HPO_4, pH 7.7 at 300 $\mu E/m^2/sec$ and other conditions as indicated.

Conditions	Photosynthesis: 7.6 mM $NaHCO_3$	μmoles $(mg\ Chl\text{-}h)^{-1}$ 0.38 mM $NaHCO_3$
No added catalase	31	9
1600 units catalase added	98	43

ACTIVATION OF RUBP CARBOXYLASE

In order to assess the roles of factors other than CO_2 in limiting photosynthesis, we have developed an assay for the active form of RuBP carboxylase in isolated chloroplasts. Activation of RuBP carboxylase involves formation of an enzyme-CO_2-Mg^{++} complex whose formation kinetics have been described (Lorimer et al., 1976). The CO_2 involved in activation is separate from that involved in the reaction (Miziorko, 1979; Lorimer, 1979). Under normal assay conditions containing 5-25 mM Mg^{++} the E-CO_2-Mg^{++} complex and any pre-existing E-CO_2 complex present contribute to the observed rate of carboxylation (Table 2). The activity of E-CO_2 is due to its conversion to E-CO_2-Mg^{++} in the assay medium. In the absence of Mg^{++} in the assay, however, E-CO_2 does not contribute to the rate. Under the Mg^{++} free conditions (excess EDTA) fully activated RuBP carboxylase remains active at about 65 percent of normal (Fig. 1). Provided the observed rate is corrected for the effect of Mg^{++} on the rate of catalysis by E-CO_2-Mg^{++}, assays carried out without Mg^{++} permit measurement of E-CO_2-Mg^{++} activity only. If such assays are conducted by lysing chloroplasts into the assay medium and are terminated at 30 sec, changes in degree of carboxylase activation during the assay will be insignificant.

The corrected activity of E-CO_2-Mg^{++} is a measure of the maximal possible rate of CO_2 fixation by the chloroplast preparation, since the assay is carried out at rate-saturating RuBP and CO_2 concentrations and at the pH optimal for RuBP carboxylase. Only two factors inside of the chloroplast can be expected to reduce the fixation rate to a value lower than the observed ocrrected carboxylase rate: sub-saturating RuBP concentrations or readily dissociable effectors of catalysis, including H^+ (stromal pH).

Table 2. Effect of Mg^{++} on apparent activity of RuBP carboxylase. Activated RuBP carboxylase was formed by incubating enzyme at 16 mM $NaHCO_3$ and 20 mM $MgCl_2$ for 6 minutes at pH 8.1 (0.2 M bicine). The E-CO_2 complex was formed by incubating enzyme 10 minutes at 94 mM $NaHCO_3$ and 1 mM EDTA at pH 8.1. Activity was measured at pH 8.1 (0.6 mM RuBP, 15 mM $NaH^{14}CO_3$, 0.05 M bicine) by transferring a 20 μl aliquot of enzyme solution at 550 μl assay medium and terminating reaction at 30 sec.

Enzyme form.	Observed activity: μmoles (mg prot.-min)$^{-1}$		
	22 mM $MgCl_2$	0.3 mM EDTA	1.0 mM EDTA
E.CO_2	0.13	0.002	-
E.CO_2-Mg^{++}	1.5	-	0.98

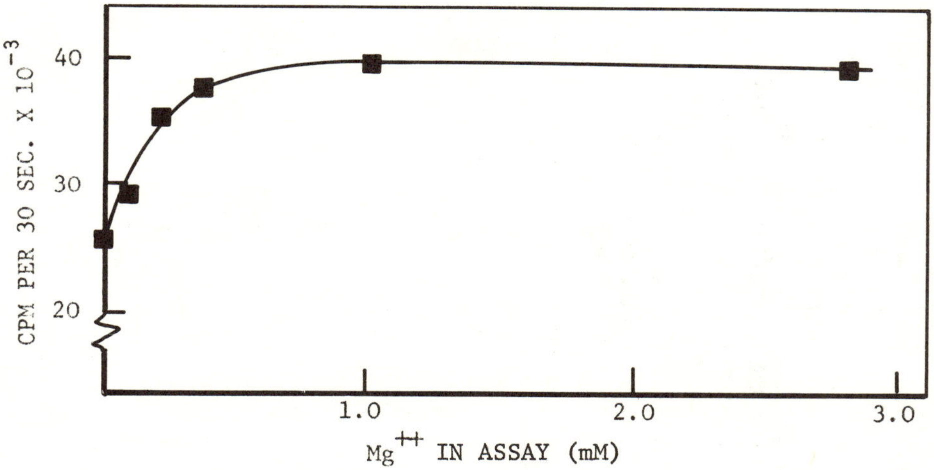

Fig. 1. Mg^{++} dependence of catalysis by active RuBP carboxylase. Purified RuBP carboxylase was activated at 16 mM $NaHCO_3$ and 20 mM $MgCl_2$ for 5 min at pH 8.1 (0.2 M bicine). An aliquot was diluted 200-fold into assay media at various final Mg^{++} concentrations determined by addition of EDTA or $MgCl_2$. Assay media contained 0.3 mM RuBP, 0.05 M bicine, 13 mM $NaH^{14}CO_3$, at pH 8.1. Assay duration was 30 sec.

The activation state of RuBP carboxylase has been measured in isolated chloroplasts and is presented in Table 3. The activation state increases upon illumination, although it does not decrease significantly upon re-darkening. From 20 to 50 percent (average 32 percent) of the total RuBP carboxylase present is in the active form. A substantial amount is also present as the E-CO_2 complex. Recently this assay has been extended to illuminated thin slices of leaf tissue, with qualitatively similar results.

For most of these chloroplast preparations the rate of CO_2 fixation at saturating CO_2 was also measured. Comparison of the rate of fixation with the catalytic capability of the active RuBP carboxylase (Table 4), allows calculation of the degree of utilization of the carboxylation potential of the chloroplasts. The degree of utilization varied from a low of 43 percent to a high of 81 percent (average value 64 percent). Most values were between 60 and 70 percent.

Table 3. Activation of RuBP carboxylase in isolated chloroplasts. Chloroplasts were incubated in the medium of Table 1 at 7.6 mM $NaHCO_3$ and 300 $\mu E/m^2/sec$, or taken directly from storage at 0 C in the dark. An appropriate aliquot of chloroplasts was transferred to RuBP carboxylase assay medium containing excess EDTA to measure active carboxylase only, or 25 mM $MgCl_2$ to measure active carboxylase plus $E.CO_2$. Total carboxylase was measured by activating at high $MgCl_2$ and CO_2 in vitro prior to initiating the assay with RuBP. Carboxylase activity is in moles (mg Chl.h)$^{-1}$ corrected for effect of Mg^{++}.

Chloroplast preparation	As prepared			Illuminated at sat. CO_2 for 6 min			
	Active enzyme	Total enzyme	Percent active	Active enzyme	Total enzyme	Percent Active	$E.CO_2$
1	39	305	13	78	347	22	148
2	27	192	14	81	294	28	78
3	--	--	--	71	162	44	80
4	30	300	10	--	--	--	--
5	30	250	12	78	222	35	71
6	36	363	10	77	405	19	130
7	56	600	9	168	--	--	--
8	52	288,366	18,15	288	582	49	155
9	25	396,348	6,7	116	324,468	36,25	182

Table 4. Relationship of active RuBP carboxylase and CO_2 fixation in isolated chloroplasts. Active RuBP carboxylase assayed with excess EDTA at saturating RuBP and CO_2. Photosynthesis measured on same preparation of saturating CO_2 and light intensity in the presence of catalase.

Preparation	Age (min)	Intactness (%)	Observed CO_2 fixation µmoles $(mg\ Chl \cdot h)^{-1}$	Active RuBP CO_2ase µmoles $(mg\ Chl \cdot h)^{-1}$	Ratio
1	-	66	101 ± 4	148 ± 7	0.68
2	-	73	165 ± 8	204 ± 7	0.81
3	60	87	164 ± 9	246 ± 6	0.67
4	40	-	91 ± 2	157 ± 4	0.58
5	60	-	170 ± 3	257 ± 4	0.66
6	150	79	81 ± 3	189 ± 10	0.43
7	30	82	141 ± 3	217 ± 7	0.65
8	45	70	140 ± 7	214 ± 3	0.65

On the basis of these data, the active carboxylase in isolated chloroplasts is not fully utilized; either because RuBP is not rate-saturating, or because soluble effectors have reduced the catalytic rate. Rate limitations might be activation of fructose bisphosphattase, phosphoribulokinase or stomal pH. Activation of RuBP carboxylase is not a rate-limiting process in isolated chloroplasts.

ACTIVATION OF FRUCTOSE-1,6-BISPHOSPHATASE

This conclusion suggested that investigation of factors controlling the thioredoxin-regulated enzymes would be useful. Fructose 1,6-bisphosphatase was selected in light of its lower over-all activity in isolated chloroplasts and convenient assay. Qualitatively, similar conclusions should apply to phosphoribulokinase.

With proper selection of assay pH the degree of activation of fructose bisphosphatase in intact chloroplasts can be assayed (Fig. 2). The "inactive" form has a pH optimum more alkaline than the "active" form at the Mg^{++} and FBP concentrations chosen. These relationships were described a few years ago by Baier and Latzko (1975). Fructose bisphosphatase undergoes a large increase in activation upon illumination of isolated chloroplasts and a similar decrease in activation upon darkening (Fig. 3).

Changes in activation of fructose bisphosphatase have been implicated in regulation of CO_2 fixation by several factors, including H_2O_2 and nitrite (Hiller and Bassham, 1965; Heldt et al., 1978), on the basis of changes in steady state concentrations of fructose 1,6-bisphosphate and fructose-6-phosphate. The results of Table 5 show changes in FBPase activation when catalase is omitted and when nitrite or methyl viologen are added to the illuminated chloroplast suspension.

The effect of methyl viologen is relatively easy to understand. It is an acceptor of reducing equivalents from photosystem I and can apparently compete effectively with thioredoxin. The response of omitting catalase is less clear. At first we thought that the build-up or H_2O_2 might lead to reoxidation of fructose bisphosphtase SH groups reduced by thioredoxin. However, H_2O_2 does not directly inactivate FBPase <u>in vitro</u> and it does not accumulate in chloroplast suspensions in the absence of catalase. Rather, H_2O_2 serves as an acceptor of electrons from photosystem I, probably via a soluble low molecular weight donor and a peroxidase. In intact chloroplast suspensions H_2O_2 supports net O_2 evolution in the light under CO_2-free conditions. Light is required, indicating that catalase is not responsible for the O_2 evolution. Inhibition is observed with DCMU and a number of other compounds which suggest involvement of a hemeprotein peroxidase. No O_2 evolution is observed if the chloroplasts are broken prior to measurement (Fig. 4). Omission of catalase forces intact

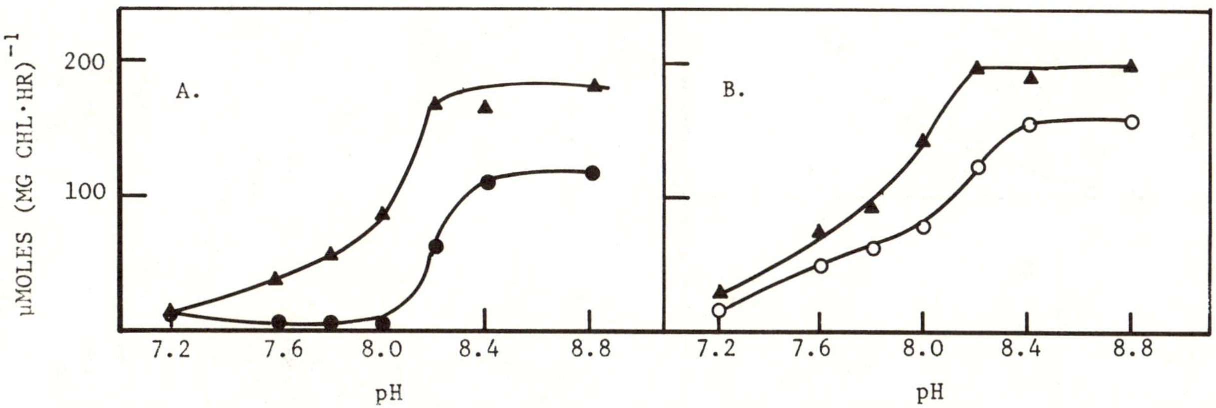

Fig. 2. pH dependence of chloroplast fructose bisphosphatase. Assay for fructose bisphosphatase contained 0.05 M Hepes, 0.05 M bicine, 7 mM $MgCl_2$, 2 mM EDTA, 1.0 mM fructose-1,6-bisphosphate, 0.1 mM reduced glutathione, 1.0 units glucose 6-phosphate dehydrogenase, 2.5 units phosphoglucoisomerase, 0.20 mM NADP, at pH as indicated. Chloroplasts (30-40 µg Chl) were incubated in 500 µl of medium of Table 1 for 5 min in light or dark and an aliquot added to assay for FBPase. A) (▲) illuminated chloroplasts with 5 mM dithioerythritol in FBPase assay; (●) dark-adapted chloroplasts. B) (▲) same as in A); (o) illuminated chloroplasts.

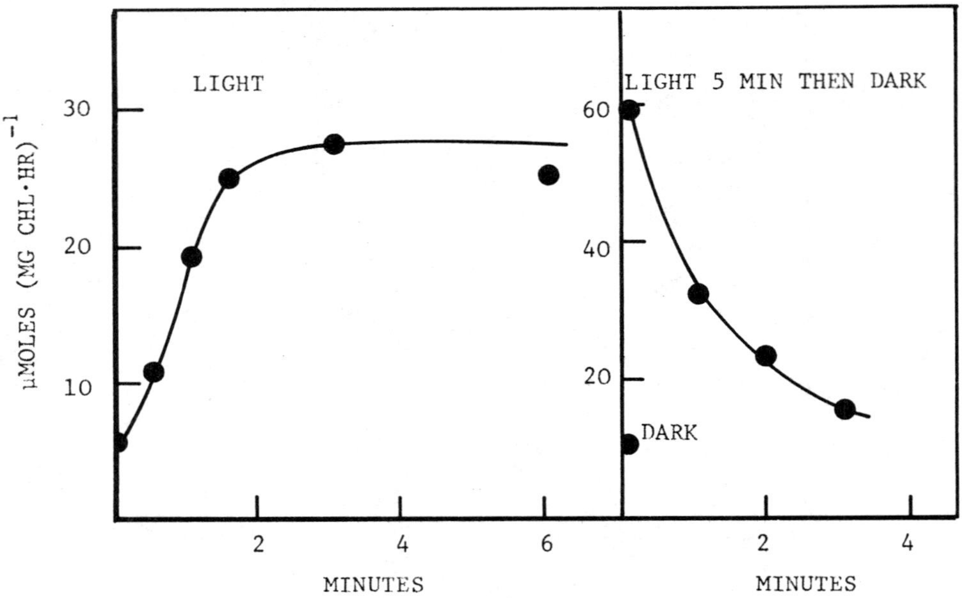

Fig. 3. Reversible activation of chloroplast fructose bisphosphatase. Chloroplasts (30 μg Chl) were incubated in medium of Table 1 with catalase present for indicated times in light or dark, and an aliquot transferred to FBPase assay at pH 8.0.

chloroplasts to reductively metabolize H_2O_2 at the expense of thioredoxin and FBPase activation. Steady state RuBP levels are reduced (Sicher and Jensen, 1979) and CO_2 fixation declines accordingly. Surprisingly, catalase has no effect on PGA reduction (Table 6). Reduction of H_2O_2 does not compete effectively with NADP reduction required to support PGA reduction.

The situation with nitrite addition is more complex. The reduction in activation of FBPase may well result from diversion of electrons from thioredoxin reduction to nitrite reduction. Photosynthetic CO_2 fixation also declines because nitrite lowers the stomal pH (Purczeld et al., 1978). Both nitrite ion and nitrous acid may traverse the chloroplast envelope and this results in a collapse of the trans-envelope proton gradient in the light. The inhibition by nitrite is sharply dependent on medium pH.

Table 5. Effects of catalase, methyl viologen and nitrite on chloroplast enzyme activation and photosynthesis. Chloroplasts were illuminated in the medium of Table 1 at high CO_2. Active forms of FBPase and RuBPCase were assayed as described above. All rates are in μmoles (mg Chl·h)$^{-1}$.

Experiment	Conditions	FBPase	RuBPCase	Photosynthesis
1	control	25	-	49
	omit catalase	16	-	7
2	control	-	58	49
	omit catalase	-	52	11
3	control	31	-	153
	plus 1 mM $NaNO_2$	15	-	48
4	control	27	-	90
	plus 10^{-6} M MV	13	-	13

Table 6. Effect of catalase on 3-PGA-dependent O_2 evolution. Oxygen evolution was monitored polarographically with chloroplasts suspended in the medium of Table 1. Suspending medium contained either 10 mM $NaHCO_3$ or 2 mM 3-PGA. Catalase was present at 1600 units in 0.5 ml.

Substrate	O_2 evolution: μmoles (mg Chl·h)$^{-1}$	
	with catalase	no catalase
$NaHCO_3$	150	40
3-PGA	70	64

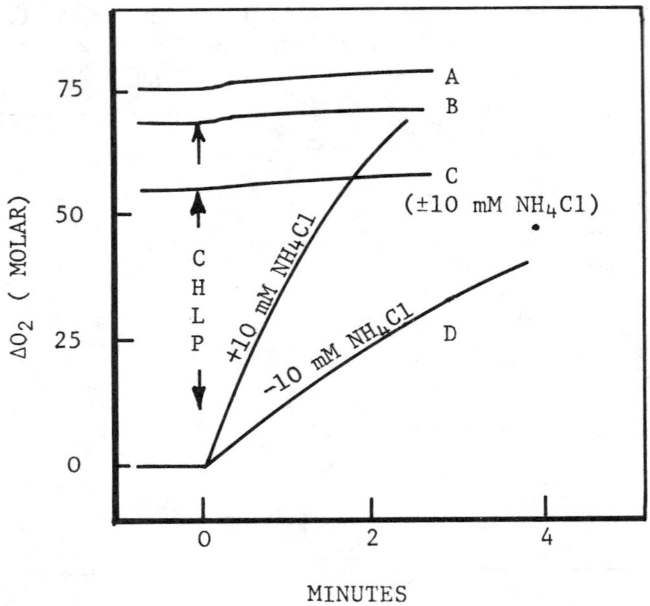

Fig. 4. H_2O_2-dependent O_2 evolution. Chloroplasts (13.8 g Chl) were added to 1 ml medium of Table 1 without catalase or CO_2, but containing 0.17 mM H_2O_2 and NH_4Cl when indicated. In a separate experiment the Km (H_2O_2) was determined to be 60 M. A) No H_2O_2, illuminated; B) No H_2O_2, dark; C) 0.17 mM H_2O_2, dark; D) 0.17 mM H_2O_2, illuminated.

If the external pH is low enough, stromal pH values in the presence of nitrite enter the pH range where photosynthesis is pH dependent and inhibition occurs. At equilibrium stromal pH is more acidic than medium pH due to Donnan potential effects. Whether it is stromal pH changes or competition for electrons between nitrite reductase and thioredoxin which reduces FPBase activation is not clear.

Changes in stromal pH in the presence of H_2O_2 (omission of catalase) would not be expected since H_2O_2 has a pKa around 12. Competition of H_2O_2 and thioredoxin for electrons appears to adequately explain the observed changes in FBPase activation.

There would appear therefore to be a ranking of acceptors of photosystem I, in which NADP reduction is favored over H_2O_2 reduction and H_2O_2 reduction favored over thioredoxin reduction. As methyl viologen also fails to inhibit PGA reduction, it can be placed at a level similar to H_2O_2. One possible mechanism for such a ranking would be differences in affinities of the appropriate reductases and acceptors for reduced ferredoxin. NADP reductase has a Km for ferredoxin of about 0.3 μM (Shin et al., 1963) whereas nitrite reductase has a Km for ferredoxin of about 10 μM (Betts and Hewitt, 1966). The fact that nitrite inhibits fructose bisphosphatase activation may put the Km for ferredoxin of ferredoxin-thioredoxin reductase even higher.

CONCLUSION

The results presented here have demonstrated several mechanisms involved in regulation of photosynthetic CO_2 fixation and suggested one way in which photosynthesis and nitrogen metabolism might interact. Although the mechanisms themselves are certain to be present *in vivo*, the quantitative relationships and importance of any particular regulatory process may be very different in isolated chloroplasts and *in vivo*. *In vivo* chloroplasts function in a particular environment which we cannot reproduce in the laboratory. As indicated by the large effects seen with catalase, our choice of suspending media for isolated chloroplasts can have major effects on the quantitative details of regulation. Further progress in regulation of photosynthesis will require both elucidation of chemical details of the regulatory processes and quantitative evaluation of these processes under *in vivo* conditions.

REFERENCES

Baier, D., and Latzko, E., 1975, Properties and regulation of C-1-fructose-1,6-diphosphatase from spinach chloroplasts, *Biochim. Biophys. Acta*, 396:141.

Betts, G. F., and Hewitt, E. J., 1966, Photosynthetic nitrite reductase and the significance of hydroylamine in nitrite reduction in plants, *Nature*, 210:1327.

Heldt, H. W., Chon, C. J., Lilley, R. McC., and Portis, A. R., 19 , The role of fructose- and sedoheptolosebisphosphatase in the control of CO_2 fixation. Evidence from the effects of Mg^{++} concentration, pH, and H_2O_2, *in*: Proc. Fourth Intern. Cong. Photosyn.

Hiller, R. G., and Bassham, J. A., 1965, Inhibition of CO_2 fixation by nitrous acid, *Biochim. Biophys. Acta*, 109:607.

Lorimer, G. H., 1979, Evidence for the existence of discrete activator and substrate sites for CO_2 on ribulose-1,5-bisphosphate carboxylase, *J. Biol. Chem.*, 254:5599.

Miziorko, H. M., 1979, Ribulose-1,5-bisphosphate carboxylase, *J. Biol. Chem.*, 254:270.

Purczeld, P., Chon, C. J., Portis, A. R., Heldt, H. W., and Heber, U., 1978, The mechanism of the control of carbon fixation by the pH in the chloroplast stroma studies with nitrite-mediated proton transfer across the envelope, Biochim. Biophys. Acta, 501:488.

Shin, M., Kagawa, K., and Arnon, D. I., 1963, Crystallization of ferredoxin-TPN reductase and its role in the photosynthetic apparatus of chloroplasts, Biochem. Zeits., 338:1305.

Sicher, R. C., and Jensen, R. G., 1979, Photosynthesis and ribulose 1,5-bisphosphate levels in intact chloroplasts, Plant Physiol., 64:880.

HYDROLYSIS OF RIBULOSE-1,5-BISPHOSPHATE CARBOXYLASE BY PARTIALLY PURIFIED ENDOPROTEINASES OF SENESCING PRIMARY BARLEY LEAVES

B. L. Miller and R. C. Huffaker

Plant Growth Laboratory and the Department of Agronomy
and Range Science, University of California,
Davis, CA 95616

INTRODUCTION

Ribulose-1,5-bisphosphate carboxylase (RuBPCase) is the protein responsible for fixation of CO_2 in photosynthetic organisms. In many higher plants RuBPCase also appears to serve as a storage protein that is hydrolyzed during leaf senescence (Huffaker et al., 1978). This provides a source of reduced N that can be transported to newly developing leaves or fruits (Dalling et al., 1976). RuBPCase is a large protein (MW $\sim$ 550 kD) consisting of 8 large subunits (50-57 kD) and 8 small subunits (13-15 kD). The large subunit is coded on chloroplast DNA and is synthesized within the chloroplast (Blair et al., 1973; Chan et al., 1972; Criddle et al., 1970; Kung, 1976). The small subunit is coded on nuclear DNA (Kung, 1976), synthesized as a precursor protein in the cytoplasm (Criddle et al., 1970; Highfield et al., 1978), and then processed at or in the chloroplast. The native protein is assembled in the chloroplast and the active enzyme is localized in the stroma. The synthesis of RuBPCase occurs predominantly during the greening of etiolated leaf tissue (Kleinkopf et al., 1970; Smith et al., 1974) or leaf expansion (Friedrich and Huffaker, 1980). The cellular concentration of RuBPCase (which can constitute 50-70% of the total soluble leaf protein) then remains nearly constant for several days; little or no apparent turnover takes place (Huffaker, 1979; Peterson et al., 1973). During senescence, protein is rapidly degraded, and RuBPCase is the predominant protein lost during the initial stages (Friedrich and Huffaker, 1980; Peterson and Huffaker, 1975). Although there is much data concerned with the synthesis of RuBPCase, information about the control of its degradation and turnover is lacking. Exo- and endoproteinases in green and senescing leaf tissue have been described (Dalling et al., 1976; Huffaker and Miller, 1978; Martin and Thimann, 1972; Peterson and Huffaker, 1975; Sopanen and Lauriere, 1976; Thomas, 1978; Wittenbach, 1978) but very little is known about their role in senescence or in normal cellular protein turnover.

Endoproteinases have been purified or partially purified from leaf tissue (Drivdahl and Thimann, 1977; Frith et al., 1978; Ragster and Chrispeels, 1979) and these will degrade several protein substrates in addition to RuBPCase (Huffaker and Miller, 1978; Peoples et al., 1979; Wittenbach, 1978). No RuBPCase-specific proteinases have been reported. Three endoproteinases from intact senescing barley leaves and their reaction with purified barley RuBPCase are described below. Analysis of the reaction kinetics and breakdown products may help to elucidate the potential role they may play in the *in vivo* breakdown of RuBPCase.

MATERIALS AND METHODS

Purification of Endoproteinases

Twelve-day-old primary barley leaves were detached and homogenized in a Waring blender with 0.1 M KP_i pH 6.0 containing 2 mM DTT and 1 mM EDTA (grinding ratio was 1 g to 3 ml buffer). Insoluble polyvinyl pyrrolidone was added at the rate of 0.1 g to 1 g of tissue. The homogenate was filtered through 8 layers of cheesecloth and centrifuged at 27,000 x g for 20 min. Solid ammonium sulfate was added to the resulting supernatant and the 35-70% pellet was recovered. This pellet was resuspended in 50 mM KP_i, pH 6.0, and 1 mM DTT, and dialyzed for 20 hrs against the same buffer. The large amount of protein that precipitated from the solution during dialysis was removed by centrifugation. Recovery of endoproteolytic activity was 80%. Apparently much of the protein that had precipitated was RuBPCase because gel electrophoresis of the supernatant showed a complete absence of RuBPCase.

The proteolytic activity was resolved into two separate activity peaks by chromatography on a Sephadex G-100 column equilibrated with 50 mM KP_i, pH 6.0, and 1 mM DTT. Any cross-contamination of the two activities was removed by DEAE chromatography. The serine proteinase EP_2 (Table 1) did not bind to a DEAE cellulose column, which had been equilibrated with 50 mM KP_i, pH 6.0, and 1 mM DTT. The thiol proteinase, EP_1 (Table 1), however, did bind and was eluted from the column with a 0.0-to-0.2 M NaCl gradient. EP_1 was then dialyzed against 50 mM KP_i, pH 6.0, and 1 mM DTT. A major non-active protein was removed from the DEAE-purified EP_1 by gel chromatography on a Sephadex G-75 superfine column equilibrated with 50 mM KP_i, pH 6.0, and 1 mM DTT. The purification procedure described resulted in an approximately 5800-fold purification of EP_1 and ∿50-fold purification of EP_2. Details of the purification and characterization of these proteinases will be described in a separate publication.

Purification of ^{14}C-Labeled RuBPCase

Barley seeds (Hordeum vulgare v. Numar) were planted in 6-inch pots containing vermiculite. Nutrient solution was continuously supplied by cotton wicks that linked the pot to a reservoir jar below. Seedlings were grown in continuous darkness for 6 days. These etiolated plants were then placed in a self-contained chamber under light (∿550 µE) and allowed to green in the presence of $^{14}CO_2$ using a method devised by Dr. H. Thomas (personal communication). After 48 hr, the green leaves were harvested and homogenized in a Waring blender using a 1 g/5 ml buffer-grinding ratio and 0.1 g insoluble polyvinyl pyrrolidone per g of tissue. Homogenization buffer was 0.2 M Tris-SO_4, pH 8, containing 10 mM $MgCl_2$, 10 mM $NaHCO_3$, 2 mM DTT, and 1 mM EDTA. The crude homogenate was filtered through 8 layers of cheesecloth and centrifuged at 27,000 x g for 20 min. Solid ammonium sulfate was added to the resulting supernatant and the 35-65% pellet was recovered. This pellet was resuspended in 50 mM Tris-SO_4, pH 8.0, 2 mM DTT, and 1 mM EDTA, and 15 ml was chromatographed on a 3 x 50-cm Sephadex G-100 column equilibrated with the same buffer. Only the first half of the RuBPCase peak was kept so as to prevent possible contamination by the major leaf proteinases (unpublished information). The pooled fractions, ∿45 ml, was applied to a 2 x 20-cm DEAE cellulose column equilibrated with 50 mM Tris-SO_4, pH 8.0, 2 mM DTT, and 1 mM EDTA. After the nonbinding protein had been washed through the column, the RuBPCase was eluted with a 0.0-0.2 M ammonium sulfate gradient. SDS-polyacrylamide electrophoresis of the final protein indicated it to be homogenous. The RuBPCase had a specific radioactivity of 4.0911×10^6 cpm/mg protein.

Hydrolysis of RuBPCase by Barley Leaf Proteinases

Added to 250 µl of reaction mixture was 1.0 µg of EP_1, DEAE prep (Table 1), or 5-15 µg of EP_2, DEAE prep (Table 1). Final reaction was 0.09 M KP_i, pH 5.7, 1 mM DTT, and 2-4 mg/ml ^{14}C-RuBPCase. The assay was run at 40 C. At designated times one aliquot was removed and added to an equal volume of 10% TCA. After ½ hour on ice, the precipitated protein was pelleted and the supernatant used to determine TCA-soluble counts. A second aliquot of reaction mixture was added to an equal volume of SDS denaturation buffer containing 4% SDS, 20 µM Leupeptin, and 2 mM PMSF. The presence of Leupeptin and PMSF, respectively, was required to inhibit EP_1 and EP_2. These enzymes had residual activity in SDS even after the samples were boiled for 2 min.

SDS Electrophoresis and Fluorography

SDS electrophoresis of hydrolysis products was done in 10% polyacrylamide slabs (1 mm) using the buffer system of Laemmli (1970) and 15 ma/slab. Each sample well included a small amount of SDS standard proteins (BioRad) as internal MW markers. The gels were stained according to Fairbanks et al. (1971). After destaining the gels were treated with EN^3HANCE (New England Nuclear) for 30 min, washed with distilled H_2O for several hours, and then placed on top of Whatman 3 MM paper and dried. The dried gels were exposed to Kodak NS-2T No-Screen X-ray film for 7-14 days at -80 C. The film was developed and scanned with a Cary 219 recording spectrophotometer.

RESULTS

Barley Leaf Proteinases

The proteinases used in this study were partially purified as described above. The thiol proteinase, EP_1, has been purified ∿5800-fold. This is 10-fold greater than previously reported for a plant leaf proteinase (Drivdahl and Thimann, 1977; Firth et al., 1978). Table 1 describes these enzymes and some of their properties. EP_1 appears to be the major endoproteinase in senescing barley primary leaves. It contributed approximately 85% of the activity measured in an *in vitro* assay of crude extracts. It is a thiol proteinase that requires reduced sulfhydryls for maximum activity and is inhibited 100% by 10 μM Leupeptin. EP_1 and EP_2

Table 1. Barley leaf proteinases.

Proteinase	MW	Type	pH opt for hydrolysis of ^{14}C-RuBPCase
EP_1	28,300	Thiol	5.7
EP_2	67,000	Serine	5.7
EP_3	--	--	∿5.2 broad

both have pH optimums for RuBPCase hydrolysis of 5.7. In contrast with other reports using partially purified proteinases (Peoples et al., 1979; Wittenbach, 1978) or crude extracts (H. Thomas, unpublished), EP_1 and EP_2 have the same pH optimums for both RuBPCase and azocasein. EP_2 appears to be a serine proteinase and is inhibited 50% with 1 mM PMSF. EP_2 contributed about 15% of the in vitro activity in crude extracts.

EP_3 is a minor enzyme that was detected when purified ^{14}C-RuBPCase (which appeared homogeneous after SDS-polyacrylamide electrophoresis) showed a very low but significant release of TCA-soluble counts when incubated at 40 C and pH 5.7. EP_3 was not inhibited by 10 μM Leupeptin, 1 mM PMSF, 10 μM pepstatin, or 1 or 10 mM EDTA. However, it was possible to competitively inhibit EP_3 by adding casein, hemoglobin, or myoglobin. BSA was not a good competitive inhibitor for EP_3, EP_1, or EP_2. Attempts to remove EP_3 from the ^{14}C-RuBPCase by affinity chromatography on hemoglobin-Separaose 4B were unsuccessful.

Hydrolysis of ^{14}C-RuBPCase by EP_1

Hydrolysis of ^{14}C-RuBPCase by EP_1 exhibited an initial rapid rate of hydrolysis followed by a linear rate beginning at about 30 minutes (Fig. 1). Preincubation of the substrate for various times before adding EP_1 gave the same kinetics. When the reaction was allowed to proceed for 30 minutes, and then additional EP_1 was added, the new rate from the time of addition was linear. These results imply the presence of 2 classes of peptide bonds, one class being particularly susceptible to hydrolysis by EP_1. Indeed, SDS-polyacrylamide electrophoresis of the degradation products showed a very rapid conversion of the large subunit of RuBPCase (57.5 kD) to a major fragment of 54.5 kD. After this fragment formed, other fragments of lesser MW appeared (Miller, unpublished). Table 2 shows the rapid decline in the large subunit to small subunit ratio that reflects this initial hydrolysis. Apparently the large and small subunits are degraded at nearly equal rates after approximately 70% of the large subunit has been converted to the 54.5 kD fragment.

Hydrolysis of ^{14}C-RuBPCase by EP_2

Compared with EP_1, there is very rapid rise in the large to small subunit ratio when EP_2 reacts with ^{14}C-RuBPCase (Table 3). During this time there is rapid loss of the small subunit (14.7 kD) and a corresponding rise in a protein fragment of 13.7 kD. During this initial 15 min there is little or no change in the amount of large subunit. Apparently EP_2 initially cleaves a sensitive region of the small subunit, but then hydrolyzes the large subunit at a slightly faster rate when approximately 70% of the small subunit has been processed to the 13.7 kD fragment.

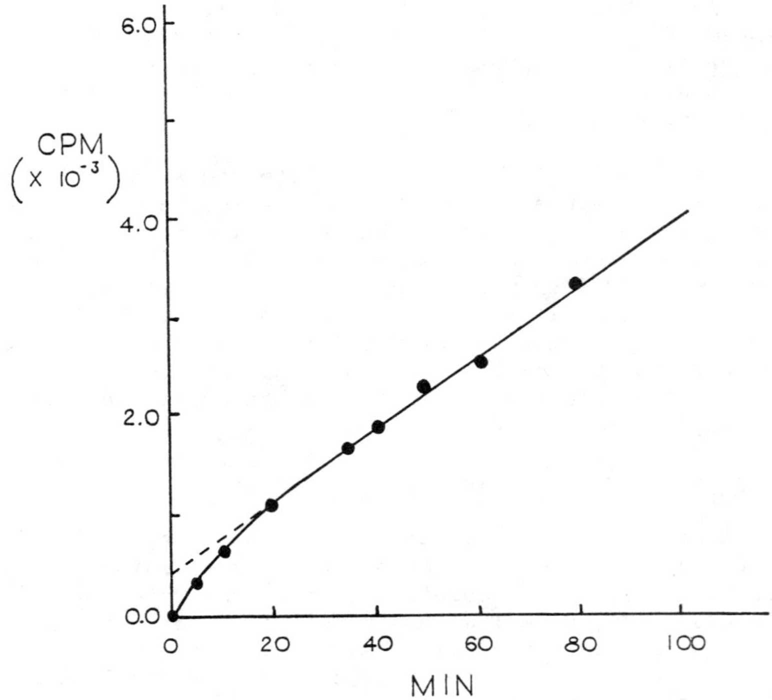

Fig. 1. Time course for ^{14}C-RuBPCase hydrolysis by EP_1. The hydrolysis of ^{14}C-RuBPCase was run and TCA soluble counts determined as described in Materials and Methods.

Hydrolysis of ^{14}C-RuBPCase by EP_3

This endoproteinase was first detected when purified ^{14}C-RuBPCase was incubated at pH 5.7 and 40 C in the absence of any added endoproteinase. There was a minor amount of TCA-soluble counts released (∿4% of the rate when EP_1 was added) at a linear rate during an 8-hr incubation. When the degradation products were separated by SDS electrophoresis, it was discovered that quite a significant amount of hydrolysis had occurred. The large subunit had declined by 50% and the small subunit by 30% after 2 hrs. However, most of these products were larger than 5000 kD and were therefore not soluble in 5% TCA (Miller, unpublished). This enzyme degrades both large and small subunits, but has a

Table 2. Hydrolysis of ^{14}C-RuBPCase by EP_1.

	% Initial		
Min*	Large subunit (LS)	Small subunit (SS)	LS/SS
0	100	100	4.01
0.5	66	99	2.37
1	59	96	2.19
2	45	85	1.94
4	37	77	1.60
8	29	68	1.57
15	23	56	1.58
30	15	38	1.61

*Hydrolysis of ^{14}C-RuBPCase was at 40 C using 1.0 μg of EP_1 as described in Materials and Methods. The amounts of LS and SS were determined after SDS electrophoresis and fluorography.

Table 3. Hydrolysis of ^{14}C-RuBPCase by EP_2.

	% Initial			Relative peak area	
Min*	Large subunit (LS)	Small subunit (SS)	LS/SS	Small subunit	13.7 kD Fragment
0	100	100	4.01	108	0
5	99	40	10.0	50	70
15	92	30	11.5	40	73
30	62	23	10.2	28	50
60	47	17	8.2	22	38
90	39	15	--	20	40

*Hydrolysis of ^{14}C-RuBPCase was at 40 C using 5-15 μg EP_2 as described in Materials and Methods. The amounts of LS and SS were determined after SDS electrophoresis and fluorography.

preference for the large subunit (Table 4). EP_3 appears to be a minor enzyme because it was not detected during purification of EP_1 and EP_2. It is present in very minor amounts in the ^{14}C-RuBPCase prep because it was not seen as a contaminating protein. Alternatively it could have a MW equivalent to either the large or small subunit and therefore would be indistinguishable from these. EP_3 is apparently a different enzyme from EP_1 and EP_2 because of its response to inhibitors, pH optimum for ^{14}C-RuBPCase hydrolysis, and its mode of ^{14}C-RuBPCase hydrolysis (Table 1, Table 4).

Table 4. Hydrolysis of ^{14}C-RuBPCase by EP_3.

	% Initial		
Min*	Large subunit (LS)	Small subunit (SS)	LS/SS
0	100	100	4.01
30	82	85	3.85
60	64	74	3.73
120	52	67	3.50
240	42	63	3.00
480	38	56	2.05

*Hydrolysis of purified ^{14}C-RuBPCase was at 40 C using no added EP_1 or EP_2, as described in Materials and Methods. The amount of LS and SS were determined after SDS electrophoresis and fluorography.

CONCLUSION

Two major endoproteinases isolated from senescing barley leaves have been partially characterized with respect to their hydrolysis of RuBPCase. The large subunit of this important protein apparently has a site particularly susceptible to hydrolysis by EP_1, the thiol proteinase. This leads to unusual kinetics (Fig. 1) for hydrolysis. When azocasein is used as the substrate, EP_1 shows linear hydrolysis with time and normal Michaelis-Menten kinetics (Miller, unpublished). Therefore, direct comparisons of different protein substrates in *in vitro* assays may not clarify the role of an endoproteinase in the *in vivo* hydrolysis of cellular protein.

The interpretation of *in vitro* assays are complicated by the possibility of hypersensitive regions of a protein. A simple cleavage could lead to enzyme inactivation and possible sensitivity to further proteolysis. Further, a particular protein substrate (for example RuBPCase) may not be available to proteolysis by EP_1 or EP_2 *in vivo* because of compartmentation. The serine proteinase EP_2 shows an initially rapid cleavage of the small subunit of RuBPCase, and in this respect it behaves differently from EP_1.

It remains unknown what role EP_1 and EP_2 play in protein turnover or in the rapid loss of protein during senescence. An increase in crude extract endoproteolytic activity is not required for senescence to occur (Huffaker and Miller, 1978; Ragster et al., 1978; Storey and Beevers, 1977; van Loon and Haverkort, 1977). In *Escherichia coli*, not all the proteinases present in the cell have the same role. Some appear to be involved in breakdown of protein during starvation, some in protein turnover, and others in degradation of aberrant proteins (Beck et al., 1980; Cheng et al., 1979). Also, some proteins may be entirely degraded by exopeptidases (Holzer and Heinrich, 1980). In view of the very significant breakdown of RuBPCase by EP_3 and the fact that it was not detected during purification of barley leaf endoproteinases suggests that EP_3 or some other minor protease not detected by standard proteolytic assays may play important roles in the turnover and rapid hydrolysis of RuBPCase during senescence.

ACKNOWLEDGMENT

The authors would like to thank H. Thomas, J. W. Friedrich, R. C. Tait, K. E. Yamamoto, and S. E. Taylor for helpful discussion and technical assistance. This work was supported by NSF Grant PFR 77-07301 and the Department of Agronomy and Range Science.

REFERENCES

Beck, I., Fink, G. R., and Wolf, D. H., 1980, The intracellular proteinases and their inhibitors in yeast, *J. Biol. Chem.*, 255:4821.

Blair, G. E., and Ellis, R. J., 1973, Protein synthesis in chloroplasts. I. Light-driven synthesis of the large subunit of fraction 1 protein by isolated pea chloroplasts, *Biochim. Biophys. Acta*, 319:223.

Chan, P.-H., and Wildman, S. G., 1972, Chloroplast DNA codes for the primary structure of the large subunit of Fraction 1 protein, *Biochim. Biophys. Acta*, 277:677.

Cheng, Y.-S. E., Zipser, D., Cheng, C.-Y., and Rolseth, S. J., 1979, Isolation and characterization of mutations in the structural gene for protease III (ptr), *J. Bacteriol.*, 140:125.

Criddle, R. S., Dau, B., Kleinkopf, G. E., and Huffaker, R. C., 1970, Differential synthesis of ribulose diphosphate carboxylase subunits, Biochem. Biophys. Res. Commun., 41:621.

Dalling, M. J., Boland, G., and Wilson, J. H., 1976, Relation between acid proteinase activity and redistribution of nitrogen during grain development in wheat, Aust. J. Plant Physiol., 3:721.

Drivdahl, R. H., and Thimann, K. V., 1977, The proteases of senescing oat leaves. I. Purification and general properties, Plant Physiol., 59:1059.

Fairbanks, G., Steck, J. L., and Wallach, D. F. B., 1971, Electrophoretic analysis of the major polypeptides of the human erythrocyte membrane, Biochemistry, 10:2606.

Feller, U. K., Soong, J.-S. J., and Hageman, R. H., 1977, Proteolytic activities and leaf senescence during grain development of field-grown corn (*Zea mays* L.), Plant Physiol., 59:290.

Friedrich, J. W., and Huffaker, R. C., 1980, Photosynthesis, leaf resistance, and ribulose-1,5-bisphosphate carboxylase degradation in senescing barley leaves, Plant Physiol., 65:1103.

Frith, G. J. T., Swinden, L. B., and Dalling, M. J., 1978, Proteolytic enzymes in green wheat leaves. II. Purification by affinity chromatography and some properties of proteinases with acid pH optima, Plant Cell Physiol., 19:1029.

Highfield, P. E., and Ellis, R. J., 1978, Synthesis and transport of the small subunit of chloroplast ribulose bisphosphate carboxylase, Nature, 271:420.

Holzer, H., and Heinrich, P., 1980, Control of proteolysis, Ann. Rev. Biochem., 49:63.

Huffaker, R. C., 1979, Protein metabolism, *in*: "Plant Physiology: A Treatise," F. C. Steward, ed., Academic Press, New York.

Huffaker, R. C., and Miller, B. L., 1978, Reutilization of ribulose bisphosphate carboxylase, *in*: "Photosynthetic Carbon Assimilation," H. W. Siegelman and G. Hind, eds., Plenum Press, New York.

Kleinkopf, G. E., Huffaker, R. C., and Matheson, A., 1970, Light-induced *de novo* synthesis of ribulose-1,5-diphosphate carboxylase in greening leaves of barley, Plant Physiol., 46:416.

Kung, S., 1976, Tobacco fraction 1 protein: a unique genetic marker, Science, 191:429.

Laemmli, U. K., 1970, Cleavage of structural proteins during the assembly of the head of bacteriophage T4, Nature, 227:680.

Martin, C., and Thimann, K. V., 1972, The role of protein synthesis in the senescence of leaves. I. The formation of protease, Plant Physiol., 49:64.

Peoples, M. B., Frith, G. J., and Dalling, M. J., 1979, Proteolytic enzymes in green wheat leaves. IV. Degradation of ribulose-1,5-bisphosphate carboxylase by acid proteases isolated on DEAE cellulose, Plant Cell Physiol., 20:253.

Peterson, L. W., and Huffaker, R. C., 1975, Loss of ribulose-1,5-diphosphate carboxylase and increase in proteolytic activity during senescence of detached primary barley leaves, Plant Physiol., 55:1009.

Peterson, L. W., Kleinkopf, G. E., and Huffaker, R. C., 1973, Evidence for lack of turnover of ribulose-1,5-diphosphate carboxylase in barley leaves, Plant Physiol., 51:1042.

Ragster, L., and Chrispeels, M. J., 1979, Azocoll-digesting proteinases in soybean leaves. Characteristics and changes during leaf maturation and senescence, Plant Physiol., 64:857.

Ragster, L., Chrispeels, M. J., Anderson, J., and Hageman, R. H., 1978, Temporal changes in proteolytic activities in the leaves of field-grown soybean plants, Plant Physiol. Suppl., 61:27.

Smith, M. A., Criddle, R. S., Peterson, L., and Huffaker, R. C., 1974, Synthesis and assembly of ribulosebisphosphate carboxylase enzyme during greening of barley plants, Arch. Biochem. Biophys., 165:494.

Sopanen, J., and Lauriere, C., 1976, Activities of various peptidases in the first leaf of wheat, Physiol. Plant., 36:251.

Storey, R., and Beevers, L., 1977, Proteolytic activity in relationship to senescence and cotyledonary development in *Pisum sativum* L., Planta, 137:37.

Thomas, H., 1978, Enzymes of nitrogen mobilization in detached leaves of *Lolium temulentum* during senescence, Planta, 142:161.

van Loon, L. C., and Haverkort, A. J., 1977, No increase in protease activity during senescence of oat leaves, Plant Physiol. Suppl., 59:113.

Wittenbach, V. A., 1978, Breakdown of ribulose bisphosphate carboxylase and changes in proteolytic activity during dark induced senescence of wheat seedlings, Plant Physiol., 62:604.

THE RELATIONSHIP BETWEEN RIBULOSE BISPHOSPHATE CARBOXYLASE CONCENTRATION AND PHOTOSYNTHESIS

James W. Friedrich and R. C. Huffaker

Plant Growth Laboratory and the Department of Agronomy and Range Science, University of California, Davis, CA 95616

Ribulose 1,5-bisphosphate carboxylase (RuBPCase) has a central role in the C and N economy of many crop plants. In C_3 plants, RuBPCase catalyzes the primary rate-limiting step in CO_2 fixation (Jensen and Bahr, 1977; Bahr and Steffens, this volume). RuBPCase also functions as a storage protein and is rapidly degraded during leaf senescence (Friedrich and Huffaker, 1980; Wittenbach et al., 1980). The catalytic properties of RuBPCase and other Calvin-cycle enzymes have been thoroughly reviewed (Jensen and Bahr, 1977; Bahr and Steffens, this volume; Buchanan, this volume). Likewise, the regulation of RuBPCase degradation has been discussed (Huffaker and Miller, 1978; Miller and Huffaker, this volume; Thomas and Stoddart, 1980). Therefore, this article will focus mainly on the role of RuBPCase as a storage protein and on the relationship between RuBPCase concentration and photosynthesis.

Changes in RuBPCase During Leaf Senescence

The concentration of total soluble protein in the primary leaf of barley (*Hordeum vulgare* L.) steadily declines as a consequence of leaf senescence (Fig. 1). The non-RuBPCase protein fraction changes little. The proportion of total soluble protein constituted by RuBPCase is quite high, ranging from 52 to 81% in this experiment (Fig. 1). This proportion declines during the later stages of senescence. Under conditions of low N supply, RuBPCase declines during senescence until it constitutes only 20% of the total soluble protein (Friedrich and Huffaker, unpublished data). Approximately 85% of the loss of soluble leaf protein during senesence in barley is due to degradation of RuBPCase (Fig. 1). Similar observations have been made in wheat (*Tritium aestivum* L.)

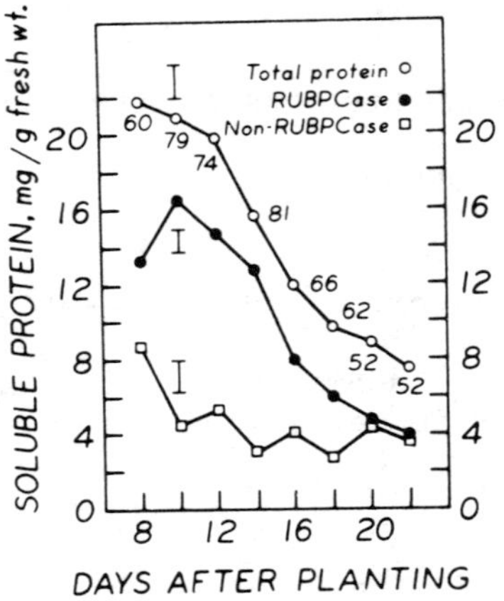

Fig. 1. Total protein, RuBPCase and non-RuBPCase protein in senescing barley primary leaves. Proportion of total protein constituted by RuBPCase is depicted for each total protein value. Bars represent LSD (0.05 level) for each parameter. Leaf fresh weight was constant (0.14 g/leaf) throughout the experiment. See Friedrich and Huffaker (1980) for experimental details.

and soybeans (*Glycine max* (L.) Merr.) (Wittenbach, 1979: Wittenbach et al., 1980) and illustrate the role of RuBPCase as a storage protein.

The *in vitro*-specific activity of RuBPCase (μmoles CO_2/mg RuBPCase·min) changes very little during senescence in barley (Friedrich and Huffaker, 1980), wheat (Wittenbach, 1979), and soybeans (Wittenbach et al., 1980). Accordingly, the observed decline in RuBPCase concentration is highly correlated with a decrease in *in vitro* RuBPCase activity (r = 0.95).

The decline in RuBPCase protein during senescence is highly correlated (r = 0.96) with gross photosynthesis (Fig. 2). Mesophyll resistance to CO_2 diffusion (Friedrich and Huffaker, 1980) increases three-fold during the period when photosynthesis declines. Thus, a cause-and-effect relationship between the decline

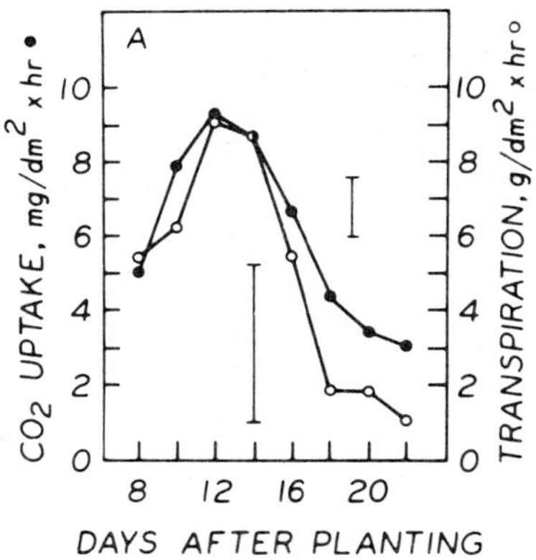

Fig. 2. Gross photosynthesis or CO_2 uptake and transpiration in senescing barley primary leaves. Bars represent LSD (0.05 level) for each parameter. See Friedrich and Huffaker (1980) for experimental details.

in RuBPCase and the decline in photosynthesis is suggested. However, the rate of photosynthesis per unit RuBPCase increases with leaf age (Table 1). Also, RuBPCase (Fig. 1) declines before the photosynthesis begins to decrease (Fig. 2). During senescence in soybeans, the decline in photosynthesis is not closely coupled with a loss in *in vitro* RuBPCase activity or RuBPCase protein (Mondal et al., 1978; Wittenbach et al., 1980).

Clearly, RuBPCase concentration is not the only determinant of photosynthesis in senescing leaves. It is possible that at high concentrations of RuBPCase, light harvesting and photosynthetic electron transport limit photosynthesis. In senescing leaves of *Phaseolus vulgaris*, electron transport between photosystem II and photosystem I is inhibited (Jenkins and Woolhouse, 1979). Photosynthesis is also affected by the *in vivo* regulation of RuBPCase and other Calvin cycle enzymes (Bahr and Steffens, this volume; Buchanan, this volume). However, the activation of RuBPCase by CO_2 and Mg^{++} is not involved in the decline in photosynthesis during barley leaf senescence (Friedrich and Huffaker, unpublished

Table 1. Photosynthesis per mg RuBPCase, the ratio of photosynthesis to transpiration, and the concentration of CO_2 in the intercellular air spaces CO_2^{IAS} in senescing barley primary leaves. External CO_2 concentration CO_2^{Ext} was 268 l/l. CO_2^{IAS} = CO_2^{Ext} minus (photosynthesis x stomatal resistance). See Friedrich and Huffaker (1980) for experimental details.

Time after planting	Photosynthesis per mg RuBPCase	Photosynthesis to transpiration	CO_2^{IAS}
days	nmol mg^{-1} min^{-1}	mg/g	µl/l
8	88	1.08	250
10	110	1.46	244
12	144	1.20	248
14	156	1.30	245
16	193	1.66	237
18	166	2.61	222
20	168	2.02	232
22	178	3.12	210
LSD, 0.05	34	1.07	20

results). Another possibility is that the measured RuBPCase concentration does not reflect the amount of catalytically active RuBPCase. The number of titratably active sites per mole of RuBPCase protein does not change during senescing in wheat (M. Dalling, personal communication). Finally, stomatal resistance to CO_2 uptake also changes during leaf senescence (Friedrich and Huffaker, 1980; Wittenbach et al., 1980). Stomatal resistance (Fig. 2) increases nine-fold during the same period when transpiration and photosynthesis are decreasing (Fig. 2). Moreover, the intercellular CO_2 concentration (Table 1) diminishes during senescence, which indicates that stomatal aperature is imposing some constraint on CO_2 uptake. However, as evidenced by the increasing photosynthesis to transpiration ratio (Table 1), the primary effect of stomatal closure is a decrease in water loss.

Effect of N Supply on RuBPCase

The concentration of nitrate in the rooting medium has a pronounced effect on RuBPCase concentration and photosynthesis. RuBPCase in mature, nonsenescent leaves of barley grown on low N is less than one-half that of high-N plants (Table 2). RuBPCase

Table 2. Concentration of RuBPCase per unit leaf area or as a percentage of the total soluble protein, rate of photosynthesis per unit leaf area (P) or per mg RuBPCase (P/RuBPCase), stomatal resistance (R_S), and mesophyll resistance (R_M) in barley primary leaves as affected by switching from a low-N regime to a high-N regime and vice versa. Plants were switched one day after collars had formed on the primary leaves. Physiological measurements were made two days after switching. Data represent the mean ± standard error of the mean.

Parameter	N supply, μmoles NO_3^-/plant/day			
	2.5/2.5	2.5/25	25/2.5	25/25
RuBPCase, Mg/dm^2	9.8±0.1	13.4±0.1	24.0±0.8	26.2±0.7
RuBPCase, %	51±2	49±3	71±4	68±2
P, Mg CO_2/dm^2/hr	5.4±0.2	6.6±0.5	7.6±0.2	8.1±0.5
P/RuBPCase, μmoles CO_2/Mg/hr	12.6±0.5	11.2±1.1	7.2±0.3	7.1±0.6
R_S, sec/cm	23.6±2.8	12.4±0.3	8.8±0.6	7.9±1.6
R_M, sec/cm	12.7±1.8	17.7±2.3	16.9±0.7	16.2±1.0

increases substantially when low-N plants are switched to a high-N regime. In contrast, switching plants from a high-N to a low-N regime does not result in any appreciable loss of RuBPCase. The proportion of total soluble protein constituted by RuBPCase is not affected by any short-term alteration in N supply. However, the RuBPCase proportion is much higher in high-N plants (68%) compared to low-N plants (51%). N supply appears to have little effect on the non-RuBPCase protein fraction (data not shown). These observations are consistent with the role of RuBPCase as a storage protein.

Photosynthesis is higher in high-N plants compared to low-N plants (Table 2). Plants grown on alternating N regimes exhibit intermediate rates of photosynthesis. The relative increase in photosynthesis is not as great as the relative increase in RuBPCase. This translates into a declining rate of photosynthesis per mg RuBPCase as N supply increases. Low-N plants also exhibit a lower mesophyll resistance compared to high-N plants. It appears that the increase in photosynthesis with increased N supply is due to a decrease in stomatal resistance. N supply has a pronounced effect on stomatal behavior in cotton (Gossypium hirsutum L.) (Radin and Parker, 1979). The mechanism by which N supply affects stomatal aperature is not understood.

Effect of Partial Defoliation

The interaction between a senescing leaf (assimilate source) and younger, developing leaves (assimilate sinks) is probably involved in senescence. Removal of the emerging second and third leaves partially prevents the usual decline in photosynthesis and RuBPCase and the increase in stomatal resistance, photosynthesis per mg RuBPCase, and mesophyll resistance (Table 3). Apparently, this type of source-sink manipulation does not affect any one physiological process in particular. Rather, partial defoliation

Table 3. Percent change in several physiological parameters in the senescing primary leaves of intact (I) and partially defoliated barley (D). Partial defoliation was effected by a daily removal of the emerging second and third leaves, beginning one day after collars had formed on the primary leaves. See Table 2 for explanation of abbreviations. Data represent the mean of 2-3 observations.

Parameter		Days after collar formation				
		1	3	6	9	12
				%		
RuBPCase/leaf area	I	100	78	72	39	39
	D		82	78	64	71
RuBPCase percentage	I	100	86	92	72	92
	D		86	95	88	98
P/leaf area	I	100	94	84	85	68
	D		94	91	99	78
P/RuBPCase	I	100	128	120	220	166
	D		123	128	164	109
R_S	I	100	104	122	135	139
	D		110	96	63	110
R_M	I	100	108	118	108	151
	D		107	118	118	136

seems to slow the entire senescence process. In contrast, during a short time period (less than 48 hr), source-sink manipulations affect stomatal resistance and photosynthesis without affecting mesophyll resistance (Setter et al., 1980a). Hormonal influences are probably involved in this phenomenon (Setter et al., 1980b). During senescence, it is likely that alterations in photosynthesis, RuBPCase, stomatal resistance, and mesophyll are at least partly due to hormonal influences (Thomas and Stoddart, 1980).

ACKNOWLEDGMENTS

This work was supported by a grant from the National Science Foundation (PFR 77-07301).

REFERENCES

Friedrich, J. W., and Huffaker, R. C., 1980, Photosynthesis, leaf resistances, and ribulose-1,5-bisphosphate carboxylase degradation in senescing barley leaves, Plant Physiol., 65:1103.

Huffaker, R. C., and Miller, B. L., 1978, Reutilization of ribulose bisphosphate carboxylase, in: "Photosynthetic Carbon Metabolism," H. W. Siegelman and G. Hind, eds., Plenum Press, New York.

Jenkins, G. I., and Woolhours, H. W., 1979, A study of photosynthetic electron transport during senescence of the primary leaves of *Phaseolus vulgaris*, Plant Physiol. Suppl., 63:74.

Jensen, R. G., and Bahr, J. T., 1977, Ribulose 1,5-bisphosphate carboxylase-oxygenase, Ann. Rev. Plant Physiol., 28:379.

Mondal, M. H., Brun, W. A., and Brenner, M. L., 1978, Effects of sink removal on photosynthesis and senescence in leaves of soybean (*Glycine max* L.) plants, Plant Physiol., 61:394.

Radin, J. W., and Parker, L. L., 1979, Water relations of cotton plants under nitrogen deficiency. II. Environmental interactions on stomata, Plant Physiol., 64:499.

Setter, T. L., Brun, W. A., and Brenner, M. L., 1980a, Stomatal closure and photosynthetic inhibition in soybean leaves induced by petiole girdling and pod removal, Plant Physiol., 65:884.

Setter, T. L., Brun, W. A., and Brenner, M. L., 1980b, Effect of obstructed translocation on leaf abscissic acid, and associated stomatal closure and photosynthesis decline, Plant Physiol., 65:111.

Thomas, H., and Stoddart, J. L., 1980, Leaf senescence, Ann. Rev. Plant Physiol., 31:83.

Wittenbach, V. A., 1979, Ribulose bisphosphate carboxylase and proteolytic activity in wheat leaves from anthesis through senescence, Plant Physiol., 64:884.

Wittenbach, V. A., Ackerson, R. C., Giaquinta, R. T., and Herbert, R. R., 1980, Changes in photosynthesis, ribulose bisphosphate carboxylase, proteolytic activity, and ultrastructure of soybean leaves during senescence, Crop Sci., 20:225.

IV. NITROGEN FIXATION BY NONLEGUMES

NITROGEN FIXATION BY CYANOBACTERIAL HETEROCYSTS

C. Peter Wolk

MSU-DOE Plant Research Laboratory
Michigan State University
East Lansing, MI 48824

This paper details current knowledge about nitrogen fixation by cyanobacterial (blue-green algal) heterocysts. I will first review the most direct evidence that aerobic nitrogen fixation is localized in heterocysts; will then recount the results of enzymatic studies which provide examples of specializations apparently supporting this localization; and finally, will consider the little that is known about the mechanisms by means of which heterocysts protect nitrogenase from inactivation by atmospheric oxygen. A broader look at the biology of heterocysts is presented in recent reviews (Haselkorn, 1978; Carr, 1979; Wolk, 1979b, 1980).

LOCALIZATION OF NITROGENASE

Diverse types of indirect evidence (see Fay et al., 1968); Stewart et al., 1969; Stewart and Lex, 1970; van Gorkom and Donze, 1971; Nielson et al., 1971; Weare and Benemann, 1973; Fleming and Haselkorn, 1973) have been interpreted as supporting the idea that all of the nitrogenase activity of aerobically grown, heterocyst-forming cyanobacteria is localized in the heterocysts. Particularly persuasive that heterocysts are foci of nitrogen fixation were observations made by van Gorkom and Donze (1971). During nitrogen starvation, the blue pigment-proteins of cyanobacteria are degraded. These authors showed that upon re-exposure of filaments to N_2 in the presence of oxygen, the pigmented proteins reformed first adjacent to heterocysts. Reformation of these proteins showed that nitrogen was no longer deficient. However, none of the indirect experiments ruled out the hypothesis suggested by the fact that nitroblue tetrazolium is reduced predominantly in vegetative cells adjacent to heterocysts (Fay and Kulasooriya, 1972). That is, much of the N_2 fixation might take place in vegetative cells adjacent to heterocysts, perhaps as a result of an interaction with the heterocysts. Direct measurements of nitrogenase activity by isolated heterocysts resolved the issue.

Heterocysts isolated from *Anabaena variabilis* have been shown to account for 60% of the nitrogenase activity of the unharvested filaments from which they were derived, and 70-75% of the activity of centrifuged filaments (Peterson and Wolk, 1978b). When proteins of *A. variabilis* were subjected to electrophoresis on anaerobic, non-denaturing gels, a protein band corresponding to nitrogenase was identified on the basis of a number of criteria, including its content of molybdenum and non-heme iron, lability to oxygen, suppression by ammonium, and subunit-polypeptide molecular weights. Quantitation by incorporation of ^{55}Fe into the protein showed that an average of 91% of the nitrogenase of intact filaments remained in the purified, isolated heterocysts (Peterson and Wolk, 1978b; see Fig. 1). Thus, all or nearly all of the nitrogenase and nitrogenase activity of whole, aerobically grown filaments of *A. variabilis* are associated with the heterocysts. It would clearly be desirable to ascertain whether similarly high recoveries are possible in work with other cyanobacteria.

HETEROCYST METABOLISM

Heterocysts of *Anabaena cylindrica* lack RuBP carboxylase (RuBPC: Winkenbach and Wolk, 1973; Codd and Stewart, 1977; see Table 1), therefore fix little or no CO_2 (Wolk, 1968; Stewart et al., 1969), and in consequence conserve energy and reductant which can be used to fix N_2. They also lack phosphoribulokinase, the enzyme which generates the substrate for RuBPC (Codd et al., 1980); polyhedral inclusions thought to contain much of the RuBPC in vegetative cells (Stewart and Codd, 1975); the oxygenase activity of RuBPC; and two enzymes, phosphoglycollate phosphatase and glycolate dehydrogenase, which could metabolize the phosphoglycollate produced by RuBP oxygenase (Codd et al., 1980). Another enzymatic constituent of the reductive pentose phosphate pathway, glyceraldehyde-3-phosphate dehydrogenase, is also essentially undetectable in heterocysts (Winkenbach and Wolk, 1973; Lex and Carr, 1974). This enzyme is also a constituent of the glycolytic and gluconeogenic pathways, so that carbohydrate neogenesis in heterocysts is presumably very limited. However, about 43% of the dry weight of a heterocyst is envelope carbohydrate (Dunn and Wolk, 1970; Lambein and Wolk, 1973; Cardemil and Wolk, 1976). If neogenesis falls short of deposition there must be a transfer of carbohydrate from vegetative cells to heterocysts. In fact, such a transfer of products of $^{14}CO_2$ assimilation takes place (Wolk, 1968; and see below).

Glucose-6-phosphate dehydrogenase and 6-phosphogluconate dehydrogenase, in contrast to the enzymes considered thus far, appear

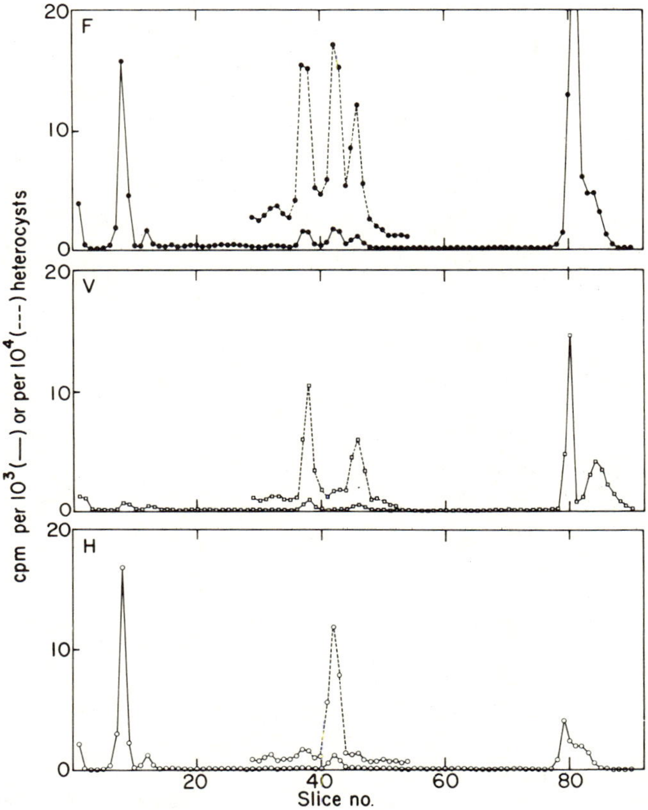

Fig. 1. Distribution of radioactivity in nondenaturing electrophoretograms of soluble extracts of filaments (F), heterocysts (H), and a vegetative cell fraction (V), from an ^{55}Fe-labeled culture of Anabaena variabilis. Counts per minute of ^{55}Fe are expressed on the basis of equal numbers of heterocysts; i.e., they are calculated for equal amounts of filaments. In this experiment, the extract of heterocysts accounted for 103%, and the extract of vegetative cells for 6%, of the nitrogenase (the band centered at slice no. 8), calculated on the basis of the extract from filaments that had not been incubated with lysozyme. The band at slice no. 42 is nitrogenase reductase, the extremely oxygen-labile Fe-protein of the nitrogenase complex, which was recovered with about 22% lower efficiency in the isolated heterocysts. Reproduced from Peterson and Wolk (1978b), with permission.

Table 1. Relative activities of enzymes of heterocysts and vegetative cells of aerobically grown *Anabaena cylindrica* (from Wolk, 1980, with permission).

Enzymes of nitrogen metabolism	Specific activity in heterocysts much greater than in vegetative cells	Relative specific activity close to 1:1	Specific activity in heterocysts much lower than in vegetative cells
Nitrogenase	Reference a		
Glutamine synthetase		b,c,d	
Glutamate synthase			c
Alanine dehydrogenase (NADH)		b,e	
Glutamic acid dehydrogenase (NADPH)		b	
Glutamate-oxaloacetate transaminase		b,e	
Glutamate-pyruvate transaminase		b	
Enzymes of carbohydrate metabolism			
Reductive pentose phosphate cycle			
RuBP carboxylase			f,g,h
RuBP oxygenase			h
Ru5P kinase			h
F-1,6-diphosphate aldolase			i
Phosphoglycollate phosphatase			h
Glycollate dehydrogenase			h
GAP dehydrogenase			f,i

Table 1. (continued)

Oxidative pentose phosphate cycle			
G6P dehydrogenase	f,i,j		
6PG dehydrogenase	f,i		
Other			
Hexokinase		f,i	
Phosphofructokinase			i
Electron transfer pathways			
"Uptake" hydrogenase	k		
Ordinary hydrogenase		l	
PS I		m,n	
PS II			n,o
Fd-NADP$^+$ oxidoreductase		j	
NADPH oxidase		o	
Miscellaneous			
Superoxide dismutase		p,q	
Catalase		q	

[a] Peterson and Wolk, 1978b
[b] Stewart et al., 1975
[c] Thomas et al., 1977
[d] Dharmawardene et al., 1973
[e] Scott and Fay, 1972
[f] Winkenbach and Wolk, 1973
[g] Stewart and Codd, 1975; Codd and Stewart, 1977
[h] Codd et al., 1980
[i] Lex and Carr, 1974
[j] Apte et al., 1978
[k] Peterson and Wolk, 1978a
[l] Eisbrenner et al., 1978; Tel-Or et al., 1978
[m] Wolk and Simon, 1969
[n] Bradley and Carr, 1971; Donze et al., 1972; Tel-Or and Stewart, 1977
[o] Bradley and Carr, 1976
[p] Daday et al., 1977
[q] Henry et al., 1978

to be present at much higher specific activity in heterocysts than in vegetative cells (Winkenbach and Wolk, 1973; Lex and Carr, 1974; Apte et al., 1978). The observed ratio of specific activities for glucose-6-phosphate dehydrogenase ranges from 6- to 7-fold, uncorrected for possible leakage from damaged heterocysts (Lex and Carr, 1974; Apte et al., 1978), to 60- to 70-fold, with correction (Winkenbach and Wolk, 1973). However, redetermination of this ratio is warranted because it has been ascertained that the regulation of cyanobacterial glucose-6-phosphate dehydrogenase involves slow relaxation between states of very different activity (Schaeffer and Stanier, 1978); that the kinetics of that enzyme differ in heterocysts and vegetative cells (Lex and Carr, 1974); and that the activity of the enzyme is also controlled by NADPH, ATP and RuBP, so that the *in vivo* and *in vitro* activities may differ. No such complications are known for 6-phosphogluconate dehydrogenase. High activities of both enzymes may serve to maintain a very high ratio of reduced to oxidized $NADP^+$, so that NADPH can -- via ferredoxin -- reduce nitrogenase. Thus, *in vitro* fluxes (Winkenbach and Wolk, 1973) catalyzed by these enzymes may greatly exceed fluxes *in vivo*. At least one enzyme of the oxidative pentose phosphate cycle, fructose-1,6-diphosphate aldolase, appears to be absent from heterocysts (Lex and Carr, 1974), so that ribulose-5-phosphate (or a product of its metabolism) may return to vegetative cells rather than being recycled within heterocysts.

Hexokinase activities in heterocysts (Winkenbach and Wolk, 1973) were inadequate to account for the electron flow to N_2 calculated from the growth rates of the cultures. Thus, if the oxidative pentose phosphate cycle is the source of electrons for nitrogenase, the mediator of electron flow from vegetative cells to heterocysts (see below) would appear not to be glucose or fructose. It remains to be determined whether the above results pertaining to carbohydrate metabolic enzymes are valid for heterocysts of cyanobacteria other than *A. cylindrica*.

As already mentioned, and as will be discussed further, later, electrons and carbon move from vegetative cells to heterocysts. What compound or compounds mediate this transport? Jüttner and Carr (1976) predigested filaments of *A. cylindrica* with lysozyme, exposed them to brief pulses of $^{14}CO_2$, and then isolated the heterocysts rapidly in the cold after rupture of the vegetative cells by passage through a French pressure cell. After a 10-s pulse, the shortest pulse reported, the principal compound labeled in heterocysts was a glucose-containing disaccharide tentatively identified as maltose, whereas the principal compounds labeled in vegetative cells were Calvin cycle intermediates. If the disaccharide, once identified with certainty, can be shown to support a high level of nitrogenase activity by isolated heterocysts (see Wolk, 1979a), it is presumably a principal substance transferring reducing equivalents and carbon into heterocysts.

Anabaena variabilis fixes nitrogen during growth in the dark at about 25% of the corresponding rate in the light (Wolk and Shaffer, 1976), and even obligately photoautotrophic *A. cylindrica* can fix nitrogen in the dark (Donze, 1973). By what pathways are electrons and energy supplied to nitrogenase in heterocysts in the dark? In the presence of $NADP^+$, substrates of $NADP^+$-linked dehydrogenases -- glucose-6-phosphate, 6-phosphogluconate and isocitrate -- can serve as electron donors to nitrogenase in heterocysts isolated by probe cavitation (Lockau et al., 1978), and provided that ferredoxin is also present, to cyanobacterial nitrogenase *in vitro* (Smith et al., 1971). In heterocysts in the dark, ferredoxin-$NADP^+$ reductase may therefore serve to reduce not $NADP^+$, but ferredoxin. However, which pathway of electron transfer to nitrogenase predominates in the dark is unknown.

At low intensities of light or in the dark, oxygen stimulates the nitrogenase activity of whole filaments (Wolk, 1970; Weare and Benemann, 1973; Benemann and Weare, 1974; Donze et al., 1974; Bothe et al., 1977) and -- if H_2 is present -- of isolated heterocysts (Peterson and Wolk, 1978b), suggesting that oxidative phosphorylation supplies the requisite ATP. Indeed, coupling of ATP formation to the oxidation of hydrogen (Peterson and Burris, 1978; see also Bothe et al., 1977) and of other substrates (Scott and Fay, 1972; Carr and Bradley, 1973; Tel-Or and Stewart, 1977; Peterson and Burris, 1976) has been demonstrated directly. However, the *in vivo* substrates of oxidative phosphorylation in heterocysts are unknown.

Isolated heterocysts lack photosystem-II activity (Bradley and Carr, 1971, 1976; Donze et al., 1972; Tel-Or and Stewart, 1977). Electrophoresis of proteins from detergent-disrupted membranes from isolated heterocysts shows two bands of chlorophyll-bearing protein which are found also in vegetative cells; a third band, associated with photosystem II in the latter cells, is missing (Reinman and Thornber, 1979). Presumably, therefore, heterocysts within filaments do not generate oxygen which might inactivate nitrogenase (Fay et al., 1968), so that reducing equivalents used by heterocyst nitrogenase must be derived from the vegetative cells.

The amounts of chlorophyll and numbers of photosystem-I reaction centers per cell are not greatly dissimilar in heterocysts and vegetative cells (Wolk and Simon, 1969; Donze et al., 1972; Tel-Or and Stewart, 1977). Action spectra of light-stimulated acetylene reduction by filaments (Fay, 1970) and isolated heterocysts (Peterson and Ke, 1979) implicate photosystem I. Even the small amount of biliproteins remaining in isolated heterocysts appears to energize photosystem I (Peterson and Ke, 1979). In the light, isolated heterocysts express nitrogenase activity when supplied with neither ATP nor reductant (Lockau et al., 1978;

Peterson and Wolk, 1978b; and see also Wolk and Wojciuch, 1971). Whether the heterocysts generate only ATP, only reductant, or both in response to illumination, is unknown. Certainly, heterocysts can photophosphorylate in the presence of phenazine methosulphate, an artificial electron carrier (Tel-Or and Stewart, 1977). However, it remains undetermined whether non-cyclic electron flow to nitrogenase via photosystem I normally occurs in heterocysts; if so, what the identities of the electron donor and electron transport pathway are to photosystem I; and whether cyclic or non-cyclic photophosphorylation, or both, takes place.

Superoxide may be generated in both vegetative cells and heterocysts by photosystem-I driven reduction of O_2. Buildup of this toxic anion is prevented by superoxide dismutase, the specific activity of which is approximately the same in vegetative cells and heterocysts (Daday et al., 1977; Henry et al., 1978).

In the light, hydrogen gas is the most effective known electron donor to nitrogenase in isolated heterocysts (Bothe et al., 1977; Peterson and Wolk, 1978b; Eisbrenner and Bothe, 1979; see also Wolk and Wojciuch, 1971; and Benemann and Weare, 1974). Some isolated heterocysts express nitrogenase activity in the dark in response to ATP and dithionite but not to ATP and H_2 (Wolk and Wojciuch, 1971; Eisbrenner et al., 1978), suggesting -- but not proving -- that H_2 donates its electrons to photosystem I in heterocysts.

An ordinary hydrogenase is present in heterocysts and vegetative cells of *A. cylindrica* and *N. muscorum*, with similar specific activity (Tel-Or et al., 1978; Eisbrenner et al., 1978). However, the uptake of hydrogen (at up to 1.74 μmol/(mg Chl·hr)) in the dark by three aerobically grown strains of *Anabaena* is -- together with heterocysts -- greatly reduced or eliminated when cultures are grown with ammonium (Peterson and Wolk, 1978a; Eisbrenner et al., 1978; Daday et al., 1977; see also Tel-Or et al., 1978). Moreover, 86 and 84% of the total *in vitro* "uptake" hydrogenase activity of filaments of *Anabaena* 7120 and *Cylindrospermum licheniforme*, respectively, were recovered in isolated heterocysts, with very little or no activity observed in extracts of vegetative cells (Peterson and Wolk, 1978a; Hirosawa and Wolk, 1979). However, the apparent implication of these results, that all or virtually all of the *uptake* hydrogenase is localized in the heterocysts, is called into question by the observation that a non-heterocystous strain of *Anabaena*, grown and assayed aerobically, consumed H_2 at a rate of 3 μmol/(mg Chl·hr) (Eisbrenner et al., 1978). The uptake hydrogenase activity present in heterocysts can recycle reducing equivalents "lost" by nitrogenase, can couple to phosphorylation (Peterson and Burris, 1978), and can help to maintain the activity of nitrogenase in the presence of oxygen (Bothe et al., 1977; Peterson and Wolk, 1978b).

Many enzymes of nitrogen metabolism, glutamine synthetase in particular (Scott and Fay, 1972; Dharmawardene et al., 1973; Stewart et al., 1975; Thomas et al., 1977), differ little in specific activity between heterocysts and vegetative cells (Table 1). As shown by use of ^{13}N, the N_2-derived nitrogen incorporated into glutamine is further metabolized principally by glutamate synthase (Wolk et al., 1976; Meeks et al., 1978). However, isolated heterocysts metabolize ^{13}N-labeled N_2 and $^{13}NH_4^+$ essentially only to glutamine; and as measured in vitro, glutamate synthase appears to be restricted to vegetative cells (Thomas et al., 1977). Thus, glutamine formed -- principally in heterocysts (Thomas et al., 1977; Meeks et al., 1977) -- from N_2-derived ammonium must move into vegetative cells in order to undergo amide transfer. It also follows that most of the glutamate from which the glutamine is synthesized is derived from the vegetative cells (Fig. 2; Thomas et al., 1977).

OXYGEN PROTECTION

Upon addition of ammonium, the nitrogenase activity of filaments of *A. cylindrica* decreases in air but not under anaerobic conditions (Ohmori and Hattori, 1974; see also Weare and Benemann, 1973). This observation suggests that together with dinitrogen, some oxygen enters heterocysts. The heterocysts presumably have to protect nitrogenase from the oxygen which enters them.

When filaments are subjected to nitrogen stepdown under aerobic conditions (see Fogg, 1949), nitrogenase activity appears from 0 to 12 h after the appearance of mature heterocysts (Neilson et al., 1971; Kulasooriya et al., 1972; Fleming and Haselkorn, 1973; Bradley and Carr, 1976; Stacey et al., 1979). However, under anaerobic conditions, nitrogenase can be active even in the absence of heterocysts (Rippka and Stanier, 1978). These results suggest that the appearance of nitrogenase activity during normal development may be controlled in part by the attainment of an adequately low intracellular partial pressure of oxygen. (However, in one mutant (PM-10) of *Anabaena* CA, nitrogenase is synthesized before the oxygen-protection system develops fully (Grillo et al., 1979).)

The final stages of heterocyst maturation, seemingly required for nitrogenase activity, involve completion of the laminated, glycolipid layer of the heterocyst envelope, and proliferation of lamellae (possibly related to oxidases: see below; Lang, 1965; Kulasooriya et al., 1972; Wilcox et al., 1973; see also Fogg, 1951, and Wahal et al., 1974). The glycolipid layer encompasses the protoplast of the heterocyst except at its junctions to adjacent vegetative cells (Winkenbach et al., 1972). If this layer prevents lipophobic substances from passing directly between the external milieu and the inside of the heterocyst, it may constrain

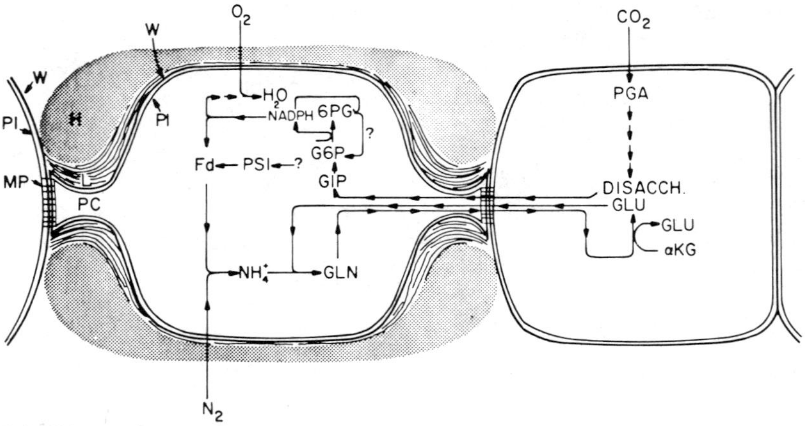

Fig. 2. Diagram illustrating the principal structural differences and known interactions between a heterocyst (at left) and a vegetative cell (at right). Outside of the wall (W) of the heterocyst is an envelope consisting principally of a laminated, glycolipid layer (L) and a homogeneous, polysaccharide layer (H). Microplasmodesmata (MP) join the plasma membranes (Pl) of the two types of cells at the end of the pore channel (PC) of the heterocyst. A disaccharide formed by photosynthesis in the vegetative cells moves into heterocysts, and may then be metabolized to glucose-6-phosphate and oxidized by the oxidative pentose phosphate pathway. Pyridine nucleotide (NADPH) reduced by this pathway can donate electrons to O_2 to maintain reducing conditions within the heterocysts, and can reduce ferredoxin (Fd). Ferredoxin can also be reduced by photosystem I. Reduced ferredoxin can donate electrons to nitrogenase, which reduces N_2 to NH_4^+. Glutamate produced principally by vegetative cells reacts with the NH_4^+ to form glutamine. The glutamine moves into the vegetative cells, where it reacts with α-ketoglutarate to form two molecules of glutamate. Reproduced from Wolk (1979b), with permission.

heterocysts to interact with adjacent vegetative cells (Winkenbach et al., 1972), possibly via plasmodesmata-like structures present at the intercellular junctions (Wildon and Mercer, 1963; Lang and Fay, 1971; Giddings and Staehelin, 1978).

The envelope glycolipid which is most abundant in heterocysts of *A. cylindrica*, 1-(O-α-D)-glycopyranosyl-3,25-hexacosanediol (Bryce et al., 1972; Lambein and Wolk, 1973), is present, but not always as the most plentiful glycolipid, in heterocysts of other taxa (Lorch and Wolk, 1974; see also Nichols and Wood, 1968, and Wahal et al., 1973). These glycolipids appear to be absent from vegetative cells (Walsby and Nichols, 1969; Wolk and Simon, 1969). The glycolipid layer has a quasi-crystalline, laminated ultrastructure with a periodicity *in situ* of 7-8 nm (Golecki and Drews, 1974; Granhall, 1976; Giddings and Staehelin, 1978; and see Winkenbach et al., 1972).

The envelope glycolipids are deficient in certain mutants of *A. variabilis* which have nitrogenase activity under microaerobic, but not aerobic, conditions, and are formed by all revertants of those mutants which are capable of aerobic fixation of dinitrogen (Haury and Wolk, 1978). Thus, the glycolipid layer may normally so greatly reduce the rate of penetration of oxygen into heterocysts that oxidases present in the heterocysts can scavenge most of that which does enter. Concordantly, the glycolipid layer of heterocysts formed by *Anabaena* strain 7120 grown under strictly anaerobic conditions appears to be irregularly formed or absent - perhaps the hydroxylated aglycones of the glycolipids cannot be synthesized in the total absence of oxygen -- and the nitrogenase activity of the organism is oxygen-sensitive (Rippka and Stanier, 1978). However, it remains possible that the effects of the mutations, and of the transition from anaerobiosis to aerobiosis, on nitrogenase activity may be independent of the quantities of glycolipid present.

Outside of the glycolipid layer of the heterocyst envelope is a homogeneous layer consisting largely of polysaccharide (Dunn and Wolk, 1970; Winkenbach et al., 1972). The polysaccharides from the envelopes of heterocysts of two species of *Anabaena* and one species of *Cylindrospermum* have been analyzed. They consist of long chains of similar but not identical subunits, in which each subunit has a β-1,3-linked backbone of the form mannosyl-glucosyl-glucosyl-glucose. The sugars in the backbone bear xylose, galactose and sometimes mannose, glucose and/or arabinose, in mono- or disaccharide substituents (Cardemil and Wolk, 1976, 1979, and unpublished observations).

The polysaccharide layer of the envelope may protect the glycolipid layer. In certain single-site mutants which are incapable of aerobic dinitrogen fixation, the carbohydrate layer of the envelope of all heterocysts develops irregularly, so that some regions remain very thin; N_2-fixing revertants form normal walls (Currier et al., 1977). These results suggest that the polysaccharide layer, like the glycolipid layer, may normally impede the entrance of oxygen.

Oxidases capable of reducing oxygen which penetrates the envelope may constitute a "second line of defense". Both NADPH and NADH are oxidized (Scott and Fay, 1972; Carr and Bradley, 1973; Peterson and Burris, 1976); and the respiratory rate of isolated heterocysts of Anabaena strain 7120 is increased by provision of glucose, fructose-6-phosphate, glucose-6-phosphate, 6-phosphogluconate, pyruvate, or isocitrate, if $NADP^+$ is present (Peterson and Burris, 1976). Nonetheless, the *in vivo* substrate for heterocyst respiration has not been identified, and it is not known with certainty whether there is a competition between oxygen and substrates of nitrogenase for electrons, within heterocysts. Evidence for such a competition, which could explain the fully reversible loss for nitrogenase activity by mutant GM-9 of *Anabaena* CA in the presence of oxygen (Gotto et al., 1979), *has* been presented for whole filaments (Lex et al., 1972; Lex and Stewart, 1973).

REFERENCES

Apte, S. K., Rowell, P., and Stewart, W. D. P., 1978, Electron donation to ferredoxin in heterocysts of the N_2-fixing alga *Anabaena cylindrica*, *Proc. R. Soc. Lond.*, *B* 200:1.

Benemann, J. R., and Weare, N. M., 1974, Nitrogen fixation by *Anabaena cylindrica*. III. Hydrogen-supported nitrogenase activity, *Arch. Microbiol.*, 101:401.

Bothe, H., Tennigkeit, J., and Eisbrenner, G., 1977, The utilization of molecular hydrogen by the blue-green alga *Anabaena cylindrica*, *Arch. Microbiol.*, 114:43.

Bradley, S., and Carr, N. G., 1971, The absence of a functional photosystem II in heterocysts of *Anabaena cylindrica*, *J. Gen. Microbiol.*, 68:xiii.

Bradley, S., and Carr, N. G., 1976, Heterocysts and nitrogenase development in *Anabaena cylindrica*, *J. Gen. Microbiol.*, 96:175.

Bryce, T. A., Welti, D., Walsby, A. E., and Nichols, B. W., 1972, Monohexoside derivatives of long-chain polyhydroxy alcohols; a novel class of glycolipid specific to heterocystous algae, *Phytochem.*, 11:295.

Cardemil, L., and Wolk, C. P., 1976, The polysaccharides from heterocyst and spore envelopes of a blue-green alga. Methylation analysis and structure of the backbones, *J. Biol., Chem.*, 251:2967.

Cardemil, L., and Wolk, C. P., 1979, The polysaccharides from heterocyst and spore envelopes of a blue-green alga. Struc-

ture of the basic repeating unit, J. Biol. Chem., 254:736.
Carr, N. G., 1979, Differentiation in filamentous cyanobacteria, in: "Developmental Biology of Prokaryotes," J. H. Parish, ed., Blackwell Scient. Publ., Oxford.
Carr, N. G., and Bradley, S., 1973, Aspects of development in blue-green algae, Symp. Soc. Gen. Microbiol., 23:161.
Codd, G. A., Okabe, K., and Stewart, W. D. P., 1980, Cellular compartmentation of photosynthetic and photorespiratory enzymes in the heterocystous cyanobacterium Anabaena cylindrica, Arch. Microbiol., 124:149.
Codd, G. A., and Stewart, W. D. P., 1977, Ribulose-1,5-diphosphate carboxylase in heterocysts and vegetative cells of Anabaena cylindrica, FEMS Microbiol. Lett., 2:247.
Currier, T. C., Haury, J. F., and Wolk, C. P., 1977, Isolation and preliminary characterization of auxotrophs of a filamentous cyanobacterium, J. Bacteriol., 129:1556.
Daday, A., Platz, R. A., and Smith, G. D., 1977, Anaerobic and aerobic hydrogen gas formation by the blue-green alga Anabaena cylindrica, Appl. Env. Microbiol., 34:478.
Dharmawardene, M. W. N., Haystead, A., and Stewart, W. D. P., 1973, Glutamine synthetase of the nitrogen-fixing alga Anabaena cylindrica, Arch. Mikrobiol., 90:281.
Donze, M., 1973, Pigment systems in algae. Electron transport to nitrogenase in Anabaena, Thesis, Rijksuniversiteit, Leiden.
Donze, M., Haveman, J., and Schiereck, P., 1972, Absence of photosystem 2 in heterocysts of the blue-green alga Anabaena, Biochim. Biophys. Acta, 256:157.
Donze, M., Raat, A. J. P., and van Gorkom, H. J., 1974, Supply of ATP and reductant to nitrogenase in the blue-green algae Anabaena cylindrica, Plant Sci. Lett., 3:35.
Dunn, J. H., and Wolk, C. P., 1970, Composition of the cellular envelopes of Anabaena cylindrica, J. Bacteriol., 103:153.
Eisbrenner, G., and Bothe, H., 1979, Modes of electron transfer from molecular hydrogen in Anabaena cylindrica, Arch. Microbiol., 123:37.
Eisbrenner, G., Distler, E., Floener, L., and Bothe, H., 1978, The occurrence of the hydrogenase in some blue-green algae, Arch. Microbiol., 118:177.
Fay, P., 1970, Photostimulation of nitrogen fixation in Anabaena cylindrica, Biochim. Biophys. Acta, 216:253.
Fay, P., and Kulasooriya, S. A., 1972, Tetrazolium reduction and nitrogenase activity in heterocystous blue-green algae, Arch. Mikrobiol., 87:341.
Fay, P., Stewart, W. D. P., Walsby, A. E., and Fogg, G. E., 1968, Is the heterocyst the site of nitrogen fixation in blue-green algae?, Nature, 220:810.
Fleming, H., and Haselkorn, R., 1973, Differentiation in Nostoc muscorum: nitrogenase is synthesized in heterocysts, Proc. Natl. Acad. Sci. USA, 70:2727.
Fogg, G. E., 1949, Growth and heterocyst production in Anabaena

cylindrica Lemm. II. In relation to carbon and nitrogen metabolism, Ann. Bot. n.s., 13:241.

Fogg, G. E., 1951, Growth and heterocyst production in Anabaena cylindrica Lemm. III. The cytology of heterocysts, Ann. Bot. n.s., 15:23.

Giddings, T. H., Jr., and Staehelin, L. A., 1978, Plasma membrane architecture of Anabaena cylindrica: occurrence of microplasmodesmata and changes associated with heterocyst development and the cell cycle, Cytobiol., 16:235.

Golecki, J. R., and Drews, G., 1974, Zur Struktur der Blaualgen-Zellwand. Gefrierätzuntersuchungen an normalen und extrahierten Zellwänden von Anabaena variabilis, Cytobiol., 8:213.

Gotto, J. W., Tabita, F. R., and van Baalen, C., 1979, Mutants of Anabaena strain CA altered in their ability to grow under nitrogen-fixing conditions, J. Bacteriol., 140:327.

Granhall, U., 1976, The presence of cellulose in heterocyst envelopes of blue-green algae and its role in relation to nitrogen fixation, Physiol. Plant., 38:208.

Grillo, J. F., Bottomley, P. J., van Baalen, C., and Tabita, F. R., 1979, A mutant of Anabaena sp. CA with oxygen-sensitive nitrogenase activity, Biochem. Biophys. Res. Commun., 89:685.

Haselkorn, R., 1978, Heterocysts, Ann. Rev. Plant Physiol., 29: 319.

Haury, J. F., and Wolk, C. P., 1978, Classes of Anabaena variabilis mutants with oxygen-sensitive nitrogenase activity, J. Bacteriol., 136:688.

Henry, L. E. A., Gogotov, I. N., and Hall, D. O., 1978, Superoxide dismutase and catalase in the protection of the proton-donating systems of nitrogen fixation in the blue-green alga Anabaena cylindrica, Biochem. J., 174:373.

Hirosawa, T., and Wolk, C. P., 1979, Factors controlling the formation of akinetes adjacent to heterocysts in the cyanobacterium Cylindrospermum licheniforme Kütz., J. Gen. Microbiol., 114:423.

Jüttner, F., and Carr, N. G., 1976, The movement of organic molecules from vegetative cells into the heterocysts of Anabaena cylindrica, in: "Proc. Second Internat. Symp. Photosynth. Prokaryotes," G. A. Codd and W. D. P. Stewart, eds., Dundee.

Kulasooriya, S. A., Lang, N. J., and Fay, P., 1972, The heterocysts of blue-green algae. III. Differentiation and nitrogenase activity, Proc. R. Soc. Lond., B 181:199.

Lambein, F., and Wolk, C. P., 1973, Structural studies on the glycolipids from the envelope of the heterocyst of Anabaena cylindrica, Biochem., 12:791.

Lang, N. J., 1965, Electron microscopic study of heterocyst development in Anabaena azollae Strasburger, J. Phycol., 1:127.

Lang, N. J., and Fay, P., 1971, The heterocysts of blue-green algae. II. Details of ultrastructure, Proc. Roy. Soc. Lond., B 178:193.

Lex, M., and Carr, N. G., 1974, The metabolism of glucose by heterocysts and vegetative cells of Anabaena cylindrica, Arch. Microbiol., 101:161.

Lex, M., Silvester, W. B., and Stewart, W. D. P., 1972, Photorespiration and nitrogenase activity in the blue-green alga, Anabaena cylindrica, Proc. R. Soc. Lond., B 180:87.

Lex, M., and Stewart, W. D. P., 1973, Algal nitrogenase, reductant pools and photosystem I activity, Biochim. Biophys. Acta, 292:436.

Lockau, W., Peterson, R. B., Wolk, C. P., and Burris, R. H., 1978, Modes of reduction of nitrogenase in heterocysts isolated from Anabaena species, Biochim. Biophys. Acta, 502:298.

Lorch, S. K., and Wolk, C. P., 1974, Application of gas-liquid chromatography to study of the envelope lipids of heterocysts, J. Phycol., 10:352.

Meeks, J. C., Wolk, C. P., Thomas, J., Lockau, W., Shaffer, P. W., Austin, S. M., Chien, W.-S., and Galonsky, A., 1977, The pathways of assimilation of $^{13}NH_4^+$ by the cyanobacterium, Anabaena cylindrica, J. Biol. Chem., 252:7894.

Meeks, J. C., Wolk, C. P., Lockau, W., Schilling, N., Shaffer, P., W., and Chien, W.-S., 1978, Pathways of assimilation of $(^{13}N)N_2$ and $^{13}NH_4^+$ by cyanobacteria with and without heterocysts, J. Bacteriol., 134:125.

Nielson, A., Rippka, R., and Kunisawa, R., 1971, Heterocyst formation and nitrogenase synthesis in Anabaena sp. A kinetic study, Arch. Mikrobiol., 76:139.

Nichols, B. W., and Wood, B. J. B., 1968, New glycolipid specific to nitrogen-fixing blue-green algae, Nature, 217:767.

Ohmori, M., and Hattori, A., 1974, Effect of ammonia on nitrogen fixation by the blue-green alga Anabaena cylindrica, Plant Cell Physiol., 15:131.

Peterson, R. B., and Burris, R. H., 1976, Properties of heterocysts isolated with colloidal silica, Arch. Microbiol., 108:35.

Peterson, R. B., and Burris, R. H., 1978, Hydrogen metabolism in isolated heterocysts of Anabaena 7120, Arch. Microbiol., 116:125.

Peterson, R. B., and Ke, B., 1979, Pigment composition and fluorescence properties of heterocysts isolated from Anabaena variabilis, in: "Abst. Third Internat. Symp. Photosynth. Prokaryotes," J. M. Nichols, ed., Dept. of Biochem., Univ. of Liverpool, Liverpool.

Peterson, R. B., and Wolk, C. P., 1978a, Localization of an uptake hydrogenase in Anabaena, Plant Physiol., 61:688.

Peterson, R. B., and Wolk, C. P., 1978b, High recovery of nitrogenase activity and of ^{55}Fe-labeled nitrogenase in heterocysts isolated from Anabaena variabilis, Proc. Natl. Acad. Sci. USA, 75:6271.

Reinman, S., and Thornber, J. P., 1979, The electrophoretic isolation and partial characterization of three chlorophyll-protein complexes from blue-green algae, Biochem. Biophys. Acta, 547:188.

Rippka, R., and Stanier, R. Y., 1978, The effects of anaerobiosis on nitrogenase synthesis and heterocyst development by nostocacean cyanobacteria, J. Gen. Microbiol., 105:83.

Schaeffer, F., and Stanier, R. Y., 1978, Glucose-6-phosphate dehydrogenase of Anabaena sp. Kinetic and molecular properties, Arch. Microbiol., 116:9.

Scott, W. E., and Fay, P., 1972, Phosphorylation and amination in heterocysts of Anabaena variabilis, Br. Phycol. J., 7:283.

Smith, R. V., Noy, R. J., and Evans, M. C. W., 1971, Physiological electron donors to the nitrogenase of the blue-green alga Anabaena cylindrica, Biochim. Biophys. Acta, 253:104.

Stacey, G., Bottomley, P. J., van Baalen, C., and Tabita, F. R., 1979, Control of heterocyst and nitrogenase synthesis in cyanobacteria, J. Bacteriol., 137:321.

Stewart, W. D. P., and Codd, G. A., 1975, Polyhedral bodies (carboxysomes) of nitrogen-fixing blue-green algae, Br. Phycol. J., 10:273.

Stewart, W. D. P., Haystead, A., and Dharmawardene, M. W. N., 1975, Nitrogen assimilation and metabolism in blue-green algae, in: "Nitrogen Fixation in Free-living Microorganisms," W. D. P. Stewart, ed., Cambridge Univ. Press, Cambridge, U.K.

Stewart, W. D. P., Haystead, A., and Pearson, H. W., 1969, Nitrogenase activity in heterocysts of blue-green algae, Nature, 224:226.

Stewart, W. D. P., and Lex, M., 1970, Nitrogenase activity in the blue-green alga Plectonema boryanum strain 594, Arch. Mikrobiol., 73:250.

Tel-Or, E., and Stewart, W. D. P., 1977, Photosynthetic components and activities of nitrogen-fixing isolated heterocysts of Anabaena cylindrica, Proc. R. Soc. Lond., B 198:61.

Tel-Or, E., Luijk, L. W., and Packer, L., 1978, Hydrogenase in N_2 fixing cyanobacteria, Arch. Biochem. Biophys. 185:185.

Thomas, J., Meeks, J. C., Wolk, C. P., Shaffer, P. W., Austin, S. M., and Chien, W.-S., 1977, Formation of glutamine from (^{13}N)-ammonia, (^{13}N)dinitrogen, and (^{14}C)glutamate by heterocysts isolated from Anabaena cylindrica, J. Bacteriol., 129:1545.

van Gorkom, H. J., and Donze, M., 1971, Localization of nitrogen fixation in Anabaena, Nature, 234:231.

Wahal, C. K., Reddy, P. M., and Talpasayi, E. R. S., 1973, Lipid composition of some blue-green algae, Ind. J. Exp. Biol., 11:588.

Wahal, C. K., Bhattacharya, N. C., and Talpasayi, E. R. S., 1974, Lipid synthesis and heterocyst development in Anabaena ambigua Rao, Ind. J. Exp. Biol., 12:92.

Walsby, A. E., and Nichols, B. W., 1969, Lipid composition of heterocysts, Nature, 221:673.

Weare, N. M., and Benemann, J. R., 1973, Nitrogen fixation by Anabaena cylindrica. I. Localization of nitrogen fixation in the heterocysts, Arch. Mikrobiol., 90:323.

Wilcox, M., Mitchison, G. J., and Smith, R. J., 1973, Pattern formation in the blue-green alga Anabaena. II. Controlled proheterocyst regression, J. Cell Sci., 13:637.

Wildon, D. C., and Mercer, F. V., 1963, The ultrastructure of the heterocyst and akinete of the blue-green algae, Arch. Mikrobiol., 47:19.

Winkenbach, F., and Wolk, C. P., 1973, Activities of enzymes of the oxidative and the reductive pentose phosphate pathways in heterocysts of a blue-green alga, Plant Physiol., 52:480.

Winkenbach, F., Wolk, C. P., and Jost, M., 1972, Lipids of membranes and of the cell envelope in heterocysts of a blue-green alga, Planta, 107:69.

Wolk, C. P., 1968, Movement of carbon from vegetative cells to heterocysts in Anabaena cylindrica, J. Bacteriol., 96:2138.

Wolk, C. P., 1970, Aspects of the development of a blue-green alga, Ann. N.Y. Acad. Sci., 175:641.

Wolk, C. P., 1979a, Developmental biology of cyanobacteria: advances and challenges, in: "Abst. Third Internat. Symp. Photosyn. Prokaryotes," J. M. Nichols, ed., Dept. of Biochem., Univ. of Liverpool, Liverpool.

Wolk, C. P., 1979b, Intercellular interactions and pattern formation in filamentous cyanobacteria, in: "Determinants of Spatial Organization," S. Subtelny and I. R. Konigsberg, eds., Academic Press, New York.

Wolk, C. P., 1980, Heterocysts, in: "The Biology of Cyanobacteria," N. G. Carr and B. A. Whitton, eds., Blackwell Scient. Publ., Oxford.

Wolk, C. P., and Shaffer, P. W., 1976, Heterotrophic micro- and macrocultures of a nitrogen-fixing cyanobacterium, Arch. Microbiol., 110:145.

Wolk, C. P., and Simon, R. D., 1969, Pigments and lipids of heterocysts, Planta, 86:92.

Wolk, C. P., Thomas, J., Shaffer, P. W., Austin, S. M., and Galonsky, A., 1976, Pathway of nitrogen metabolism after fixation of ^{13}N-labeled nitrogen gas by the cyanobacterium, Anabaena cylindrica, J. Biol. Chem., 251:5027.

Wolk, C. P., and Wojciuch, E., 1971, Photoreduction of acetylene by heterocysts, Planta, 97:126.

PHOTOSYNTHATE LIMITATION OF NITROGEN FIXATION IN THE BLUE-GREEN ALGA, *ANABAENA VARIABILIS*

H. Spiller, J. F. Haury[1], and K. T. Shanmugam[2]

Plant Growth Laboratory
University of California
Davis, CA 95616

Nitrogen fixation in nodulated legumes can be increased by CO_2 enrichment of the air surrounding leaves of soybean and peanut plants (Hardy and Havelka, 1974). In the above symbiotic associations organic compounds from CO_2 fixation of the plant are transported to the root nodules to supply energy for N_2 fixation. In heterocystous blue-green algae under conditions of nitrogen limitation filaments differentiate into heterocysts and vegetative cells (see Wolk, this volume and references therein for discussion of the physiology and biochemistry of heterocysts). The resultant two cell types carry different functions. Heterocysts contain only photosystem I producing ATP by cyclic photophosphorylation and lack photosystem II. The localization of nitrogenase in heterocysts in connection with an efficient O_2 uptake system provides protection for nitrogenase against denaturation by oxygen. The absence of oxygen evolution in heterocysts makes them highly dependent on the neighboring vegetative cells that contain both photosystems and can fix CO_2. Thus vegetative cells are equipped to deliver reductant supplied as an organic compound to the heterocyst (see Wolk, 1968). This raises the possibility that the generation of reductant by photosynthesis and CO_2 fixation as well as the transport of reductant (photosynthate) may limit nitrogen fixation and nitrogenase-mediated H_2 evolution in heterocystous

[1] ORE-IDA Foods, Inc., P. O. Box 10, Ontario, OR 97914

[2] Department of Microbiology & Cell Science, 1059 McCarty Hall, Institute of Food and Agricultural Sciences, University of Florida, Gainesville, FL 32611

blue-green algae. The experiments discussed in this communication show that nitrogenase activity and growth of Anabaena variabilis is limited by photosynthate availability.

MATERIALS AND METHODS

A. variabilis was axenically grown in 125 ml Erlenmeyer flasks with 50 ml culture volume in the medium of Allen and Arnon (1955) modified by dilution to ½ strength and containing 6 mM phosphate pH 7 as buffer. Cultures were shaken using a New Brunswick gyrotory shaker (model G10) at 100 rpm, under low intensity fluorescent light (3,000 Lux). The cultures were sparged with either pure N_2 or air, both gas phases enriged with 0.4% or 2% CO_2 in gas mixing chambers from Matheson or supplied with 10 mM fructose. Growth temperature was 24 C. The optical density of cells was measured after sonicating 1 ml cell suspension for 16-20 sec to generate 3-6 cell filament pieces, and after proper dilution of the suspension to OD_{750} nm $\leq$ 0.55. Nitrogenase activity was assayed by placing 1.5 ml of cell suspensions diluted with growth medium in 10 ml reaction vials, which were degassed and flushed with argon before adding C_2H_2 to establish a 10% C_2H_2 gas phase. Incubation under 5,000 Lux light was carried out for 1.5 h after which C_2H_4 formation was measured in a Varian model 1400 gas chromatograph.

RESULTS

Anabaena variabilis depends on the atmospheric CO_2 for production of cell material under phototrophic growth conditions. Production of organic compounds by the vegetative cells becomes much more important under conditions in which N_2 is the sole nitrogen source, since the organic compounds serve an additional function as the carriers of reducing power for nitrogenase activity in the heterocysts. This raises the possibility that the low concentrations of CO_2 present in the air (0.033%) may limit the growth under N_2-fixing conditions. To study this possibility, the cultures were exposed to higher concentrations of CO_2 and their growth was followed. These results are presented in Figure 1. Upon transfer to a fresh medium containing N_2 as the sole nitrogen source, A. variabilis culture showed a lag period of varying length of time depending on the CO_2 concentration in the sparging gas mixture. After this initial lag period, a rapid increase in cell mass was observed during the early exponential phase of the culture, for about one or two generations, presumably at the expense of endogenous reserves. After about 5 days, the culture reached a specific growth rate that was a reflection of the CO_2 concentration. Doubling time of the culture in the presence of air (CO_2 concentration, 0.033%) is 6.5 days at 25 C. This value decreases to 3.2 days in the presence of 0.4% CO_2 and to 2.1 days in the presence of 2% CO_2. Since cell growth requires both fixed nitrogen and carbon, NH_4^+ was

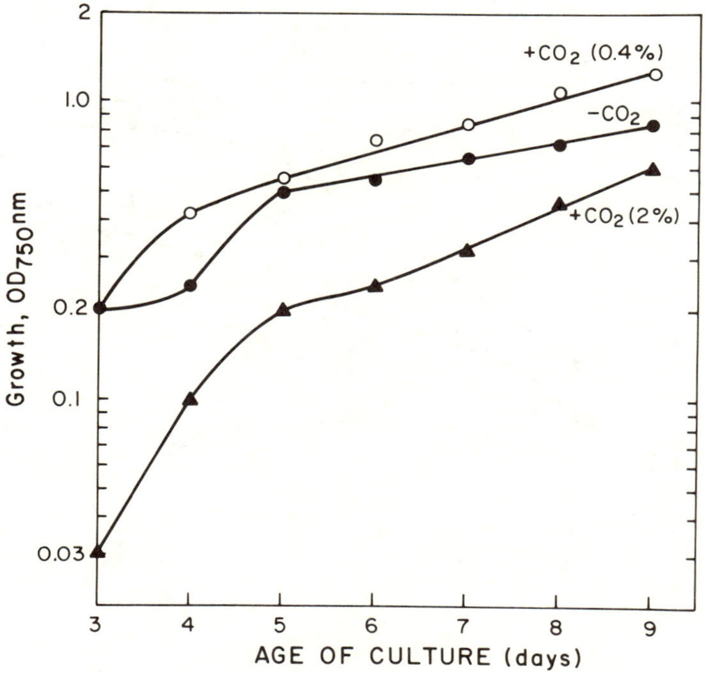

Fig. 1. Effect of CO_2 on growth of Anabaena variabilis in air. For conditions, see Methods.

added to determine the limiting effect of photosynthate on growth without the complications of nitrogen fixation. With NH_4^+ as the nitrogen source instead of N_2, the generation time of the culture in air, was found to be 2.8 days (Fig. 2). Carbon dioxide supplementation to a final concentration of 0.4% increased this growth rate lowering the generation time to 1.4 days. Two percent of CO_2 completely inhibited the growth in the NH_4^+-containing medium. The reason for this growth inhibition is unkonwn at this time. These results show that the growth of A. variabilis is limited by the availability of carbon (CO_2). This limitation appears to be more severe under N_2 fixing conditions when doubling times increase by more than two-fold to 3.2 days in air/0.4% CO_2.

Wolk and Shaffer (1976) demonstrated that A. variabilis is capable of growing heterotrophically at the expense of added

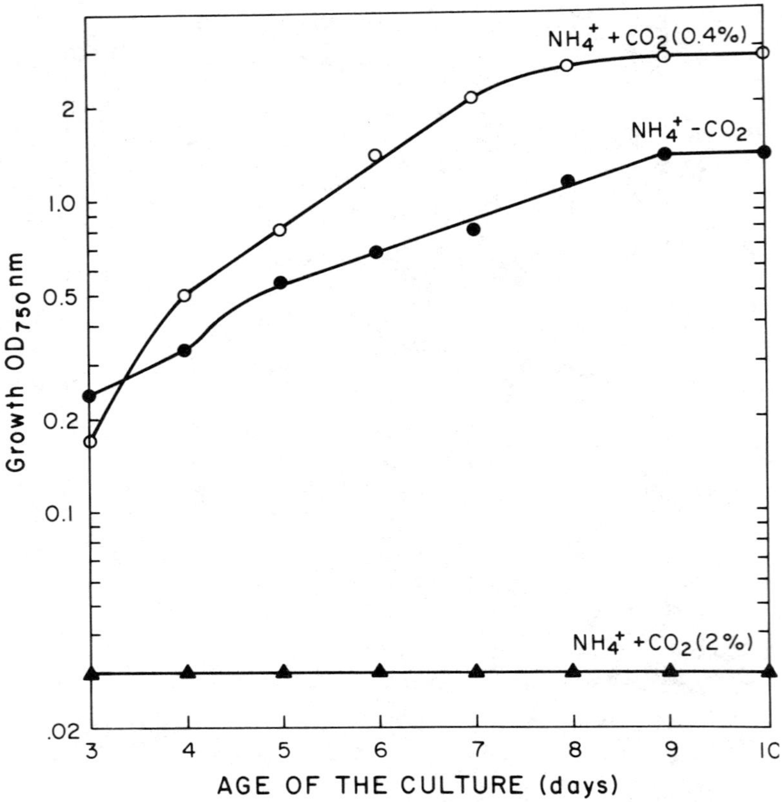

Fig. 2. Effect of CO_2 on growth of Anabaena variabilis in presence of NH_4^+ (8 mM). For conditions, see Methods.

fructose. If growth of A. variabilis is stimulated by adding CO_2, it is possible that fructose is also capable of enhancing the growth rate of the culture. The results presented in Figure 3 show that this is indeed the case. Upon addition of fructose (10 mM) the generation time of a nitrogen-fixing culture in air decreased to 1.2 days from 4.2 days for the control. Even small amounts of fructose (0.5 mM) stimulated growth but the effect was transient and lasted only until all the added fructose was utilized (data not shown). If fructose and CO_2 were added together, the growth rate although greater than the control was not additive and less than with fructose supplemented cultures but comparable to a CO_2-culture (data not presented).

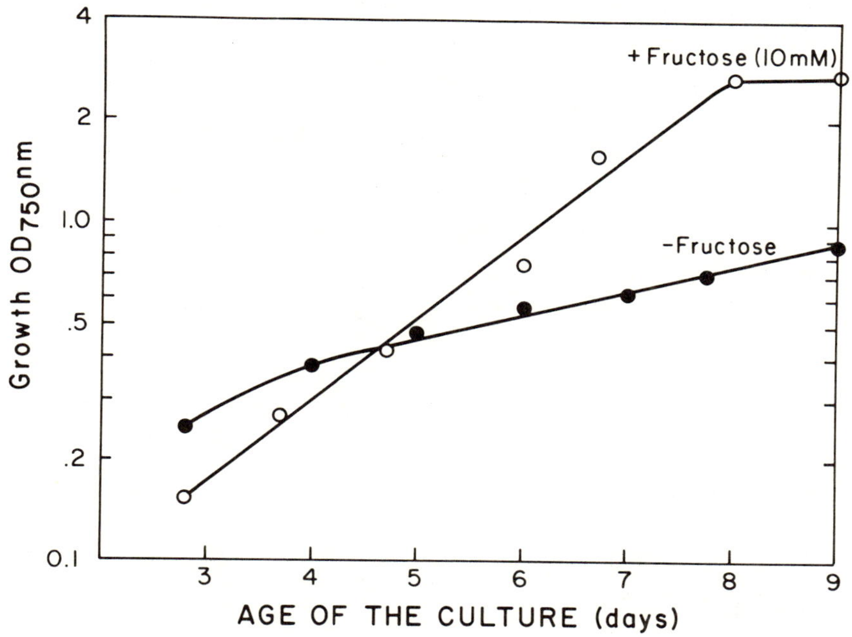

Fig. 3. Effect of fructose on growth of *Anabaena variabilis*. For conditions, see Methods.

The increase in growth rate of *A. variabilis* in the presence of added CO_2 or fructose under N_2-fixing conditions was found to be due to an increase in nitrogen activity (Table 1). If an aerobic culture of *A. variabilis* previously grown without any supplementation was enriched with CO_2 (0.4%) nitrogenase activity, under air, was found to be 51 units. However, if the new growth medium contained fructose (10 mM) instead of CO_2, the nitrogenase activity increased to 173 units under similar conditions. It was observed in other experiments (data not shown) that the catabolism of fructose requires oxygen. This raises the possibility that the increase in nitrogenase activity in the presence of fructose is due to a decrease in the O_2 concentration of the medium, achieved by fructose catabolism. The low levels of nitrogenase activity observed in the presence of CO_2 (0.4%) may be a consequence of elevated O_2 evolution due to CO_2 fixation. To test this possibility,

Table 1. Effect of fructose or CO_2 supplementation on nitrogenase activity.

New growth medium	Previous growth history	Nitrogenase activity (nmoles C_2H_4 produced·hr^{-1} OD_{750}^{-1} nm)	
		Air	Argon
CO_2 (0.4%)	Air	51	256.0
Fructose (10 mM)	Air	173.0	330.0
CO_2 (0.4%)	Fructose	22.0	102.0
Fructose* (10 mM)	Fructose	60	290.0
N_2 + CO_2 (0.4%)	Air	ND	220.0
N_2 + CO_2 (2.0%)	Air	ND	17.0
Air	Air	ND	120
Air	Fructose	120	202

* Culture without sparging under air.
ND = Not determined.

All cultures were sparged with either air, air + CO_2 or N_2 + CO_2. Nitrogenase activity of the culture was determined when the culture was 6 days old.

nitrogenase activity was determined in the presence of argon as the gas phase instead of air (20% O_2). As presented in Table 1, both fructose as well as CO_2 (0.4%) grown cultures stimulated nitrogenase activity to a level of 256 and 300 units, respectively, in an argon atmosphere. Cultures grown in ambient air (0.033% CO_2) exhibited rather low activity (120 units). However, CO_2 supplementation (0.4%) after growth in fructose failed to stimulate nitrogenase maximally (Table 1, line 3). When a fructose culture was allowed to grow on a shaker without constant bubbling with air, nitrogenase activity also increased indicating a negative effect of high oxygen concentration on nitrogenase activity of a fructose-grown culture (Table 1, line 4).

Results presented in Table 1 show that the supplementation of the medium with fructose or CO_2 at varying levels increased the nitrogenase activity leading to increased growth rate of the culture. The inhibitory effect of O_2 described above raises the question whether complete absence of O_2 in the gas phase (N_2) would dramatically alter nitrogenase and growth. However, this is not the case since N_2 enriched by 0.4% CO_2 yields an intermediate value of 220 units for nitrogenase and a slow growth rate (not shown) whereas 2% CO_2 in N_2 severely inhibits nitrogenase (Table 1, line 6).

DISCUSSION

Biological nitrogen fixation, an energy intensive process requires large inputs of energy and reducing power for maximal production of NH_4^+. In blue-green algae, where H_2O is the source of reductant, O_2 evolution is an integral part of generation of reducing power. In order to achieve the simultaneous presence of O_2 evolution and nitrogenase, filamentous blue-green algae developed heterocysts, a morphological entity for localizing nitrogenase (Stewart, 1977). The reducing power (NADPH) generated by light-dependent reactions is transferred via organic compounds derived from CO_2 fixation in vegetative cells to the heterocysts where it is utilized to drive nitrogenase activity. This leads to a dependence of heterocysts on vegetative cells for supply of reductants and as such can be compared to photosynthetic bacteria which are dependent on the medium for supply of reductant while using light for generating ATP.

This raises an interesting question: can vegetative cells supply enough reducing power to maintain high nitrogenase activity? The results presented in this paper show that in fact the vegetative cells are severely limited in supplying maximum amounts of reducing power for nitrogenase activity.

Supplementation of the medium with increasing levels of CO_2 did stimulate growth of Anabaena variabilis. This stimulation occurred irrespective of whether N_2 or NH_4^+ served as nitrogen source (Figs. 1 and 2). The enhancement of growth in the NH_4^+-containing medium by CO_2 suggests that ambient CO_2 concentration is too low to supply all the organic compounds required for biosynthesis of cell materials needed for rapid growth. The requirement for CO_2 is greater however under N_2-fixing conditions since a fraction of the total organic material produced by the cell is used as a carrier of reducing power for nitrogenase activity in the heterocysts.

It was shown that fructose supports heterotrophic growth of A. variabilis in the dark (Wolk and Shaffer, 1976). Fructose supplementation of the culture medium provided the maximal growth rates in light observed in our experiments (Fig. 3). In contrast, carbon dioxide supplementation failed to provide these maximal growth rates and higher levels of CO_2 actually inhibited growth (Figs. 1 and 2). This difference observed may be related to nitrogenase activity in the presence of these compounds. The difference in the nitrogenase activity (Table 1) is probably due to the amount of O_2 present in the medium, since nitrogenase activity of a CO_2 grown culture was higher if the O_2 in the gas phase was replaced by argon during the assay. This suggests that in the presence of CO_2, nitrogenase protein is present at maximal levels in an aerobic culture and the presence of O_2 in the gas phase and the O_2 produced by the cells during CO_2 fixation inhibits nitrogenase activity. In contrast addition of fructose to the culture medium actually reduces the amount of O_2 in the medium since fructose metabolism consumes O_2 (Spiller, unpublished data).

The results discussed above show that N_2 fixation is limited by the availability of reductant in the heterocystous blue-green alga, A. variabilis and in order to enhance nitrogen fixation, it is essential to generate sufficient carbohydrate to supply reducing power to nitrogenase. However, generation of reducing power by photosystem II at high rates seems to have a negative effect on nitrogen fixation because of its byproduct, oxygen. Supplementation of the medium with fructose, a preformed reductant, seems to be an alternative at present for enhancing nitrogenase activity and growth of A. variabilis. These results further suggest that although the blue-green alga, A. variabilis developed heterocysts for O_2 protection of nitrogenase, oxygen still inhibits the production of maximal nitrogenase activity allowing only slow growth under aerobic conditions.

It is interesting to speculate that the high levels of N_2 fixation activity achieved by the Azolla-Anabaena symbiosis (see papers in this volume by Peters and Talley and Rains) may be due in large part to the availability of photosynthate supplied by the host plant.

ACKNOWLEDGMENT

We thank R. C. Valentine for encouragement and continuous support. This investigation was supported by a grant from the National Science Foundation (PFR 77-07301) and a fellowship to H.S. from the Deutsche Forschungsgemeinschaft.

REFERENCES

Allen, M. B., and Arnon, D. I., 1955, Studies on nitrogen-fixing blue-green algae. I. Growth and nitrogen fixation by Anabaena cylindrica Lemm., Plant Physiol., 30:366.

Hardy, R. W. F., and Havelka, U. D., 1975, Nitrogen fixation research: a key to world food?, Science, 188:633.

Stewart, W. D. P., 1977, Blue-green algae, in: "A Treatise on Dinitrogen Fixation," Sect. III, E. W. S. Silver, ed., John Wiley and Sons, New York.

Wolk, C. P., 1968, Movement of carbon from vegetative cells to heterocysts in Anabaena cylindrica, J. Bact., 96:2138.

Wolk, C. P., and Shaffer, P. W., 1976, Heterotrophic micro- and macrocultures of a nitrogen-fixing Cyanobacterium, Arch. Microbiol., 110:145.

PHYSIOLOGICAL STUDIES ON N_2-FIXING AZOLLA*

G. A. Peters, O. Ito, V. V. S. Tyagi, and D. Kaplan

Charles F. Kettering Research Laboratory
150 East South College Street
Yellow Springs, OH 45387

Azolla, a genus of heterosporous aquatic ferns generally included in the *Salviniaceae*, is widely distributed in tropical and temperate fresh-water ecosystems. Members of the genus are capable of growth in environments deficient in combined nitrogen since they invariably contain an N_2-fixing cyanophyte which can provide their total N requirements. The current interest and potential of these N_2-fixing associations as an alternative N source in rice culture, as well as their long time usage for this purpose in the Far East, is well documented (Moore, 1969; Tuan and Thuyet, 1979; Liu, 1979; Singh, 1979; Rains and Talley, 1979; Talley and Rains, 1980; Watanabe et al., 1977).

The genus is generally considered to contain four new world species in the subgenus *Euazolla*, *A. caroliniana* Willd., *A. filiculoides* Lam., *A. mexicana* Presl. and *A. microphylla* Kaulfaus, and two old world species in the subgenus Rhizosperma, *A. pinnata* R.Br. and *A. nilotica* DeCaisne (Moore, 1969). The *A. nilotica* is atypical in that it is much larger than the other species (Moore, 1969; Lumpkin, pers. commun.) and the *A. pinnata* R.Br. is reported to have two morphologically distinct forms; *A. pinnata* var. *pinnata* R. Brown and *A. pinnata* var. *imbricata* (Roxb.) Bonap. (Becking, 1979). Furthermore, the *A. filiculoides* introduced to the People's Republic of China from E. Germany appears to be larger and morphologically distinct from other *A. filiculoides* (Shi et al., 1980 and Peters, unpublished observations). Based on the current interest in these associations their taxonomy would seem to merit a reassessment, and possibly a revision, using new technologies available to taxonomists.

---

*Contribution No. 713 from C. F. Kettering Research Laboratory

We have recently shown that when growth conditions are optimized, populations of A. caroliniana, A. filiculoides, A. mexicana and A. pinnata can double their biomass in two days or less and contain 5-6% N on a dry weight basis (Peters et al., 1980b). Results of comparative studies of various physiological processes in the four species are currently in preparation. These include: rates of growth, photosynthesis, light and dark respiration, N_2 fixation ($^{15}N_2$ and C_2H_2), %C, %N, etc., as a function of 12 hr, 16 hr and continuous light; the contributions of light and dark N_2 fixation; and, the effect of combined N as ammonium, nitrate and urea on physiological processes and the input from combined N and N_2 fixation. However, our most indepth studies have been, and are currently, conducted with A. caroliniana Willd. and the following account is restricted to this species.

PHOTOSYNTHESIS

The Azolla sporophyte consists of a branched floating stem bearing deeply bilobed leaves and true roots. The endophytic Anabaena occupies a specialized chamber which is formed in the aerial dorsal leaf lobes during their development (Peters, 1977; Peters et al., 1978). The Azolla contains chlorophylls a and b as well as carotenoids while the endophytic Anabaena contains chlorophyll a, phycobilins and carotenoids. The phycobilin complement is comprised of phycoerythrocyanin, phycocyanin, and allophycocyanin (Tyagi et al., 1980b). Action spectra have been obtained for photosynthesis by the association, cyanophyte-free plants and the endophyte removed from the leaf cavities (Ray et al., 1979). While the relative quantum yield is highest in the region of phycobilin absorption with the isolated endophyte, any contribution of the endophyte to the association's action spectrum is effectively masked by the preponderance of the Azolla pigments. Other studies have shown that the endophyte accounts for less than 20% of the association's chlorophyll (Peters and Mayne, 1974a; Ray et al., 1978) and the action spectra for photosynthesis by the association and endophyte-free Azolla are similar to one another and to those obtained with other green plants.

The association, fern and endophyte exhibit Calvin cycle intermediates of photosynthetic CO_2 fixation with phosphoglyceric acid being the initial product followed by hexose phosphates. Sucrose is a primary end product in the Azolla but not in the isolated endophyte (Ray et al., 1979). CO_2 fixation in the association is saturated at approximately 400 $\mu E \cdot m^2 \cdot sec$ (Ray et al., 1979) as is the growth rate (Peters et al., 1980b). The association and endophyte-free Azolla exhibit a 40% inhibition of photosynthesis (CO_2 fixation) at atmospheric O_2 as compared to 2% O_2 and an O_2 dependent CO_2 compensation point. The isolated endophyte exhibits the same, low CO_2 compensation point of about 4 ppm CO_2 at 2% and 20% O_2 and its rate of photosynthesis is also constant at both O_2 tensions (Ray et al., 1979). Results obtained with the isolated endophyte are based on experiments in which it was isolated indiscriminately from all stages of leaf cavity formation (Peters and Mayne, 1974b; Peters et al., 1980a). As noted previously (Ray et al., 1979), since the endophyte undergoes morphological and physiological changes as a function of leaf cavity age, and since the microenvironment within the leaf cavity is still unresolved, the results obtained with the isolated endophyte demonstrate its attributes but do not necessarily extrapolate to how it functions within the Azolla leaves. The actual contributions of the individual partners to the association's total photosynthesis and photosynthesis by the association and individual partners as a function of leaf age are subjects of current investigations.

PHOTOSYNTHESIS AND N_2 FIXATION

N_2 fixation in the association and isolated endophyte has been investigated using C_2H_2 reduction, ATP-dependent H_2 production, $^{15}N_2$ fixation and their relationships (Peters and Mayne, 1974b; Peters et al., 1976, 1977). Photosynthesis is the ultimate source of all the ATP and reductant utilized in these nitrogenase-catalyzed reductions. Dark, aerobic reductions are dependent upon endogenous reserves of photosynthate, as a substrate for respiration. Rates of dark, aerobic reductions are always less than half of those obtained aerobically with light intensities saturating for N_2 fixation (Peters and Mayne, 1974b; Peters et al., 1979). This has been taken to imply that as with free-living cyanobacteria (Bottomly and Stewart, 1977; Stewart et al., 1979), respiratory driven reductions are ATP limited (Peters et al., 1980a). Dark, anaerobic reductions are negligible.

Studies in which the endogenous reserves of photosynthate have been either maintained or depleted by preincubation in the light or dark, respectively, followed by the simultaneous measurements of photosynthesis (CO_2 fixation) and N_2 fixation (C_2H_2 reduction) in the presence of DCMU (an inhibitor of Photosystem II activity) indicated the following: Photosystem II is required to provide photosynthate for reducing power but non-cyclic photophosphorylation is not a principle source of ATP for nitrogenase; CO_2 fixation can be completely inhibited by DCMU with no more than a 30% inhibition of C_2H_2 reduction if the endogenous reserves of reductant have not been depleted. Thus, the role of Photosystem II is indirect and cyclic photophosphorylation associated with Photosystem I is clearly implicated as the primary source of ATP for nitrogenase-catalyzed reductions in the light.

Action spectra of C_2H_2 reduction in the association and isolated endophyte have further demonstrated the interaction of photosynthesis with N_2 fixation (Tyagi et al., 1980b and unpublished results). Phycobilins are generally considered to be accessory pigments for PS II, as indicated by the action spectrum for photosynthesis in the endophyte (Ray et al., 1979) and to be depleted or absent in heterocysts (Thomas, 1970; Tel-Or and Stewart, 1977; Haskelkorn, 1978). Although the above studies demonstrated that N_2 fixation is a PS I linked process not directly dependent on PS II, in the action spectra for C_2H_2 reduction in the association and the isolated endophyte, the relative rate of C_2H_2 reduction per incident quantum was essentially equal in the region of absorption by the phycobilins and that of the chlorophyll. Further, there was no appreciable affect of DCMU on the action spectrum of the endophyte and the action spectra for C_2H_2 reduction and photosynthesis differed appreciably only in that the action spectrum of photosynthesis decreased much more markedly in the region of chlorophyll absorption. At present we have no corroborating evidence for phycobilins being associated with PS I in heterocysts of the endophyte. However, there is evidence for the association of phycobilins with PS I (Wang et al., 1977) and heterocysts isolated from *Anabaena variabilis* not only contain phycobilins but the light absorbed by them (600-650 nm) is as effective in driving H_2 supported C_2H_2 reduction as the light absorbed by chlorophyll a (Peterson and Ke, 1979, and personal communication). In the case of the endophyte, attempts to isolate pure preparations of heterocysts have been frustrated by contamination with akinetes, the fact that there is a large variation in size and state of differentiation, and the simple problem of obtaining sufficient quantities of material. Current studies using fluorescence microspectroscopy of filaments of the endophyte removed from leaves of varying age have indicated that, in accord with the studies on *Anabaena variabilis* (Peterson and Ke, 1979), heterocysts of the endophyte exhibit phycobilin fluorescence (Calvert and Peters, unpublished observation).

UNIDIRECTIONAL HYDROGENASE

Previous studies (Peters et al., 1976, 1977; Peters, 1977) have shown that when the association was grown in the absence of combined nitrogen, nitrogenase-catalyzed H_2 production under Ar was generally appreciably less than the rates of C_2H_2 reduction, that H_2 production was greatest under an atmosphere of C_2H_2 and CO (see Smith et al., 1976) and that H_2 production under air was extremely low. These findings were all strongly suggestive that these associations contained a unidirectional hydrogenase as was the observation that they were efficient, in the terminology of Schubert and Evans (1976). Recent studies (Tyagi, unpublished) have shown that H_2 production is maximal under an atmosphere of 25% C_2H_2 and 0.5% CO; that preincubation under C_2H_2 for 6 hr in the light followed by assays of H_2 production and C_2H_2 reduction under the appropriate gas phase increased H_2 production up to

5-fold and C_2H_2 reduction by only 20%. The latter observation is similar to that reported for a free-living heterocystous cyanobacterium (Scherer et al., 1980) and suggestive of C_2H_2 inhibition of the unidirectional hydrogenase. H_2 production under Ar $\pm$ O_2 was also higher when photosynthesis was inhibited by DCMU, indicating that uptake may occur via the oxygen-hydrogen reaction (Peterson and Burris, 1978). Studies of the actual uptake of H_2, using 0.1% H_2 in air, have shown that after a lag period of several hours, there is appreciable H_2 uptake by the association and none in the endophyte-free Azolla and the uptake by the association is diminished by preincubation with C_2H_2. In accord with previous studies (Peters et al., 1976) there is no evidence for the occurrence of a reversible hydrogenase. These studies are continuing and details will be presented elsewhere.

N_2 FIXATION AND TRANSFER FROM THE ENDOPHYTE TO THE AZOLLA

Studies with $^{15}N_2$ have shown that the endophyte Anabaena, isolated from Azolla leaf cavities in all stages of development, releases approximately half of the N_2 it fixes into the incubation medium as ammonium with only small amounts of organic N (Peters, 1977; Peters et al., 1980a). It should be noted that more than half of the N_2 fixed may actually be released since undifferentiated filaments in these preparations may well have assimilated some of the released ammonium. Additional studies comparing the supernatants after incubations of the endophyte under $^{14}CO_2$ in Ar and in N_2 atmospheres showed ammonium under N_2 but not Ar and no difference in the ^{14}C content of the supernatants. Both contained less than 5% of the incorporated ^{14}C, further indicating that the symbiont released few or no organic N compounds such as amino acids (Ray et al., 1978; Peters et al., 1980a). In the association, incubations under $^{15}N_2$ in air followed by chase periods with air showed that there is a rapid assimilation of newly fixed N_2 into ethanol-soluble and then ethanol-insoluble fractions with a low level of free ammonia (Peters et al., 1979).

Both the fern and endophytic Anabaena were found to have the capacity to metabolize ammonium, based on the activities of glutamine synthetase (GS), glutamate synthase (GOGAT) and glutamate dehydrogenase (GDH) in each. However, about 90% of the association's GS activity and 80% of its GDH activity were attributable to the fern (Ray et al., 1978; Peters et al., 1980a). Further, since the contributions by the endophyte to the association's total activity reflected an average of its activity in all developmental stages, and a developmental gradient was known to exist (Hill, 1975, 1977; Peters et al., 1980a), it was suggested (Ray et al., 1978; Peters et al., 1979, 1980a) that activities attributed to the Anabaena might be associated with specific developmental stages, notably the undifferentiated filaments.

It has been suggested by others (Stewart, 1977; Haselkorn, 1978) that decreased GS activity in symbiotic cyanobacteria could provide an explanation for the release of ammonia with the possibility that the host modifies the endophyte's ability to assimilate ammonia by inhibiting its GS activity or synthesis. In the Azolla-Anabaena association there is evidence indicating that the endophyte's GS activity is associated with filaments in young leaf cavities and that as a function of leaf development, and increasing heterocyst frequency, the Azolla plant may actually prevent GS synthesis in the endophyte (Orr, Toan, and Haselkorn, personal communication, Third Int. N_2 Symposium, Madison, Wis.).

DEVELOPMENTAL MORPHOLOGY AND PHYSIOLOGY

General aspects of the morphology and ultrastructure of the A. caroliniana association have been presented (Duckett et al., 1975; Peters, 1977; Peters et al., 1978, 1980a). Recent studies have dealt with the developmental morphology taking advantage of the sequential leaf development. Reconstruction of serial sections of leaves of various age, as defined by their position on the stem axis (Hill, 1977), and the subsequent use of cleared and stained whole mounts, have provided new insights into the development and organization of the leaf cavity (Calvert and Peters, 1979). Of particular interest is the finding that the epidermal hairs which line the leaf cavity are of two types, simple and branched, and that whereas the simple hairs increase in number with leaf development, the branched hairs are not only differentiated early in leaf development, but their number is restricted to two per cavity. Moreover, while the simple hairs are randomly positioned around the cavity, except for the lower quadrant, the branched hairs are always found in the same position and in close proximity to the path of the foliar trace. A detailed account of these and other aspects of leaf cavity development will be presented elsewhere (Calvert et al., in preparation).

Physiological studies have been conducted in parallel with the morphological studies. Using main stem axes containing 12 or more leaves, these studies have shown that the N content is highest in the apical region, decreasing in progressively older leaves and that the dry weight as a percent of the fresh weight exhibits a similar pattern. Moreover, the gradient in nitrogenase activity as a function of leaf age shown using C_2H_2 reduction (Hill, 1977; Peters et al., 1980a) has been confirmed using $^{15}N_2$ (Fig. 1). Employing $^{15}N_2$ followed by varying chase periods with air, and a rather involved experimental protocol which will be presented in detail elsewhere (Kaplan and Peters, in preparation), it has also been demonstrated that N_2 fixed in the older leaf cavities is transported to the apical region. Studies to determine the transported compound(s), the relationship of the hairs to the physiological processes involved, and the interaction of host-

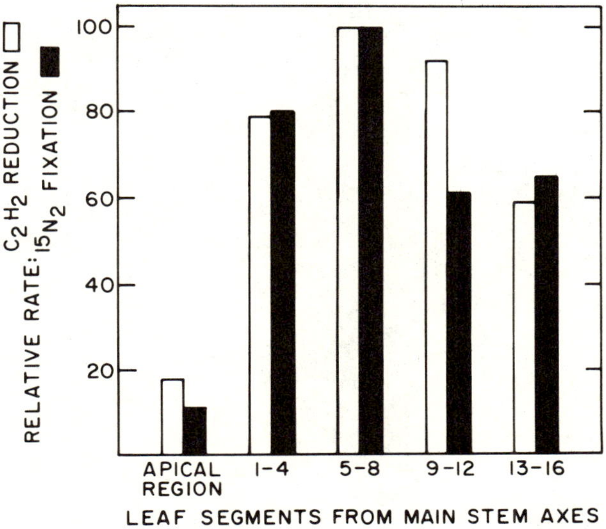

Fig. 1. Gradient in nitrogenase activity (C_2H_2 and $^{15}N_2$ reduction) as a function of leaf age in A. caroliniana.

symbiont carbon metabolism as a function of leaf age are in progress.

PHYSIOLOGICAL STUDIES USING THREE PHOTOPERIODS

Table 1 summarizes the results of studies on photosynthesis, N_2 fixation, and associated processes after two weeks growth with weekly samplings under optimized growth conditions for A. caroliniana (Peters et al., 1980b) as a function of 12 hour, 16 hour and continuous light regimes. Although the doubling times in Table 1 are slightly longer than those obtained previously (Peters et al., 1980b), they are indicative of healthy, actively growing plants.

An analysis of the data shown in Table 1 (other than that for growth and C_2H_2 reduction and H_2 production in vials), using the Duncan multiple range test at p = .05, indicated few statistically significant differences in the physiological process as a function of photoperiod. Photosynthesis in air, with and without C_2H_2, under continuous light was significantly different from the

Table 1. Photosynthesis, N_2 fixation and associated processes in A. carolininana during the light interval of three photoperiods.

	Light Interval 12 hr	16 hr	24 hr
Doubling Time (days)	3.57± .50	2.44± .12	2.33± .12
Photosynthesis (μmoles CO_2/g fr wt·hr)			
Air	123±6	109±13	71±3
Air + 15% C_2H_2	113±5	106±14	69±5
2% O_2 - .03% CO_2 in N_2	141±4	170±21	93±10
"Photorespiration" (μmoles CO_2/g fr wt·hr)			
Photosyn. in 2% O_2 - Photosyn in air	18±8	61±27	22±11
Extrapolation to zero CO_2 *	35±8	29±6	18±3
Aerobic CO_2 Compensation Point (ppm CO_2)	41±2	44±2	42±7
Dark Respiration (μmoles CO_2/g fr wt·hr)			
Air, during light period	15±3	19±6	16±1
C_2H_2 Reduction (nmoles C_2H_4/g fr wt·min)			
light, aerobic in vials	45±4	57±6	27±5
light, aerobic simultaneously with determination of photosynthesis	36±8	34±12	27±10
dark, aerobic simultaneously with respiration during light period	15±1	11±1	12±4
H_2 Production (nmoles H_2/g fr wt·min)			
light, 20% O_2 - .03% CO_2 in Ar	4.3±1.6	11.6±3.2	7±2
light, air	.2±.05	.3±.02	.0
Relative Efficiency in Light $\left(1 - \frac{H_2(O_2\text{-}CO_2\text{-}Ar)}{C_2H_4\ (C_2H_2 \text{ in air})}\right)$	.90±.01	.80±.00	.74±.00
μmoles C_2H_2 reduced/μmole CO_2 fixed	.019	.019	.023
C to N ratio	7.7±.3	-	8.8±.8

Data is the average ± S.D. of duplicate analyses in two experiments after 7 days and 14 days of growth under specified photoperiod. Growth as described in Peters et al., 1980b: IRRI medium buffered at pH 6 with 10 mM MES; constant temperature of 25°C; Light at 400 $\mu E \cdot m^2 \cdot sec$.

* Method of Decker (1957).

rates at 16 hr and 12 hr. The value obtained for "photorespiration" at the 16 hr photoperiod using the rate of photosynthesis in 2% O_2 minus that in air, was significantly different from that at 12 hr or continuous light. However, when using the method of Decker (1957) with extrapolation to zero CO_2, the value obtained for the plants under 16 hr of illumination was not significantly different from that obtained with either 12 hr or continuous illumination.

In general the data in Table 1 are entirely consistent with earlier comments in this manuscript in regard to photosynthesis, N_2 fixation, their interaction, and associated processes. For example, the O_2 inhibition of photosynthesis and values obtained for the aerobic CO_2 compensation point are consistent with a C3 type of photosynthesis and the contribution of photosynthetically-generated ATP is clearly indicated by comparison of the rates of C_2H_2 reduction in the light and dark simultaneously with measurements of photosynthesis and respiration. As seen in Table 1, the dark, aerobic rate of C_2H_2 reduction is only 25-40% of the rate obtained in the light. Furthermore, the data on C_2H_2 reduction versus H_2 production and the relative efficiency expressions, determined here by

$$1 - \frac{H_2(Ar\text{-}O_2\text{-}CO_2)}{C_2H_4(C_2H_2 \text{ in air})}$$

are consistent with the occurrence of a unidirectional hydrogenase. The data also show that there is no significant effect of 15% C_2H_2 on photosynthesis during the 10-15 minute assay period.

In order to assess the contributions of light and dark N_2 fixation to the plant's nitrogen budget, rates of $^{15}N_2$ fixation and C_2H_2 reduction were determined at the midpoint of the light and dark periods of the 12 hr - 12 hr and 16 hr - 8 hr light-dark photoperiods in separate experiments. As shown in Table 2, the ratio of C_2H_2 reduced to $^{15}N_2$ fixed is approximately 4 and, using the rates of $^{15}N_2$ fixation, dark, aerobic N_2 fixation is estimated to contribute 27% of the total daily N with a 12 hr dark period and 19% with an 8 hr dark period. (These studies were conducted several months after those shown in Table 1 and the rates of C_2H_2 reduction at 12 hr and 16 hr were lower and higher, respectively, than those obtained in the previous studies. Except for invoking biological variability we have no explanation for the differences. Nevertheless the data are indicative of the relative inputs during the light and dark cycles.) Although not shown, dark respiration and C_2H_2 reduction were found to remain constant throughout the 8 hr dark period following 16 hr of light but to begin to decline after about 7 hr of darkness after 12 hr of light when plants had been maintained under optimal conditions with these photoperiods for at least a week.

Table 2. Aerobic C_2H_2 reduction and $^{15}N_2$ fixation at midpoint of light and dark cycles for two photoperiods.

	12 hr - 12 hr		16 hr - 8 hr	
	light	dark	light	dark
nmoles C_2H_4/g fr wt·min	17.2±2.1	-	88.1±8.7	43.9±5.4
nmoles N_2/g fr wt·min	4.2±0.8	1.6±0.3	21.6±2.2	10.0±0.6
C_2H_2 reduced/N_2 fixed	4.07	-	4.08	4.40
% of total N fixed during 24 hr period	73	27	81	19

Growth conditions as for Table 1 except 600 $\mu E \cdot m^2 \cdot sec$. C_2H_2 was 15% in air, $^{15}N_2$ was approximately 40 atom % in air. Incubation period was 30 or 60 min under optimal growth conditions.

A. CAROLINIANA GROWN ON COMBINED N SOURCES (AMMONIUM, NITRATE, OR UREA)

Although Peters and Mayne (1974b) were unable to obtain growth of *A. caroliniana* on ammonium, this was the result of inadequate buffering capacity and pH drift of the growth medium (Ito, unpublished observation). Employing media buffered at pH 6 with 10 mM MES (Peters et al., 1980b) there is good growth of both the association and endophyte-free plants.

As shown in Table 3, growth rates on up to 2.5 mM ammonium, and in subsequent studies up to 5 mM, are comparable to those of plants grown in its absence. There is a gradual increase in chlorophyll content, free ammonia-N, soluble N and protein N and a gradual decrease in C_2H_2 reduction activity - 75% of the control at 2.5 mM - with increasing concentrations of ammonium in the growth medium. (In subsequent studies 2.5 mM NH_4^+ has resulted in up to a 50% inhibition of C_2H_2 reduction.) Table 4 shows that the ammonium absorption rate (AAR) determined with $^{15}NH_4^+$, increases modestly with increasing ammonium concentrations. While the nitrogen fixation rate (see legend to Table 4 for method of determination) decreases with increasing ammonium concentrations, there is generally a recovery of nitrogen fixation activity upon transfer to N-free medium and the ratios of C_2H_2 reduced to the estimated N_2 fixed are entirely reasonable. The ratio of AAR/(AAR + 2(NFR)), an index of the relative contribution of ammonium absorption to the total N input, indicates that

Table 3. Effect of NH_4^+ on growth, chlorophyll content, C_2H_2 reduction and the relationships of free ammonium-N, soluble N and protein N.

mM NH_4^+	Doubling Time (days)	mg Chl*	ARA†	μg NH_4-N*	μg soluble N*	μg protein N*
0	2.35±.26	.57±.02	61±13	15±9	333±103	1550±309
0.25	2.19±.22	.60±.02	57±12	16±12	334±65	1824±336
0.50	2.31±.19	.62±.03	53±11	19±10	366±71	1820±449
1.00	2.23±.29	.64±.03	48±12	24±15	385±96	1985±475
2.50	2.27±.30	.65±.04	46±13	34±20	417±104	2000±442

† nmoles C_2H_2 reduced/g fr wt·min

* per g fr wt

Protein N is Lowry protein ÷ 6.25

Growth as in Table 1 using a 16 hr light - 8 hr dark cycle. Values are average ± S.D. of duplicate or triplicate samples from 6 sampling periods during 28 days with frequent transfer to maintain ammonium concentrations at desired level. The variation between sampling periods was much greater than that within the duplicate or triplicate samples at the same sampling period.

Table 4. Ammonium absorption ($^{15}NH_4^+$) and N_2 fixation in A. caroliniana.

$mMNH_4^+$ for Growth and ^{15}N labelling	$^{14}NH_4^+$ in medium after $^{15}NH_4$ labelling	AAR[1]	NFR[2]	ARA[3]	ARA/NFR	AAR/AAR+2(NFR[4]
0.25	+	6.2	21.9	82.2	3.86	.12
	-		21.3	75.0	3.52	
0.50	+	7.5	22.7	77.8	3.80	.14
	-		15.4	76.1	4.94	
1.00	+	8.7	15.8	69.5	4.96	.22
	-		20.7	66.2	3.20	
2.50	+	11.8	13.4	54.5	4.43	.30
	-		18.6	74.8	4.02	

1. Ammonium absorption rate (nmoles $^{15}NH_4^+$/g fr wt·min)
2. Nitrogen fixation rate (nmoles N_2 fixed/g fr wt·min).
3. Acetylene reduction activity (nmoles C_2H_4/g fr wt·min).
4. Index of contribution of ammonium absorption to N input. NFR is x2 since nmoles N_2 fixed = 2 nmoles NH_4^+.

After 36 days on specified NH_4^+ concentration using conditions as per Table 3, plants were transferred to medium containing the same concentration of $^{15}NH_4^+$ for two days. After removing material for ^{15}N analysis, half of the remaining material was transferred to medium with the same concentration of $^{14}NH_4$ and the other half to N free medium for two days. NFR was calculated from the dilution of ^{15}N in the plant material, correcting for ammonium absorption when ammonium was present.

at 2.5 mM ammonium, after five weeks of growth with frequent transfers, 70% of the plant's N is still derived from N_2 fixation. This is in reasonably good agreement with the finding that the average value for C_2H_2 reduction over the five week period was 75% of the control (Table 3).

The growth rate of A. caroliniana is comparable when grown on N_2 alone or on medium containing up to 25 mM NO_3^-. Table 5 shows that the chlorophyll content and percent dry matter remain quite constant with increasing nitrate concentration in the growth medium. While photosynthesis remains quite constant with increasing concentrations of nitrate, as do the %C and %N, there is a decrease in C_2H_2 reduction activity, especially between 10 mM and 25 mM, with the latter resulting in about a 60% decrease in C_2H_2 reduction activity relative to the control after five weeks of growth. This value is the average obtained from assays at weekly sampling periods which ranged from 45% to 65% inhibition relative to the control plants. Of note is the observation that H_2 production does not decrease in parallel with C_2H_2 reduction as a function of nitrate concentration and there is a marked decrease in the relative efficiency expression for plants grown on 25 mM nitrate. Previous studies have indicated that nitrate resulted

Table 5. Effect of NO_3^- on growth, photosynthesis, N_2 fixation and associated processes and other physiological parameters in *A. caroliniana*.

mM NO_3^-	Doubling Time (days)	mgChl / g fr wt	% Dry Matter	%C	%N	Photosyn[1]	ARA[2]	HPA[3]	RE[4]
0	2.07±.10	.45±.08	5.2±.5	43.9±.6	4.9±.6	75.9±1.1	56.6±28	10.4±6.5	.81
2.5	1.91±.13	.44±.03	5.5±.6	43.4±.6	4.7±.5	74.9±6.3	46.9±20	9.2±6.6	.80
10	2.03±.02	.46±.09	5.3±.3	43.4±.5	4.8±.3	73.8±14	42.0±16	11.7±6.4	.72
25	2.24±.02	.52±.05	5.8±.2	43.2±1.0	4.7±.2	68.0±2.4	23.0±6.5	15.8±8.3	.32

1 Photosynthesis (µmoles CO_2 fixed/g fr wt·hr) in air.

2 Acetylene reduction activity (nmoles C_2H_4/g fr wt·min) in air.

3 H_2 production activity (nmoles H_2/g fr wt·min) in Ar-20% O_2 - .03% CO_2.

4 Relative efficiency (1 - HPA/ARA).

Plants grown as in Table 1 except for 600 $\mu E \cdot m^{2} \cdot sec$ light intensity, 16 hr - 8 hr photoperiod. Data is average ± S.D. of duplicate determinations at weekly sampling periods and transfers to new media over a five-week period.

in an increase in H_2 production both relative to C_2H_2 reduction and under an air atmosphere (Newton, 1976; Peters et al., 1976) and it was suggested that a product of nitrate metabolism might be altering the occurrence or activity of a unidirectional hydrogenase (Peters, 1977).

A study analogous to that shown in Table 4, using NH_4^+, is shown for nitrate in Table 6. The nitrate absorption rate, determined with $^{15}NO_3$, increases with increasing nitrate concentrations in the medium, and, as with ammonium, there is a recovery of nitrogenase activity in plants which are transferred from nitrate-containing solution to N-free medium. Although the ratio of C_2H_2 reduced to N_2 fixed in Table 6 are slightly higher than obtained with ammonium (Table 4), they are again quite reasonable. The index analogous to that employed with ammonium indicates that after five weeks of growth on 25 mM nitrate, 46% of the N input to the plants is derived from nitrate and 54% from N_2 fixation. In accord with this it was shown in Table 5 that the %N content was essentially constant at all concentrations of nitrate, including 25 mM where C_2H_2 reduction was inhibited 60%. Thus, the decrease in N_2 fixation is compensated by nitrate utilization, maintaining a constant N content in the tissue. Furthermore, the increase in H_2 production relative to C_2H_2 reduction as a function of nitrate concentration did not diminish the N content of the plants.

As with nitrate and ammonium, the growth of *A. caroliniana* is comparable when grown on N_2 alone or on medium containing up to 12.5 mM urea (Table 7). Increasing concentrations of urea had no obvious effect on chlorophyll content or rates of photosynthetic CO_2 fixation but resulted in an increase in the percent dry matter and a decrease in C_2H_2 reduction. Relative to the average value of the control at the weekly sampling intervals over the three week study, 2.5 mM urea resulted in a 38% decrease, and 12.5 mM a 53% decrease, in C_2H_2 reduction activity. (Although not shown in Table 7, there was no significant effect of increasing urea concentrations on the %C and %N compared to the control.) There was considerable variation in the nitrogenase-catalyzed H_2 production under Ar-O_2-CO_2 between sampling periods as evidenced by the standard deviations (Table 7). However, within an individual sampling period the variability was much less. In contrast to the results obtained with nitrate, the average value for H_2 production decreased relative to C_2H_2 reduction with increasing urea resulting in a higher value for the relative efficiency. Studies are currently in progress to ascertain the consistency and significance of this observation in regard to unidirectional hydrogenase activities. It should also be noted that while the effect of NH_4^+ on H_2 production was not determined in conjunction with the studies shown in Tables 3 and 4, subsequent studies

Table 6. Nitrate absorption ($^{15}NO_3^-$) and N_2 fixation in *A. caroliniana*.

mM NO_3^- for Growth and ^{15}N labelling	$^{14}NO_3$ in medium after $^{15}NO_3$ labelling	NAR[1]	NFR[2]	ARA[3]	ARA/NFR	NAR/(NAR + 2(NFR))[4]
2.5	+	3.11	9.74	42.8	4.40	0.14
	-		11.60	49.3	4.20	
10.0	+	5.77	9.43	41.0	4.37	0.23
	-		9.84	48.7	4.95	
25.0	+	10.10	5.97	25.3	4.24	0.46
	-		8.21	42.0	5.11	

1. Nitrate absorption rate (nmoles $^{15}NO_3^-$/g·fr wt·min)
2. As in Table 4.
3. As in Table 4.
4. Index of contribution of nitrate absorption to N input.

Growth conditions as for Table 5. Plants had been grown on specified NO_3^- concentration for 35 days prior to transferring to media with equivalent concentration of $^{15}NO_3$ for 4 days followed by division of material as in Table 4 for 4 more days to determine input from N_2 fixation ± NO_3 uptake.

Table 7. Effect of urea on growth, photosynthesis, N_2 fixation and associated processes and other physiological parameters.

mM Urea	Doubling Time (days)	Mg Chl/g·fr wt	% Dry Matter	Photosyn[1]	ARA[2]	HPA[3]	RE[4]
0	1.96±.09	.51±.03	5.4±.4	72.1±5.0	61.6±17	13.2±4.7	.79
1.25	1.88±.08	.54±.03	5.3±.3	75.4±9.6	41.9±12	8.3±3.4	.81
2.50	1.85±.08	.49±.06	5.8±.3	70.4±9.4	38.4±8	7.2±4.3	.82
5.00	1.88±.09	.50±.06	6.5±.6	77.9±5.3	30.7±8	4.9±2.5	.85
12.50	1.89±.09	.50±.07	6.8±.7	72.3±4.7	28.8±7	3.3±5.1	.88

1. As in Table 5.
2. As in Table 5.
3. As in Table 5.
4. As in Table 5.

Growth conditions are in Table 5. Data is average ± S.D. of triplicate analyses at each of three weekly samplings.

which will be presented elsewhere have shown that the effect of ammonia is basically similar to that of nitrate.

Table 8. Urea absorption (^{15}N urea) and N_2 fixation in A. carolininana.

mM Urea for Growth and ^{15}N Labelling	^{14}N Urea After ^{15}N Labelling	UAR[1]	NFR[2]	ARA[3]	ARA/NFR	UAR/(UAR + 2(NFR))[4]
1.25	+	10.0	11.1	62.5	5.63	.31
	-		11.6	76.5	6.59	
2.50	+	11.3	9.9	50.5	5.13	.37
	-		10.7	68.1	6.36	
5.0	+	14.4	9.0	51.9	5.77	.44
	-		11.2	71.8	6.41	
12.5	+	18.7	7.4	42.0	5.71	.56
	-		8.7	77.2	8.91	

1. Urea absorption rate (nmoles ^{15}N-urea/g fr wt·min).
2. As in Table 4.
3. As in Table 4.
4. Index of contribution of urea absorption to N input.

Growth conditions as for Table 5. Plants had been grown on specified urea concentration for 21 days prior to transferring to medium with equivlent concentration of ^{15}N-urea for 4 days followed by division of material as in Table 4 for 4 more days to determine in part from N_2 fixation $\pm$ urea uptake.

Table 8 shows the urea absorption rate, using ^{15}N-urea, and other data analogous to that presented for ammonium (Table 4) and nitrate (Table 6). As with the other combined N sources there is a rapid recovery of nitrogenase activity upon transfer of the plant material from medium containing urea to N-free medium. The values obtained for C_2H_2 reduced to N_2 fixed, based on dilution of the ^{15}N-labeled plant material by either N_2 fixation alone or N_2 fixation plus urea absorption, using the urea absorption rate determined with ^{15}N urea, are somewhat high, especially upon transfer to N-free medium. The reason for this is not known at present. The index analogous to that employed with ammonium and nitrate indicates that after 3 weeks growth on 2.5 and 12.5 mM urea, N_2 fixation provides 73% and 44%, respectively, of the plants total N input. In regard to the urea studies it should be noted that after a week's growth on medium containing a given urea concentration, up to 10% of the urea N is found in the medium as free ammonium-N. This breakdown requires the presence of *Azolla* but we do not as yet know whether it is associated with an extracellular urease activity, with breakdown of urea before uptake, or if the ammonium is being released from the plant after uptake of urea and the action of an intracellular urease.

CONCLUDING REMARKS

The studies of physiological processes as a function of photoperiod and the utilization of combined N sources, as ammonium, nitrate and urea, and their effect on physiological processes presented here were restricted to *A. caroliniana*. This association was employed as the test organism to work out experimental procedures for use in conjunction with comparative studies using *A. filiculoides*, *A. mexicana* and *A. pinnata* in addition to *A. caroliniana*. The results of these comparative studies are currently being prepared for publication.

ACKNOWLEDGMENTS

The authors are indebted to Drs. W. R. Evans, B. C. Mayne, and H. E. Calvert, and R. E. Toia, Jr., M. K. Pence, D. K. Crist, R. E. Poole, and V. L. Kaiser for their efforts in conjunction with the studies presented here. Portions of this work were supported by NSF grant DEB 74-11679 A01, USDA grant 7900263 and NSF/ASRA PFR 77-27269.

REFERENCES

Becking, J. H., 1979, Environmental requirements of *Azolla* for use in tropical rice production, *in*: "Proc. Intern. Rice Res. Inst., Nitrogen and Rice," IRRI, Los Baños, Laguna, Philippines.

Bottomley, P. J., and Stewart, W. D. P., 1977, ATP and nitrogenase activity in nitrogen fixing heterocystous blue-green algae, *New Phytol.*, 79:625.

Calvert, H. E., and Peters, G. A., 1979, New morphological observations on the *Azolla caroliniana*-*Anabaena* symbiosis relevant to physiological interaction, *Plant Physiol*, S5:113.

Decker, J. P., 1957, Further evidence of increased carbon dioxide production accompanying photosynthesis, *J. Solar Energy Sci. Eng.*, 1:30.

Duckett, J. R., Toth, R., and Soni, S. L., 1975, An ultrastructural study of the Azolla-Anabaena azollae relationship, New Phytol., 75:111.

Haselkorn, R., 1978, Heterocysts, Ann. Rev. Plant Physiol., 29:319.

Hill, D. J., 1975, The pattern of development of Anabaena in the Azolla-Anabaena symbiosis, Planta, 122:179.

Hill, D. J., 1977, The role of Anabaena in the Azolla-Anabaena symbiosis, New Phytol., 78:611.

Liu, C. C., 1979, Use of Azolla in rice production in China, in: "Proc. Intern. Rice Res. Inst., Nitrogen and Rice," IRRI, Los Baños, Laguna, Philippines.

Moore, A. W., 1969, Azolla: Biology and agronomic significance, Bot. Rev., 35:17.

Newton, J. W., 1976, Photoproduction of molecular hydrogen by a plant-algal symbiotic system, Science, 191:559.

Peters, G. A., 1977, The Azolla-Anabaena azollae symbiosis, in: "Genetic Engineering for Nitrogen Fixation," A. Hollaender, ed., Plenum Press, New York.

Peters, G. A., Evans, W. R., and Toia, R. E., Jr., 1976, Azolla-Anabaena azollae relationship. IV. Photosynthetically driven nitrogenase-catalyzed H_2 production, Plant Physiol., 58:119.

Peters, G. A., and Mayne, B. C., 1974a, The Azolla-Anabaena azollae relationship. I. Initial characterization of the association, Plant Physiol., 53:813.

Peters, G. A., and Mayne, B. C., 1974b, The Azolla-Anabaena azollae relationship. II. Localization of nitrogenase activity as assayed by acetylene reduction, Plant Physiol., 53:820.

Peters, G. A., Mayne, B. C., Ray, T. B., and Toia, R. E., Jr., 1979, Physiology and biochemistry of the Azolla-Anabaena symbiosis, in: "Proc. Intern. Rice Res. Inst., Nitrogen and Rice," IRRI, Los Baños, Laguna, Philippines.

Peters, G. A., Ray, T. B., Mayne, B. C., and Toia, R. E., Jr., 1980a, Azolla-Anabaena association: Morphological and physiological studies, in: "Nitrogen Fixation, Vol., II, Symbiotic Associations and Cyanobacteria," W. E. Newton and W. H. Orme-Johnson, eds., University Park Press, Baltimore.

Peters, G. A., Toia, R. E., Jr., and Lough, S. M., 1977, Azolla-Anabaena azollae relationship. V. $^{15}N_2$ fixation, acetylene reduction and H_2 production, Plant Physiol., 59:1021.

Peters, G. A., Toia, R. E., Jr., Raveed, D., and Levine, N. J., 1978, The Azolla-Anabaena azollae relationship. VI. Morphological aspects of the association, New Phytol., 80:583.

Peters, G. A., Toia, R. E., Jr., Evans, W. R., Crist, D. K., Mayne, B. C., and Poole, R. E., 1980b, Characterization and comparisons of five N_2 fixing Azolla-Anabaena associations. I. Optimization of growth conditions for biomass increase and N content in a controlled environment, Plant, Cell Environ., in press.

Peterson, R. B., and Burris, R. H., 1978, Hydrogen metabolism in isolated heterocysts of Anabaena 7120, Arch. Microbiol., 116: 125.

Peterson, R. B., and Ke, B., 1979, Presence of phycobilins in heterocysts isolated from Anabaena variabilis, Plant Physiol., S5:28.

Rains, D. W., and Talley, S. N., 1979, Use of Azolla in North America, in: "Proc. Intern. Rice Res. Inst., Nitrogen and Rice," IRRI, Los Baños, Laguna, Philippines.

Ray, T. B., Peters, G. A., Toia, R. E., Jr., and Mayne, B. C., 1978, The Azolla-Anabaena relationship. VII. Distribution of ammonia-assimilating enzymes, protein, and chlorophyll between host and symbiont, Plant Physiol., 62:463.

Ray, T. B., Mayne, B. C., Toia, R. E., Jr., and Peters, G. A., 1979, The Azolla-Anabaena relationship. VIII. Photosynthetic characterization of the association and individual partners, Plant Physiol., 64:791.

Scherer, S., Kerfin, W., and Böger, P., 1980, Regulatory effect of hydrogen on nitrogenase activity of the blue-green alga (cyanobacterium) Nostoc muscorum, J. Bacteriol., 141:1037.

Schubert, K. R., and Evans, H. J., 1976, Hydrogen evolution: a major factor affecting the efficiency of nitrogen fixation in nodulated symbionts, Proc. Natl. Acad. Sci. USA, 73:1207.

Shi, T. C., Li, C. K., Wang, F. C., Chung, C. P., Chu, L. P., and Peters, G. A., 1980, Studies on nitrogen fixation and photosynthesis in Azolla imbricata (Roxb.) and Azolla filiculoides Lam., Acta Botanica Sinica, in press (in Chinese).

Singh, P. K., 1979, Use of Azolla in rice production in India, in: "Proc. Intern. Rice Res. Inst., Nitrogen and Rice," IRRI, Los Baños, Laguna, Philippines.

Smith, L. A., Hill, S., and Yates, M. G., 1976, Inhibition by acetylene of conventional hydrogenase in nitrogen-fixing bacteria, Nature, 262:209.

Stewart, W. D. P., 1977, A botanical ramble among the blue-green algae, Br. Phycol. J., 12:89.

Stewart, W. D. P., Rowell, P., Ladha, J. K., and Sampaio, M. J. A. M., 1979, Blue-green algae (cyanobacteria)-some aspects related to their role as sources of fixed nitrogen in paddy soils, in: "Proc. Intern. Rice Res. Inst., Nitrogen and Rice," IRRI, Los Baños, Laguna, Philippines.

Talley, S. N., and Rains, D. W., 1980, Azolla filiculoides Lam. as a fallow-season green manure for rice in temperate climate, Agronomy J., 72:11.

Tel-Or, E., and Stewart, W. D. P., 1977, Photosynthetic components and activities of nitrogen-fixing isolated heterocysts of Anabaena cylindrica, Proc. R. Soc. London Ser. B., 198:61.

Thomas, J., 1970, Absence of the pigments of photosystem II of photosynthesis in heterocysts of a blue-green alga, Nature, 258:715.

Tuan, D. T., and Thuyet, T. Q., 1979, Use of Azolla in rice production in Vietnam, in: "Proc. Intern. Rice Res. Inst., Nitrogen and Rice," IRRI, Los Baños, Laguna, Philippines.

Tyagi, V. V. S., Mayne, B. C., and Peters, G. A., 1980a, Action spectra of acetylene reduction in the Azolla-Anabaena association and in the isolated endophyte, Plant Physiol., S

Tyagi, V. V. S., Mayne, B. C., and Peters, G. A., 1980b, Purification and initial characterization of phycobiliproteins from the endophytic cyanobacterium of Azolla, Arch. Microbiol., in press.

Wang, T. C., Stevens, C. R. L., and Myers, J., 1977, Action spectra for photoreactions I and II of photosynthesis in the blue-green alga Anacystis nidulans, Photochem. Photobiol., 25:103.

Watanabe, I., Espinas, C. R., Berja, N. S., and Alimagno, C. B., 1977, Utilization of the Azolla-Anabaena complex as a nitrogen fertilizer for rice, International Rice Research Institute Research Paper Series, 11:1.

APPLICATION OF *AZOLLA* IN CROP PRODUCTION

S. N. Talley, E. Lim, and D. W. Rains

Department of Agronomy & Range Science and The
Plant Growth Laboratory, University of California
Davis, CA 95616

INTRODUCTION

Rice paddies in Vietnam and southern China have been fertilized for centuries utilizing biological nitrogen fixation of the *Azolla pinnata*-*Anabaena azollae* association (Liu, 1979; Mishustin and Shil-Nikova, 1971). Within the past 25 years, research has expanded the areas of *Azolla* cultivation in China to 1,340,000 ha and to 400,000 ha in Vietnam (Liu, 1979; Tran and Dao, 1973), and the practice may spread to areas of high rainfall on the east coast of India (Singh, 1977; 1979). The principle advantage of cultivating *Azolla* rather than leguminous green manures is that the former can be cultivated simultaneously with rice (Liu, 1978).

In Vietnam, *Azolla* is cultivated on fallow-flooded rice fields from November through January, between the early summer and spring rice crops. *Azolla* for inoculation onto paddies is propagated during late summer in the nursery beds used for early summer rice (Dao and Tran, 1979). Individual paddies are 0.2 ha.[1] Ferns are placed onto rice paddies at 2.5 to 5.0 MT fr wt$\cdot$ha^{-1} and cultivated until they weigh about 16 MT fr wt$\cdot$ha^{-1}. At this time the ferns may be incorporated into soil, or two-thirds of the biomass may be transferred to other paddies or dry fields and the remainder left to reestablish the *Azolla* cover (Tran and Dao, 1973; Dao and Tran, 1979). The transfer is usually done by manual labor.

---

[1]Bui, H. D., and Nguygen, T. N., 1976, Practice of *Azolla* fertilization in Vietnam. Unpublished IRRI Memo based on a visit to IRRI between 15 and 24 April.

Two or three *Azolla* crops, each amounting to approximately 25 kg $N \cdot ha^{-1}$, are possible using this procedure (Dao and Tran, 1979), which provides 50 to 100% of the nitrogen needed to produce 5 $MT \cdot ha^{-1}$ of paddy rice by the spring rice crop (Tran and Dao, 1973).[1] Fallow-season *Azolla* cultivation competes with other winter crops: corn, soybean, potatoes, vegetables (Dao and Tran, 1979). Phosphorus must be added to paddies at 2.3-4.5 kg $P \cdot ha^{-1}$ as superphosphate every 5 days to maintain optimum *Azolla* growth. The ratio of nitrogen fixed to phosphorus applied is nearly 5:1 (Tran and Dao, 1973). Potassium must also be applied to some paddy soils (Dao and Tran, 1979).

In both temperate and subtropical regions of China, *A. pinnata* is released onto rice paddies one day before transplanting, the same day as transplanting, or 5-7 days after transplanting. Inoculation of *Azolla* at transplanting utilizes 3.6 to 4.5 MT fr $wt \cdot ha^{-1}$. Seven to 8 MT fr $wt \cdot ha^{-1}$ of *Azolla* are used if paddies are inoculated after transplanting. In either case, 15 to 22 MT fr $wt \cdot ha^{-1}$ of *Azolla* possessing 37-45 kg $N \cdot ha^{-1}$ exists on the paddy 15 to 20 days after transplanting. Results are an additional 600- to 700-kg paddy $yield \cdot ha^{-1}$, which represents an approximate 18% increase over paddies in which *Azolla* is not cultivated (Anonymous, 1975; Liu, 1979).

Chinese researchers stress the importance of temperature in determining the fertilizer requirements of *A. pinnata*. If water temperature is below 10 C nutrient additions have little effect upon *Azolla* growth. When water temperature ranges between 10 and 15 C ammonium sulfate (2-4 kg $N \cdot ha^{-1}$), phosphorus (5.7-9.4 kg $P \cdot ha^{-1}$), and pig manure (20-40 $kg \cdot ha^{-1}$), are supplied each time the *Azolla* cover is thinned. Thinning takes place at approximate weekly intervals. In water over 20 C phosphorus alone is needed but the application rate is increased to 9.5-13.14 kg $P \cdot ha^{-1}$ every 5-7 days. The most effective means of fertilizing *Azolla* with phosphorus is to apply Na_2HPO_4 as a fine mist to the frond surface. This method is followed in phosphorus use efficiency by applying P as a dust, applying it to water, and applying it to soil (Anonymous, 1975).

Recent research in Fukien, China demonstrates that two rice crops may yield as much as 13,230 $kg \cdot ha^{-1}$ if the rice is planted in double rows 13 cm apart and separated from the next double row by 53 to 66 cm (Liu, 1979). This practice, used in conjunction with short-stature rice, permits *Azolla* cultivation to extend an extra month into the rice cultivation period before shade limits growth (Liu, 1978).

Azolla culture similar to that developed in Vietnam and China should be possible in regions in which the mean temperature ranges between 22 and 27 C, average maximum temperature does not exceed

32 C before or during the early stages of rice growth, and adequate water is available to keep paddies flooded. Regions fitting these criteria are northern Thailand, Burma, and central India (Watanabe, 1978). Year-round field and lab experiments at the Central Rice Research Institute, Cuttack, India, confirm that Azolla increases growth, tiller number, and grain and straw yield of rice. Incorporating 5, 10, and 15 MT fr wt$\cdot$ha^{-1} of Azolla into paddy soil one month after transplanting increased grain yield 0.50, 0.62 and 1.29 MT$\cdot$ha^{-1}, respectively. Subsequent studies confirm the Chinese experience that early planting of Azolla gives higher yields of transplant rice, but even higher yields are possible if Azolla is incorporated 7-15 days before transplanting and again one month after transplanting. Inoculation density of Azolla used in these experiments was 0.1 kg fr wt$\cdot$m^{-2} -- less than that used in Vietnamese or Chinese Azolla culture -- but a longer cultivation period (20-30 days) permitted the fern cover to reach maturity in flooded fields (Singh, 1979).

Chinese agronomists estimate that 24 to 33 man-days are required to grow one crop of Azolla over 1 hectare. Capital cost is approximately $16-21 per hectare (Anonymous, 1975). Manual tasks associated with Azolla cultivation include maintaining of small paddies for Azolla cultivation, collecting and transporting large quantities of fresh Azolla from one field to the next, spraying for pests, caring for Azolla during periods of high- or low-temperature stress, incorporating Azolla into wet paddy soil, and frequent application of phosphorus to Azolla paddies (Anonymous, 1975). Phosphorus inputs to Azolla are equivalent to 20% of the weight of nitrogen fixed under the most favorable circumstances and apparently account for much of the capital cost associated with cultivating Azolla. For comparison, approximately 16 man-hours are required to produce one hectare of rice using complete mechanization (Stout, 1966). It is therefore unlikely that Azolla culture in its present form will be incorporated into rice-production practices in the United States or in other countries which use highly mechanized cultivation procedures, even if there are severe shortages of inorganic nitrogen fertilizer.

In this article we review the problems and progress of efforts at the University of California, Davis, to integrate biological nitrogen fixation via Azolla-Anabaena with the highly mechanized rice production methods and temperate climate of California. This is being accomplished by cultivating Azolla as a green manure in fallow-flooded fields during fall after rice has been harvested, or during late-winter-early-spring, before fields are prepared for the rice crop. In preliminary experiments Azolla has also been cultivated as a dual crop with rice during summer. Objectives of both fallow-season and dual-culture studies have been to maximize rice yields by Azolla-Anabaena nitrogen fixation using propagation methods amenable to mechanization.

AZOLLA AS A FALLOW-SEASON GREEN MANURE IN TEMPERATE CLIMATE

Temperature

Azolla filiculoides and *A. caroliniana* are both moderately frost-tolerant, but *A. filiculoides* has higher rates of growth and nitrogen fixation at low temperatures (Table 1). *A. filiculoides* can be observed throughout the Sacramento Valley during December and January, when freezing is common and average maximum temperatures are 12 C or less (U. S. Department of Commerce, 1967-1976). However, growth and nitrogen fixation at these low temperatures are nil (Talley and Rains, 1980a). Between mid-February and late April, mean maximum air temperatures in the Sacramento Valley increase from 15 C to about 25 C. Throughout this period exponential growth rate and nitrogen content of *A. filiculoides* increase according to a positive logarithmic correlation with maximum air temperature. Longer-term climatic records for the Sacramento Valley suggest that *A. filiculoides* could grow in fallow rice paddies after mid-February (Fig. 1). Controlled-environment studies confirm a primary temperature limitation upon nitrogen fixation by *A. filiculoides*-*Anabaena azollae* between 12 hr day/night regimes of 10/1 C and 20/10 C. Light saturation for growth and N-fixation of *A. filiculoides* is sufficiently low that ferns growing in fallow rice fields should seldom be limited by low light (Fig. 2).

Inoculation of *Azolla filiculoides* at 0.05 kg fr wt$\cdot$m^{-2} (equivalent to 1.2 kg N$\cdot$ha^{-1}) onto fallow flooded rice paddies during spring and fall, respectively, has produced crops containing between 50 and 102 kg N$\cdot$ha^{-1} within 35-45 days (Table 2). Low temperature decreases the daily nitrogen-fixation capacity of *A. filiculoides* (Fig. 2). However, when average maximum and minimum temperature, respectively, for the *Azolla* growth period are between 34 and 20 C and 13 and 8 C, the decline in daily nitrogen fixation resulting from decreasing temperature is overcompensated by an increase in the development time of the fern mat such that nitrogen yield per *Azolla* crop increases as growth temperature decreases (Table 2). When maximum and minimum temperatures are above or below these limits, *A. filiculoides* does not form a spore-bearing multiple-layered mat. Rather, the fern cover persists as a compact single-layered mat which has a maximum nitrogen content of 35-45 kg N$\cdot$ha^{-1}. Crops of *Azolla filiculoides* that are arrested in development by low temperatures in late fall will mature during the coming late winter and early spring. Specific temperature tolerance levels of *A. filiculoides* also appear to vary according to population and whether the seasonal temperature is increasing or decreasing.

The 250-500 kg fr wt$\cdot$ha^{-1} of *A. filiculoides* used to inoculate fallow-flooded rice paddies in the experiments cited in Table 2 represents 5 to 10% of the biomass used in Vietnamese and Chinese

Table 1. Exponential growth rate and nitrogen content[1], in relation to temperature, of populations representing three *Azolla* species. Thermoperiod and photoperiod were 12 hr, and light was provided by metal halide and self-ballasted mercury lamps at 1000 $\mu E \cdot m^{-2} \cdot sec^{-1}$. *Azolla* was inoculated at 1.2 g fr wt onto the 240-cm^2 surface of each pot, which contained 2 l of nitrogen-free nutrient media described in Talley and Rains (1980a). Nitrogen was determined by Kjeldahl analysis. Experiments were carried out in triplicate.

EXPONENTIAL GROWTH RATE ($g \cdot g \cdot day^{-1}$)

	A. filiculoides Sacramento Co., CA	*A. caroliniana* Dane Co., WI	*A. mexicana* Sutter Co., CA
15/5	0.136±0.014	0.109±0.009	0.070±0.007
20/10	0.222±0.001	0.140±0.009	0.165±0.011
25/15	0.286±0.006	0.267±0.011	0.275±0.011
30/20	0.233±0.028	0.299±0.006	0.323±0.013
35/25	0[2]	0.278±0.022	0.288±0.006

NITROGEN CONTENT (% dry weight)

	A. filiculoides Sacramento Co., CA	*A. caroliniana* Dane Co., WI	*A. mexicana* Sutter Co., CA
15/5	3.40±0.08	1.98±0.07	2.43±0.12
20/10	5.22±0.01	3.39±0.23	4.27±0.07
25/15	6.09±0.07	3.62±0.02	5.54±0.14
30/20	4.82±0.07	--	--
35/25	0[2]	3.76±0.09	4.37±0.08

[1] Average ± standard deviation.

[2] *A. filiculoides* died after 2 months at this temperature.

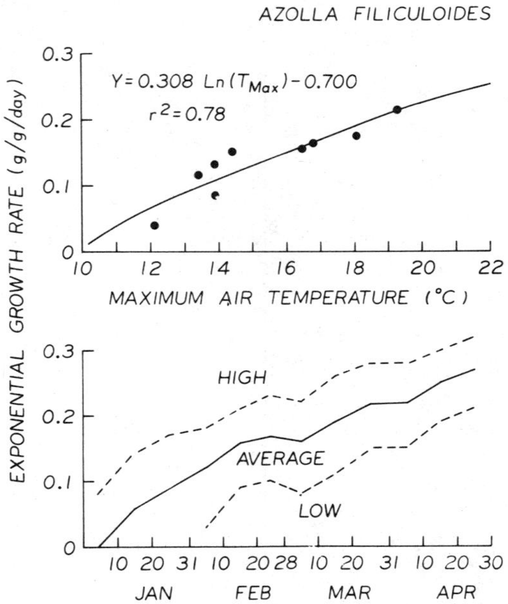

Fig. 1. The potential relationship between temperature and growth rate of A. filiculoides during late winter and early spring in the Sacramento Valley, California. Growth of A. filiculoides was monitored between 4 February and 8 May 1975 in a small agricultural drainage ditch (Tisdale Weir Quadrangle, Lat 39° 1' 27", Long 121° 47' 14"). Maximum temperature data were obtained from the Sutter Basin Reclamation District No. 1660 headquarters 0.7 km northeast of the study area. Exponential growth rate projections were derived from the U. S. Weather Bureau climatological station at Davis, California, between 1967 and 1976 (U. S. Dept. Commerce, 1967-1976).

Azolla culture. Use of such small amounts of Azolla for inoculating paddies approximately doubles the time necessary to obtain a mature Azolla cover but these lighter inoculations are well within the carrying capacities of agricultural aircraft currently used to fertilize, seed, and control pests in California rice fields. Small "monolayer" A. filiculoides fronds used for inoculation should readily disperse over a field when dropped from a low-flying aircraft.

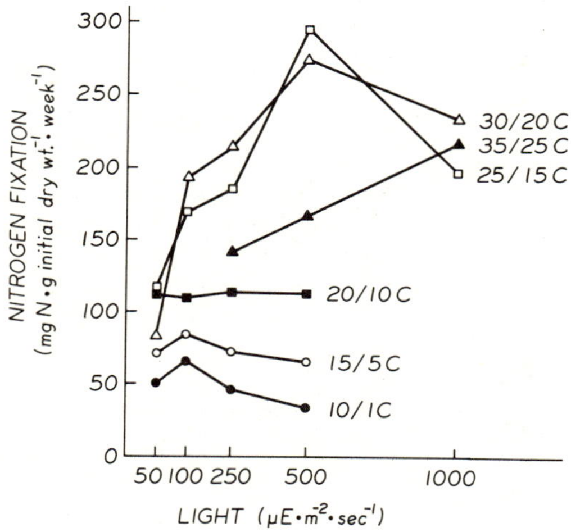

Fig. 2. Nitrogen fixation by A. filiculoides as functions of light and day/night temperature for controlled-environment studies using a 12-hour thermoperiod and photoperiod. Reprinted from Talley and Rains (1980a).

Nutrients

Phosphorus and iron must be added to waters available to rice fields in California if Azolla growth and nitrogen fixation are to be significant (Fig. 3). A. filiculoides in exponential-growth phase can contain phosphorus equal to 1% of tissue dry weight. When tissue phosphorus content is below 0.5%, phosphorus and nitrogen content appear to be positively correlated (Table 3). When Azolla tissue phosphorus content falls below 0.2% of dry weight, growth and nitrogen content are severely affected. Net phosphorus uptake in these same populations is negligible in solutions containing 0.1 mg $P \cdot l^{-1}$ or less. Maximum growth and nitrogen fixation are obtained with 0.8 mg $P \cdot l^{-1}$. At this rate of application, 38.4 mg P applied to a water surface area of 240 cm^2 over a 35-day growth period yielded 193 mg N, or a nitrogen-to-phosphorus ratio of 5:1 (Fig. 4).

Batch culture experiments conducted in a greenhouse under low light (250-700 $\mu E \cdot m^{-2} \cdot sec^{-1}$) produced A. filiculoides, A. caroliniana, and A. mexicana that had high nitrogen content. When differences in phosphorus concentration of inoculum tissue are accounted for, the A. filiculoides population from Mineral Co., Nevada, and A. caroliniana from Dane Co., Wisconsin, appear to

Table 2. A comparison of nitrogen yield of *Azolla filiculoides* populations cultivated in a fallow-flooded rice field in the Sacramento Valley, CA, and the temperature during the growth period.[1] All experiments commenced with a 50-g-fr wt·m^{-2} *Azolla* inoculum (equivalent to approximately 1.2 kg·N·ha^{-1}), except in the June-August 1980 studies, which used 25 g fr wt·m^{-1} of *Azolla*. Population maturity was indicated by abundant orange-brown sporocarps. The 10 October-27 November population did not mature but possessed smaller number of yellow-green sporocarps.

Population[2]	Growth Period	Nitrogen Yield (kg N·ha^{-1})	Average temperatures (^{o}C)[3] Maximum	Mean	Minimum
Tisdale Road	17 Aug - 7 Sept 77	34±11	33.4±3.4	23.12±2.6	12.9±2.8
Benton Crossing	13 June-29 July 80	52±3	31.1±4.7	31.8±3.2	11.7±2.4
Walker Lake	13 June-17 July 80	49±9	30.1±3.7	20.8±2.4	11.2±2.1
Sherman Island	13 June-17 July 80	54±8	30.1±3.7	20.8±2.7	11.2±2.1
Tisdale Road	19 May -23 June 77	42±12	28.5±5.8	19.5±4.1	10.7±3.3
Tisdale Road	9 May -13 June 77	53±12	25.5±6.4	17.5±4.4	9.7±3.1
Tisdale Road	4 May -13 June 77	48±11	24.7±6.7	17.4±5.4	8.9±5.8
Sherman Island	1 Oct -14 Nov 79	102±18	22.8±5.2	15.6±3.8	8.1±3.0
Sherman Island	4 Apr -19 May 79	68±5	22.3±7.4	15.3±3.8	7.7±3.1
Benton Crossing	24 Mar - 8 May 78	93±6	20.9±4.2	14.3±3.4	7.9±3.1
Sherman Island	10 Oct -27 Nov 79	72±5	19.7±5.0	13.6±4.8	7.0±3.2

[1]All figures are average ± standard deviation.

[2]Population locations are as follows: Tisdale Road, agricultural drainage ditch adjacent to Tisdale Road, Sutter Co., California, elevation 10m; Benton Crossing, on the Owens River, Mono Co., California, elevation 2550m; Walker Lake, small pond within the northwest portion of the old Walker Lake bed, Mineral Co., Nevada, elevation 1440m; Sherman Island, a drainage canal on the western end of Sherman Island, Sacramento Co., California, elevation -3m.

[3]Temperature data are from 2 m above ground in an open field at the U.S. Weather Bureau Station at the University of California, Davis. This weather facility is 3 km east of the Rice Research Facility where *Azolla* is cultivated.

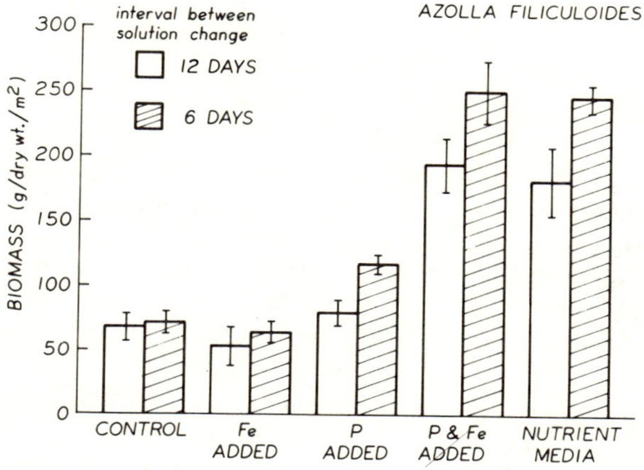

Fig. 3. Biomass of A. filiculoides growing in plastic pots (240-cm^2 surface area) filled with 2 l of Sacramento River water and various nutrient supplements. Phosphorus was added as KH_2PO_4 (12.4 mg $P \cdot l^{-1}$). Iron was added at 2.8 mg $Fe \cdot l^{-1}$ (as 0.02 M Fe·EDTA). Nutrient media consisted of 100 ml of a 20x concentration of N-free nutrient media described in Talley and Rains (1980a) and 10 ml of 0.02 M Fe EDTA prepared after Olsen (1958). The experiment utilized a 1.2 g fr $wt \cdot m^{-2}$ inoculation of A. filiculoides equivalent to 22 kg dry $wt \cdot ha^{-1}$, 1.2 kg $N \cdot ha^{-1}$, and 0.09 kg $P \cdot ha^{-1}$ and was conducted in triplicate outdoors between 11 March and 18 April 1978.

exhibit superior growth and nitrogen fixation under moderate phosphorus concentration. Although all populations appear to exhibit reduced growth when tissue phosphorus concentration is between 0.11 and 0.15% of dry wt, the Mineral Co. A. filiculoides population was able to take more phosphorus out of solution and could therefore maintain a high nitrogenase activity for a longer period (Table 4). Preliminary field studies under phosphorus-limiting conditions indicate that the Mineral Co. A. filiculoides population has better growth and nitrogen-fixation rates than do the Sacramento Co. or Mono Co. populations (unpublished data).

The iron requirement for A. filiculoides may be as low as 20 μg $Fe \cdot l^{-1}$ as Fe-EDTA (Fig. 5). Field studies have shown that 0.15 kg $Fe \cdot ha^{-1}$ as Fe-EDTA or Fe-citrate, or 1.0 kg $Fe \cdot ha^{-1}$ as $FeSO_4 \cdot 7H_2O$ are sufficient to meet the iron requirements of a crop of A. filiculoides, A. caroliniana, or A. mexicana if added in 3 split applications every 10 days (unpublished data). Fertilization with

Table 3. Average and standard deviation for tissue phosphorus, nitrogen content, and biomass of *Azolla filiculoides* growing on a range of phosphorus concentrations in a greenhouse for 18 days during August 1978. *Azolla* was inoculated at 1.2 g fr wt per pot (50 g fr wt or 3.31 g dry wt·m^{-2}). Nutrient media was changed every 70-90 hrs. Samples were carried out in triplicate.

Nutrient Media P conc. (mg/l)	Tissue P content (% dry wt)	Tissue N content (% dry wt)	Biomass (g dry wt·m^{-2})
0	0.106±0.001	3.77±0.03	29.07±4.41
0.1	0.107±0.002	3.65±0.04	31.27±1.32
0.2	0.129±0.004	4.03±0.01	37.44±3.52
0.4	0.179±0.003	4.39±0.04	46.25±3.50
0.8	0.276±0.000	5.11±0.14	49.78±4.85
1.6	0.426±0.003	5.43±0.20	55.94±6.61
3.1	0.664±0.003	5.38±0.16	59.03±0.88
6.2	1.087±0.040	5.44±0.16	61.67±3.08
12.4	1.046±0.024	5.55±0.16	63.43±2.64
49.6	--	5.59±0.22	--
99.2	0.961±0.012	5.53±0.16	--
198.4	0.997±0.030	5.49 0.12	--

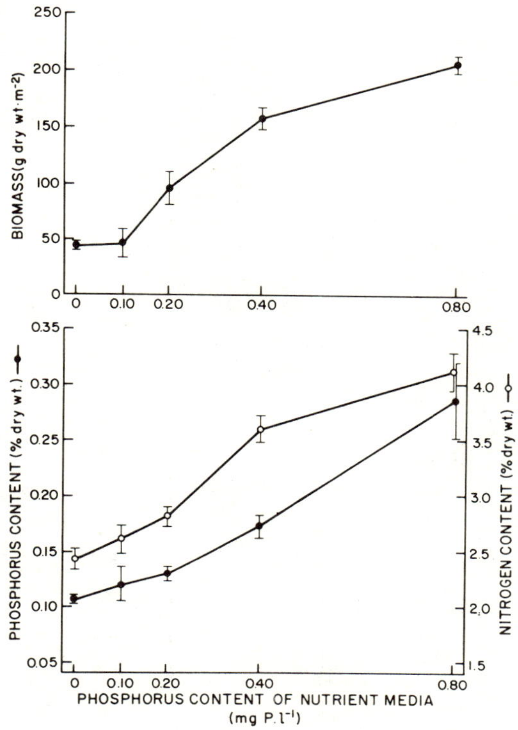

Fig. 4. Biomass, tissue nitrogen content and tissue phosphorus content of A. filiculoides growing on a phosphorus gradient for 36 days in a glass-enclosed greenhouse. Azolla was inoculated at 1.2 g fr wt onto the 240-cm^2 surface area of plastic pots containing 2 l of N-free nutrient media described in Talley and Rains (1980a), except that phosphorus was added as 0.4 M KH_2PO_4 to produce the desired concentration. Pots were changed every 3 days (6 March-11 April 1978).

both iron and phosphorus could be accomplished by airplane, particularly if large areas were fertilized at once.

Rice Yields

Azolla filiculoides incorporated into dry soil of fallow rice fields at 40 and 90 kg $N{\cdot}ha^{-1}$, respectively, increased paddy rice yield 1.5 and 2.6 $MT{\cdot}ha^{-1}$ over unfertilized controls. When the 40- and 90-kg-$N{\cdot}ha^{-1}$ Azolla treatments were augmented with equal amounts of ammonium sulfate, rice yields were statistically equivalent to results obtained by incorporating, respectively, 80 and

Table 4. Growth, nitrogen fixation, and phosphorus uptake of *Azolla filiculoides* Lam. (A.f.), *A. caroliniana* (A.c.), and *A. mexicana* Presl. (A.m.), in batch culture with various phosphorus concentrations. Experiments were conducted within a glass-enclosed greenhouse at Davis, California, between 21 January and 29 February 1980. One- and two-tenths grams fresh weight of *Azolla* was inoculated onto the 240-cm^2 surface of round plastic pots containing 2 l of nitrogen-free nutrient media described in Talley and Rains (1980a), except that phosphorus (KH_2PO_4) was deleted or added to make a solution of 0, 0.8, or 1.6 mg $P \cdot l^{-1}$. Assays were conducted over 10-day periods, nutrient media was changed every 48 hrs. Data represent averages of three replicates.

	Azolla Population					
	A.f. Mineral Co. CA	A.f. Mono Co. CA	A.f. Sacramento Co. CA	A.c. Dane Co. WI	A.m. Sutter Co. CA	A.m. Butte Co. CA
Mean day temperature (°C)	28	28	26	28	26	28
Mean night temperature (°C)	21	21	21	21	21	21
	Exponential Growth Rate ($g \cdot g^{-1} \cdot day^{-1}$)					
Phosphorus in media ($mg\ P \cdot l^{-1}$)						
0	0.24	0.17	0.20	0.16	0.21	0.17
0.8	0.24	0.20	0.20	0.23	0.26	0.21
1.6	0.29	0.21	0.21	0.22	0.25	0.21
	Total Acetylene Reduction ($\mu M\ C_2H_4 \cdot m^{-2} \cdot min^{-1}$)					
0	9.1	15.3	9.7	14.0	28.5	10.0
0.8	44.1	38.7	33.8	43.0	32.4	57.3
1.6	47.8	39.6	32.5	45.3	42.4	55.0

Table 4 continued.

			Total Tissue Nitrogen (mg N)			
0	16.6	23.5	25.5	21.7	19.5	22.4
0.8	37.2	33.9	28.3	43.0	26.6	33.8
1.6	36.5	33.2	31.8	40.1	25.3	33.4
			Total Tissue Phosphorus (mg P)			
0	0.38	0.50	0.66	0.34	0.73	0.34
0.8	3.93	3.68	3.33	3.62	3.81	3.72
1.6	5.78	4.72	4.20	4.72	4.44	4.91
			Tissue Phosphorus Content (% dry wt)			
0	0.13	0.17	0.19	0.11	0.22	0.12
0.8	0.66	0.86	0.88	0.62	0.88	0.86
1.6	1.00	1.12	1.07	0.86	1.10	1.11
			Total Phosphorus Uptake From Pots (mg P)			
0	0	0	0	0	0	0
0.8	4.87	4.32	3.64	3.32	3.89	4.39
1.6	7.41	7.05	5.72	6.61	5.23	7.32

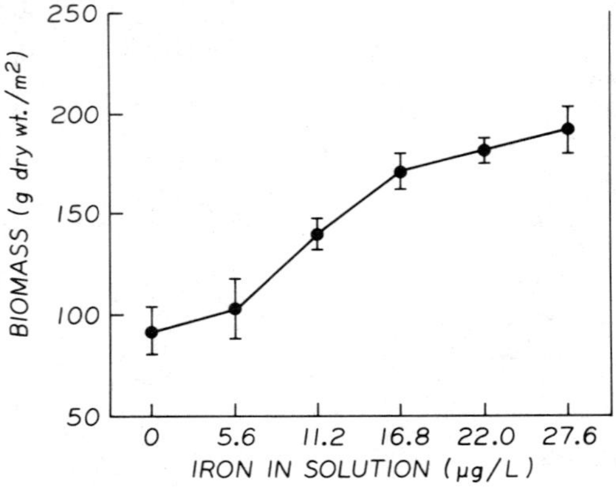

Fig. 5. Biomass of A. filiculoides maintained on an iron gradient for 36 days in a glass-enclosed greenhouse. Azolla was inoculated at 1.2 g fr wt onto the 240-cm^2 surface of plastic pots containing 2 l of N-free nutrient media described in Talley and Rains (1980a), except that iron was added as 0.002 M Fe EDTA to produce the desired concentrations. Pots were changed every 3 days (6 March-11 April 1978).

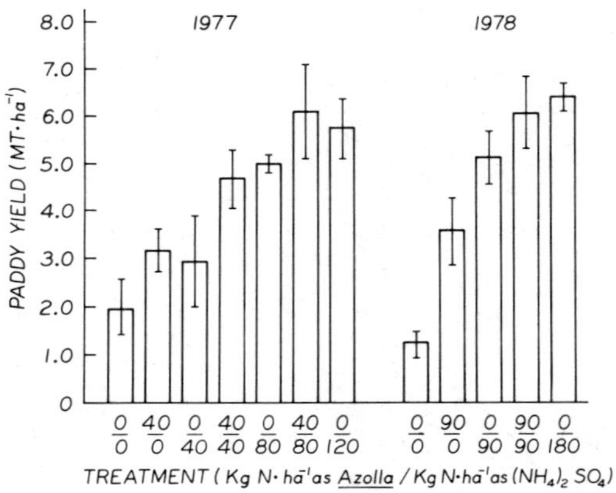

Fig. 6. Average and standard deviations of paddy yield for rice variety ED7 as functions of Azolla and ammonium sulfate fertilization during 1977 and 1978 at the U.C. Davis Rice Research Facility. Data for 1977 and 1978, respectively, represent averages of 3 and 4 22.4-m^2 plots. Planting dates were 22 May 1977 and 2 June 1978. Reprinted from Talley and Rains (1980a).

180 kg $N \cdot ha^{-1}$ as ammonium sulfate. However, yields obtained from a 90-kg$N \cdot ha^{-1}$ *Azolla* treatment not supplemented with inorganic fertilizer produced only 70% of the paddy yield obtained by incorporating an equal amount of ammonium sulfate (Fig. 6).

Azolla and ammonium sulfate nitrogen have equal effect upon rice yield when small amounts (25-30 kg $N \cdot ha^{-1}$) are used (Singh, 1977; Govindarajan et al., 1980). Experiments using large amounts of *Azolla* N (250-500 kg $N \cdot ha^{-1}$) report significantly lower rice yields than are obtained using similar amounts of ammonium sulfate nitrogen (Watanabe et al., 1977). Decreased effectiveness of high *Azolla* N could be the result of difficulty of thorough incorporation of large amounts of *Azolla*, denitrification of *Azolla* nitrogen when the soil profile is flooded and returned to an anaerobic condition after spring field preparation, and/or incomplete decomposition of *Azolla* nitrogen. The quantitative significance of any of these losses would be expected to increase if larger amounts of *Azolla* nitrogen are used. *Azolla* decomposition products may also be slightly toxic to rice. We know of no detailed studies dealing with these factors.

AZOLLA AS A COMPANION CROP WITH RICE

A. filiculoides can be cultivated as a companion crop with rice during summer. High temperatures produce high growth and N-fixation rates. A 50-g-fr·wt·m^{-2} inoculation of *A. filiculoides*, representing slightly under 10% cover, will cover the water surface in 10 to 14 days. *A. filiculoides* mats begin multilayering immediately after attaining 100% cover, spore induction occurs simultaneously with multilayering or within the next week. Within approximately 21-27 days, the *Azolla* cover contains 30-50 kg $N \cdot ha^{-1}$ and is mature. Growth and nitrogen fixation cease and cannot be stimulated in the field by thinning and fertilization.

Azolla mexicana Presl. is also native to California's Sacramento Valley and possesses several traits that make it superior to *A. filiculoides* for growth in summer. *A. mexicana* tolerates higher temperatures than does *A. filiculoides* (Table 1). When *A. mexicana* mats attain 100% cover, they form a compact monolayer that exhibits low growth and nitrogen fixation rates as biomass increases. If left unthinned, this mat will senesce, but thinning and fertilizations with phosphorus and iron will usually stimulate formation of a new cover. Paddies cultivated with *A. mexicana* as a companion crop also produce grain yields that are significantly greater than those obtained with *A. filiculoides* (Talley et al., 1977; Talley and Rains, 1980b). More detailed evaluation of *A. mexicana*-rice dual culture indicates that rice yield is equivalent to that which would be obtained by incorporating between 40 and 80 kg $N \cdot ha^{-1}$ as ammonium sulfate into paddy soil. Nitrogen effect of the *A. mexicana* companion crop is additive with ammonium sulfate when the latter is applied at 40 kg $N \cdot ha^{-1}$ (Table 5).

The mode(s) of yield enhancement for rice grown with Azolla are unknown. Observations suggest that at least five processes may be involved: 1) the pH of paddy water is less alkaline under Azolla cover, resulting in a more stable environment for ammoniacal nitrogen; 2) Azolla cover shades and subsequently kills the autotrophic aquatic flora of the paddy, subjecting the nitrogen to ammonification; 3) Azolla nitrogen may be lost to the paddy via death and decay; 4) Azolla nitrogen may be lost as ammonium when nitrogen fixation is very high; 5) late, usually sterile, tillers in rice are suppressed by shading from the Azolla cover, resulting in a conservation of nitrogen and carbon within tillers producing fertile panicles.

Macrophytic algae and cyanobacteria are not well developed under rice fertilized with 80 or 120 kg $K \cdot ha^{-1}$ as ammonium sulfate, but may represent 14 to 20 kg $N \cdot ha^{-1}$ in unfertilized or lightly fertilized paddies (Table 5). Indirect evidence for excretion of ammonia from rapidly growing A. mexicana has been outlined in Talley and Rains (1980b). Total daily acetylene reduction activity and average daily ammoniacal nitrogen concentration in the paddy water are positively correlated in A. mexicana but not in A. filiculoides, although significant ammonia is released from senescing A. filiculoides (Fig. 7). The lower ammoniacal nitrogen concentration in paddies with A. mexicana is probably due to uptake by rice, which can utilize ammoniacal nitrogen in concentrations as low as 35 $\mu g \cdot l^{-1}$ (Fried et al., 1965). Significant excretion of ammoniacal nitrogen between 24 and 72 hrs after fixation has recently been demonstrated with $^{15}N_2$ for A. pinnata (Liu et al., 1980).

When rice is grown as a dual crop with Azolla, a greater portion of the total above-ground nitrogen and biomass appears as paddy yield than when rice is grown with ammonium sulfate as the nitrogen source. For each additional metric ton of rice straw produced in Azolla-rice companion-crop paddies the paddy yield increased 3.6 metric tons. By contrast, each 1-ton increase in straw was correlated with an average increase in paddy yield of only 1.7 tons when ammonium sulfate was the nitrogen source (Table 5). These experiments suggest that the high rice grain/straw ratio in Azolla dual-culture paddies was due to suppression of late tillers by the fully developed fern mat.

DISCUSSION

The most productive periods for cultivating A. filiculoides in fallow rice paddies of the Sacramento Valley, California, are from mid-March to mid-May and mid-September to mid-November, when

Table 5. Average and standard deviation for biomass and nitrogen content of rice and *Azolla mexicana* obtained from 1977 field studies.

	TREATMENT							
	Azolla[a]	0 N[b]	0 N + Azolla[c]	40 kg N/ha	40 kg N/ha + Azolla	80 kg N/ha	80 kg N/ha + Azolla	120 kg N/ha
COMPONENT								
Azolla								
July 15-Aug 10								
Max. biomass (kg dry wt/ha)	--	--	317±197	--	371±174	--	86±58	--
Max. N-content (kg/ha)	--	--	17.3±3.6	--	15.4±7.2	--	3.4±2.3	--
Aug 17-Sept 30								
Max. biomass (kg dry wt/ha)	996±160	--	613±70	--	603±68	--	323±89	--
Max. N-content (kg/ha)	36.6±4.0	--	29.2±3.3	--	27.1±3.1	--	14.6±4.0	--
N-content at experiment termination	36.6±4.0	--	13.3±1.3	--	10.5±3.7	--	9.4±3.0	--
Macrophytic algae N (kg/ha)[d]	27.1	19.5	14.2	15.7	14.3	4.0	5.1	Trace
Rice (ESD-7)								
Straw (MT/ha)	--	1.80±0.85	2.79±0.45	2.32±0.20	2.82±0.50	3.32±0.77	3.74±0.54	5.20±0.60
Yield (MT/ha)	--	1.59±0.75	4.14±0.27	3.15±0.55	5.78±0.32	5.55±0.16	7.55±0.62	6.04±0.54
Straw corr. w/Azolla (MT/ha)[e]	--	--	0.99	--	0.50	--	0.42	--
Yield corr. w/Azolla (MT/ha)	--	--	2.55	--	2.63	--	2.00	--
Straw N (kg/ha)	--	10.7±4.8	16.9±1.8	13.5±0.1	14.9±1.6	17.9±2.9	19.2±2.5	24.1±2.5
Yield N (kg/ha)	--	15.2±0.75	39.8±4.6	28.9±3.8	52.5±0.5	53.1±3.3	68.8±3.3	54.4±3.7
Straw N corr. w/Azolla (kg/ha)	--	--	6.2	--	1.4	--	1.3	--
Yield N corr. w/Azolla (kg/ha)	--	--	24.6	--	23.6	--	15.8	--
Rice Yield N/Rice Straw N	--	1.42	2.36	2.14	3.52	2.97	3.58	2.26
Rice Yield/Rice Straw	--	0.88	1.48	1.36	2.05	1.67	2.02	1.16

[a] *Azolla mexicana* grown from 1 Sept. to 2 Oct. 1977. [b] Nitrogen in kg/ha incorporated into dry soil as ammonium sulfate.
[c] *Azolla mexicana* companion crop with rice. [d] Macrophytic algae determined on 5 Aug. 1977. [e] Corr. = correlated.

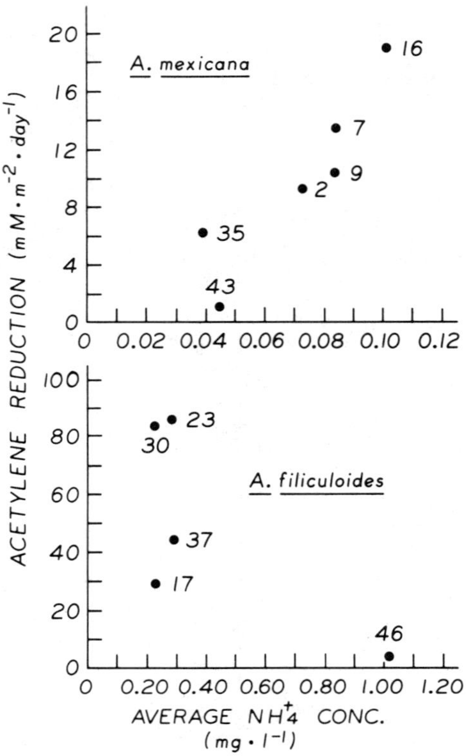

Fig. 7. The relationship between total diurnal acetylene reduction activity of A. mexicana and A. filiculoides per unit area of paddy surface and average concentration of ammoniacal nitrogen in paddy water for the same diurnal periods. Numbers beside points indicate days elapsed since Azolla was inoculated at 50 g fr wt$\cdot$m^{-2} onto the paddy. Culture period was 1 Sept.-20 Oct. 1977 for A. mexicana, and 7 April-19 May 1978 for A. filiculoides.

average maximum, mean, and minimum temperatures, respectively, are about 23-26 C, 15-17 C, and 7-8 C (Talley et al., 1977) (Table 2). Contemporary California rice-culture practices require significant field activity during both periods. Paddies are dried, cultivated, leveled, fertilized, flooded, and directly seeded with rice during spring. In fall, fields are dried and harvested, and, if weather permits, the stubble is burned and the residue plowed under before the onset of significant winter rain -- usually in November. Azolla cultivation is, therefore, precluded immediately before or after the rice crop, when climate is best for production of Azolla N.

We estimate that within a decade, the combined effects of using early maturing short-stature rice (to minimize cold injury (Board et al., 1980)) and the need to develop a cost-effective alternative to burning rice stubble in the fall will result in fallow rice fields free of straw (excepting basal stubble) by mid-October of most years. This will be due to early harvest dates possible with new rice varieties and the necessity to get dry rice straw out of the fields by early October before the onset of significant precipitation in fall[2]. If fields could be planted with Azolla by mid-October, there would be sufficient time to fix 40-60 kg $N \cdot ha^{-1}$ before low temperature halts growth through December and January (Table 2) (Talley et al., 1977). Preliminary studies conducted in fall, winter, and spring of 1979-1980 suggest that an A. filiculoides crop grown during fall can overwinter in flooded or wet rice paddies.

Cultivation of Azolla during late winter and early spring requires either that fields be prepared for rice when they are still wet or that rice be transplanted if it is to be planted early enough to minimize cold injury and if dry stubble is to be available in early October. We know of no attempts to either develop wet field preparation or transplant rice culture in the United States. However, the successful development of small-scale machinery to puddle rice soil and the widespread mechanization of transplanting in Japanese rice paddies (Torijama, 1976) suggest transplant rice culture may warrant long-term research priority in California. Japanese rice transplanters, which plant 2-4 rows at a time, can complete 1 ha of paddy in 6 hrs (Kanei, 1977). Development of large-scale transplanters for the much larger fields in the United States should further reduce the labor and time requirements for transplanting. Potential benefits other than spring Azolla culture, which would accompany transplant rice culture, would be: 1) conservation of seed rice; 2) conservation of perhaps 20% of the water currently used to cultivate rice; 3) the capacity to reliably double crop rice with winter crops such as wheat or barley.

If the substantial yield increases attributable to Azolla-rice dual culture (Table 5) are confirmed by future studies, this

---

[2] Climatic records for Marysville, California, which is approximately at the middle of latitudinal extent of the Sacramento Valley rice-growing region, show that storms producing 2.5 cm or greater precipitation end the summer dry period by the 2nd week of October in approximately one year out of five (U.S. Dept. of Commerce, 1951-1978).

method of cultivation could be the first to be tested on a large scale, because few changes in current rice culture practices would be required. Azolla-rice dual culture with partial or complete incorporation of the Azolla cover would require systematic placement of rice (through either transplanting into saturated soil or drill seeding into dry soil) and development of the required machinery.

The principal inorganic nutrient requirement for Azolla growth in rice paddies is phosphorus. Even if fields are available for Azolla culture at the proper time and the rice-yield-enhancing capacity of the Azolla crop is assured, it will still be difficult to justify use of phosphorus (a nonrenewable resource) to produce nitrogen (a renewable resource), unless the phosphorus requirements for the Azolla crop can be held within the 20-30 kg $P \cdot ha^{-1} \cdot yr^{-1}$ already needed to cultivate rice in California (Mikkelsen and Evatt, 1973). Thirty to 60 kg $P \cdot ha^{-1}$ and about 45 days are required to produce an A. filiculoides crop containing 90-100 kg $N \cdot ha^{-1}$ (Talley and Rains, 1980a). However, phosphorus-efficient A. filiculoides populations can fix 40 to 60 kg $N \cdot ha^{-1}$ in fallow rice fields within 35 days through application, respectively, of 6 to 12 kg $P \cdot ha^{-1}$ in three split applications (unpublished data). The diminishing return to rice incurred when more than about 60 kg $N \cdot ha^{-1}$ as Azolla green manure is incorporated into soil (Talley and Rains, 1980a) helps justify growing a sub-optimal (40-60 kg $N \cdot ha^{-1}$) Azolla crop. Additional phosphorus needed to grow rice but not needed to grow a spring or fall Azolla crop could be applied along with supplemental inorganic nitrogen fertilizer during spring field preparation for rice or it could be added during summer to fertilize an Azolla companion crop and the rice.

ACKNOWLEDGMENT

This work was supported in part by a grant from the National Science Foundation (NSF PFR 77-07301).

REFERENCES

Anonymous, 1975, Cultivation, propagation and utilization of Azolla, Institute of Soils and Fertilizers, Chekiang Academy of Agricultural Sciences, Agricultural Publishing House, Peking Distributing Office of the Hsin-Hua Shu-Tien.

Board, J. E., Peterson, M. L., and Ng, E., 1980, Floret sterility in rice in a cool environment, Agron. J., 72:483.

Dao The Tuan, and Tran Quang Thuyet, 1979, Use of Azolla in rice production in Vietnam, in: "Nitrogen and Rice," International Rice Research Institute, Los Banos, Laguna, Philippines.

Fried, M., Zsoldos, F., Vose, P. B., and Shatokhin, I. L., 1965, Characterizing the NO_3^- and NH_4^+ uptake process of rice roots by use of ^{15}N labeled NH_4NO_3, Physiol. Plant., 18:313.

Govindarajan, K., Kannaiyan, S., Jagannathan, R., Palaniyandi, V. G., and Ramachandran, M., 1980, Effects of Azolla inoculation on rice yields, International Rice Research Newsletter, 5(1):20.

Kanei, K., 1977, New trends on the development of rice transplanters examined by the (Japanese) government, Farming Mechanization March, 1977:28.

Liu Chung-Chu, 1978, Studies on cultivation of Azolla in the rice field for most of the year, Soil and Fertilizer Institute, Fukien Academy of Agricultural Sciences, People's Republic of China.

Liu Chung-Chu, 1979, Use of Azolla in rice production in China, in: "Nitrogen and Rice," International Rice Research Institute, Los Banos, Laguna, Philippines.

Liu Chung-Chu, Chen Bing-Huan, Wei Wen-Shong, You Chung Biao, Li Jiang-Wei, and Song Wei, 1980, Studies on Nitrogen Excretion by Azolla. Draft paper for a Special Workshop on Nitrogen Fixation and Utilization in Rice Fields 20-30 April, The International Rice Research Institute, Los Banos, Laguna, Philippines.

Mikkelsen, D. S., and Evatt, N. S., 1973, Soils and Fertilizers, in: "Rice in the United States: Varieties and Production," U.S.D.A. Agricultural Handbook No. 289.

Mishustin, E. N., and Shil-Nikova, U. K., 1971, Biological Fixation of Atmospheric Nitrogen, London, MacMillan.

Olsen, C., 1958, Iron absorption in different plant species as a function of pH value of the solution, Physiol. Plant., 11:889.

Singh, P. K., 1977, Multiplication and utilization of fern Azolla containing nitrogen fixing algal symbiont as green manure in rice cultivation, Il Riso, 26:125.

Singh, P. K., 1979, Use of Azolla in rice production in India, in: "Nitrogen and Rice," International Rice Research Institute, Los Banos, Laguna, Philippines.

Stout, B. A., 1966, Equipment for rice production, FAO Development Paper 87, United Nations, Rome.

Talley, S. N., Talley, B. J., and Rains, D. W., 1977, Nitrogen fixation by Azolla in rice fields, in: "Genetic Engineering for N Fixation," A. Hollaender, ed., Plenum Press, New York.

Talley, S. N., and Rains, D. W., 1980a, Azolla filiculoides Lam. as a fallow-season green manure for rice in a temperate climate, Agron. J., 72:11.

Talley, S. N., and Rains, D. W., 1980b, Azolla as a nitrogen source for temperate rice, in: "Nitrogen Fixation, Vol. II, Symbiotic Associations and Cyanobacteria," W. E. Newton and W. H. Orme-Johnson, eds., University Park Press, Baltimore, Maryland.

Torijama, K., 1976, Rapid progress of mechanical transplanting in Japan. Rice, Newsletter of the International Rice Commission, 25(1/2):35.

Tran Quan Thuyet, and Dao The Tuan, 1973, Azolla: A green compost, in: "Vietnamese Studies No. 38, Agricultural Problems (Vol. 4), Agronomical Data," Nguyen Khac Vien, ed., Xunhasaba, Hanoi, Vietnam.

U. S. Department of Commerce, 1951-1978, Climatological data, California, V 55-82, U.S. Gov. Printing Office, Washington, D.C.

Watanabe, I., 1978, Azolla and its use in lowland rice culture, Tsuchi To Biseibutsu (Soil and Microbe, Japan), 20:1.

Watanabe, I., Espinas, C. R., Berja, N. S., and Alimagno, V. B., 1977, Utilization of the Azolla-Anabaena complex as a N fertilizer for rice, IRRI Research Paper Series No. 11.

A NEW WOODY PLANT WHICH FIXES NITROGEN: *CHAMAEBATIA FOLIOLOSA* (ROSACEAE)

Rod M. Heisey and C.C. Delwiche

Department of Land, Air, and Water Resources
University of California
Davis, California 95616

INTRODUCTION

Root nodules, and presumably nitrogen fixation, occur in at least 18 genera of nine angiosperm families outside of the Leguminosae (Bond, 1976; Akkermans and Houwers, 1979; Righetti and Munns, 1980). Nearly all the nodulated non-legumes are woody shrubs or trees. Except for one genus in the Ulmaceae (Akkermans et al., 1978), which is nodulated by *Rhizobium* (Trinick, 1973), all the other such species whose endophytes have been investigated are thought to be nodulated by actinomycetes (Akkermans and Houwers, 1979).

In the Rosaceae, nodulation has been reported for *Dryas* (Lawrence et al., 1967), *Purshia* (Webster et al., 1967), and *Cowania* (Righetti and Munns, 1980), all members of the tribe Dryadeae (Hutchinson, 1964), as well as for *Cercocarpus* (Vlamis et al., 1964) and *Rubus* (Bond, 1976). We have recently discovered nodules on *Chamaebatia foliolosa* Benth. (Rosaceae, Dryadeae) and have obtained direct evidence for their nitrogen fixation (Heisey et al., 1980).

Chamaebatia foliolosa (Fig. 1) is an evergreen shrub 2 to 6 dm high having numerous leafy branches and glandular-pubescent young twigs (Munz and Keck, 1973). The leaves, which are mostly thrice pinnate and 2 to 10 cm long, are covered with a glutinous, aromatic material giving the plant the common name of mountain misery. *Chamaebatia* occurs between 600 and 2100 m on the western slope of the Sierra Nevada from Shasta to Kern counties, California (Munz and Keck, 1973) in ponderosa pine, white fir-mixed conifer, and Jeffrey pine forests (Rundel et al., 1977). Reproduction is primarily vegetative by means of extensive creeping rhizomes (Anon., 1976), but viable seeds are also produced.

Fig. 1. Nodulated plant of *Chamaebatia foliolosa*.

Chamaebatia is favored by fire and becomes less vigorous when heavily shaded (Rundel et al., 1977). It often forms extensive, dense stands in semi-open forest areas which hinder the growth of seedling trees (Munns, 1922; Magill, 1974).

A second member of the genus, *C. australis* (Bdg.) Abrams. occurs in chaparral of southern San Diego County, California, and northern Baja California (Munz and Keck, 1973). It is 6 to 20 dm high, but otherwise very similar to *C. foliolosa* in appearance. No nodules have been found upon examination of a small number of *C. australis* plants.

We first discovered nodulated *Chamaebatia* at an elevation of 1100 m near Pollock Pines, El Dorado County, California. The site had been burned about 10 years previously and was revegetated mainly by *Chamaebatia* and *Ceanothus integerrimus* H. & A.; few overstory trees survived the fire. The soil, a cobbly to very rocky loam derived from volcanic parent material, had little O horizon and was severely eroded in places. Nodulated *Chamaebatia* was subsequently found at a second site about 6 km from the first, at elevation 900 m, on a very rocky loam overlying metamorphic rocks. Although an open ponderosa pine overstory was present, the second site also had poorly developed soil organic horizons and showed evidence of recent fire.

MATERIALS AND METHODS

For ^{15}N studies, nodulated Chamaebatia plants were dug in December 1978, potted in habitat soil, and maintained in a greenhouse at Davis. Nodulated and non-nodulated root segments were removed in March 1979, placed immediately into 23 ml glass-stoppered tubes which were evacuated to the vapor pressure of water and refilled with 20% O_2, 70% Ar, 10% $^{15}N_2$, and 0.04% CO_2, and incubated 5 hr at 22 C. The incubated tissues, as well as non-incubated foliage and soil from the nodule site, were digested by a modified Kjeldahl procedure (Bremner, 1965) and analyzed for ^{15}N on a mass spectrometer.

Chamaebatia was also tested for nitrogen-fixing activity with the acetylene reduction technique (Hardy et al., 1968). Tests were done in the field in May, a time when Chamaebatia was showing vigorous new shoot growth. Nodules or non-nodulated roots were excavated, placed in 13 ml stoppered vials, and incubated in 10% acetylene in air. Incubation was begun within 35 minutes after nodules and roots were detached from the plants. To simulate natural conditions, the vials were buried 15 cm deep in the soil under the canopy of a Chamaebatia plant during incubation. Soil temperatures near the vials ranged from 15 to 18 C. Gas samples of 1.45 ml were removed at 1, 3, and 5 hr and stored in evacuated 13 ml vials. An equivalent amount of 10% acetylene in air was added to the incubation vials after withdrawing samples to maintain a constant pressure around the nodules and roots. At the lab, the storage vials were equilibrated to atmospheric pressure with air and 0.1 ml aliquots were analyzed for ethylene on a gas chromatograph. Ethylene peaks of samples were quantified by comparison to standard curves.

Nodule tissues were examined and photographed with a phase-contrast microscope.

RESULTS AND DISCUSSION

The results of $^{15}N_2$ uptake (Table 1) and acetylene reduction (Fig. 2) are very similar (the reduction of 3 μmoles of acetylene is theoretically equivalent to the fixation of 1 μmole of N_2) and provide positive evidence of nitrogen fixation. The rates of nitrogen fixation by Chamaebatia are comparable to those of other non-legumes such as Alnus (Fleschner et al., 1976), Ceanothus (Delwiche et al., 1965), Myrica (Bond, 1971; Schwintzer, 1979), and Purshia (Webster et al., 1967; Dalton and Zobel, 1977). Acetylene reduction by

Table 1. Total % N, isotopic composition, and rates of nitrogen fixation (µmoles g^{-1} fresh wt hr^{-1}) for excised nodules and non-nodulated roots of Chamaebatia foliolosa incubated 5 hr in $^{15}N_2$ compared to % N and isotopic composition of non-incubated foliage and soil from the nodule site (from Heisey et al., 1980).

Sample	Total % N[a]	Atom % excess ^{15}N[b]	Rate of N_2 fixation
Nodular tissue	.53	.369	.130
Nodular tissue	.55	.030	.011
Non-nodulated roots	.51	-.0003	-
Non-nodulated roots	.56	-.0011	-
Non-incubated foliage from nodule site	1.38	-.0004	-
Site soil (0 to 5 cm)	.23	.0030	-
Site soil (10 to 15 cm)	.13	.0033	-

[a]On fresh weight basis for nodules and roots, dry weight basis for non-incubated foliage and soil.

[b]Referred to atmospheric N_2.

Chamaebatia nodules continues actively for at least 5 hr after their detachment from the plants, but the rate declines slightly with time.

The coralloid, di- or trichotomously lobed Chamaebatia nodules are spherical, ovoid, disc, or fan-shaped and grow to at least 20 mm long and 15 mm thick (Fig. 3). Individual lobes are about 0.8 to 1 mm in diameter. Young nodules are light brown, turn darker brown with age, and eventually become almost black during senescence. We do not know when nodulation first occurs in the field, but newly emerged seedlings collected at both sites in April and May 1979 and grown in habitat soil in a greenhouse were well nodulated by the following March.

Internally, Chamaebatia nodules are very much like those of other actinorhizal nodules (Furman, 1959; Lawrence et al., 1967; Hoeppel and Wollum, 1971; Krebill and Muir, 1974; Strand and Laetsch, 1977). Many cells of the cortex, especially the medial region, are enlarged and nearly filled with the endophyte (Fig. 4). Higher magnification of the endophyte clusters therein shows a central mass of filamentous hyphae surrounded on the periphery by hyphae bearing terminal vesicles (Fig. 5).

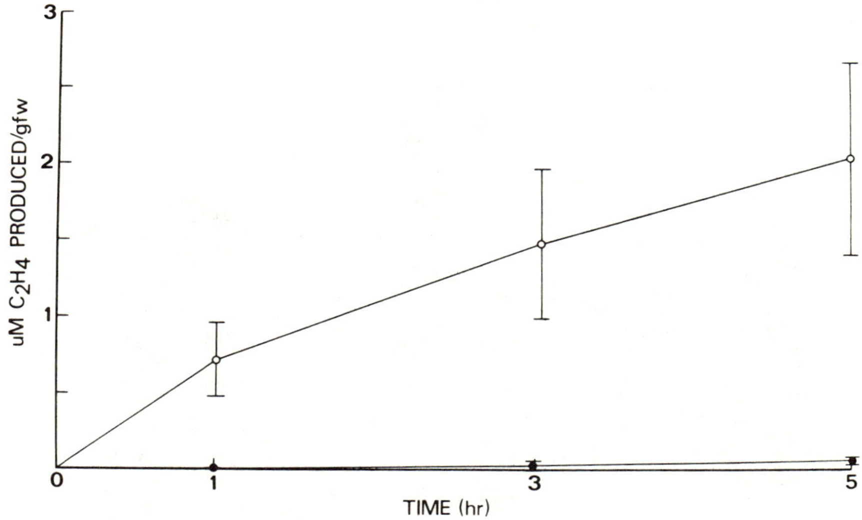

Fig. 2. Acetylene reduction, on fresh-weight basis, by excised nodules (open dots) and non-nodulated roots (closed dots) of Chamaebatia foliolosa. Bars are ± $s_{\bar{x}}$; n=3 for nodules, 2 for roots.

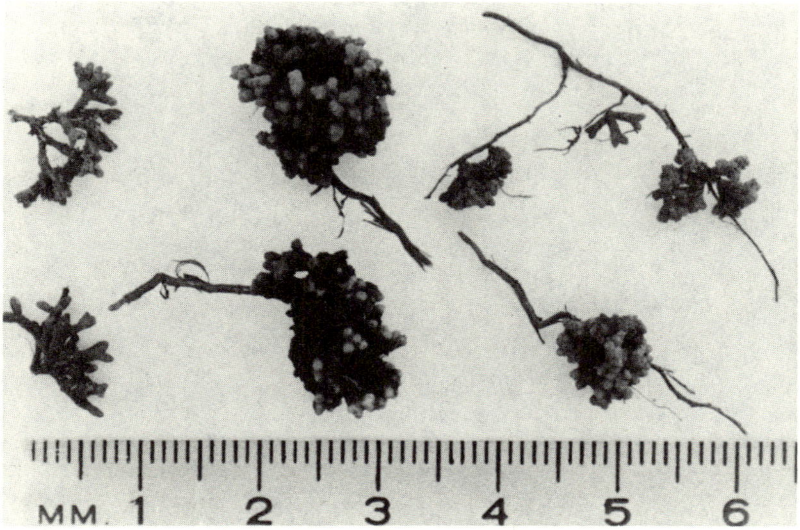

Fig. 3. Root nodules of Chamaebatia foliolosa.

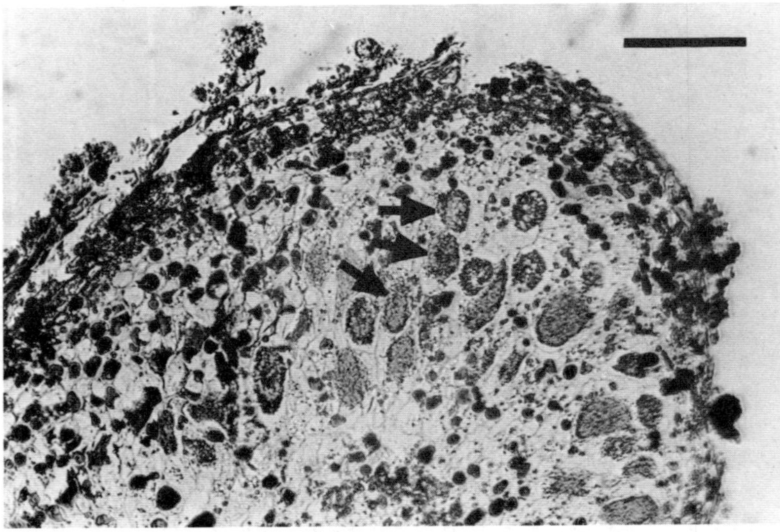

Fig. 4. Oblique section of Chamaebatia nodule lobe showing hypertrophied cortical cells filled with the endophyte (arrows). Scale bar indicates 100 µm.

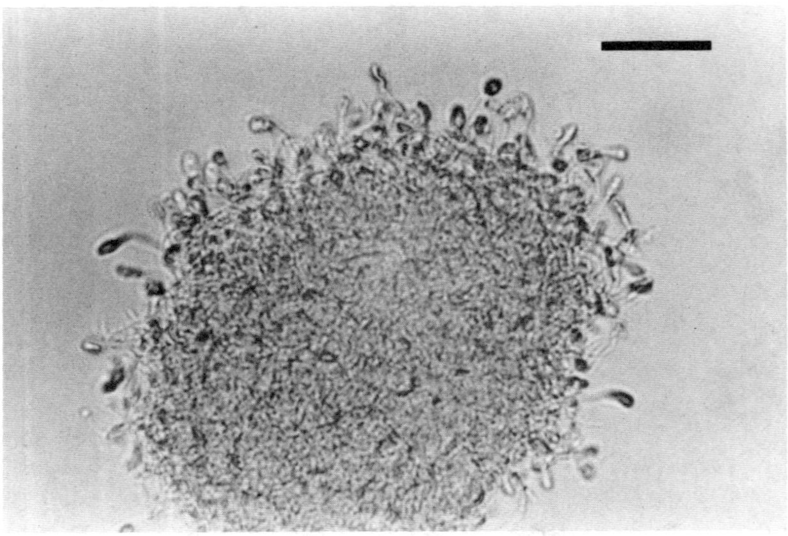

Fig. 5. Endophyte cluster from Chamaebatia nodule showing filamentous central region surrounded by hyphae bearing club-shaped terminal vesicles. Scale bar indicates 20 µm.

The vesicles are primarily club-shaped and are similar to those found in nodules of other members of the Rosaceae, as well as outside the family in Casuarina, Myrica, and Comptonia (Akkermans et al., 1979).

Nodulation of Chamaebatia is not ubiquitous. We have examined plants at a number of sites with thick soil organic horizons and found no nodules. Since Chamaebatia roots extend as deep as 1.8 m (Anon., 1976) it is possible that nodules at these sites were below the 0 to 30 cm depth we sampled. Like many other nitrogen fixers, Chamaebatia frequently inhabits disturbed or nutrient-deficient sites. One factor which often influences the occurrence of nodules on nitrogen-fixing plants is the nitrogen status of the soil. Nodulation is commonly greatest at low concentrations and reduced or inhibited at high concentrations (Vlamis et al., 1964; Zavitkovski and Newton, 1968; Hoeppel and Wollum, 1971). Losses of soil nitrogen by volatilization, leaching,and erosion as a result of forest fire, may explain why Chamaebatia nodules have been found only on recently burned areas.

Chamaebatia has heretofore been regarded as a forest nuisance. Its recently discovered capacity to fix nitrogen makes it deserving of new consideration. More information is needed concerning the geographical extent of nodulation, the factors which affect nodulation, and the contribution made by Chamaebatia to the nitrogen input of the forest.

REFERENCES

Akkermans, A. D. L., Abdulkadir, S., and Trinick, M. J., 1978, Nitrogen-fixing root nodules in Ulmaceae, Nature, 274:190.

Akkermans, A. D. L. and Houwers, A., 1979, Symbiotic nitrogen fixers available for use in temperate forestry, in: "Symbiotic Nitrogen Fixation in the Management of Temperate Forests," J. C. Gordon, C. T. Wheeler, and D. A. Perry, eds., Forest Research Laboratory, Oregon State University, Corvallis.

Akkermans, A. D. L., Roelofsen, W., and Blom, J., 1979, Dinitrogen fixation and ammonia assimilation in actinomycetous root nodules of Alnus glutinosa, in: "Symbiotic Nitrogen Fixation in the Management of Temperate Forests," J. C. Gordon, C. T. Wheeler, and D. A. Perry, eds., Forest Research Laboratory, Oregon State University, Corvallis.

Anonymous, 1976, Proceedings of the Bear Clover Conference, Stanislaus National Forest, March 17, 1976.

Bond, G., 1971, Fixation of nitrogen in two African species of Myrica, New Phytol., 70:1.

Bond, G., 1976, The results of the IBP survey of root-nodule formation in non-leguminous angiosperms, in: International Biological Programme, Vol. 7, "Symbiotic Nitrogen Fixation in Plants," P. S. Nutman, ed., Cambridge University Press, Cambridge.

Bremner, J. M., 1965, Isotope-ratio analysis of nitrogen in nitrogen-15 tracer investigations, in: Agronomy, No. 9, "Methods of Soil Analysis," (part 2), American Society of Agronomy, Madison.

Dalton, D. A. and Zobel, D. B., 1977, Ecological aspects of nitrogen fixation by Purshia tridentata, Plant and Soil, 48:57.

Delwiche, C. C., Zinke, P. J., and Johnson, C. M., 1965, Nitrogen fixation by Ceanothus, Plant Physiol., 40:1045.

Fleschner, M. D., Delwiche, C. C., and Goldman, C. R., 1976, Measuring rates of symbiotic nitrogen fixation by Alnus tenuifolia, Amer. J. Bot, 63:945.

Furman, T. E., 1959, The structure of the root nodules of Ceanothus sanguineus and Ceanothus velutinus, with special reference to the endophyte, Amer. J. Bot., 46:698.

Hardy, R. W. F., Holsten, R. D., Jackson, E. K., and Burns, R. C., 1968, The acetylene-ethylene assay for N_2 fixation: Laboratory and field evaluation, Plant Physiol., 43:1185.

Heisey, R. M., Delwiche, C. C., Virginia, R. A., Wrona, A. F., and Bryan, B. A., 1980. A new nitrogen-fixing non-legume: Chamaebatia foliolosa (Rosaceae), Amer. J. Bot., 67:429.

Hoeppel, R. E. and Wollum, A. G., II, 1971, Histological studies of ectomycorrhizae and root nodules from Cercocarpus montanus and Cercocarpus paucidentatus, Can. J. Bot., 49:1315.

Hutchinson, J., 1964, " The Genera of Flowering Plants," Vol. 1, Oxford University Press, London.

Krebill, R. G. and Muir, J. M., 1974, Morphological characterization of Frankia purshiae, the endophyte in root nodules of bitterbrush, Northwest Sci., 48:266.

Lawrence, D. B., Schoenike, R. E., Quispel, A., and Bond, G., 1967, The role of Dryas drummondii in vegetation development following ice recession at Glacier Bay, Alaska, with special reference to its nitrogen fixation by root nodules, J. Ecol., 55:793.

Magill, A. W., 1974, Chamaebatia foliolosa Benth., in: "Seeds of Woody Plants in the United States," Agricultural Handbook 450, Forest Service, U. S. Department of Agriculture, Washington, D. C.

Munns, E. N., 1922, Bear clover and forest reproduction, J. Forestry, 20:745.

Munz, P. A. and Keck, D. D., 1973, "A California Flora and Supplement," University of California Press, Berkeley.

Righetti, T. L. and Munns, D. N., 1980, Nodulation and nitrogen fixation in cliffrose (Cowania mexicana var. stansburiana (Torr.) Jeps.), Plant Physiol., 65:411.

Rundel, P. W., Parsons, D. J., and Gordon, D. T., 1977, Montane and subalpine vegetation of the Sierra Nevada and Cascade ranges, in: "Terrestrial Vegetation of California," M. G. Barbour and J. Major, eds., John Wiley & Sons, New York.

Schwintzer, C. R., 1979, Nitrogen fixation by Myrica gale root nodules Massachusetts wetland, Oecologia (Berl.), 43:283.

Strand, R. and Laetsch, W. M., 1977, Cell and endophyte structure of the nitrogen-fixing root nodules of Ceanothus integerrimus H.

and A. I. Fine structure of the nodule and its endosymbiont, Protoplasma, 93:165.

Trinick, M. J., 1973, Symbiosis between Rhizobium and the non-legume, Trema aspera, Nature, 244:459.

Vlamis, J., Schultz, A. M., and Biswell, H. H., 1964, Nitrogen fixation by root nodules of western mountain mahogany, J. Range Manage., 17:73.

Webster, S. R., Youngberg, C. T., and Wollum, A. G., II, 1967, Fixation of nitrogen by bitterbrush (Purshia tridentata (Pursh) D. C.), Nature, 216:392.

Zavitkovski, J. and Newton, M., 1968, Ecological importance of snowbrush Ceanothus velutinus in the Oregon Cascades, Ecology, 49: 1134.

SOIL FACTORS LIMITING NODULATION AND NITROGEN FIXATION IN *PURSHIA*

Timothy L. Righetti and Donald N. Munns

Land, Air and Water Resources
University of California
Davis, CA 95616

INTRODUCTION

Approximately 50 species in 10 genera capable of actinomycete-induced nodulation and nitrogen fixation are found in the western United States. Their ecological importance has recently been reviewed (Klemmedson, 1979). Nitrogen fixation may enhance colonization of natural or man-induced disturbances. Some of these genera may have applications in revegetation and rehabilitation of drastically nitrogen deficient disturbances brought on by increasing energy and mineral demands. There is also interest in forestry applications utilizing nitrogen-fixing trees such as alder, or a nitrogen-fixing understory (Gordon and Dawson, 1979).

Endophyte soil populations seem to be adequate and environmental conditions usually permit nodulation for actinorhizal systems inhabiting moist ecosystems (Bond, 1976b). In more xeric climates nodule biomass may be suboptimal. The lack of knowledge on the occurrence, effectiveness and duration of nodules on actinorhizal plants in semiarid areas makes estimates of their ecologial importance and the level of nitrogen inputs difficult. It appears that many sites are sparsely nodulated, but the question of nodulation incidence is largely unanswered. In some nodulated species, nodules may be situated too deeply to be revealed by conventional excavation (Bond, 1976b). Factors associated with soil depth may be important.

Factors limiting nodulation are not clear. Moisture and temperature stresses may be severe and fertility problems may exist. Until environmental limitations are better understood, the opportunities to utilize actinorhizal systems in semiarid regions will remain speculative.

Purshia was selected for study as an example of the nodulating genera in semiarid regions because of its exceptional browse value and wide distribution (Nord, 1965). A management handbook (Giunta et al., 1978) and extensive reviews on the autecology (Nord, 1965; Stanton, 1959) are in the literature. Both species in the genus, P. tridentata and P. glandulosa, exhibit pioneer capabilities on semiarid plains, foothills and mountain slopes, and have been successfully employed in revegetation efforts (McArthur et al., 1974; Plummer, 1976).

Speculation that Purshia contributes substantial nitrogen into the environment (Silvester, 1977) has been disputed. An estimate based on seasonal acetylene reduction rates and a determination of nodule biomass on volcanic soils in Oregon was only 0.057 kg $ha^{-1}yr^{-1}$ (Dalton and Zobel, 1977). Insufficient nodule biomass may have been caused by moisture constraints. However, some soils collected at Purshia sites fail to nodulate Purshia seedlings even under favorable greenhouse conditions indicating limiting factors other than moisture (Klemmedson, 1979; Wagle and Vlamis, 1961).

In addition to differences between the two Purshia species, there are ecotype variations within them (Nord, 1965). Some ecotypes are better adapted to specific soil conditions than others (Giunta et al., 1978; Plummer et al., 1968). It is not clear how ecotype or species differ in nodulation capabilities.

Where their ranges overlap, hybridization and backcrosses make differences between the two Purshia species indistinct (Nord, 1965). Cowania mexicana hybridizes with both Purshia species and introgression occurs between the two genera (Blauer et al., 1975; Stebbins, 1959; Stutz and Thomas, 1964). Putative C. mexicana x Fallugia paradoxa hybrids have been reported (Blauer et al., 1975).

This genetic exchange among these rosaceous shrubs suggests that genetic manipulation to meet nodulation constraints may be possible if considerable variability exists.

This study consists of a series of greenhouse experiments designed to assess the occurrence of nodulation in soils collected at Purshia sites, nodulation limitations due to insufficient endophyte populations, effects of phosphorus and sulfur amendments, differences between the two Purshia species, and differences between ecotypes of the same species. When nodulation was difficult to demonstrate in some desert soils, speculation that other genera capable of nitrogen fixation have escaped detection was verified with the demonstration of nitrogen fixation in Cowania. Preliminary studies to evaluate if nodulation occurs in Fallugia paradoxa were attempted.

MATERIALS AND METHODS

General Greenhouse Procedure

In each greenhouse trial, a surplus of the appropriate seed was planted with subsequent thinning to a uniform number of seedlings. Containers were watered by weight to maintain a moisture content slightly below field capacity and kept in the greenhouse under natural light. After the growth period, seedlings were separated from the soil and the roots gently washed. Shoot dry weights, root dry weights, nodule number, nodule fresh weights, and ethylene production were measured. Details for most experiments are presented elsewhere (Righetti, 1980).

Occurrence of Nodulation

Purshia tridentata seedlings were grown in soil samples collected from the 10 Purshia sites listed in Table 1. Soil samples will be referred to by soil numbers corresponding to site identification numbers in the table. Samples were collected from 0-30 cm and 30-60 cm depths. Soils were air-dried and sieved (5 mm) to remove larger stones and organic material before planting seeds in small plastic cups. In vitro nitrogen mineralization-nitrification measurements were made using a modified method of Stanford and Smith (1972) to determine if common soil nitrogen levels inhibit nodulation.

RESULTS AND DISCUSSION

Figure 1a presents nodule numbers on 3-month-old P. tridentata seedlings. Only four of the ten surface soils produced nodulated seedlings. Subsurface soils from three of these four sites produced seedlings with several-fold more nodules. Nodulation was also observed in other subsurface soils. No nodules were found on seedlings grown in soils 2, 9 and 4 (subsoil at site 4 not sampled). These latter soils were from P. glandulosa shrub communities.

Nodule masses display the same relationships as nodule number, but differences between sites and sampling depth are relatively smaller. Where both surface and subsoils had measurable nodule mass, surface samples had fewer nodules with a larger mass per nodule. The compensation does not preclude the importance of nodule number. Increased nodulation would be required to increase nodule biomass in the field. Thus it is important to determine if the number of successful infections can be increased.

Table 1. Site identification, location, elevation and estimated precipitation[1].

Site identification		Location	Elevation	Estimated precipitation
1. Burcham Flat	PT[2]	NW¼Sec23 T6N R23E	2190 m	350 mm
2. Cactus Flat	PG[3]	SW¼Sec19 T3N R2E	1780 m	350 mm
3. Chilcoot	PT	NW¼Sec25 T23N R16E	1580 m	405 mm
4. Independence	PG	NE¼Sec27 T13S R34E	1580 m	305 mm
5. Sherwin Hill	PT	SW¼Sec18 T5S R30E	2120 m	350 mm
6. Shingle Mill Flat	PT	SE¼Sec16 T7N R23E	1830 m	635 mm
7. Silver Lake	PT	SE¼Sec8 T2S R26E	2220 m	500 mm
8. Truckee	PT	NW¼ Sec1 T17N R16E	1780 m	765 mm
9. Valyermo	PG	SE¼Sec7 T4N R9W	1150 m	275 mm
10. Walker	PT	SE¼Sec19 T8N R23E	1630 m	380 mm

[1] Elevations were estimated using topographical map lines, and precipitation was estimated using a California rainfall map.

[2] PG = _P_. _tridentata_ dominate shrub at site.

[3] PT = _P_. _glandulosa_ dominate shrub at site.

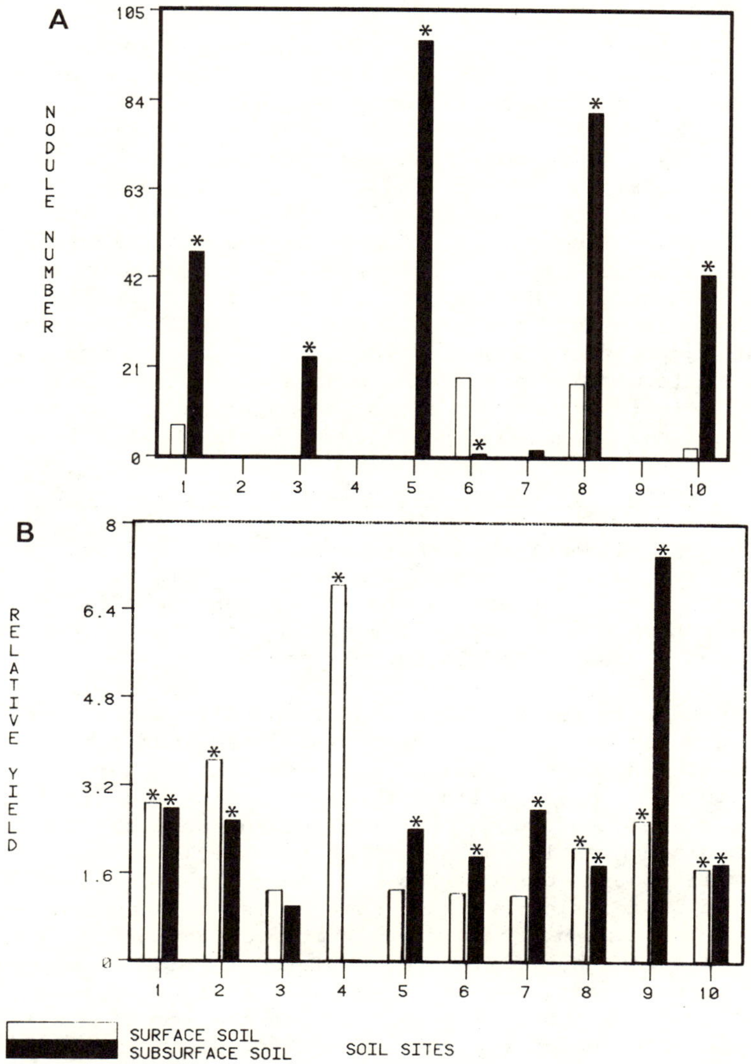

Fig. 1a. Nodule numbers for P. tridentata seedlings grown in unamended soils from Purshia stands. Asterisks indicate numbers for seedlings in the 0-30 cm depth are significantly (p = .05) different than numbers for seedlings in the 30-60 cm depth from the same soil site.

Fig. 1b. Relative yields (nitrogen treatment:unamended treatment) for P. tridentata seedlings grown in soils from Purshia stands. Asterisks indicate that relative yields are significantly (p = .05) greater than 1.0.

Nitrogen inhibits nodulation in actinorhizal and other symbiotic systems. In this trial, none of the nitrogen-amended soils produced nodulated seedlings, and it is possible that common soil nitrogen levels inhibit nodulation, especially in surface samples. However, it is unlikely that nitrogen would be inhibitory in soils where seedlings respond well to added nitrogen, and only a few soils showed no nitrogen response (Fig. 1b). The lack of a nitrogen response in site 3 soils and surface soil 6 is probably due to other limiting factors. Growth was poor and available nitrate levels low in these samples. Subsurface soil 3 produced nodulated seedlings, but surface soil 3 did not, even though available nitrate was approximately the same. Surface soils 7 and perhaps 5 have inhibitory nitrate levels. The lack of nodules produced in these two soils was accompanied by high available nitrate levels and no dry weight response to nitrogen additions. In other samples, sparse nodulation cannot be explained by nitrogen inhibition.

Nodulation Limitations Due to Insufficient Endophyte Populations

Another sampling of the same surface soils was used for an experiment with both *P. tridentata* and *P. glandulosa*. Seedlings were grown in six-inch plastic pots. Treatments consisted of a control with the soils unaltered, a six mmole nitrogen addition as NH_4NO_3 and a crushed nodule inoculation.

Nodule numbers for the 3½-month-old seedlings are presented in Figure 2. The results in unuinoculated treatments are similar to those reported above. Soil 5 was an exception. Nitrogen was not inhibitory in this sampling, thus nodulation and growth responses to combined nitrogen occurred. When inoculated, two of five nonnodulating soils and all three sparsely nodulating soils produced well-nodulated seedlings. In some cases, nodulation increases with inoculation on these same soils were accompanied by increases in percent nitrogen, total nitrogen and dry weight. The three desert soils that failed to nodulate *P. tridentata* above, failed to nodulate either species when uninoculated. Two of these soils did not respond appreciably to the inoculum. In both nonresponding cases the nodules that developed reduced acetylene. The barrier is in nodulation rather than nodule function. Perhaps it could be overcome with a better inoculum. Soil 7 failed to nodulate in all treatments. Nitrogen data and lack of growth response to nitrogen suggest that, as above, it was not nitrogen deficient.

Nitrogen fertilizer completely prevented nodulation in all but soil 1 which had the lowest nitrogen status. It appears that these seedlings nodulated because they had exhausted the added nitrogen, not because the endophyte population had any special properties. The seedling dry masses and nitrogen contents support this contention. Increased early growth from a nitrogen

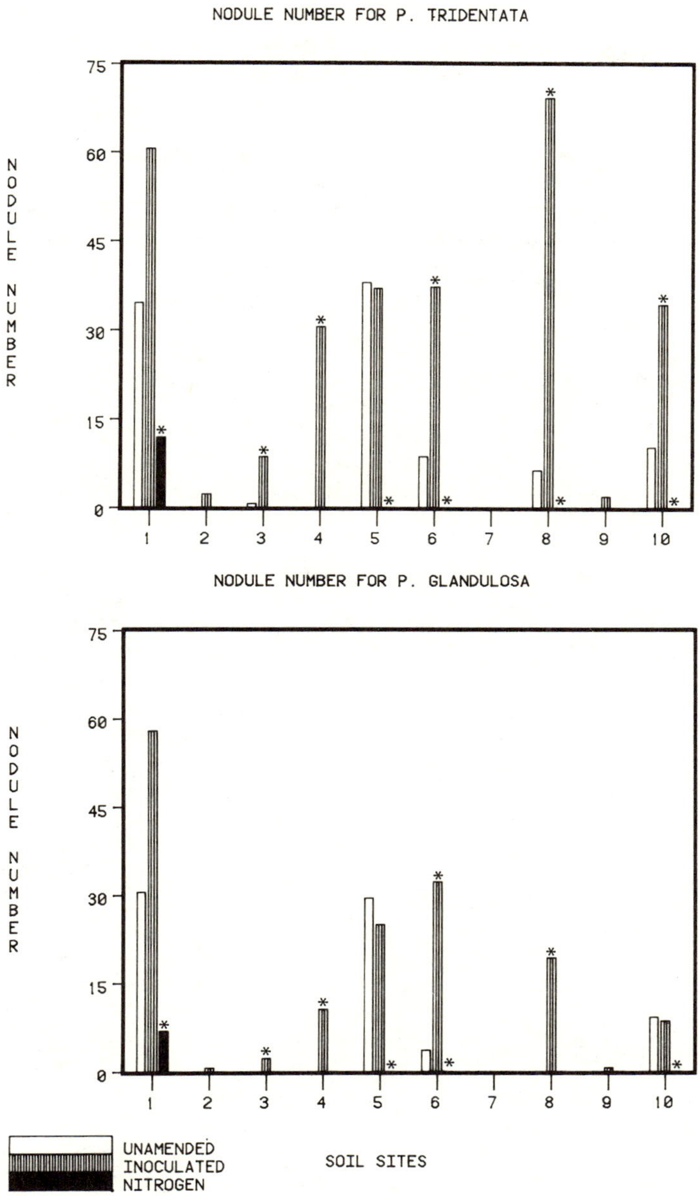

Fig. 2. Nodule numbers for P. tridentata and P. glandulosa seedlings grown in ten soils. Asterisks indicate that numbers are significantly (p = .05) different than numbers for control seedlings grown in the same soil. Differences between species are significant for inoculated treatments in soils 4, 8 and 10.

application, followed by nodulation and nitrogen fixation could prove valuable in revegetation efforts considering the substantial lag in symbiosis establishment (Bond, 1976a).

The two species had similar trends in nodulation and response to inoculation with a few exceptions. In particular, nodule numbers were often lower for P. glandulosa, and significantly so for inoculated treatments in soils 4, 8 and 10. Soil 10 responded to inoculation with an increase in nodule number for P. tridentata but not for P. glandulosa.

Fertility Factors Limiting Nodulation

Responses of P. tridentata and P. glandulosa to sulfur (80 ppm) and phosphorus (50 ppm) amendments under symbiotic conditions were explored. A nitrogen-, phosphorus-, and sulfur-deficient sample that produced well-nodulated seedlings was used for the greenhouse trial. Treatments included a control, phosphorus and sulfur applied alone, and the two nutrients applied together.

Fertility constraints may prevent an optimum nodule percentage of root. Nodule percentages were similar to previous studies (less than 5%) in unamended subsoil, but approached solution culture values of 11% (Bond, 1976a) when the soil was treated with phosphorus and sulfur. The increase is due to increased nodulation or larger nodules; root weights did not appreciably change with fertility treatments. Purshia glandulosa responded to phosphorus alone but P. tridentata did not. Purshia tridentata had significantly more nodules in all treatments and appears to be more capable of maintaining high numbers of successful infections with low fertility. This superiority at low fertility is apparent in ethylene production (Fig. 3). Treatments did not significantly differ from the unamended control for P. tridentata, but P. glandulosa responded significantly to phosphorus and the phosphorus-sulfur combination. Specific activities varied slightly, but were not significantly different for treatment or species.

Ecotype Differences

Seedlings grown from P. glandulosa seeds collected from three locations and seeds collected from a putative P. glandulosa x C. mexicana hybrid were compared. Plants were grown in six-inch plastic pots containing the same bulk soil sample used for fertility experiments. A basal phosphorus and sulfur addition was made before planting.

Differences between ecotypes are apparent in Table 2. It remains to be seen if a fertility x ecotype interaction can be demonstrated similar to the fertility x species interaction described above. The P. glandulosa x C. mexicana hybrid was

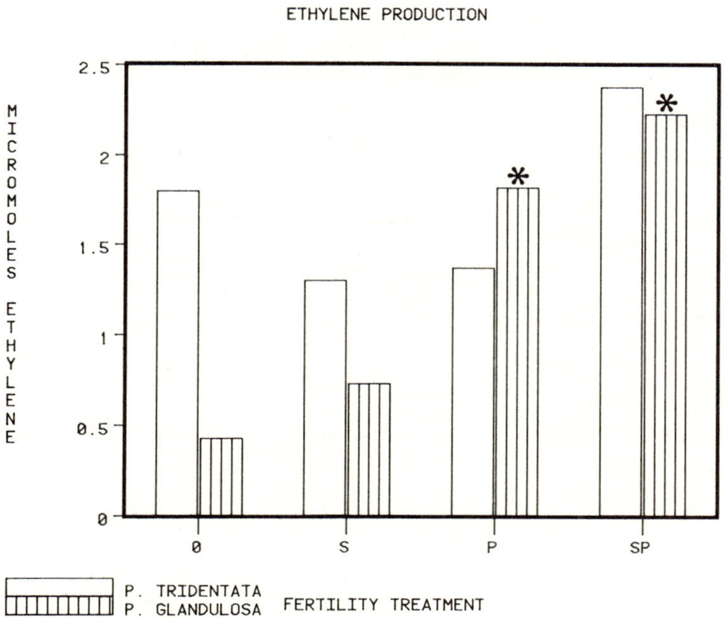

Fig. 3. Ethylene production for P. tridentata and P. glandulosa seedlings grown in soils with different fertility treatments. Asterisks indicate production is significantly ($p = .05$) greater than production for control seedlings for the same species. Differences between species are significant for the unamended treatment.

well nodulated and nodules reduced acetylene. The hybrid's growth was similar to growth observed in one P. glandulosa ecotype, but slower than growth in two others. It compared to growth observed for C. mexicana in other experiments. Differences in growth may be due to differences in vigor rather than a direct result of nitrogen fixation capacity.

INVESTIGATIONS ON "NEW" NITROGEN-FIXING GENERA

In the experiment with Cowania seedlings, seeds collected from three Southern California Cowania populations were sown into soil known to nodulate both Purshia species to which a crushed nodule suspension had been added. In the experiments with F. paradoxa, seedlings grown from seeds collected in Richfield, Utah were transplanted into a mixture of seven soils known to nodulate P. tridentata, P. glandulosa, C. mexicana, Chamaebatia foliolosa,

Table 2. Shoot dry weight, nodule number, nodule mass, and total N in shoot of three P. glandulosa ecotypes and a P. glandulosa x C. mexicana hybrid.

Seed source	Shoot Dry wt GM	Nodule #	Nodule Wt MG	Total N in shoot MG
2. Cactus Flat	1.79 a	136 a	491 ab	35.3 a
4. Independence	0.91 b	103 a	362 ab	15.8 b
9. Valyermo	2.20 a	107 a	545 a	35.5 a
11. White Mtn (Hybrid)	0.60 b	108 a	354 b	11.1 b

Values followed by the same letter are not significantly different from other values in the same column (p = .05).

and Ceanothus integerrimus. Purshia tridentata seedlings of the same age were transplanted into the same soil mix.

It is not surprising that nodulation and nitrogen fixation was demonstrated for Cowania. Although there was considerable variation, all cliffrose seedlings were nodulated. Nodules were coralloid and visually indistinguishable from those on Purshia. Light microscopy revealed a nodule and endophyte structure similar to that reported in Purshia nodules (Krebill and Muir, 1974). Direct evidence for nitrogen fixation was apparent from enrichment in ^{15}N-incubated Cowania nodules. The specific activity was approximately 4.0 µmoles ethylene gm^{-1} hr^{-1} and is similar to values observed for Purshia seedlings grown in this and other greenhouse trials.

Unlike Cowania, F. paradoxa did not nodulate and remained nitrogen deficient and stunted, although Purshia seedlings nodulated and grew well in the same soil mix. The nodulated Purshia seedlings were six-fold larger than non-nodulated Fallugia seedlings. When nitrogen fertilized, Fallugia responded well and grew more than Purshia. It is possible that Fallugia would nodulate and fix nitrogen with other endophytes or in another soil. If it does not, a thorough study of Cowania x Fallugia hybrids would be interesting. Fallugia seedlings should be grown in a wider variety of soils known to nodulate other actinorhizal systems. When air dry, soils have been stored for 18 months without loss of infectivity or change in nodulation capability relative to other samples. A collection of soils would be helpful in investigating Fallugia and other arid adapted rosaceous shrubs for a nitrogen-fixing capability.

CONCLUSIONS

Nodulation does not always occur in soils collected from Purshia stands. In greenhouse trials, increases in nodule number, total nodule mass, total nitrogen, nitrogen content and shoot dry weight can be obtained through inoculation, especially in surface samples. There may be room for improvement, even in the better nodulated subsurface samples, because the percentage of root system (dry mass) consisting of nodule tissue can be increased with fertility treatments. Nodulation did not occur in some soils even in the presence of an inoculum. Soil nitrogen was rarely inhibitory.

Nodule mass showed the same trends as nodule number, but varied less with treatment and depth. Seedlings compensated for sparse nodulation with an increase in mass per nodule. Ethylene production usually corresponded to nodule mass and the average specific activity was about 4.0 µmoles ethylene gm^{-1} hr^{-1}.

Although important exceptions exist, P. tridentata and P. glandulosa were similar in nodulation, nitrogen fixation and growth. When nitrogen fixation parameters are pooled for all soils, P. tridentata is superior. Purshia tridentata appears to be more capable of maintaining nodulation and nitrogen fixation with low fertility. Differences in nitrogen fixation have been demonstrated among ecotypes of the same Purshia species.

Species of genera not known to fix nitrogen may be capable of nitrogen fixation, but escape detection because of soil and environmental constraints or insufficient numbers of the appropriate endophyte reduce nodulation in the field. Nodulation and nitrogen fixation in Cowania has been demonstrated. A preliminary study indicates that F. paradoxa does not nodulate. If this is true, a study of C. mexicana x F. paradoxa hybrids would be interesting.

Although moisture may limit nodulation in the field, the greenhouse data suggest other factors are also important. If a soil is incapable of optimum nodulation in the greenhouse, more severe difficulty can be expected in the field. The biggest questions are whether the results for Purshia can be extrapolated to other semiarid actinorhizal shrubs and whether inoculation and fertilizers could make the symbiosis useful in revegetation efforts on marginal land. Water may be the limiting factor with symbiosis inputs and endophyte populations unimportant. The answer will require further research culminating in field trials.

REFERENCES

Blauer, A. C., Plummer, A. P., McArthur, E. D., Stevens, R., and Guinta, B. C., 1975, Characteristics and hybridization of important intermountain shrubs. I. Rose Family, USDA For. Serv. Res. Pap. INT-169, Intermtn. For. and Range Exp. Stn., Ogden, Utah.

Bond, G., 1976a, Observations on the root nodules of Purshia tridentata, Proc. Roy. Soc. Lond., Ser B, 193:127.

Bond, G., 1976, The results of the IBP survey of root-nodule formation in nonleguminous angiosperms, in: "Symbiotic Nitrogen Fixation in Plants," P. S. Nutman, ed., Int. Biol. Prog. No. 7, Cambridge Univ. Press, Cambridge.

Dalton, D. A., and Zobel, D. B., 1977, Ecological aspects of nitrogen fixation by Purshia tridentata, Plant Soil, 48:57.

Guinta, B. C., Stevens, R., Jorgensen, K. P., and Plummer, A. P., 1978, Antelope bitterbrush. An important wildland shrub, Utah St. Div. of Wildlife Resources Pub. No. 78-12.

Gordon, J. C., and Dawson, J. O., 1979, Potential use of nitrogen-fixing trees and shrubs in commercial forestry, Bot. Gaz., 140(Suppl):588.

Klemmedson, J. O., 1979, Ecological importance of actinomycete-nodulated plants in the western United States, Bot. Gaz., 140(Suppl):591.

Krebill, R. G., and Muir, J. M., 1974, Morphological characterization of Frankia purshiae, the endophyte in root nodules of bitterbrush, NW Sci., 48(4):266.

McArthur, E. D., Guinta, B. C., and Plummer, A. P., 1974, Shrubs for restoration of depleted ranges and distrupted areas, Utah Sci., 35:28.

Nord, E. C., 1965, Autecology of bitterbrush in California, Ecol. Monogr., 35(3):307.

Plummer, A. P., 1976, Revegetation of disturbed intermountain area sites, in: "Reclamation and Use of Disturbed Land in the Southwest," J. L. Thomas, ed., Univ. Ariz. Press, Tucson.

Plummer, A. P., Christensen, D. R., and Monsen, S. B., 1968, Restoring big game range in Utah, Utah Div. Fish and Game Pub. 68-3.

Righetti, T. L., 1980, Soil factors limiting nodulation and nitrogen fixation in bitterbrush (Purshia), Ph.D. Thesis, Univ. of California, Davis.

Silvester, W. B., 1977, Dinitrogen fixation by plant associations excluding legumes, in: "A Treatise on Dinitrogen Fixation," Section IV, Agronomy and Ecology, R. W. F. Hardy and A. H. Gibson, eds., John Wiley and Sons, New York.

Stanford, G., and Smith, 1972, Nitrogen mineralization potentials of soils, Soil Sci. Soc. Amer. Proc., 36:465.

Stanton, F. W., 1959, Autecological studies of bitterbrush (*Purshia tridentata* (Pursh) D.C.), Ph.D. Thesis, Oregon St. Univ., Corvallis.

Stebbins, G. L., 1959, The role of hybridization in evolution, *Amer. Phil. Soc. Proc.*, 103:231.

Stutz, H. C., and Thomas, L. K., 1964, Hybridization and introgression in *Cowania* and *Purshia*, *Evolution*, 18:183.

Wagle, R. F., and Vlamis, J., 1961, Nutrient deficiencies in two bitterbrush soils, *Ecology*, 42:745.

V. CONSERVATION OF FIXED NITROGEN

OVERVIEW OF DENITRIFICATION

W. J. Payne and M. A. Grant

Department of Microbiology
University of Georgia
Athens, Georgia 30602

INTRODUCTION

Despite its restriction to a relatively small number of procaryotes, denitrification exerts a pervasive effect on the metabolism of organisms at many levels of biology. In the development of their Gaia, or Mother Earth, hypothesis of the Earth's atmosphere as a circulatory system of biological origin, Margulis and Lovelock (1974) attributed a major role to denitrifying bacteria. They concluded that the release of the dinitrogen that formed the major component of the gaseous envelope more than 3000 million years ago greatly influenced the establishment of the control systems necessary for the subsequent evolution of eucaryotes. Today, on a series of lesser but still gigantic scales, denitrification impinges critically upon the nitrogen cycle; heavily upon carbon and energy cycling in both heterotrophic and autotrophic bacteria; destructively upon the fertility of the soil; to an unknown but probably controlling degree, upon the productivity of marshes, bottom sediments and marine upwelling regions; helpfully upon wastewater treatment; and, for now, acceptably upon stratospheric gas composition. Therefore, instead of examining the component parts of the phenomenon, as we have done before (Payne et al., 1980), it might be useful to consider as many of these areas of impact as we can profitably analyze to gain some insights into the total force of denitrification.

GENERAL BIOLOGICAL AND BIOCHEMICAL ASPECTS

The Denitrifiers

A group of morphologically and, except for this one property, largely unrelated bacteria comprise the denitrifiers (Payne et al., 1980). Most are prototrophic, heterotrophic, Gram-negative rods such as Acinetobacter (Focht and Joseph, 1974), Alcaligenes, Achromobacter (probably now subsumed under Alcaligenes, taking along with it the presumed Corynebacterium nephridii), Agrobacterium, Chromobacterium, Flavobacterium, Kingella, Moraxella, Pseudomonas and (remarkably, because they also fix nitrogen) Rhizobium species. A few are Gram-positive rods such as Bacillus, Corynebacterium and Propionibacterium. Others are spirals in the genus Spirillum (one species perhaps better accommodated as Aquaspirillum) or vibrioid (and spiralled only before the cells fully separate) and placed in the genus Azospirillum (another nitrogen fixer). . A certain number of Gram-negative, presumably denitrifying cocci are loged in the genus Neisseria, whereas two remarkably versatile species of Gram-negative coccoid bacteria are placed in the genus Paracoccus -- after many years of reposing uncomfortably in the genus Micrococcus.

If single reports prove accurate after further examination, prosthecate and other complex denitrifiers reside in the genera Arthrobacter, Cytophaga and Thermothrix, whereas a well-known group of budding denitrifiers is found in the genus Hyphomicrobium.

Turning to the chemolithotrophic denitrifiers, we find that a Gram-negative, rod-shaped, sulfur and sulfide oxidizer is assigned to the genus Thiobacillus and a closely kindred spirillum to a fairly new genus called Thiomicrospira. The paracocci have one foot in either camp. Not only can they grow heterotrophically as denitrifiers at the expense of a truly extensive range of organic compounds as sole sources of carbon and energy (i.e., methanol to protein) but autotrophically as hydrogen-dependent denitrifiers as well.

Then, to complete the scatter of morphologic types, a number of photoorganotrophic denitrifiers have reportedly been described among the members of another genus of nitrogen fixers, Rhodopseudomonas (Satoh, 1977).

Physiology of Denitrification

Denitrification is, of course, one type of anaerobic respiration -- a property shared out unevenly among the bacteria. Propionibacteria ferment and denitrify; some bacilli and chromobacteria can respire aerobically, ferment and denitrify; but the majority of the species either respire aerobically or denitrify, and do not ferment. The presence of molecular oxygen represses synthesis of denitrifying enzymes in all the denitrifiers (possibly excepting the propionibacteria), but sensitivity varies with the species. In one exemplary study, Sias and Ingraham (1979) found that Pseudomonas aeruginosa grew as a denitrifier only if the oxygen-flow value (K_LaC^*, the term that should be universally adopted for studies of oxygen effects) was not equal to or greater than 0.4 mM O_2/liter/min in nitrate-amended liquid medium. No other studies with cultures have addressed this question with such precision.

Even in the same species, initiation and continuance differ in sensitivity. Once denitrification has begun, the intrusion of oxygen at any point aborts the process, but with varying degrees of rapidity in the different species. The effect is achieved by inactivation of enzymes and blockage of replacement synthesis.

But, it is behavior in the absence of oxygen that interests us most. Thus we note that, upon sensing the absence or near absence of oxygen, cells close out production of ordinary respiratory c-type and $a+a_3$ cytochromes but apparently leave the formation of the pyridine nucelotides, flavins, quinones and cytochrome b unchanged (Payne, 1973). Most denitrifiers then find it necessary to sense the presence of nitrate or nitrite to trigger the synthesis and emplacement of: 1) a molybdoprotein (nitrate reductase), 2) either a c-d-cytochrome or, in a few species, a copper protein (nitrite reductase), 3) a c_{554} (probably nitric oxide reductase), and 4) what appears from minimal observation (Matsubara, 1975) to be a c-type cytochrome comprising nitrous oxide reductase. (There is doubt about the cytochromes of nitrous oxide reductase because the enzyme has been difficult to stabilize in cell-free extracts.) A few denitrifiers require nothing more than the absence of oxygen to derepress their capacity for production of the entire array of reductases, but the overwhelming majority require the presence of one of the nitrogen oxides as well.

Denitrification does not always simply flow from nitrate to dinitrogen. A greater or lesser amount of nitrite accumulates, as nitrate reduction proceeds in all but one or two of the various species, and is then rapidly reduced. Some small quantity of nitric oxide has been seen in several experiments. Subsequently, nitrous oxide also accumulates in some quantity in the atmosphere of denitrifying cultures of most denitrifiers before it too is reduced. An example of a notable exception is *Pseudomonas perfectomarinus*, which does not release any nitrous oxide during normal denitrification. Nearly all the denitrifiers can grow with nitrite as the sole oxidant but tolerate only 0.1-0.2 as much nitrite as nitrate to begin with, and several species can grow anaerobically at the expense of nitrous oxide as sole oxidant (Payne, 1973). The presence of an elevated but not lethal concentration of nitrite intensifies the amount of nitrous oxide transiently released by many. In addition to these normal divergencies, there are several aberrations. For example, a few *Alcaligenes* species begin denitrification at nitrite and cannot reduce nitrate. Other *Alcaligenes* and *Pseudomonas* species start with nitrate but stop

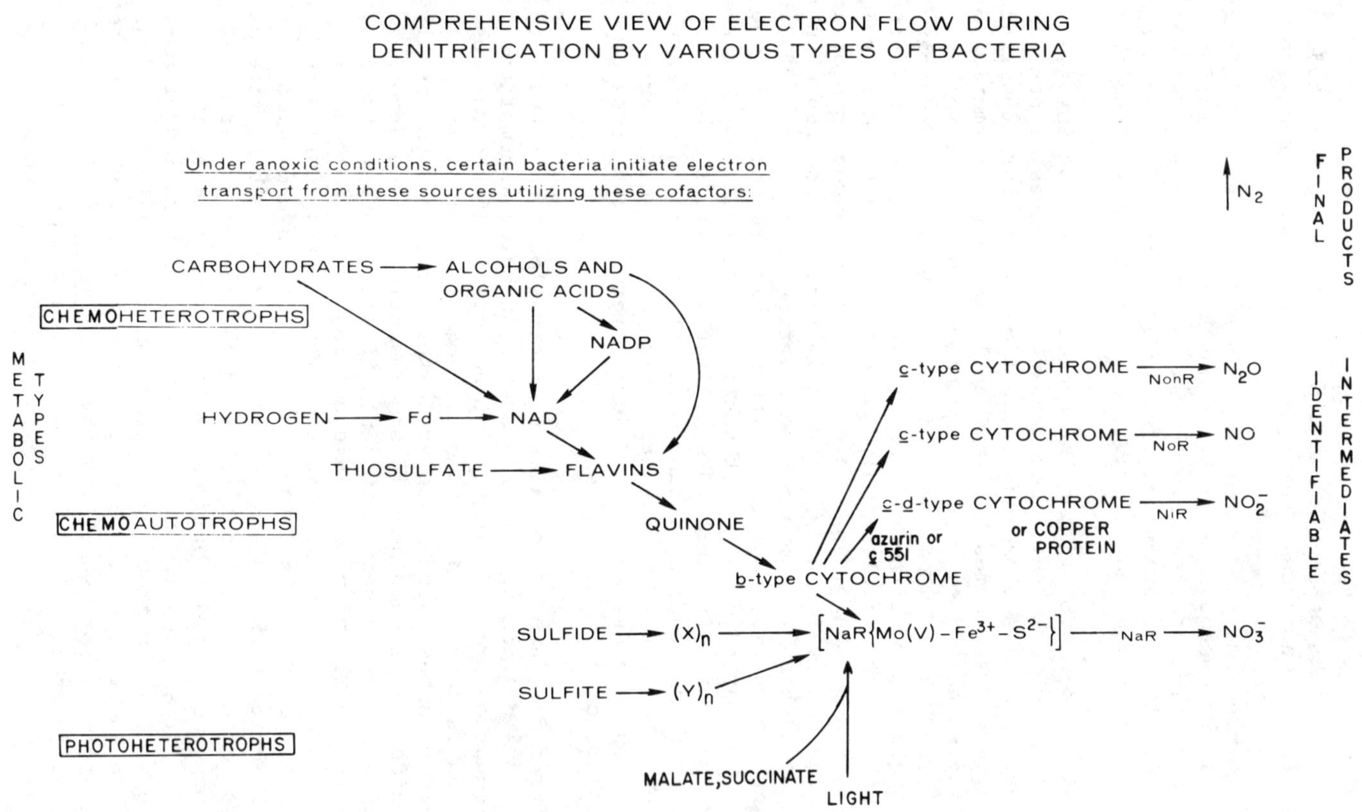

Fig. 1. Diagrammatic accounting for the events known to comprise bacterial denitrification in Nature.

the process with the release of nitrous oxide rather than continuing reduction to the release of dinitrogen.

IMPACTS ON THE CARBON AND ENERGY CYCLES

As prototrophs that can function about 67-77% as effectively when denitrifying as when they respire aerobically (Elliott and Gilmour, 1971; Justin and Kelley, 1978), many species are able to utilize a remarkable variety of carbon compounds as sole sources of carbon and energy during anaerobic growth (Fig. 1). As would be expected, all the simple and many of the complex carbohydrates, organic acids, proteins and peptides, lipids, alcohols, nucelotides and their components, and other such biochemicals will serve. But, less well expected is the capacity of certain Moraxella (Williams and Evans, 1975) and Pseudomonas species (Taylor and Heeb, 1972) to grow as denitrifiers at the expense of aromatic compounds. It had long been thought that such compounds were assailable only by oxygenases, but the denitrifiers first reduce the double bonds of the rings to render them non-aromatic and then hydrolytically (rather than oxidatively) cleave the cyclic products. They further fragment the resulting oxygenated (from water) chains into organic acids, which then serve as unexceptional carbon and energy sources for denitrifying growth.

Such a capacity does not demarcate the limits to the verastility of the denitrifiers. We have now observed that another type of compound also thought to be susceptible only to oxygenases, the ether glycol, will serve as the sole source of carbon and energy for denitrifying growth of Alcaligenes faecalis var. denitrificans (Fig. 2). Di-, tri- and tetraethylene glycols and a variety of ethoxy oligomers in mixed sizes support slow but significant denitrifying growth, as does closely related, but non-ether-bearing, ethylene glycol. In addition, a series of Tergitols^R, which are primary and secondary alkyl esters of ether glycol oligomers, also serve as substrates for denitrifying growth (Fig. 3). Release of dinitrogen accompanies growth and reflects its rate (Figs. 4 and 5). Carbon dioxide and nitrous oxide appear during denitrifying growth on all these substrates. The nitrous oxide persists for days in some, perhaps because these particular bacteria use such small fragments of the detergent molecules (i.e., the ethoxy components) and thus run short of electrons. The denitrifying A. faecalis does not grow on the alcohol residues but, abroad in Nature, other bacteria are certain to do.

The significance of such observations is the reassurance they provide that, if supplied with nitrate, bacteria are able anaerobically to degrade various types of pollutants, such as nonionic detergents that incorporate ether glycol units as their water-soluble moieties. They are thus likely to do so either at large in Nature or within the basins of treatment plants. We have

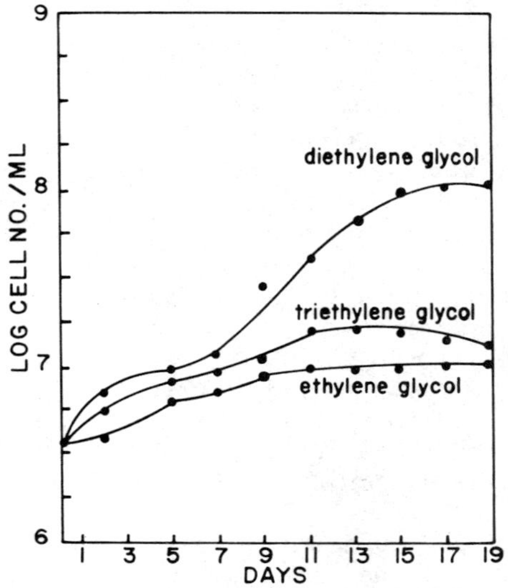

Fig. 2. Slow denitrifying growth of Alcaligenes faecalis var. denitrificans in minimal media at the expense of ether glycols and ethylene glycol.

further noted that, when supplied with nitrate, an unidentified Bacillus species from coastal Georgia salt marsh can anaerobically degrade purified lignocellulose (Browning, 1967) and use the products as electron donors for denitrification. The significance of these additional observations is the notice they provide that denitrification apparently also contributes to degradation of biologically durable, but not recalcitrant, polymers from fibrous plants and to the cycling of their carbon in an important and productive environment. It would be useful to find that denitrification could serve as an anaerobic vehicle for degradation of hydrocarbons, but repeated and impressively well designed experiments have failed to demonstrate such a capability (Swain et al., 1978).

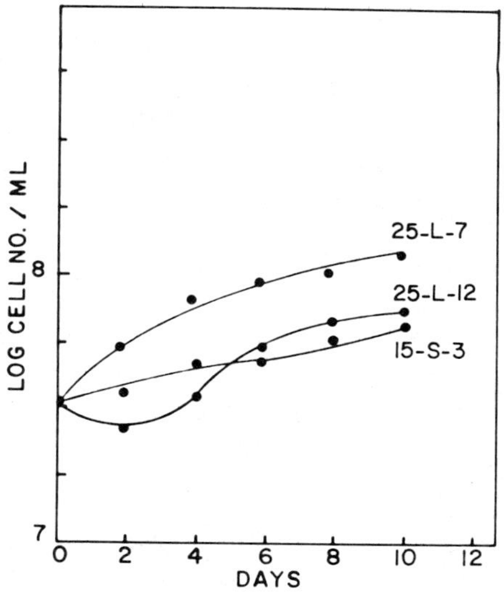

Fig. 3. Slow denitrifying growth of _A_. _faecalis_ var. _denitrificans_ in minimal media at the expense of Tergitols^{R}.

Impacts on the carbon and energy cycles must be viewed as mixed blessings, for they force upon a system the necessity of trading fixed nitrogen for the regeneration of carbon. One disagreeable result is then a dependence of the system upon renewed fixation of nitrogen for the restoration of its nitrogen balance. That in turn requires the functioning of a series of processes -- all ultimately dependent upon a massive input of solar energy. In a stable salt or fresh water marsh, forest soil or aquatic or sediment community, such dependence is acceptable; but in a highly productive marine upwelling area such as that observed in the coastal waters of Peru (Dugdale et al., 1977), intensification of denitrification in the oxygen minimum zone at depths of 100-500 m may initiate changes in the orderly relationships

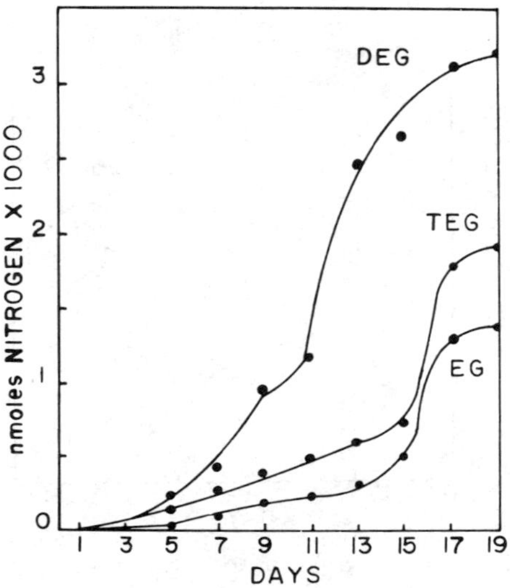

Fig. 4. Release of dinitrogen by A. faecilis var. denitrificans growing on ether glycols and ethylene glycol.

that comprise the system's food web to the detriment of production in the fisheries industry. Only time and solar input can restore the lost productivity and lead to restabilization of the prices of fish protein in the markets of the world. There are few more telling examples of the undermining effect of denitrification on our desire to enhance ammonia production than the induction of nitrate deficiency seen periodically in the oxygen minimum layers of the Peruvian coastal waters.

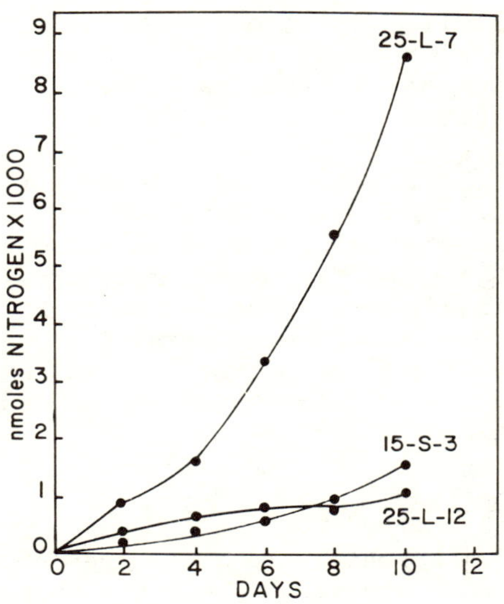

Fig. 5. Release of dinitrogen by A. faecalis var. denitrificans growing on Tergitols[R].

BASIC CONCERNS AND FUTURE PROSPECTS IN AGRICULTURE

Assessment of Impact

Impoverishment of arable land by denitrification puzzled and dismayed soil scientists of a century ago, and realization that denitrifiers are universally distributed in the soils of the world added to the shock. For a time it was feared that the destructive agents were added with the manure or hay plowed under during preparation of the land for sowing (Voorhees, 1902). But, workers active in those days were soon able to draw some comfort from learning that what manure and hay added was oxidizable organic matter that activated rather than introduced the denitrifying agents, which were already present in abundance. Such a perception formed a basis for improved management of soil nitrogen but provided no progress toward elimination of the pernicious phenomenon. We have simply learned to live with it.

And, even today, we are still ignorant of (but eager to learn about) the true dimensions of the annual loss of fertilizer and other fixed nitrogen suffered by our crop lands. Granted the volume lost will vary from soil to soil, we must still be shocked to learn that the average has been estimated at an appalling 25% -- year in and year out.

It is not difficult to catalogue the factors that influence the process in soil. Some are general, others specific. Assays reveal that all the affectors of every other sort of biological phenomenon in soil (e.g., temperature, moisture, acidity and availability of energy source) also influence rates of nitrogen loss. And, the same sorts of measurement further reveal that two additional specific affectors must be simultaneously operative (i.e., a dminished, perhaps eliminated, oxygen supply and an enriched stock of nitrate or nitrite).

Speculation on Inhibitors

Where can such information lead us? Perhaps we can focus on the latter observations and find some profitable pathways to follow. Even if we hypothesize in vain, let us presume a universal interest in finding some means of interdicting denitrification in cultivated soil, and further a universally held prospect that we can do so. The latter presumption is the more difficult to accept, but we aks indulgence for the sake of argument. It is a linkage between specificity and vulnerability that we seek, if we are to strive for a sort of "chemotherapy."

Because all the consequential denitrifiers in soil can interconvert rapidly and repeatedly between aerobic and anaerobic respiration, whereas most other significant anaerobic respirers and fermenters do not, there are two sensing mechanisms as well as an initiating mechanism peculiar (but probably not unique) to denitrifiers. The first is the process sensitive to the absence of oxygen which then directs a "shut-down" of the ordinary cytochrome production line. An agent that could specifically prevent aerobically grown, but potentially denitrifying, cells in the soil from becoming "aware" that oxygen was in short supply would prevent both the cessation of synthesis and the onset of the second (the initiating) mechanism that directs the production and implacement of the denitrifying enzymes. There are two possibilities here. Even if the anoxia sensor functioned, an agent capable of specifically blunting the force of the signal activating the initiating mechanism could still frustrate the conversion of the cells from aerobic to denitrifying respiration. Failing at either of those, we could turn to a third function peculiar to denitrifiers that might be vulnerable (i.e., sensitivity to the presence of nitrate or nitrite). An agent that specifically desensitizes cells that can produce the respiratory reductases to the availability of those two oxidants would represent a last prospect for imposing blockage on transitional biosynthesis.

Moving on in search of vulnerability to a consideration of specific functions, it is soon apparent that the denitrifying nitrate reductases (which are molybdoproteins) so resemble other molybdoproteins whose activities we wish to protect and promote that the first reaction seems best left alone. But, the nitrite reductase produced by the apparently greater number of denitrifiers is a unique two-heme (*c-d*) cytochrome (Horio et al., 1961). As the catalyst that initiates the irrevocable step in the return of fixed nitrogen to the elemental state (the agency that divests oxidized nitrogen atoms of any prospect of service as a nutrient short of reduction to the elemental stage, release and possible refixation), the *c-d* cytochrome is the bane of agriculture. No less ruinous, although perhaps less prevalent, is a copper protein that serves as a denitrifying nitrite reductase and is apparently devoid of cytochrome (Iwasaki et al., 1975; Sawada et al., 1978).

Let us consider what precisely we mean by the term, *denitrifying* nitrite reductase. Workers in various laboratories continue to argue about the identity of the product of the denitrifying nitrite reductases and draw upon results of experiments with soil or with whole cells to question the place of nitric oxide in the scheme. But, such concerns seem unproductive. Evidence from cell-free work is unequivocal. By use of gas chromatography, it has been established that crude cell-free extracts (Barbaree and Payne, 1967) and fractionated extracts (Payne and Riley, 1969) of *P. perfectomarinus* and purified *c-d* cytochrome (Table 1) from *P. aeruginosa* all generate nitric oxide from the reduction of nitrite. Additional evidence has been supplied by experiments in which $^{15}NO_2^-$ was reduced to ^{15}NO by purified *c-d* cytochrome from *P. aeruginosa* (D. C. Wharton, personal communication, 1979). An artificial electron donor, ascorbate, was used with the purified enzymes, but ascorbate alone did not reduce nitrite in the experiments reported here (Table 1) -- nor in parallel experiments did the addition of phenazine methosulfate (PMS), tetramethy-*p*-phenylene diamine, dichlorophenol indophenol or mammalian cytochrome *c* enable ascorbate to reduce nitrite in the absence of the *c-d* cytochrome. Nitrous oxide was not observed in most of the experiments reported here and only negligible amounts were seen after some time in any of the experiments. Both Zumft et al. (1979) and LeGall et al. (1979) have since shown that purified *c-d*-cytochrome from *P. perfectomarinus* and *Thiobacillus denitrificans*, respectively, reduce nitrite to nitric oxide, and Sawada et al. (1978) showed

Table 1. Nitrite reduction by purified P. aeruginosa enzyme and cofactors.*

System	Nitric oxide peak height (per 0.1 ml of headspace gas sample)	
	30 min	60 min
Control 1 (no c-d or no nitrite or no ascorbate)	0	0
Control 2 (no azurin or c_{551})	0	0
Experiment 1 (ascorbate, c_{551} and c-d)	92	104
Experiment 2 (ascorbate, azurin and c-d)	64	64
Experiment 3 (ascorbate, azurin, c_{551} and c-d)	100	128

Complete reaction mixture contained, potassium phosphate buffer, 50 mM, pH 6.5, 0.5 ml; potassium nitrite, 1 mM, 0.1 ml; ascorbate, 10 mM, 0.1 ml; azurin, 0.012 μmoles in 0.1 ml; cytochrome c_{551}, 0.024 μmoles in 0.1 ml; and (last to initiate reaction) cytochrome c-d, 0.5 mg in 0.1 ml. Each mixture was incubated at 30 C.

*Materials purified by D. C. Wharton. Experiments performed by D. C. Wharton, W. J. Payne, J. J. Rowe, and B. F. Sherr, August 1978.

that the purified copper enzyme from a photosynthetic denitrifier does so as well. All three groups used ascorbate + PMS as electron sources.

The distinctiveness of denitrification thus appears to lie in 1) the releasability of nitric oxide by cell-free enzymes, and 2) its unique fate. We are led to that conclusion by the following arguments: When Cox et al. (1971) first noted from electron paramagnetic resonance (EPR) measurements that a nitrite-reducing fraction from P. perfectomarinus (which reduced no other nitrogen oxides) formed a heme-nitric oxide complex while nitrite was actively under attack, the phenomenon was thought unique. Now we know that other dissimilatory (Liu et al., 1980) as well as assimilatory nitrite reductases from green plants (Lancaster et al., 1979) also form heme-nitric oxide intermediates. The difference is that those heme-nitric oxide complexes yield ammonia.

It has not mattered that other reductases form the complexes. We have found other useful applications for EPR analyses. Working with the purified c-d from *T. denitrificans*, LeGall et al. (1979) showed that it is the d_1 heme that forms the complex signaled by EPR.

The critical point in the impoverishment of soil for nitrogen is then precisely identified as the post-(d_1-heme-nitric oxide) instant when ammonia does not arise. In and around the cells, we see no significant amount of nitric oxide liberated, and that has led many to remain skeptical at its significance. But a c_{554} cytochrome that also forms a distinctive heme-nitric oxide compound (Cox et al., 1971) apparently contributes to reduction of the nitric oxide to nitrous oxide (Cox and Payne, 1973). Thus, an agent capable of interfering specifically with d-heme or c_{554} synthesis or function in soil denitrifiers long enough for crop plants to compete successfully for a greater share of soil nitrate or nitrite could conceivably add a gratifying measure of fertility to a treated plot of soil.

We have been led to propose that such finely tuned inhibitors may be found by previous demonstrations that acetylene specifically inhibits nitrous oxide reductase activity apparently without influencing any other enzyme in the assimilatory, fermentative or dissimilatory pathways of nitrate reduction (Payne and Balderston, 1978). Some useful techniques have been developed to exploit the effects of acetylene (Smith et al., 1978), and their further use in the study of the interactions of crop plant roots and denitrifiers (Smith and Tideje, 1979a) has been instructive and should continue to prove even more so.

There are additional reasons to believe that the nitrite reductase complex is vulnerable to selective manipulation. Megãlhaes et al. (1978) isolated a number of mutants and naturally altered strains (nir^-) of azospirilla that had lost the capacity to synthesize denitrifying nitrate reductase. Plant roots with which the nitrogen-fixing azospirilla associate in Nature appeared to exert selective pressure against the establishment of nir^+ strains within the root tissue and yielded high proportions of nir^- strains to recovery procedures even though wild-type (nir^+) strains abounded in the soil all about the roots. Identification of the selective agent or set of circumstances that the roots bring into play to effect such selectivity would provide useful insights.

It appears that the polynucleotides in the *P. aeruginosa* genome that control synthesis of the denitrifying nitrite reductase can be determined. Van Hartingsveldt et al. (1971) obtained *nir*⁻ mutants whose extracts failed to display the *d*-heme light absorption spectrum. The loci of several *nir* lesions that prevent production of the *c-d* enzyme have been mapped. It might prove useful if those sites could be characterized and subjected to vulnerability analyses.

Standardized Test Strains

If programs of evaluation of the efficacy of anti-denitrification agents ever are implemented, attention to standardization of test bacterial populations seems advisable. Smith and Tiedje (1979b) reported a first (and promising) step toward standardization by the introduction of stably marked (Rifampicin-resistant) denitrifying pseudomonads into soil and discovery that the bacteria were recoverable at a very high rate of survival for more than a month. Development and testing of an extensive array of such denitrifiers now seem doubly attractive. The alternative to use of standardized, labeled populations is dependence on assay of the activity of mixed populations in soils and thus a lack of reproducible conditions of assay.

ATMOSPHERIC INVOLVEMENT

A primary concern of atmospheric scientists is fear that nitrogen oxides may be released by man's activities in quantities that will destructively erode the stratospheric ozone layer (Crutzen, 1976). With the demonstration that nitrification contributes significantly to the production of nitrous oxide in soil (Bremner and Blackmer, 1978), and the additional observation that marine waters in various parts of the world contain unexpectedly high concentrations of nitrous oxide (Yoshinari, 1976), a decision as to whether denitrification is a source or sink for atmospheric nitrous oxide has become increasingly difficult to achieve. It is certain that, when and where soils are anaerobic and electron donors are available, nitrous oxide will be reduced (Firestone et al., 1980) -- and further that reduction of nitrous oxide takes place at a significant rate in marine systems as well (Cohen and Gordon, 1978). But without additional information that can be obtained only by extensive observation over a long period, we are unable to assess the extent to which denitrification may affect the stratospheric ozone layer. The small fraction of gas liberated as nitrous oxide during denitrification may represent only a minimal portion of the gas reaching the stratosphere, and it may be shown that active denitrification is the protective mechanism that keeps the upward surge of nitrogen oxides as low as it is.

CONCLUDING REMARKS

No commentary as brief as this one could adequately define the significant aspects of a phenomenon so monumentally consequential as denitrification. Our efforts here have been restricted to identifying those phenomena that run counter to our desire to enhance ammonia production (which, we all understand, would increase the productivity of agriculture and of fisheries). If we have accomplished nothing more in our overview than drawing forceful attention to the destructive significance of denitrifying nitrite reductase and to the need to know more about the role of nitric oxide in every sort of biological conversion of nitrite, we are convinced of the value of the exercise. Until denitrifying nitrite reduction is fully understood, Nature's most effective antagonist to our enjoyment of the benefits of enhanced production of ammonia will continue unchecked along its deleterious way.

ACKNOWLEDGMENTS

Our work is supported by NSF grant OCE-79-19550.

REFERENCES

Barbaree, J. M., and Payne, W. J., 1967, Products of denitrification by a marine bacterium as revealed by gas chromatography, Mar. Biol., 1:136.

Bremner, J. M., and Blackmer, A. M., 1978, Nitrous oxide: Emission from soils during nitrification of fertilizer nitrogen, Science, 199:295.

Browning, B. L., 1967, Preparation of extractive-free lignocellulose, in: "Methods of Wood Chemistry," Vol. II, Interscience Publ., New York.

Cohen, Y., and Gordon, L. I., 1978, Nitrous oxide in the oxygen minimum of the eastern tropical North Pacific: evidence for its consumption during denitrification and possible mechanisms for its production, Deep-Sea Res., 25:509.

Cox, C. D., and Payne, W. J., 1973, Separation of soluble denitrifying enzymes and cytochromes from *Pseudomonas perfectomarinus*, Can. J. Microbiol., 19:861.

Cox, C. D., Payne, W. J., and DerVartanian, D., 1971, Electron paramagnetic resonance studies on the nature of hemoproteins in nitrite and nitric oxide reduction, Biochim. Biophys. Acta, 253:290.

Crutzen, P. J., 1976, Upper limits on stratospheric ozone reductions following increased application of fixed nitrogen to the soil, Geophys. Res. Lett., 3:169.

Dugdale, R. C., Goering, J. J., Barber, R. T., Smith, R. L., and Packard, T. T., 1977, Denitrification and hydrogen sulfide in Peru upwelling during 1976, Deep-Sea Res., 24:601.

Elliott, L. F., and Gilmour, C. M., 1971, Growth of Pseudomonas stutzeri with nitrate and oxygen as terminal electron acceptors, Soil Biol. Biochem., 3:331.

Firestore, M. K., Firestone, R. B., and Tiedje, J. M., 1980, Nitrous oxide from soil denitrification: factors controlling its biological production, Science, 208:749.

Focht, D. D., and Joseph, H., 1974, Degradation of 1,1-diphenylethylene by mixed cultures, Can. J. Microbiol., 20:631.

Horio, T., Sekuzu, I., Higashi, T., and Okunuki, K., 1961, Cytochrome oxidases of Pseudomonas aeruginosa and ox-heart muscle, and their related respiratory components, in: "Haematin Enzymes," J. E. Falk, R. Lemberg, and R. E. Morton, eds., Pergamon Press, London.

Iwasaki, H., Noji, B., and Shidara, S., 1975, Achromobacter cycloclastes nitrite reductase. The function of copper, amino acid composition and ESR spectra, J. Biochem., Tokyo, 78:355.

Justin, P., and Kelley, D. P., 1978, Metabolic changes in Thiobacillus denitrificans accompanying the transition from aerobic to anaerobic growth in continuous chemostat culture, J. Gen. Microbiol., 107:131.

Lancaster, J. R., Vega, J. M., Kamin, H., Orme-Johnson, N. R., Orme-Johnson, W. H., Krueger, R. J., and Siegel, L. M., 1979, Identification of the iron-sulfur center of spinach ferredoxin-nitrite reductase as a tetranuclear center, and preliminary EPR studies of mechanism, J. Biol. Chem., 254:1268.

LeGall, J., Payne, W. J., Morgan, T. V., and DerVartanian, D. V., 1979, On the purification of nitrite reductase from Thiobacillus denitrificans and its reaction with nitrite under reducing conditions, Biochem. Biophsy. Res. Commun., 87:355.

Liu, M. C., Abou-Jaoude, A., Chippaux, M., Morgan, T. V., DerVartanian, D. V., Peck, H. D., and LeGall, J., 1980, The identification of the low potential cytochrome $\underline{c}_{552}$ of Escherichia coli as a nitrite reductase, Bacteriol. Proc., p. 156.

Margulis, L., and Lovelock, J. E., 1974, Biological modulation of the earth's atmosphere, Icarus, 21:471.

Matsubara, T., 1975, The participation of cytochromes in the reduction of N_2O to N_2 by a denitrifying bacterium, J. Biochem., Tokyo, 77:627.

Megalhães, L. M. S., Neyra, C. A., and Döbereiner, J., 1978, Nitrate and nitrite reductase negative mutants of N_2-fixing Azospirillum spp., Arch. Microbiol., 117:247.

Payne, W. J., 1973, Reduction of nitrogenous oxides by microorganisms, Bacteriol. Rev., 37:409.

Payne, W. J., and Balderston, W. L., 1978, Denitrification, in: "Microbiology-1978," D. Schlessinger, ed., ASM Publications, Washington.

Payne, W. J., and Riley, P. R., 1969, Suppression by nitrate of enzymatic reduction of nitric oxide, Proc. Soc. Exp. Biol. Med., 132:258.

Payne, W. J., Rowe, J. J., and Sherr, B. F., 1980, Denitrification: A plea for attention, in: "Nitrogen Fixation, Vol. I, Free-Living Systems and Chemical Models," W. E. Newton and W. H. Orme-Johnson, eds., University Park Press, Baltimore, Maryland.

Payne, W. J., Sherr, B. F., and Chalmers, A., 1980, Nitrification-denitrification associated with plant roots, in: "Associative Nitrogen Fixation," P. V. Vose, ed., CRC Press, West Palm Beach, Florida, in press.

Satoh, T., 1977, Light-activated, -inhibited and -independent denitrification by a denitrifying phototrophic bacterium, Arch. Microbiol., 115:293.

Sawada, E., Satoh, T., and Kitamura, H., 1978, Purification and properties of a dissimilatory nitrite reductase of a denitrifying phototrophic bacterium, Plant Cell Physiol., 19:1339.

Sias, S. R., and Ingraham, J. L., 1979, Isolation and analysis of mutants of Pseudomonas aeruginosa, Arch. Microbiol., 122:263.

Smith, M. S., Firestone, M. K., and Tideje, J. M., 1978, The acetylene inhibition method for short-term measurement of soil denitrification and its evaluation using nitrogen-13, Soil Sci. Soc. Amer. J., 42:611.

Smith, M. S., and Tiedje, J. J., 1979a, The effects of roots on soil denitrification, Soil. Sci. Soc. Amer. J., 43:951.

Smith, M. S., and Tiedje, J. M., 1979b, Use of antibiotic resistance to study the growth and survival of denitrifiers in soil, Agron. Abst., p. 163.

Swain, M. M., Somerville, H. J., and Cole, J. A., 1978, Denitrification during growth of Pseudomonas aeruginosa on octane, J. Gen. Microbiol., 197:103.

Taylor, B. F., and Heeb, J., 1972, The anaerobic degradation of aromatic compounds by a denitrifying bacterium. Radioisotope and mutant studies, Arch. Microbiol., 83:165.

Van Hartingsveldt, J., Marinus, M. G., and Stouthamer, A. H., 1971, Mutants of Pseudomonas aeruginosa blocked in nitrate or nitrite dissimilation, Genetics, 67:469.

Voorhees, E. B., 1902, Studies in denitrification, J. Amer. Chem. Soc., 24:785.

Williams, R. J., and Evans, C. W., 1975, The metabolism of benzoate by Moraxella species through anaerobic nitrate respiration, Biochem. J., 148:1.

Yoshinari, T., 1976, Nitrous oxide in the sea, Mar. Chem., 4:189.

Zumft, W. G., Sherr, B. F., and Payne, W. J., 1979, A reappraisal of the nitric oxide-binding protein of denitrifying Pseudomonas, Biochem. Biophys. Res. Commun., 88:1230.

THE PHYSIOLOGICAL GENETICS OF DENITRIFICATION IN *PSEUDOMONAS*

Curtis A. Carlson and John L. Ingraham

Department of Bacteriology
University of California
Davis, CA 95616

INTRODUCTION

Interest in the fate of soil nitrates is shared by a number of scientific disciplines. This is reflected by the diversity of approaches to the study of nitrate metabolism. An important element in the nitrate budget of the soil is the competition between bacteria and plants for this important and often-limiting nutrient.

Bacteria metabolize nitrate by two distinct reductive pathways: one leading to ammonium ion and the other largely to dinitrogen gas (Fig. 1). By the assimilatory pathway, bacterial nitrate reduction has a conservative function. Many microbes convert nitrate to ammonia aerobically. Others excrete ammonia during anaerobic reduction of nitrate or nitrite (Cole, 1978). Nitrogen is kept within the soil system by both of these processes. Ammonia production by bacteria eventually may be exploited as a means of conserving soil nitrogen.

Denitrification, on the other hand, leads to the loss of fixed nitrogen from soil, and can be quantitatively significant when soils become anaerobic in response to levels of water content, alkalinity, temperature, and bacterial activity (Payne, 1973; Payne, this volume; Focht and Verstraete, 1977; Focht, this volume). Our interest is the control of this process, with the goal of finding ways to modify or suppress the flow of nitrate to gaseous products.

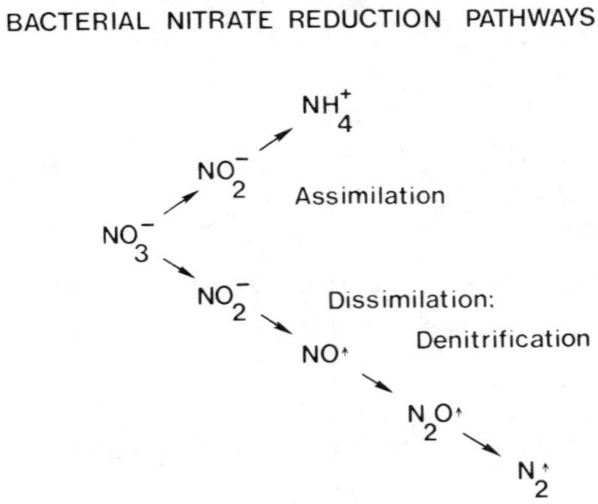

Fig. 1. Bacterial nitrate reduction pathways.

THE PHYSIOLOGICAL GENETICS APPROACH TO DENITRIFICATION

The kinds of questions we want to ask about denitrification on a molecular level include: What is the pathway? What are the intermediate chemical species? What is the physiological role of each reductive step? What are the structural and catalytic components of the enzymes and associated proteins involved?

Approaches to nitrate metabolism studies based mainly on chemistry and biochemistry have had definite limitations. A chemical approach has been employed to sort out the intermediates of denitrification. The use of tracer isotopes and identification of compounds by gas chromatography and mass spectrometry has been successfully explored by a number of workers (St. John and Hollocher, 1977; Firestone et al., 1979; Hollocher et al., 1980) However, major questions remain unanswered. Classic biochemical approaches have been slowed because the enzyme systems appear to be highly complex, oxygen-sensitive, and largely membrane associated (Stouthamer, 1976). For these and other reasons, several of the reductases are extremely unstable *in vitro* (Kristjansson and Hollocher, 1980). Thus, the purification and study of active components promises to be arduous.

The tools of physiological genetics have proven to be powerful in the study of a variety of microbial functions. They offer considerable promise to the study of denitrification, but have been largely neglected. This approach is especially well-suited to answer questions about in vivo functions. One may readily produce mutant strains of bacteria that exhibit an altered phenotype of the process under investigation. Then the loss of biochemical function can be correlated with the change of organism's metabolic capability. In this way we should be able to study each step of denitrification and find answers to the kinds of questions posed above.

THE DENITRIFIER OF CHOICE?

Meaningful genetic studies require the use of pure cultures of a bacterial strain that can be manipulated genetically, that possesses a wide range of physiological functions associated with the process to be studied, and that is typical of the sorts of bacteria that carry out the process in nature.

Most of the information available on denitrification has been derived from studies on a variety of bacterial species and genera. This is comforting in that generalizations can be drawn that cut across taxonomic lines. However, it is discouraging when one is faced with data that seemingly conflict, and with the realization that progress in the field is being slowed by parallel efforts. Admittedly, patterns of nitrate metabolism are diverse among bacteria. Some organisms reduce nitrate only for assimilation, some only via denitrification, and some do both. Further, the denitrification pathway can be truncated at the beginning, end, or middle, in a variety of species. In a sense, this diverse array of patterns of nitrate metabolism might be looked upon as an existing set of mutant strains, but we have no means of genetic comparison among them. We urge that future metabolic studies on denitrification concentrate on one or two microorganisms.

Pseudomonas aeruginosa has much to recommend it for denitrification studies. It occurs ubiquitously and possesses two complete systems for nitrate reduction: either to ammonium ion for assimilation or to nitrogen gas via denitrification. It has several well-defined systems for genetic exchange: transduction via phages F116, G101 and others; conjugation via the mating factors FP2, FP110 and R68.45; and transformation (Holloway et al., 1979). This genetic system is well established -- more than 100 discrete functions have now been mapped on the P. aeruginosa chromosome. Admittedly, there are disadvantages. For instance the objection may be made that the type strain PAO has been too long domesticated as a laboratory strain since its first isolation some 25 years ago. Furthermore, P. aeruginosa apparently is not the predominant species among denitrifying pseudomonads found in soil (Gamble et al., 1977). A more serious shortcoming is its inability to grow on nitrous oxide. This property makes it exceptional among denitrifiers metabolizing nitrous oxide, and will be addressed later.

THE SEPARATION OF NITRATE ASSIMILATION AND DISSIMILATION IN *P. AERUGINOSA*

We have used the tools of physiological genetics to establish the separateness of the two nitrate reductases in *P. aeruginosa*, and we will briefly summarize these studies.

The work of Stouthamer (van Hartingsveldt et al., 1971) made a significant start with the isolation of mutants defective in dissimilatory nitrate reductase, the first step in denitrification. Some five separate functions were identified and mapped, and given the *nar* designation. However, a number of them contained multiple mutations (Sias et al., 1980), and so were not definitive. In this laboratory, a number of new mutants have been isolated as having defects in assimilatory nitrate reduction. Table 1 shows growth phenotypes for mutants affected in growth on nitrate alone, or both nitrate and nitrite. These assimilatory mutants were designated *nas* and *nis*, respectively; 23 *nas* mutants and 21 *nis* mutants have now been isolated. Through the work of Sias and Ingraham (1979) and of Jeter and Ingraham (unpublished results) at least seven discrete functions essential for assimilatory nitrate and nitrite reduction have been identified by transductional analysis. The four *nas* loci have been mapped by conjugation (Sias and Ingraham, 1980). These data made it clear that the *nar* loci for dissimilatory nitrate reductase, and the *nas* loci for assimilatory nitrate reductase, are widely separated on the genetic map, and encode no functions common to both enzyme systems.

Table 1. Growth phenotypes of mutant strains of *P. aeruginosa* blocked in nitrate reduction.

Phenotype	Aerobic growth[a]			Anaerobic growth[b]		
	NO_3^-	NO_2^-	NH_4^+	NO_3^-	NO_2^-	$NO_2^-+ClO_4^-$
wild type	+	+	+	+	+	–
nas	–	+	+	+	+	n.d.
nis	–	–	+	+	+	n.d.
nar	+	+	+	–	+	+
nar-ts, 40 C	+	+	+	–	+	+
28 C	+	+	+	+	+	–

[a]Growth on minimal plates, supplemented with the compound as sole nitrogen source aerobically, is indicated as "+"; non-growth as "-"; not determined as "n.d.".

[b]Growth on rich medium, in an anaerobic jar, with nitrate or nitrite as terminal electron acceptor.

The mutants have also been useful in establishing a rationale for two separate nitrate reductases. A nas mutant growing vigorously under denitrifying conditions stopped when shifted to aerobic conditions. Thus, the enzyme which reduces nitrate anaerobically cannot function in the presence of oxygen. A separate assimilatory enzyme is required for growth on nitrate as the sole nitrogen source.

Finally, the nas mutants have provided a clean background in which to isolate dissimilatory nitrate reductase mutants, with no low level growth attributable to anaerobic function of the assimilatory enzyme.

NEW MUTANTS FOR DISSIMILATORY NITRATE REDUCTASE

To pursue the actual process of denitrification, we have isolated a number of mutants of *P. aeruginosa* defective in the first step of the pathway (Carlson and Ingraham, 1980). With these we can explore pleiotropic effects upon later steps in the pathway resulting from functional blockage of nitrate reductase. Such effects may reveal regulatory consequences of nitrate reduction to nitrite and features common to a number of steps. A longer range goal is continued study of the make-up of the nitrate reductase complex: what are its structural components and how are they involved in expression of its activity *in vitro* and *in vivo*?

Two types of dissimilatory mutants are being studied. Their growth phenotypes are illustrated in Table 1. Certain of the nar mutants are "nonconditional" in their lack of ability to grow anaerobically on nitrate. They grow on nitrite as an alternate electron acceptor and are resistant to the effect of the nitrate analogue, chlorate, because they cannot reduce it to chlorite which is toxic. A second group, designated nar-ts, is "heat-sensitive"; they fail to grow on nitrate and are insensitive to chlorate at 40 C but not at 28 C. Tables 2 and 3 show growth data on several of these strains. Cell-free extracts were prepared for *in vitro* assay of nitrate reductase using dithionite-reduced benzyl viologen as electron donor. Also shown is the *in vitro* response to heating at 50 C. Native enzyme extracts were slowly heat-activated several-fold in a process which at least partially involved release of the enzyme from membrane. The nonconditional mutants (Table 2) exhibited a variety of characteristics. Three strains (JM124, 126, 132) produced wild-type level of enzyme, after growth of nitrite, but their enzyme was not heat-activated. Two strains (JM129, 130) lacked activity even in unheated extracts. Growth yields of the several strains on 0.025% nitrite were similar but their rates of growth on this substrate varied greatly, especially in the case of strain JM132. This variation in growth rate suggested marked pleiotropic effects on the nitrate and nitrite reductase systems.

Table 2. Characteristics of nonconditional nar mutants.

Strain	Doubling time[a] (hr)	Yield[b] (OD_{650})	Enzyme activity[c] (%)	Heat-activation[d]
JM41 (wild type)	2.3	0.20	100	+ (1.8-fold)
JM129 (narA)	4.2	0.15	9	-
JM130 (narB)	2.3	0.14	0	-
JM126 (narC)	3.0	0.14	110	± (1.1-fold)
JM124 (narD)	5.5	0.14	140	-
JM132 (narE)	15.0	0.10	75	-

[a]Cultures were grown on tryptone-yeast extract broth with 0.1% sodium thioglycollate and 0.025% sodium nitrite in completely-filled bottles held at 40 C.

[b]The density of stationary-phase cultures.

[c]The relative nitrate reductase specific activity of crude cell-free extracts of the harvested cultures, as measured at 30 C by the dithionite-reduced benzyl viologen assay.

[d]The response of activity after heating the extracts at 50 C for up to 90 min at a protein concentration of about 5 mg/ml.

The heat-sensitive mutant strains (Table 3) produced cell yields comparable to wild-type when grown under either permissive condition; that is, at 28 C on 0.5% nitrate or at 40 C on 0.025% nitrite. Again, growth rates varied widely. Enzyme from all strains grown at 28 C appeared normal in terms of activity and heat-activation, when assayed in vitro. However, enzyme activity was absent after 40 C growth and did not appear even after heating. Possible explanations include the following: 1) a regulatory signal was blocked, preventing induction by nitrite at 40 C; 2) enzyme was synthesized but immediately heat-inactivated; 3) proper assembly of enzyme components was prevented. It was clear from the effect of heating cell-free extracts that enzyme from mutant strains was not intrinsically heat-labile, at least not in vitro. When a culture of JM152, growing on nitrite at 40 C, was shifted in mid-log phase to 28 C in the presence of chloramphenicol for several hours before harvest, no in vitro activity developed. Therefore, if improper assembly of enzyme accounted for the observed loss of activity, the process was not reversible.

We concluded that the nar-ts mutants are heat-sensitive for synthesis of nitrate reductase. A corroborative experiment is

Table 3. Characteristics of heat-sensitive nar-ts mutants.

Strain	Doubling time (hr)	Yield (OD_{650})	Enzyme activity (%)	Heat-activation
		Grown on nitrate[a]		
Wild type	2.5	0.80	100	+ (5.1-fold)
JM152	8	>0.6	78	+ (6-fold)
JM146	6.5	0.63	155	+ (7.2-fold)
JM147	7.5	>0.5	275	+ (4.2-fold)
JM193	25	>0.6	50	+ (8.7-fold)
JM195	25	>0.6	56	+ (5.6-fold)
		Grown on nitrite		
Wild type	2	0.16	100	+ (2-fold)
JM152	2	0.16	0	-
JM146	8	0.15	0	-
JM147	8	0.14	0	-
JM193	10	0.17	0	-
JM195	10	0.18	0	-

[a]Cultures were grown as in Table 2 on either 0.5% sodium nitrate at 28 C, or 0.025% sodium nitrite at 40 C.

illustrated in Figure 2. When mutant cells growing at 28 C on nitrate were shifted to 40 C, logarithmic growth ceased almost immediately, suggesting that no new enzyme was synthesized, and in vitro specific activity decreased slowly. Wild-type continued to grow and produce enzyme when subject to the same shift of growth temperature. We are continuing to study the heat-sensitive defect in terms of which enzyme components are affected and whether or not the mutant strains show differential effects.

NITROUS OXIDE REDUCTION

A second focus of our interest in denitrification has been the reduction of nitrous oxide. The metabolism of nitrous oxide (N_2O) by soil bacteria is a subject of increasing interest and importance. Its reduction to nitrogen gas is generally accepted to be the final step of denitrification. However, large amounts of N_2O are released directly into the atmosphere. The magnitude and ecological impact of this process have been subject to much discussion (CAST, 1976; Hutchinson and Mosier, 1979). The proportion of N_2O to N_2 released during denitrification in mixed soil cultures is governed by a wide variety of conditions including

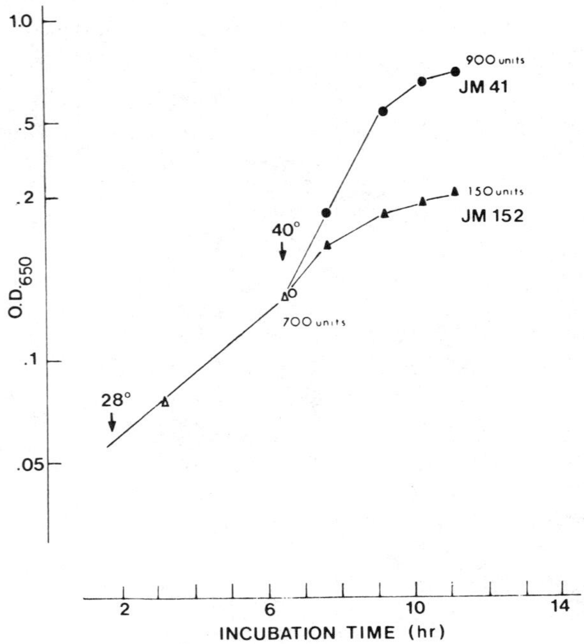

Fig. 2. Anaerobic growth of a heat-sensitive mutant on nitrate. Cultures of JM152 and JM41 (wild type) were grown in completely filled bottles containing tryptone-yeast extract, sodium thioglycollate and 0.5% sodium nitrate, held at 28 C. After 7 hours, they were placed at 40 C. Aliquots were harvested, extracts prepared, and total nitrate reductase activity assayed, after 7 and 12 hours.

aeration, nitrate concentrations and pH (Firestone et al., 1980; Sorensen et al., 1980).

Since N_2O reduction is so strongly implicated as a point of regulation in the denitrification pathway, we have examined the ratio of gases produced by anaerobic cultures of P. aeruginosa. The level of anaerobiosis of the medium was adjusted by addition of reducing agents (Meynell and Meynell, 1970). Table 4 shows that both the proportion of N_2O to N_2 produced from nitrate, and cell yield, depended upon how rigorously the culture was kept anaerobic. In medium C, with an oxidation-reduction potential of about -200 mV, the gas evolved was entirely N_2. It is evident that endogenous N_2O reduction may often be rate-limiting.

We further examined the reduction of N_2O by supplying anaerobic cells with exogenous N_2O. Both growing and resting cultures of P. aeruginosa were able to reduce it to N_2 as expected (Fig. 3).

Table 4. Gases evolved by log-phase cells in anaerobic culture.

Culture medium[a]	Doubling time (hr)	Cell yield (OD_{650})	Gas evolution[b] ($\mu l \cdot hr^{-1}$)		
			N_2	N_2O	NO
A	1.7	1.6	461	2,469	0
B	1.5	1.7	1,661	1,812	0
C	1.8	3.0	13,690	0	

[a]Medium A contained tryptone-yeast extract and 1% KNO_3. Medium B contained, in addition to A, 0.1% sodium thioglycollate. Medium C contained, in addition to B, 0.5% glucose, 0.25% sodium chloride and 0.1% L-cystine.

[b]Aliquots of the growing cultures were removed to stoppered serum vials during log-growth to follow gas evolution. Gases collected in the vial headspace were sampled and identified by gas chromatography on Porapak Q.

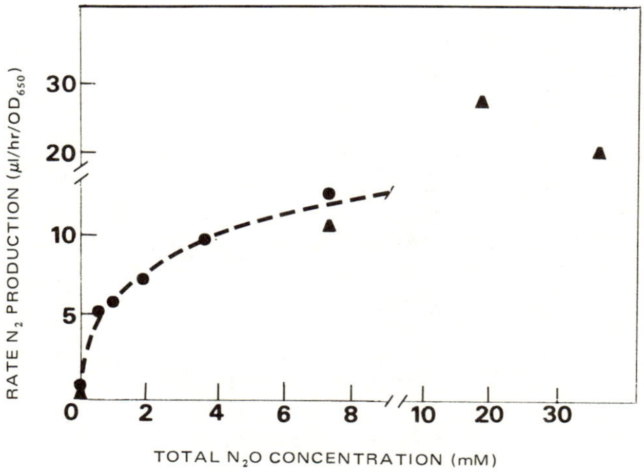

Fig. 3. Reduction of exogenous nitrous oxide by a cell suspension of P. aeruginosa. Cells in late-log growth in medium C (Table 4) with 0.1% KNO_3 were harvested, washed twice in the medium without nitrate, and resuspended at 8-fold concentration in a sealed serum vial. Nitrous oxide was supplied as N_2O-saturated water. The evolution of N_2 into the headspace was measured by gas chromatography. The two symbols (●, ▲) refer to results from separate experiments.

Table 5. Nitrogen gas production and growth yield of nitrate-limited cultures.

Electron acceptor[a]	N_2 Production ($\mu l \cdot hr^{-1}$)	Growth yield (OD_{650})
KNO_3 (18 μmol)	30.5	0.38
KNO_3 (625 μmol)	115.5	1.45
KNO_3 (18 μmol) plus N_2O (375 μmol)	48.6	0.36

[a]A culture was grown to stationary phase anaerobically on medium C (Table 4) with 0.3% KNO_3. Aliquots were then diluted to a density of 0.2 OD_{650} in sealed serum vials, and supplied with nitrate or nitrous oxide. Gas production was followed for several hours, and total growth was measured after overnight incubation.

However, the rate of N_2O reduction was slow, less than 3% of the rate of N_2 production from nitrate. Furthermore, we were unable to demonstrate anaerobic cell growth using N_2O as the terminal electron acceptor. This was true on both solid and liquid media and when N_2O was supplied alone or in conjunction with a low level of nitrate (Table 5). St. John and Hollocher (1977) also reported the inability of P. aeruginosa to grow on N_2O, but presented evidence that it is a freely-exchangeable intermediate of denitrification. Thus, there is a real dilemma with P. aeruginosa with regard to N_2O utilization, since it appears to be exceptional among denitrifiers thus far known to metabolize N_2O. Table 6 lists a dozen strains which reduce nitrate to N_2 and that grow on exogenous N_2O. There are also a number of denitrifiers that do not reduce N_2O to N_2 but simply release N_2O as the final product of nitrate reduction (Table 6). P. aeruginosa is the only example, thus far, of N_2 production but lack of growth on N_2O.

Recent data of Bryan and Delwiche (Bryan, 1980) only heighten the dilemma. They found that P. stutzeri (which does grow on N_2O) and P. aeruginosa produced similar cell yields when grown on limiting nitrate (or nitrite). Also, when acetylene was used to block endogenous N_2O reduction, the yields of P. aeruginosa and P. stutzeri decreased by the same factor. It thus appeared that the reduction of N_2O generated internally from nitrate was energy-yielding for both organisms. This seemingly refutes the idea that lack of growth by P. aeruginosa on N_2O lies simply in an inability to generate energy from N_2O.

Table 6. N_2 utilization or production by denitrifiers.

Strains capable of growing on N_2O and producing N_2 from NO_3^- or NO_2^-	
Pseudomonas perfectomarinus	Payne et al., 1971
P. denitrificans	Matsubara, 1971
P. stutzeri	Allen and van Niel, 1952
P. lemoignei	Pichinoty et al., 1977c
P. fluorescens	Greenberg and Becker, 1977
P. chlororaphis B561[a]	Greenberg and Becker, 1977
P. pickettii	Garcia et al., 1977
Paracoccus denitrificans	Pinchinoty et al., 1977b
Agrobacterium tumefaciens	Pinchinoty et al., 1977a
A. radiobacter	Pinchinoty et al., 1977a
Bacillus azotoformans	Pinchinoty et al., 1976a
Flavobacterium sp.	Pinchinoty et al., 1976b
Alcaligenes sp.	Pinchinoty et al., 1975
Strains incapable of growth on N_2O	
A. Do not produce N_2	
Pseudomonas fluorescens PJ185	Greenberg and Becker, 1977
P. chlororaphis	Greenberg and Becker, 1977
Spirillum itersonii	Hansen, 1972
S. lipoferum	Neyra et al., 1977
Corynebacterium nephridii	Hart et al., 1965
B. Produce N_2	
Pseudomonas aeruginosa	St. John and Hollocher, 1977

[a]Although N_2O growth was not established, N_2 was produced from nitrate.

The dilemma of *P. aeruginosa* and N_2O poses difficulty in using the tools of physiological genetics to study the latter steps of denitrification. Although *P. aeruginosa* represents a fascinating phenotype for biochemical studies, we are stymied in attempts to select for mutants lacking N_2O reductase, for mapping studies, or for mutants that overproduce N_2O reductase. We are also limited in study of the induction and regulation of N_2O reductase *in vivo*.

THE CASE FOR *P. STUTZERI*

With the limitations imposed by the N_2O metabolism of *P. aeruginosa*, we have turned to *P. stutzeri*. In its favor are vigorous denitrification and rapid growth on N_2O. In fact, it was one of the first denitrifiers known to grow on N_2O (Allen and van Niel, 1962). We have found that a "smooth colony" variant handles easily for plating and transfer. Furthermore, *P. stutzeri* is apparently closely related to *P. aeruginosa* and might share several mechanisms of genetic exchange. The plasmid R68.45, for example, has been shown to act as a sex factor in a wide range of bacterial strains (Holloway et al., 1979; Paraskeva, 1979). To begin the exploration of genetic manipulation in *P. stutzeri*, we have been working out procedures for mutagenesis and counterselection and have isolated a number of auxotrophs. We have been able to demonstrate, by conjunction, the interspecies transfer of plasmid R68.45 from *P. aeruginosa* to *P. stutzeri*, as well as between strains of *P. stutzeri*, using these auxotrophic markers and antibiotic resistance markers. With the isolation of mutants unable to grow on N_2O we will be able to initiate a study on the make-up and regulation of the N_2O reductase.

Mutants of *P. stutzeri* that no longer utilize exogenous N_2O should reveal the position of N_2O as an intermediate of denitrification. If such mutants do not evolve N_2, and reduce nitrate stoichiometrically to N_2O, with decreased cell yield, then the biochemical defect must be at site (1) and N_2O must be an obligate intermediate as has been proposed:

$$NO_3^- \rightarrow \rightarrow \rightarrow N_2O \xrightarrow[(1)]{(4)\ \sim ATP} N_2$$

However, if such mutants produce normal amounts of N_2O or N_2 from nitrate then we must consider another intermediate (N), to which N_2O is normally converted by an enzymatic step that is blocked here (site (2)).

$$\begin{array}{lll} NO_3^- \rightarrow \rightarrow \rightarrow & (N) & \xrightarrow[(3)]{(4)\ \sim ATP} N_2 \\ & \uparrow\downarrow (2) & \\ & N_2O & \end{array}$$

This would be phenotypically similar to *P. aeruginosa*. In this postulation, a cell suspension would be unable to reduce exogenous N_2O, and the evolution of N_2 would still be sensitive to inhibition by acetylene. However, cell yield from nitrate would still be normal, if site (3) is the energy-yielding reduction.

A third possibility, arising from a mutational defect at site (3), would result in the production of N_2O exclusively. N_2O could still be reduced, but cell yield would be decreased.

Finally, ineffective N_2O utilization could involve a block in electron transport during the reductive step to N_2 (site (4)). The above schemes could not be distinguished in this class of mutants, since in either case we would expect normal N_2O and N_2 production, decreased yield from nitrate, and exogenous N_2O reduction.

CONCLUSION

We believe the time has come when laboratory studies of denitrification can progress quickly through the application of physiological genetics. With a focus upon one or two species, we should gain access to answers not otherwise available up to this time. This information can then be profitably applied to mixed cultures and the field situations which are the ultimate concern.

ACKNOWLEDGMENTS

This work was supported by a grant from NSF (PFR 77-07301).

REFERENCES

Allen, M. B., and van Niel, C. B., 1952, Experiments on bacterial denitrification, J. Bact., 64:397.

Bryan, B., 1980, "Cell Yield and Energy Characteristics of Denitrification with *Pseudomonas stutzeri* and *Pseudomonas aeruginosa*," Ph.D. Thesis, Univ. of California, Davis.

Carlson, C. A., and Ingraham, J. L., 1980, Structural mutations in nitrate reductase of denitrifying *Pseudomonas aeruginosa*, Amer. Soc. Microbiol. Abstr., 148.

CAST (Council for Agric. Science and Technology), 1976, "Effect of Increased Nitrogen Fixation on Stratospheric Ozone," Rep. 53, Iowa State Univ., Ames.

Cole, J. A., 1978, The rapid accumulation of large quantities of ammonia during nitrite reduction by *Escherichia coli*, FEMS Microbiol. Lett., 4:327.

Firestone, M. K., Firestone, R. B., and Tiedje, J. M., 1979, Nitric oxide as an intermediate in denitrification: evidence from nitrogen-13 isotope exchange, Biochem. Biophys. Res. Comm., 91:10.

Firestone, M. K., Firestone, R. B., and Tiedje, J. M., 1980, Nitrous oxide from soil denitrification: factors controlling its biological production, Science, 208:749.

Focht, D. D., and Verstraete, W., 1977, Biochemical ecology of nitrification and denitrification, Adv. Microbiol. Ecol., 1:135.

Gamble, T. N., Betlach, M. R., and Tiedje, J. M., 1977, Numerically dominant denitrifying bacteria from world soils, Appl. Environ. Microbiol., 33:926.

Garcia, J.-L., Pichinoty, F., Mandel, M., and Greenway, B., 1977, A new denitrifying saprophyte related to Pseudomonas pickettii, Ann. Microbiol. (Inst. Pasteur), 128B:243.

Greenberg, E. P., and Becker, G. E., 1977, Nitrous oxide as end product of denitrification by strains of fluorescent pseudomonads, Can. J. Microbiol., 23:903.

Hansen, E. J., 1972, "Denitrification by Spirillum itersonii," M.Sc. Thesis, University of Iowa, Iowa City.

Hart, L. T., Larson, P. D., and McCleskey, C., 1965, Denitrification by Corynebacterium nephridii, J. Bact., 89:1104.

Hollocher, T. C., Garber, E., Cooper, A. J. L., and Reiman, R. E., 1980, ^{13}N, ^{15}N isotope and kinetic evidence against hyponitrite as an intermediate in denitrification, J. Biol. Chem., 255:5027.

Holloway, B. W., Krishnapillai, V., and Morgan, A. F., 1979, Chromosomal genetics of Pseudomonas, Microbiol. Rev., 43:73.

Hutchinson, G. L., and Moiser, A. R., 1979, Nitrous oxide emissions from an irrigated cornfield, Science, 205:1125.

Kristjansson, J. K., and Hollocher, T. C., 1980, First practical assay for soluble nitrous oxide reductase of denitrifying bacteria and a partial kinetic characterization, J. Biol. Chem., 255:704.

Matsubara, T., 1971, Studies on denitrification, XIII, Some properties of the N_2O-anaerobically grown cell, J. Biochem., 69: 991.

Meynell, G. G., and Meynell, E., 1970, "Theory and Practice in Experimental Bacteriology," Cambridge Univ. Press, Cambridge.

Neyra, C. A., Döbereiner, J., Lalande, R., and Knowles, R., 1977, Denitrification by N_2-fixing Spirillum lipoferum, Can. J. Microbiol., 23:300.

Paraskeva, C., 1979, Transfer of kanamycin resistance mediated by plasmid R68.45 in Paracoccus denitrificans, J. Bacteriol., 139:1062.

Payne, W. J., 1973, Reduction of nitrogenous oxides by microorganisms, Bact. Rev., 37:489.

Payne, W. J., Riley, P. S., and Cox, C. D., 1971, Separate nitrite, nitric oxide, and nitrous oxide reducing fractions from Pseudomonas perfectomarinus, J. Bact., 106:356.

Pichinoty, F., de Barjac, H., Mandel, M., Greenway, B., and Garcia, J.-L., 1976a, Une novelle bactérie sporulée, dénitrificante, mésophile: Bacillus azotoformans n. sp., Ann. Microbiol., 127B:351.

Pichinoty, F., Bigliardi-Rouvier, J., Mandel, M., Greenway, B., Méténier, G., and Garcia, J.-L., 1976b, The isolation and properties of a denitrifying bacterium of the genus Flavobacterium, Antonie van Leeuwenhoek, 42:349.

Pichinoty, F., Mandel, M., and Garcia, J.-L., 1977a, Étude de six souches de Agrobacterium temefaciens et A. radiobacter, Ann. Microbiol. (Inst. Pasteur), 128A:303.

Pichinoty, F., Mandel, M., and Garcia, J.-L., 1977b, Étude physiologique et taxonomique de Paracoccus denitrificans, Ann. Microbiol. (Inst. Pasteur), 128B:243.

Pichinoty, F., Mandel, M., Greenway, B., and Garcia, J.-L., 1977c, Isolation and properties of a denitrifying bacterium related to Pseudomonas lemoignei, Int. J. System. Bacteriol., 27:346.

Pichinoty, F., Mandel, M., Greenway, B., and Garcia, J.-L., 1975, Isolement à partir du sol et étude d'une bactérie dénitrifiante appartenant au genre Alcaligenes, C. R. Acad. Sc. Paris, T. 381 (Série D):1273.

St. John, R. T., and Hollocher, T. C., 1977, Nitrogen-15 tracer studies on the pathway of denitrification in Pseudomonas aeruginosa, J. Biol. Chem., 252:212.

Sias, S. R., and Ingraham, J. L., 1979, Isolation and analysis of mutants of Pseudomonas aeruginosa unable to assimilate nitrate, Arch. Microbiol., 122:263.

Sias, S. R., and Ingraham, J. L., 1980, Chromosomal location in Pseudomonas aeruginosa of genes encoding assimilatory nitrate reductase, J. Bact., in press.

Sias, S. R., Stouthamer, A. H., and Ingraham, J. L., 1980, Assimilatory and dissimilatory nitrate reductases in Pseudomonas aeruginosa are encoded by different genes, J. Gen. Microbiol., 118:229.

Sorsensen, J., Tiedje, J. M., and Firestone, R. B., 1980, Inhibition by sulfide of nitric and nitrous oxide reduction by denitrifying Pseudomonas fluorescens, Appl. Environ. Microbiol., 39:105.

Stouthamer, A. H., 1976, Biochemistry and genetics of nitrate reductase in bacteria, Adv. Microbiol. Physiol., 14:315.

van Hartingsveldt, J., Marinus, M. G., Stouthamer, A. H., 1971, Mutants of Pseudomonas aeruginosa blocked in nitrate or nitrite dissimilation, Genetics, 67:469.

ON REGULATING THE SYNTHESIS OF NITRATE REDUCTASE

Curtis A. Carlson

Bacteriology Department
University of California
Davis, CA 95616

INTRODUCTION

Denitrification plays an important role in the nitrogen budget of soil and water. It is carried out only by bacteria and provides a means for them to generate metabolic energy anaerobically, by utilizing various nitrogenous oxides as terminal electron acceptors in place of oxygen. Reduction of nitrate to nitrite, the first step in the denitrification pathway, is catalyzed by dissimilatory nitrate reductase (DNR). It is clear that study devoted to the structure and function of this enzyme, and to the regulation of its synthesis and activity, will help to form a clearer picture of the various factors controlling denitrification.

This study explored a semi-anaerobic cell suspension for characterizing the synthesis of nitrate reductase in a soil bacterium, *Pseudomonas aeruginosa*. The appearance of different levels of enzyme activity in response to different chemical inducers is shown, as well as the utility of the system to screen mutants.

METHODS

The induction system used here is based upon that devised for *Paracoccus denitrificans* by Calder et al. (1980). A 10 ml inoculum of *P. aeruginosa* PA01 in Luria broth was grown under fully aerobic conditions by shaking at 300 rpm at 37 C in a 500 ml Erlenmeyer flask covered with a milk filter. At stationary phase, this culture was transferred to a 125 ml flask containing 50 ml L-broth and capped with a foam plug. The flask was shaken at 200 rpm at 40 C for up to 8 hours in the presence or absence of chemical inducers. Cultures were harvested by centrifugation, resuspended in

0.2 M phosphate buffer pH 7.2 and broken in a French pressure cell for assay of nitrate reductase. The assay measured the production of nitrite from nitrate in the presence of dithionite-reduced benzyl viologen as the electron donor (Lowe and Evans, 1964).

RESULTS

The results of an initial survey of a variety of potential inducers are shown in Table 1. A low level of enzyme activity was present under the semi-anaerobic condition in the absence of added inducer. However, this level was greatly enhanced by sodium nitrate, nitrite, azide or chlorate. Nitrate and nitrite are metabolized during denitrification so that fairly high initial concentrations were required for them to be effective. Sodium azide was not metabolized under these conditions and was very effective at low concentrations in inducing a high level of enzyme activity. Nitrite, azide, and chlorate are all toxic to the cell at different higher concentrations, so they must be used judiciously.

The kinetics of induction of DNR by nitrate or azide were quite different (data not shown). Nitrate-induced enzyme activity increased quickly with time up to 1.5 hours, then decreased at the same time that nitrate (and nitrite) concentrations decreased due to enzymatic reduction. By contrast, azide-induced enzyme levels rose slowly but continuously for up to 8 hours. A three hour incubation was used in experiments comparing inducers.

Table 2 illustrates that enzyme appearing in response to the various inducers had different characteristics. DNR from *P. aeruginosa* cells grown anaerobically on nitrate can be activated *in vitro* by heating. The activation is a slow, irreversible, and temperature-dependent process which at least partially involves solubilization of the enzyme from a membrane-associated state (Carlson and Ingraham, 1980, and unpublished results). Enzyme from cells induced with nitrate or nitrite here could also be heat-activated (Table 2). Azide-induced enzyme, however, could not be heat-activated. In fact, its activity slowly decreased upon heating. Azide-induced DNR and nitrate-induced DNR were both 85-90% membrane-bound before heating (data not shown), so the lack of activation of the former was not the result of its existence as a soluble form already.

When azide and nitrate were added together as co-inducers, their effect on the level of enzyme activity produced was synergistic (Tables 1 and 2). This further suggests that they act by different mechanisms. Interestingly, the enzyme produced by co-induction was similar to the enzyme induced by azide alone, with respect to its response to heating. Thus, if coinduction did produce a mixed population of nitrate-induced and azide-induced enzyme molecules, their net response to heating was unexpected.

Table 1. Induction of wild-type nitrate reductase.

Inducer	Conc. (mM)	DNR Activity (nm $min^{-1}mg^{-1}$)
None	-	0.9
Nitrate	10	8.8
	20	13.9
	50	61.6
Nitrite	1.0	1.8
	10	4.2
	20	7.9
Azide	0.1	33.0
	0.5	54.1
	1.0	31.2
	5.0	0.04
Chlorate	1.0	3.3
Nitrate + Azide	20 + 0.5	154
Nitrite + Azide	20 + 0.5	5.8

Table 2. Effect of different inducers upon response of wild-type nitrate reductase to heating *in vitro*.

		DNR Activity[a] (nm $min^{-1}\cdot mg^{-1}$)		
Inducer	Conc. (mM)	Before heating	After heating	Net response
Nitrate	20	6.6	36.2	5.8-fold increase
Nitrite	20	2.6	6.2	2.4-fold increase
Azide	0.5	53.3	30.0	50% decrease
Nitrate + azide	20 + 0.5	165	91.5	50% decrease
Nitrite + azide	20 + 0.5	7.3	4.8	35% decrease

[a]Cell-free extracts were assayed for enzyme activity at 30 C, then heated (at about 3 mg ml^{-1} protein) for 90 min at 50 C before being chilled and assayed again at 30 C.

The net effect of azide and nitrate as co-inducers was complex. The initial enzyme activity induced was similar to that induced by nitrite alone. Azide and the excess nitrite reacted chemically to eventually degrade the azide under these conditions (data not shown). Nevertheless, the enzyme activity was not increased by heating, as it was when nitrite was the sole inducer.

Finally, Table 3 demonstrates the efficacy of the induction system in screening the response of a variety of mutant strains to different inducers. These strains were mutants of *P. aeruginosa* which either did not grow anaerobically on nitrate as the electron acceptor at any temperature ("nonconditional"), or did so at 28 C but not at 40 C ("heat-sensitive"). The functional defects in these strains cannot yet be correlated with specific components of the nitrate reductase complex. Nevertheless, these data illustrate a varied pattern of response to inducers by the different mutants which may help to characterize them. The azide-induced enzyme levels were generally high, and nitrate-induction and azide-induction showed very similar patterns. It is noteworthy that the *in vitro* enzyme activity could be induced in strains JM152 and JM195, even though they failed to grow on nitrate at 40 C.

Table 3. Induction of nitrate reductase in mutant strains.

	DNR activity (nm $min^{-1} \cdot mg^{-1}$) after induction at 40 C by			
Strain	Class	Nitrate	Nitrite	Azide
JM41	Wild-type	7.5	8.4	57.2
JM129	Nonconditional	5.5	4.2	37.0
JM130	Nonconditional	6.6	2.0	30.0
JM126	Nonconditional	0	4.6	0
JM124	Nonconditional	0	3.5	2.2
JM132	Nonconditional	2.4	0	7.0
JM152	Heat-sensitive	9.5	2.9	29.7
JM195	Heat-sensitive	7.3	2.9	47.7

DISCUSSION

The induction technique described here allowed DNR synthesis in the absence of net cell growth. Enough oxygen was supplied by shaking the dense cell suspension for aerobic respiration to provide low levels of ATP. This is a critical point for non-fermentative organisms like P. aeruginosa. Nevertheless, the limited aeration and high cell denisty produced a semi-anaerobic condition in which the cells were poised to initiate synthesis of DNR and the denitrification process in response to added inducers. Nitrate and nitrite appeared to act similarly. Azide, however, acted by a different mechanism, since different enzyme activity levels were produced, the kinetics of induction were different, and the enzyme itself was different as judged by its physiochemical characteristics. Moreover, azide and nitrate acted synergistically when added as co-inducers.

Azide is a respiratory electron transport inhibitor and may regulate the synthesis of nitrate reductase via the altered oxidation-reduction state of an electron-transport intermediate or of DNR itself. Such a redox control model in Proteus mirabilis has been discussed by Stouthamer (1976), and the involvement of azide in such a system in Paracoccus denitrificans has been suggested by Lascelles (Calder et al., 1980).

The present study of the synthesis of nitrate reductase is continuing with a search for new gratuitous inducers. In conjunction with a variety of mutant strains of P. aeruginosa, it may be possible to elucidate the factors regulating the synthesis and assembly of active dissimilatory nitrate reductase.

ACKNOWLEDGEMENT

This work was supported by a grant from NSF (PFR 77-07301).

REFERENCES

Calder, K., Burke, K. A., and Lascelles, J., 1980, Induction of nitrate reductase and membrane cytochromes in wild-type and chlorate-resistant Paracoccus denitrificans, Arch. Microbiol., 126:149.

Carlson, C. A., and Ingraham, J. L., 1980, Structural mutations in nitrate reductase of denitrifying Pseudomonas aeruginosa, Amer. Soc. Microbiol. Abstr., 148.

Lowe, R. H., and Evans, H. J., 1964, Preparation and some properties of a soluble nitrate reductase from Rhizobium japonicum, Biochim. Biophys. Acta, 85:377.

Payne, W. J., 1973, Reduction of nitrogenous oxides by microorganisms, Bacteriol. Rev., 37:409.

Stouthamer, A. H., 1976, Biochemistry and genetics of nitrate reductase in bacteria, Adv. Microbiol. Physiol., 14:315.

THE CONVERSION OF NO_3^- TO NH_4^+ IN KLEBSIELLA PNEUMONIAE

J. R. Thayer and R. C. Huffaker

Plant Growth Laboratory and the Department of
Agronomy & Range Science, University of California
Davis, CA 95616

INTRODUCTION

In soil environments several microbial oxidations and reductions of nitrogen (N) may occur depending on the conditions. Figure 1 shows those of interest to this study. Anaerobically, NO_3^- is reduced to NO_2^-, which can be converted to gaseous forms and the N lost, or may be converted to NH_4^+ and assimilated or excreted and the N conserved through interaction with soil particles.

Nitrogen fixation also occurs in the soil when N is limiting; thus, some of the N lost during denitrification may be regained where conditions permit. When soils become sufficiently aerobic, the NH_4^+ present is oxidized by microbes to NO_2^- and NO_3^- (nitrification), thus causing a set of cyclic N conversions in the soil.

Nitrification rates appear to be unaffected by the presence of plants, and denitrification rates are enhanced by both proximity to roots and plant development (Alexander, 1977). This may be due to the decreased oxygen tension associated with active respiration (by the roots) or may reflect the presence of organic compounds excreted by the roots. Among other organic compounds, roots of aseptically grown plants excrete essentially all naturally occurring amino acids, a number of growth factors, sucrose, glucose, fructose, acetate butyrate, propionate, valerate, succinate, glycolate, citrate, and fumarate (Alexander, 1977). These compounds would be expected to support the growth of a wide variety of heterotrophs.

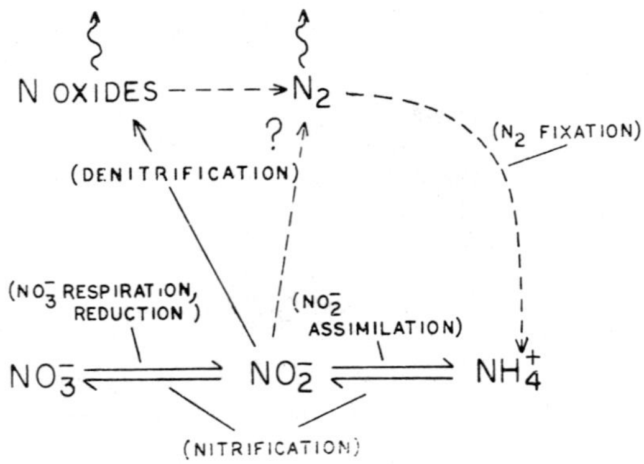

Fig. 1. The transformations of nitrogen occurring in the soil.

The purpose of this study was to obtain and characterize microbes that would reduce NO_3^- to NH_4^+ and excrete it under both aerobic and anaerobic conditions. Such organisms may be useful in converting some of the soil NO_3^- to NH_4^+, thus reducing N loss by both NO_3^- leaching and denitrification. Once such organism are obtained, they may be genetically manipulated to perform this task faster and more efficiently. In order to intelligently manipulate the genome of an organism to enhance a process, the rate-limiting step of that process must be known. Therefore, this study was undertaken to determine the rate-limiting step(s) in the conversion of NO_3^- to NH_4^+ under aerobic and anaerobic conditions.

The organism chosen was Klebsiella pneumoniae M5A1 and its derivatives. This soil and enteric bacterium is capable of N_2 fixation and NO_3^- respiration and assimilation. Being a member of the Enterobacteriaceae, it has a genetic system that is readily manipulated. Mutants of this organism lacking glutamine synthetase and/or glutamate synthase are derepressed for nitrogenase (andersen and Shanmugam, 1977) and for the activities responsible for the conversion of NO_2^- to NO_x (x = 1 or 2) in the presence of NH_4^+ (Hom et al., 1980; Shanmugam, this volume). These strains require glutamine, asparagine, or other amino acids for growth, and excrete NH_4^+ from N_2 or NO_3^- when limited for N in the presence of a suitable carbon source (Andersen and Shanmugam, 1977; Thayer and Huffaker, this communication). In fact, natural isolates of Klebsiella species grown under N limitation excrete over 60% of the NO_3^--N taken up as NH_4^+ (Herbert, 1980). The fact that glutamine, glucose and sucrose may be excreted actively by plant roots suggests that the use of a glutamine auxotroph of Klebsiella may be a reasonably good choice for the purposes of this study.

Because glutamine auxotrophs cannot assimilate NH_4^+ produced from NO_3^- or N_2, they excrete it. This and other products of metabolism that may inhibit further cell growth were minimized in the cell culture by growing the cells in a dialysis bag surrounded by a medium of the same composition into which such products could diffuse (Fig. 2). The effects of pH, temperature, and oxygen availability in the flasks were determined.

Sensitive measurements of initial rates of metabolic reactions have traditionally required the use of radioactive substrates. In this study we have used ^{13}N, the longest-lived radioisotope of N: it has a half-life of 9.96 minutes (Tiedje et al., 1979). Using ^{13}N, we have studied NO_3^- transport and metabolism as affected by NO_3^- concentration, cultural conditions, and presence of NH_4^+.

MATERIALS AND METHODS

The dialysis-culture system shown in Figure 2 was adapted from the studies of Hom et al. (1980) and Andersen and Shanmugam (1977). The mineral medium used was that of Cohen-Bazire et al. (1957); the cultures were sparged with oxygen, air, or argon as indicated. Temperature studies were conducted by immersing the flasks in a water bath maintained at the indicated temperature by a termperature-control unit (Lauda K2-R). Nitrate and nitrite were

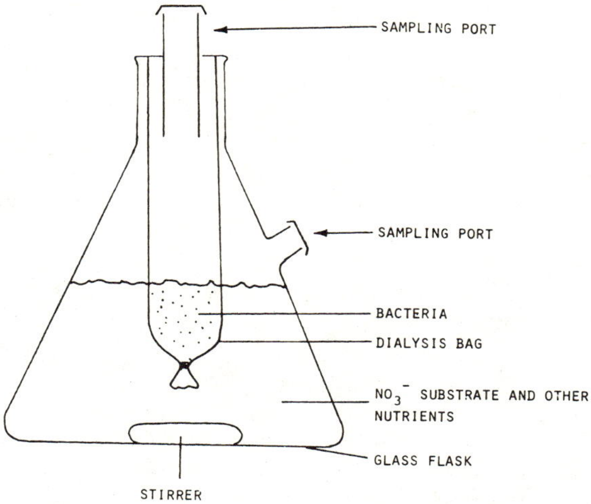

Fig. 2. Schematic representation of a dialysis-culture vessel. The gas-sparging tubes used in both chambers have been omitted for simplicity in presentation.

determined by the methods of Schrader et al. (1968) or Thayer and Huffaker (1980). NH_4^+ was determined using NH_4^+-specific electrodes from Orion Inc. (11 Blackstone St., Cambridge, MA 02139). Glucose was determined using the statzyme reagent from Worthington Diagnostics (Grandview Business Center, South San Francisco, CA 94080). Nitrogen sources were as indicated.

For studies utilizing $^{13}NO_3^-$, cultures were grown in 100- and 1,000-ml flasks half-filled with media and sparged with air through 12 C frits at flow rates of about 1 ml (minute ml of culture)$^{-1}$. $^{13}NO_3^-$ was prepared at the Crocker Nuclear Laboratory at UC Davis using the target system of Parks and Krohn (1978). It was purified from the other products of the proton bombardment of water by the method of Tiedje et al. (1979). Transport was assayed by sedimentation of cells through silicone oil (Kashket and Barker, 1977; Thayer and Huffaker, 1981). In experiments in which the products of transport and metabolism were to be analyzed, sedimented cells were subjected to extraction in 80% methanol (aqueous) about 60 seconds after sedimentation (Thayer and Huffaker, 1981). High-pressure liquid chromatographic analyses of such extracts were performed using the method of Thayer and Huffaker (1980), except that the column eluate was passed through a well γ-counting system connected to a computer for generation of half-life corrected chromatograms.

Bacteria

Klebsiella pneumoniae M5Al and its derivative SK-25 (Shanmugam et al., 1975) were obtained from K. T. Shanmugam and R. C. Valentine.

RESULTS

Steady-State Conversion of NO_3^- to NH_4^+

Table 1 shows the effect of pH on the rate, duration, and efficiency of NH_4^+ production from NO_3^- by the glutamine auxotroph SK-25. As pH increased from 6.5 to 8.5, the maximum production rate also increased. However, efficiency of NH_4^+ production, as measured by the NH_4^+ produced per glucose utilized (molar ratio) was greatest at neutrality.

The effect of temperature on the aerobic conversion of NO_3^- to NH_4^+ by SK-25 is shown in Table 2. The maximal rate observed in oxygen-sparged cultures was nearly constant from 20 C to 30 C. However, the greatest efficiency was obtained at 20 C. Van't Riet et al. (1968) reported that oxygen-sparging inactivated the assimilatory NR of *Aerobacter* (*Klebsiella*) *aerogenes*. As a control in this experiment, an air-sparged culture was also analyzed. The maximal rate of the air-sparged culture was 83% higher than the O_2-sparged culture, and its efficiency was 2.4-fold higher.

Table 1. The effect of pH on NH_4^+ production from NO_3^- in Klebsiella pneumoniae-derivative SK-25.[1]

Flask	pH	NH_4^+ production: Duration (days)	NH_4^+ production: Maximal rate[2]	Molar ratio NH_4^+/glucose: Day 3	Molar ratio NH_4^+/glucose: End of run
1	6.5	8	6.5	0.12	0.13
2	7.0	4	12.4	0.32	0.30
3	7.5	8	11.7	0.31	0.22
4	8.0	8	12.6	0.27	0.23
5	8.5	8	17.5	0.29	0.20

[1]Reactor vessel condition: T = 25 C, sparged with air, (NO_3^-) = 3 mM as KNO_3 in HMB (mineral base) 0.5 μg/ml GLN added each day.

[2]μmoles/day

Table 2. The effect of temperature on the aerobic conversion of nitrate to ammonium by SK-25.[1]

Flask	T^oC	NH_4^+ production: Duration (days)	NH_4^+ production: Maximal rate[2]	Molar ratio NH_4^+/ glucose at end of run
1	16	3.5	2.4	0.048
2	20	3.5	6.9	0.132
3	25	3.5	6.6	0.085
4	30	3.5	6.8	0.033
5[3]	25	3.7	12.4	0.32

[1]Conditions: pH 7.0, HMB sparged with O_2 gas, GLN added at 1 μg/ml each day, 5 mM KNO_3^-.

[2]μmoles/day·mg dry wt.

[3]Sparged with air.

The effect of the sparging gas was analyzed further. Cultures were sparged with O_2, air, or argon gas as indicated in Table 3. No NO_2^- was excreted into the medium in the air-or oxygen-sparged cultures. Large amounts of NO_2^- appeared in the medium in the argon-sparged cultures. Under argon, when all the NO_3^- had been consumed, the NO_2^- level dropped concomitant with the excretion of more NH_4^+.

The oxygen-sparged cultures gave rates of 37% and efficiencies 47% those of the air-sparged cultures. The argon-sparged cultures produced NH_4^+ at 2.5 times the rate of the air-sparged culture, but did so with no apparent increase in efficiency. We conclude that the conversion of extracellular NO_3^- to NO_2^- is rate-limiting in aerobic cultures and that the conversion of NO_2^- (extra- or intracellular) to NH_4^+ becomes rate-limiting under anaerobic conditions, where the increase in NR activity due to NO_3^- respiration overcomes the aerobic NR limitation. These conclusions are based on the observations that NO_2^- appears in the medium under anaerobic but not aerobic conditions, for the specific activity of the assimilatory NR is much lower than that of the respiratory enzyme, and the rate of the overall production of NH_4^+ from NO_3^- increases as the culture becomes increasingly anaerobic.

Transport

There is a controversy about the site of respiratory nitrate reduction. Both the extracellular (Garland et al., 1975; Haddock and Jones, 1977) and intracellular (John, 1977; Kristajansson et al., 1978; Kristjansson and Hollocher, 1979) sites have been proposed. Both the assimilatory and respiratory enzymes appear to be membrane-associated, so assimilatory reduction might occur on the external aspect of the cell membrane, thus eliminating transport from our consideration. If transport was required for NO_3^- assimilation, then a mutant class that grows on NO_2^-, not on NO_3^- and has ANR activity should be possible to obtain (permease$^-$). A genetic analysis of NO_3^- assimilation in Pseudomonas aeruginosa by Sias and Ingraham (1979) did not produce such a class. All NO_2^- positive, NO_3^- negative strains obtained possessed significantly less than wild-type ANR levels. However, the absence of a permease-negative class may be due to the presence of multiple-transport systems. Because mutations in 2 or more genes may be required, the probability of obtaining such a class could be vanishingly low.

In order to test this theory, we performed a kinetic analysis of initial rates of NO_3^- transport at NO_3^- concentrations between 1.0 μM and 1.0 mM. A double-log plot of initial rates versus NO_3^- concentration is shown in Figure 3. This graph indicates the presence of two saturation curves with apparent K_m values of 2 mM and 4 μM. In separate studies, analyzing velocities at NO_3^-

Table 3. Comparison of NH_4^+ production by SK-25 under oxygen-, air-, and argon-sparging.

Flask	pH	Temp	NH_4^+ production: Duration (days)	NH_4^+ production: Overall rate[1]	$\overline{X}$	Molar ratio NH_4^+/glucose: Day 3	Molar ratio NH_4^+/glucose: End of run
1[2]	7	20 C	7.5	4.4	(3.8±0.8)	0.22	0.08
2	7	25 C	7.1	3.1		0.09	0.035
3[3]	7.5	25 C	7.7	12.3	(10.3±2)	0.303	0.23
4	7	25 C	3.7	8.3		0.36	0.31
5[4]	7	25 C	5.8	30.3	(26.2±4)	0.43	0.34
6	7	25 C	5.8	22.0		0.29	0.22

[1] μmoles/day·mg dry wt.

[2] Reactor vessel condition: HMB + 5 mM KNO_3, GLN @ μg/ml culture per day with O_2-sparging.

[3] Reactor vessel condition: HMB + 3 mM KNO_3, GLN @ 0.5 μg/ml per day, air-sparging.

[4] Reactor vessel condition: HMB + 25 mM KNO_3, GLN @ 0.35 μg/ml per day, argon-sparging.

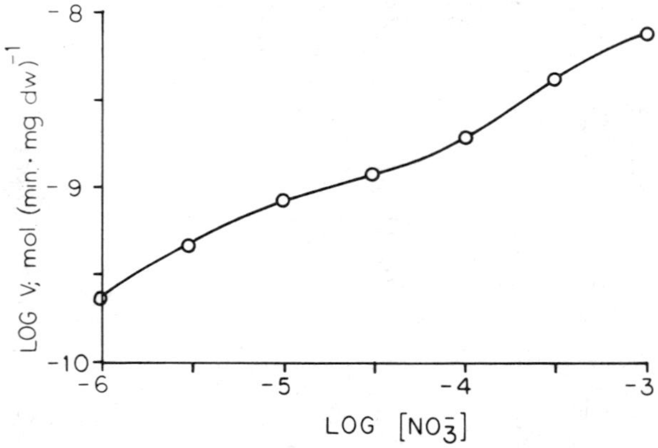

Fig. 3. A double-log plot of nitrate concentration vs. initial rate of nitrate transport between 0 and 60 seconds after addition of $^{13}NO_3^-$ to Klebsiella pneumoniae cultures grown aerobically on nitrate.

concentrations between 1.25 and 5 mM NO_3^- and between 3 and 9 μM NO_3^-, these values were found to be 2.5 mM and 7 μM, respectively.

In these transport assays, subsequent metabolism was not blocked. Therefore, the radioactivity that sedimented with the cells may have been primarily NO_2^- or NH_4^+, rather than NO_3^-. The demonstration that NO_3^- was the predominant form of N in the cellular pellet was crucial to our study. Methanoliç extracts were subjected to HPLC analysis, and the results shown in Table 4. Only NO_3^- was found after 15 seconds of incubation and 99% of the total activity was found in NO_3^- after 45 seconds. Therefore, we conclude that NO_3^- is the transported form of N in this study. It is also noteworthy that NO_3^- is accumulated against a 29-fold concentration gradient, which suggests that transport of NO_3^- is active. The fact that NO_2^- does not appear intracellularly after a 45-second exposure to NO_3^-, although NH_4^+ does, supports the conclusions of our dialysis-culture studies that NO_3^- rather than NO_2^- reduction is rate-limiting under aerobic conditions.

Finally, the observation that NO_3^- can be concentrated against a gradient without artificially blocking reduction strongly indicates that reduction, rather than transport, is rate-limiting. That this represents a specific NO_3^- transport system, rather than gratuitous transport by permeases for some other compounds (e.g.,

Table 4. Intracellular concentration of ^{13}N-labeled metabolites of NO_3^- after 15- and 45-second exposures to $^{13}NO_3^-$ followed by sedimentation through silicone oil and methanolic extraction.[1]

Time in $^{13}NO_3^-$ (seconds)	$[NO_3^-]$	$[NO_2^-]$	$[NH_4^+]$
15	9.9 μM	<2 nM	<2 nM
45	29 μM	<2 nM	0.16 μM

[1]Initial, external $[NO_3^-]$ was 1 μM.

$SO_4^=$ or $SO_3^=$), was suggested by regulation of transport by NH_4^+ (Thayer and Huffaker, manuscript in preparation).

DISCUSSION

We have now demonstrated that NO_3^- transport occurs in Klebsiella pneumoniae (Thayer and Huffaker, 1981) and that at least two transport systems are involved (Thayer and Huffaker, unpublished results). We show here that under both steady-state (i.e., dialysis-culture studies) and initial-rate (^{13}N-studies) conditions, NO_3^- reduction is limiting in the assimilatory conversion of NO_3^- to NH_4^+. Under steady-state conditions, respiratory NO_3^- reduction (anaerobic) overcomes this limitation and NO_2^- reduction (or re-entry transport) becomes rate-limiting.

Future genetic manipulations aimed at increasing the overall rate of NH_4^+-production from NO_3^- in Klebsiella pneumoniae should probably focus on the assimilatory (aerobic) NO_3^- reduction and NO_2^- transport or reduction under anaerobic conditions.

ACKNOWLEDGMENT

We thank Nelinia Henry and Suzanne Epperly for their help in preparing the manuscript. We are deeply indebted to Drs. J. Meeks, K. Krohn, and J. Parks for sharing their cyclotron time and chromatography equipment with us and to Joe Chasko for help with the computer-assisted data reductions for the HPLC analyses. This work was supported in part by the National Science Foundation (Grant Nos. PFR 77-07301 and DEB 77-01199).

REFERENCES

Alexander, M., 1977, An Introduction to Soil Microbiology, 2nd Ed., John Wiley & Sons, New York.

Andersen, K., and Shanmugam, K. T., 1977, Energetics of biological nitrogen fixation: determination of the ratio of formation of H_2 to NH_4^+ catalyzed by nitrogenase of Klebsiella pneumoniae in vivo, J. Gen. Microbiol., 103:107.

Cohen-Bazire, G., Sistrom, W. R., and Stanier, R. Y., 1957, Kinetic studies of pigment synthesis by non-sulphur bacteria, J. Cell Comp. Physiol., 49:25.

Garland, P. B., Downie, J. A., and Haddock, B. A., 1975, Proton translocation and respiratory nitrate reductase of Escherichia coli, Biochem. J., 152:549.

Haddock, B. A., and Jones, C. W., 1977, Bacterial respiration, Bacteriol. Rev., 41:47.

Herbert, R. A., 1980, Nitrogen cycling, with particular reference to nitrate dissimilation in estuarine and marine sediments, Abstracts of the Symposium on the Microbiology of Sediments at Cambridge, March 1980, Soc. Gen. Microbiol. Quart., 7:65.

Hom, S. S. M., Hennecke, H., and Shanmugam, K. T., 1980, Regulation of nitrogenase biosynthesis in Klebsiella pneumoniae: effect of nitrate, J. Gen. Microbiol., 117:169.

John, P., 1977, Aerobic and anaerobic bacterial respiration monitored by electrodes, J. Gen. Microbiol., 98:231.

Kashket, E. R., and Barker, S. L., 1977, Effects of potassium ions on the electrical and pH gradients across the membranes of Streptococcus lactis cells, J. Bacteriol., 130:1017.

Kristajansson, J. K., and Hollocher, T. C., 1979, Substrate binding site for nitrate reductase of Escherichia coli is on the inner aspect of the membrane, J. Bacteriol., 137:1227.

Kristajansson, J. K., Walter, B., and Hollocher, T. C., 1978, Respiration-dependent proton translocation and the transport of nitrate and nitrite in Paracoccus denitrificans and other denitrifying bacteria, Biochemistry, 17:5014.

Parks, N. J., and Krohn, K. A., 1978, The synthesis of ^{13}N labeled ammonia, dinitrogen, nitrite, and nitrate using a single cyclotron target system, Int. J. Appl. Radiat. Isot., 29:754.

Schrader, L. E., Ritenour, G. L., Elrich, G. L., and Hageman, R. H., 1968, Some characteristics of nitrate reductase from higher plants, Plant Physiol., 43:930.

Shanmugam, K. T., Chan, I., and Morandi, C., 1975, Regulation of nitrogen fixation. Nitrogenase-derepressed mutants of Klebsiella pneumoniae, Biochim. Biophys. Acta, 408:101.

Sias, S. R., and Ingraham, J., 1979, Isolation and analysis of mutants of Pseudomonas aeruginosa unable to assimilate nitrate, Arch. Microbiol., 122:263.

Thayer, J. R., and Huffaker, R. C., 1980, Determination of nitrate and nitrite by high-pressure liquid chromatography. Comparison with other methods, Anal. Biochem., 102:110.

Thayer, J. R., and Huffaker, R. C., 1981, Use of ^{13}N-nitrate to study nitrate transport in Klebsiella pneumoniae, in: "Short-Lived Radionuclides in Chemistry and Biology," K. A. Krohn and J. W. Root, eds., Advances in Chemistry Series, Washington, D. C., American Chemical Society (in press).

Tiedje, J. M., Firestone, M. K., Betlach, M. R., Smith, M. S., and Caskey, W. H., 1979, Methods for the production and use of nitrogen-13 in studies of denitrification, Soil Sci. Soc. Am. J., 43:709.

Van't Riet, J., Stouthamer, A. H., and Planta, R. J., 1968, Regulation of nitrate assimilation and nitrate respiration in Aerobacter aerogenes, J. Bacteriol., 96:1455.

DENITRIFICATION BY THE PHOTOSYNTHETIC BACTERIUM, <u>RHODOPSEUDOMONAS SPHAEROIDES</u> FORMA SP. <u>DENITRIFICANS</u>: MECHANISM OF ELECTRON TRANSPORT TO NITRATE AND NITRITE REDUCTION

Toshio Satoh, Eiko Sawada, and Makoto Bannai

Department of Biology
Tokyo Metropolitan University
Tokyo, Japan

INTRODUCTION

Our laboratory is concerned with water quality and purification with special emphasis on the role of photosynthetic bacteria. Recently, we have isolated a photosynthetic bacterium from polluted water which possesses an active denitrification system (Satoh et al., 1974; Satoh et al., 1976), a finding which has opened the door to the studies of the mechanism of denitrification and its relationship to photosynthesis in this organism.

The new phototroph exhibits denitrification as a biological energy-transforming system in addition to oxygen-respiration and photosynthesis common to other photosynthetic bacteria. This organism is capable of both denitrification and nitrogen fixation, two major components of the biological nitrogen cycle. The isolation of this photodenitrifier raises the following questions: 1) how is the denitrification affected by light?; 2) what kinds of relationships exist between electron transport pathways and energy-transforming systems of photosynthesis, denitrification and oxygen-respiration? In this paper we focus on these questions with emphasis on the nature of the electron transport system linking denitrification.

MATERIALS AND METHODS

Organism and Growth Conditions

A denitrifying phototrophic bacterium, <u>Rhodopseudomonas</u>

sphaeroides forma sp. denitrificans IL106 (Satoh et al., 1976) and the green mutant strain of this organism, which was isolated as a green colony under anaerobic-light conditions after UV treatment, were used. Cells were cultured anaerobically in light in the presence of 0.2% KNO_3 as described before (Satoh, 1977).

Denitrification Activity

Manometric measurements were made at 30 C using conventional Warburg vessels under a gas phase of 100% He as described before (Satoh, 1977). The light intensity was 4000 lux using Tungsten lamps which was sufficient for maximum inhibition of denitrification using lactate as the electron donor.

Spectrophotometric Measurement

Difference spectra of cytochromes *b* and *c* and the kinetics of their oxidation were determined using a Hitachi 356 spectrophotometer. The kinetics of cytochromes *b* and *c* oxidation were determined from absorption changes at 560 nm and 550 nm, respectively, using 572 nm as reference wave length. All measurements were carried out under He at 30 C.

Materials

HOQNO (2-n-heptyl-4-hydroxyquinoline-N-oxide, Sigma Chemical Co.) was dissolved in ethanol. KCN (potassium cyanide) was dissolved in distilled water.

RESULTS AND DISCUSSION

Effect of Light and Electron Donors on NO_3^- Reduction to N_2

Photosynthetic bacteria such as *R. sphaeroides* utilize a variety of organic substances for growth. In the course of studying the effect of light on the denitrification of NO_3^- it was found that light was inhibitory depending on the electron donors used. As shown in Figure 1, the denitrification of NO_3^- is linear with time. The rate of N_2 production at 30 C in the dark is 0.6-0.7 μmoles·hr^{-1}·mg cell dry $weight^{-1}$ irrespective of electron donors. Denitrification of NO_3^- using succinate, malate and pyruvate as the electron donors, respectively, is independent of illumination. However, when lactate or glucose is used denitrification is severly inhibited by illumination, but is restored in the second dark period, showing that the inhibitory effect of light is reversible.

Effect of Light and Electron Donors on NO_2^- Reduction to N_2

When NO_2^- is used as the substrate of denitrification to N_2 (Fig. 2), denitrification is found to be non-linear with time and

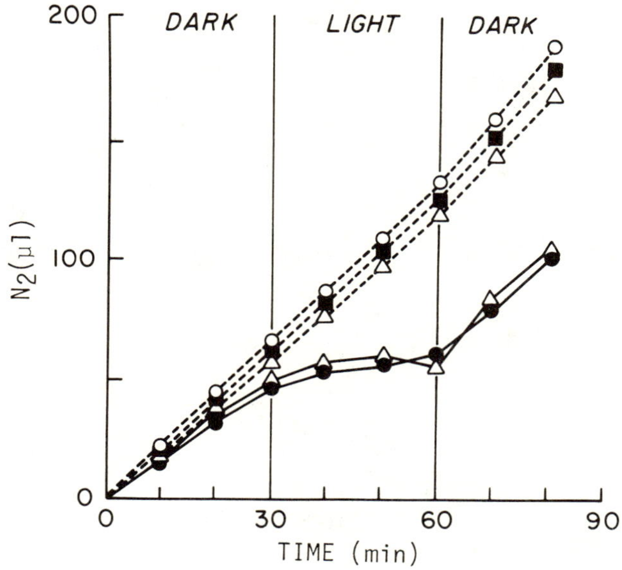

Fig. 1. Effects of illumination on the denitrification of nitrate with various electron donors under alternating dark and light phases. Cells (8 mg dry weight) at the exponential phase of growth were used. Cells were incubated in the presence of various electron donors in the Warburg vessels under He for 30 min and at zero time NO_3^- was added from the side-arm of the vessel. --O-- succinate; --Δ-- malate; --■-- pyruvate; —●— lactate; —Δ— glucose.

and greater than that of NO_3^- (the rate of N_2 formation from NO_2^- at 30 C during the first 5 min dark period is about 2.1 μmoles·hr^{-1}·mg cell dry weight^{-1} using either lactate or succinate as the electron donor). Although denitrification of NO_2^- occurs either in light or dark, it clearly shows that denitrification is inhibited by light irrespective of the electron donor.

Effect of Electron Transport Inhibitors and Energy Uncouplers on the Denitrification of NO_3^-

To elucidate the mechanism of inhibition of denitrification of NO_3^- (see Fig. 1), the effects of inhibitors of electron transport and oxidative phosphorylation were studied. Rotenon and amytal, inhibitors of NADH dehydrogenase (Izawa and Good, 1972), inhibited denitrification and utilization of NO_3^- when lactate

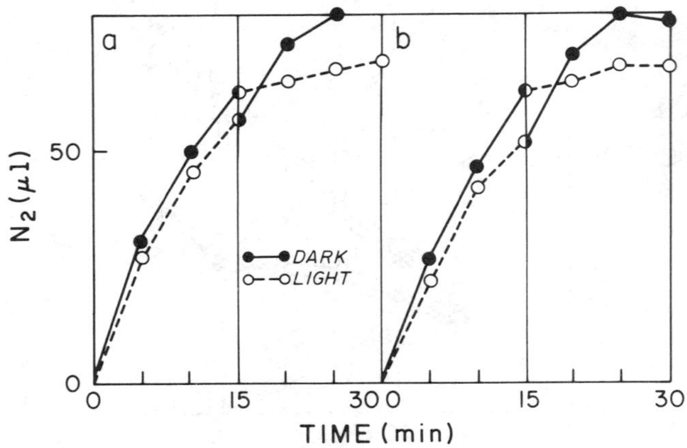

Fig. 2. Effects of illumination on the denitrification of nitrite using succinate or lactate as the electron donor. Cells (8 mg dry weight) at the exponential phase of growth were used. The experiment started with both dark and light phases (15 min) and then followed by light and dark phases (15 min), respectively. Figures are shown separately for lactate (a) and succinate (b). Other experimental procedures were the same as those described in Figure 1 except that $NaNO_2$ was used instead of KNO_3. --O-- light phase; —●— dark phase.

was present in dark, but not when succinate was used in light. Gramicidin D, an uncoupler (Izawa and Good, 1972) accelerated these reactions when lactate was used as an electron donor in dark, but partially inhibited these reactions when succinate was used in light. Antimycin A, an inhibitor of cyclic electron transport between cytochromes b and c_2 (Nishimura, 1963) strongly inhibited the denitrification of NO_3^-. The photochemical reduction of NAD by photosynthetic bacteria coupled to the oxidation of succinate or H_2 has been demonstrated and the mechanism of the photoreduction has been concluded to be an energy-linked reverse electron flow (Gest, 1972). Therefore, it is likely that NADH is a direct electron donor for electron transport from lactate or glucose and that the inhibition by light results from competition between NO_3^- reduction and energy-dependent reverse electron flow for NAD reduction as discussed by Oelze and Weaver (1971) for light inhibition by oxygen uptake in *Rhodospirillum rubrum*.

Kinetic Studies of Oxidation of Cytochromes b and c by NO_3^- and NO_2^-

In many bacteria which exhibit nitrate respiration or denitrification, a B-type cytochrome functions as the electron carrier for nitrate reduction (Stouthamer, 1976) and a C-type cytochrome is thought to be an immediate electron donor (Yamanaka and Okunuki, 1963; Miyata and Mori, 1969). The mechanism of electron transport involved with NO_3^- and NO_2^- reduction and its relationship to photosynthetic cyclic electron transport in this photodenitrifier were studied using cell suspensions. The kinetics of cytochrome b and c oxidation and reduction were monitored spectrophotometrically. In these studies, a green mutant strain of this photodenitrifier with the carotenoid absorption bands shifted to shorter wave lengths was used to avoid the disturbance caused by carotenoid-shift. The carotenoid-shift is thought to be due to the high energy state of the cell membrane which results in oxidative phosphorylation (Jackson and Crofts, 1971).

Succinate-reduced minus NO_3^- or NO_2^--oxidized difference spectra (Fig. 3) show that NO_3^- oxidizes cytochrome b (α band, 560 nm) preferentially over cytochrome c (α band, 550 nm) (compare with the difference spectrum with ferricyanide in Fig. 3a) and NO_2^- oxidizes both cytochrome b and c equally (Fig. 3b). In the next experiments the kinetics of cytochromes b and c oxidation by NO_3^- and NO_2^- in the presence of KCN was studied (Fig. 4). This inhibitor at the concentration of 2.5 mM almost completely inhibits the nitrite reductase from this strain (Sawada et al., 1978) and cytochrome oxidation (Fig. 4f). In the presence of KCN, where only NO_3^- reduction to NO_2^- occurs, NO_3^- oxidizes only cytochrome b (Fig. 4, c and d) although oxidation of cytochrome b is reduced because of an inhibitory effect of KCN on nitrate reductase. These results suggest that cytochrome b is involved in the NO_3^- reduction and that cytochromes b and c are involved in the NO_2^- reduction.

In the next experiment the effect of HOQNO (2-n-heptyl-4-hydroxyquinoline-N-oxide), an inhibitor of cytochrome b oxidation (Nishimura, 1963), on the patterns of oxydation by NO_3^- and NO_2^- was studied (Fig. 5). It is found to inhibit the oxidation of cytochrome b by NO_3^- (Fig. 5a), but not that of cytochromes b and c by NO_2^- (Fig. 5b). Since HOQNO had no inhibitory effect on the nitrate reductase of this denitrifier (assayed with benzyl viologen as an electron donor) the HOQNO effect suggests that the cytochrome b involved in the NO_3^- reduction is different from the one involved in the NO_2^- reduction.

It was reported that cells of *R. sphaeroides* grown photosynthetically had at least three kinds of cytochrome b (Dutton and

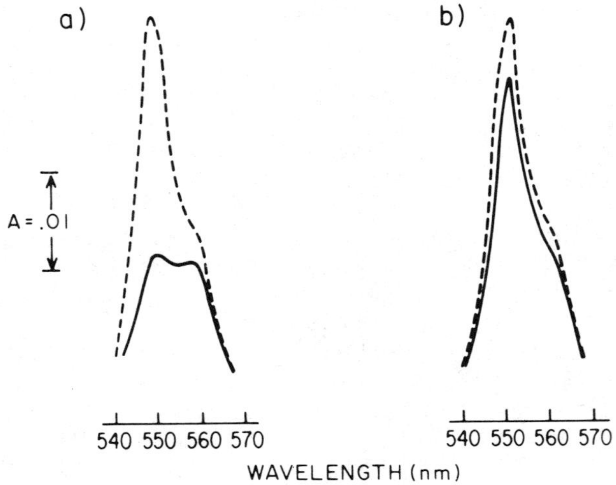

Fig. 3. Succinate reduced minus NO_3^- or NO_2^- oxidized difference spectra of a green mutant strain of R. sphaeroides forma sp. denitrificans. After cells (2 mg cell weight·ml^{-1}) were incubated with 1.5 mM succinate in the 0.1 M K-phosphate (pH 7.0) for 30 min, NO_3^- (2.9 mM) or NO_2^- (2.9 mM) were added to the cell suspension. Ferricyanide (2.6 mM) was used to see the maximum level of cytochromes oxidized (broken line). a) Succinate reduced minus NO_3^- oxidized; b) succinate reduced minus NO_2^- oxidized.

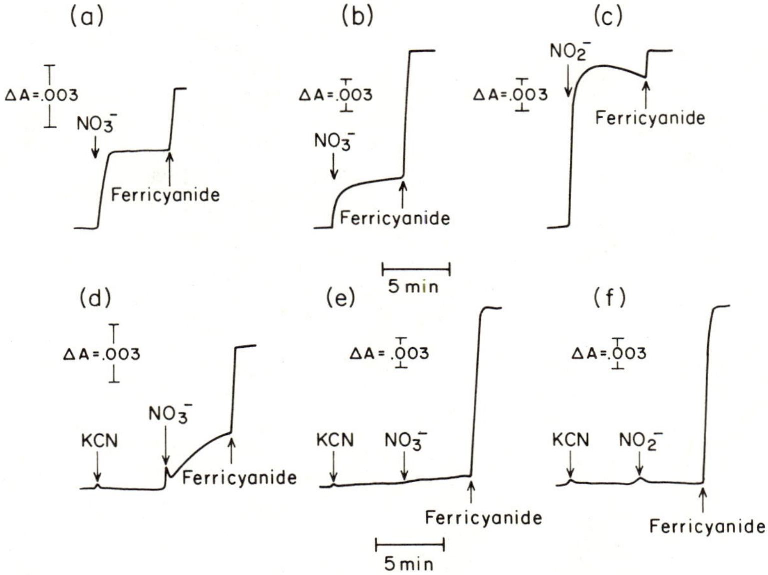

Fig. 4. Time course of oxidation of cytochrome b and c. After cells (2 mg dry weight·ml^{-1}) of a green mutant strain were incubated with 1.5 mM succinate in the 0.1 M K-phosphate (pH 7.0) for 30 min, NO_3^- (2.9 mM), NO_2^- (2.9 mM) or KCN (2.5 mM) were added to the cell suspension. Ferricyanide (2.6 mM) was used to see the maximum level of cytochromes oxidized. a) and d), Cytochrome b, NO_3^-; b) and e), cytochrome c, NO_3^-; c) and f), cytochrome c, NO_2^-.

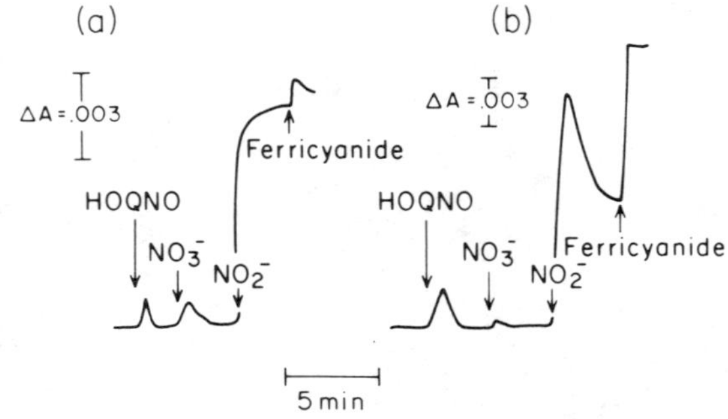

Fig. 5. Effect of NOQNO on the oxidation of cytochromes b and c by NO_3^- and NO_2^-. After cells (2 mg dry weight·ml^{-1}) of a green mutant strains were incubated with 1.5 mM succinate in the 0.1 M K-phosphate (pH 7.0) for 30 min, HOQNO (150 μM) and NO_3^- (2.9 mM) or NO_2^- (2.9 mM) were added to the cell suspension. Ferricyanide (2.6 mM) was added to see the maximum level of cytochromes oxidized. a) Cytochrome b; b) cytochrome c.

Jackson, 1972; Conelly et al., 1973). Next we plan to investigate what type of cytochrome b may be involved in NO_3^- reudction.

Role of Cytochrome c_2 in NO_2^- Reduction

Cytochrome c_2, which is common to Rhodospirillaceae and is involved in the photosynthetic cyclic electron transport as the electron donor to the reaction center bacteriochlorophyll, P870, is known to serve as an electron donor for purified nitrite reductase of this strain (Sawada et al., 1978). Both nitrite reductase and cytochrome c_2 of this photodenitrifier are found in the periplasmic space as soluble proteins (Sawada and Satoh, 1980). If cytochrome c_2 is the physiological electron donor for NO_2^- reduction, this periplasmic fraction should be able to catalyze NO_2^- reduction without any exogenous addition of electron donors. When the periplasmic fraction was incubated for 1 h with NO_2^- and a small amount of dithionite, the fraction showed nitrite reduction and the typical difference spectrum of cytochrome c_2.

These results suggest that cytochrome c_2 is the immediate electron donor for nitrite reductase. Therefore, the inhibition of denitrification of NO_2^- by light using succinate or lactase as the electron donor (Fig. 2) is postulated to be due to competition for electrons at the site of cytochrome c_2 between the reaction center bacteriochlorophyll, P870, and nitrite reductase. Also the effect of antimycin A might be explained as follows: antimycin A may inhibit at a site between cytochromes b and c_2 blocking NO_2^- reduction, thus resulting in inhibition of denitrification. Therefore, antimycin A does not inhibit utilization of NO_3^-.

SUMMARY

Electron transport systems of photosynthesis and denitrification in this photodenitrifier appear to be closely linked. Light appears to inhibit denitrification by competitively partitioning electrons between denitrification and photosynthesis. There may be at least two sites of inhibition, energy dependent NAD reduction and cytochrome $\underline{c}_2$ oxidation.

A cytochrome b appears to be involved in NO_3^- reduction and cytochrome $\underline{c}_2$ is thought to be the immediate electron donor for NO_2^- reductase.

A cytochrome b involved in NO_3^- reduction is thought to be different from the one which is involved in the NO_2^- reduction.

ACKNOWLEDGMENTS

We thank Dr. R. C. Valentine, Plant Growth Laboratory, University of California, Davis, for his advice on the manuscript (T.S.). We also thank Dr. Kitamura for his encouragement and advice through this work.

REFERENCES

Conelly, J. L., Jones, O. T. G., Saunders, U. A., and Yates, D. W., 1973, Kinetic and thermodynamic properties of membrane-bound cytochromes of aerobically and photosynthetically grown Rhodopseudomonas sphaeroides, Biochim. Biophys. Acta, 292:644.

Dutton, J. L., and Jackson, J. B., 1972, Thermodynamic and kinetic characterization of electron-transfer components in situ in Rhodopseudomonas spheroides and Rhodospirillum rubrum, Eur. J. Biochem., 30:495.

Gest, H., 1972, Energy conversion and generation of reducing power in bacterial photosynthesis, Adv. Microbial. Physiol., 7:243.

Izawa, S., and Good, N. E., 1972, Inhibition of photosynthetic electron transport and photophosphorylation, in: "Methods in Enzymology," Vol. 7, A. San Pietro, ed., Academic Press, New York-London.

Jackson, J. B., and Crofts, A. R., 1971, The kinetics of light induced carotenoid changes in Rhodopseudomonas spheroides and their relation to electric field generation across the chromatophore membrane, Eur. J. Biochem., 18:120.

Miyata, M., and Mori, T., 1971, Studies on denitrification X. The "denitrifying enzyme" as a nitrite reductase and the electron donating system for denitrification, J. Biochem., 66:463.

Nishimura, M., 1963, Studies on the electron transfer system in photosynthetic bacteria. II. The effect of heptylhydroxyquinoline-N-oxide and antimycin A on the photosynthetic and respiratory electron-transfer systems, Biochim. Biophys. Acta, 66:17.

Oelze, J., and Weaver, P., 1971, The adjustment of photosynthetically grown cells of Rhodospirillum rubrum to aerobic light conditions, Arch. Mikrobiol., 79:108.

Satoh, T., Hoshino, Y., and Kitamura, H., 1974, Isolation of denitrifying photosynthetic bacteria, Agr. Biol. Chem., 38:1749.

Satoh, T., Hoshino, Y., and Kitamura, H., 1976, Rhodopseudomonas sphaeroides forma sp. denitrificans, a denitrifying strain as a subspecies of Rhodopseudomonas sphaeroides, Arch. Microbiol., 108:265.

Satoh, T., 1977, Light-activated, -inhibited and -independent denitrification by a denitrifying phototrophic bacterium, Arch. Microbiol., 115:293.

Sawada, E., Satoh, T., and Kitamura, H., 1978, Purification and properties of a dissimilatory nitrite reductase of a denitrifying phototrophic bacterium, Plant Cell Physiol., 19:1339.

Sawada, E., and Satoh, T., 1980, Periplasmic location of dissimilatory nitrate and nitrite reductases in a denitrifying phototrophic bacterium, Rhodopseudomonas sphaeroides forma sp. denitrificans, Plant Cell Physiol., 21:205.

Stouthamer, A. H., 1976, Biochemistry and genetics of nitrate reductase in bacteria, Adv. Microbiol. Physiol., 14:315.

Yamanaka, T., and Okunuki, K., 1963, Crystalline pseudomonas cytochrome oxidase. I. Enzymic properties with special reference to the biological specificity, Biochim. Biophys. Acta, 67:379.

PRODUCTION OF NITROUS OXIDE AS A PRODUCT OF NITRITE METABOLISM BY ENTERIC BACTERIA

T. Satoh, S. S. M. Hom, and K. T. Shanmugam*

Plant Growth Laboratory
University of California
Davis, CA 95616

INTRODUCTION

Biological denitrification is part of the global nitrogen cycle in which NO_3^- and NO_2^- are converted to gaseous products such as nitrous oxide (N_2O) and N_2 (Payne 1973, and this volume; Delwiche and Bryan, 1976; Knowles, 1978). Several microorganisms are known to denitrify NO_2^- only to nitrous oxide (an environmental pollutant), while in other organisms N_2 is the final product of denitrification. Although there is a large body of literature in the area of biological denitrification, very little is known about the organization and the mechanism of regulation of the genes involved in this process (see also Carlson and Ingraham, this volume; Thayer and Huffaker, this volume).

The strategy of utilization of enteric bacteria as a tool for studying the molecular genetics of denitrification benefits greatly from the wealth of genetic knowledge about these organisms, but suffers (as seen below) from the fact that these organisms catalyze only a partial denitrification reaction terminating at N_2O. However, the steps involving the conversion of NO_2^- to N_2O may be cardinal to the process.

Several investigators reported that Escherichia coli is capable of reducing NO_2^- to NH_4^+ (see Cole and Brown, 1980, for a review). Cole and co-workers (Cole, 1978; Cole and Brown, 1980)

*Present Address: Dept. of Microbiology and Cell Science, University of Florida, Gainesville, FL 32611.

reported that this process continues rapidly even when substantial quantities of NH_4^+ have accumulated in the medium and as a result, NO_2^- can serve as the sole nitrogen source for growth. However, E. coli and other enteric bacteria are not known to produce products other than NH_4^+ from NO_2^-, although preliminary reports show that these organisms are capable of producing some nitrous oxide (Tiedje et al., 1979). In this communication we report evidence that several enteric bacteria including E. coli, Salmonella typhimurium and Klebsiella pneumoniae have evolved mechanisms for NO_2^- reduction yielding both NH_4^+ as well as N_2O and all three organisms are capable of growth using NO_2^- as the sole nitrogen source.

MATERIALS AND METHODS

Bacterial Strains

Wild type strains of Escherichia coli K10, Salmonella typhimurium LT-2 and Klebsiella pneumoniae M5A1 were used.

Bacterial Media and Growth Conditions

Culture media used were L-broth which contained per liter, 10 g Bactotrypton (Difco), 5 g Bacto yeast extract (Difco), and 10 g NaCl, and minimal medium which contained per liter, 6.25 g Na_2HPO_4, 0.75 g KH_2PO_4, 2.0 g NaCl, 0.01 g $FeSO_4 \cdot 7H_2O$, 0.01 g $Na_2MoO_4 \cdot 2H_2O$, 0.2 g $MgSO_4 \cdot 7H_2O$, and 15 g glucose. For K. pneumoniae sucrose (15 g) replaced glucose. As a nitrogen source, $NaNO_3$ or $NaNO_2$ (0.05%) was used. For anaerobic growth experiments, 20 ml of growth medium in a 71 ml serum bottle was inoculated with 0.2 ml of a starter culture of the appropriate organism and the gas phase was replaced with argon. For aerobic growth conditions, 20 ml of growth medium in a 125 ml flask was inoculated with 0.2 ml of a starter culture, and incubated aerobically in a waterbath shaker. Cultures for inoculation were grown aerobically containing a limiting concentration of NH_4^+ (50 $\mu g \cdot ml^{-1} (NH_4)_2SO_4$) and incubated overnight to exhaust the nitrogen source. This procedure eliminated any carry over of organic or inorganic nitrogen with the inoculum.

Analytical Methods

Ammonium ion was detected as described previously (Hom et al., 1980). N_2O was determined using a gas chromatograph equipped with a porapak R (80-100 mesh) column and ^{63}Ni-electron capture detector. The carrier gas was 5% methane in argon at a flow rate of 30 $ml \cdot min^{-1}$. The amount of N_2O produced was estimated by combining the N_2O in the gas phase and the appropriate (calculated) value of N_2O solublized in the aqueous phase. The amount of NH_4^+ utilized for growth was calculated from a previously determined relationship between cell yield and NH_4^+ concentration.

Identification of the Gaseous Product of NO_2^- Reduction as N_2O

Two different methods were used to identify the gaseous product of NO_2^- reduction as N_2O. The retention times of pure samples of N_2O (Matheson Gas Products) were compared to the retention times of the gas samples from the cultures in a gas chromatograph. The production of both NO and N_2O in cultures of *K. pneumoniae* exposed to iostopic nitrogen ($^{15}NO_3^-$ or $^{15}NO_2^-$) were verified by mass spectrometry (Hom et al., manuscript in preparation). Upon exposure to $^{14}NO_3^-$ or $^{14}NO_2^-$, *K. pneumoniae* strain M5A1 produced N_2O which was identified in the mass spectrometer by the characteristic peaks for N_2O at masses 44, 45, and 45 and at 28, 29, and 30. When $^{15}NO_3^-$ or $^{15}NO_2^-$ was added to the culture medium, the major abundances at masses 46 and 30 were further verified by the presence of N_2O

RESULTS

Kinetics of the Anaerobic Conversion of NO_3^- and NO_2^- to NH_4^+ and N_2O

Two types of NO_3^- and NO_2^- reduction systems have been recognized in several bacteria, the assimilatory and dissimilatory pathways (Payne, 1973). The assimilatory pathway utilizes NO_3^- or NO_2^- as the sole nitrogen source for the biosynthesis of all the nitrogen containing compounds of the cell. The assimilatory process functions under both aerobic and anaerobic conditions and is regulated by NH_4^+. During this process NO_2^- or NH_4^+ do not accumulate in the medium. On the other hand, dissimilatory reduction is a rapid process yielding NO_2^-, N_2O or N_2 which accumulates in the culture medium. This process is regulated by O_2. The experiments described in this section focus on the kinetics and subsequent appearance of the various reduction products of NO_3^- and NO_2^- produced by cultures of *E. coli*, *S. typhimurium* and *K. pneumoniae*.

All three organisms grew under anaerobic conditions using NO_3^- and NO_2^- as the sole nitrogen source. Moreover, each organism reduced NO_3^- and NO_2^- and accumulated both NH_4^+ in the medium and N_2O in the gas phase. The growth rate, final cell yield, and rates of production of NH_4^+ and N_2O at 30 C and 37 C, both in the presence and absence of yeast extract are summarized in Table 1. When *E. coli* and *S. typhimurium* were grown using NO_3^- as the nitrogen source, the growth rates declined as NO_2^- accumulated in the medium. However, an increase in the growth rate was observed upon initiation of NO_2^- reduction to NH_4^+ and N_2O. Therefore the growth of *E. coli* and *S. typhimurium* utilizing NO_3^- as a nitrogen source was found to be biphasic.

The production of NH_4^+ and N_2O from NO_3^- by all three strains commenced following NO_2^- accumulation to a maximal level of 4

Table 1. Rate of NH_4^+ and N_2O formation from NO_2^- by enteric bacteria. *E. coli* strain K-10, *S. typhimurium* strain LT-2 and *K. pneumoniae* strains M5A1 were grown anaerobically as described in Materials and Methods. Samples were removed at time intervals with syringes for cell density (OD_{420}), NH_4^+ and N_2O determinations. Yeast extract was added where indicated at the initial concentration of 1 mg/ml. The reported values were calculated from the data at the exponential phase of growth.

Strain	Growth conditions			Final cell yield (OD_{420})	Growth rate constant (h^{-1})	Rate of product formation ($\mu mol \cdot h^{-1} \cdot mg$ cell protein^{-1})		
	N-source	Temperature (C)	Yeast extract			NH_4^+ in the medium	NH_4^+ in the cell	N_2O
E. coli	NO_3^-	30	-	2.0	.044			
		37	-	1.7	.091			
		30	+	5.3	.69			
	NO_2^-	30	-	1.6	.037	1.1	.42	.065
		37	-	1.6	.095	.94	1.04	.077
		30	+	4.6	.41	4.4	4.6	.060
S. typhimurium	NO_3^-	30	-	2.9	.18			
		37	-	3.5	.21			
		30	+	4.8	.63			
	NO_2^-	30	-	2.6	.067	3.4	.76	.13
		37	-	2.6	.23	3.1	2.6	.62
		30	+	4.5	.49	6.8	5.5	.069
K. pneumoniae	NO_3^-	30	-	5.4	.61			
		37	-	4.8	.73			
		30	+	8.0	1.14			
	NO_2^-	30	-	5.4	.47	5.5	5.0	.25
		37	-	4.3	.50	3.2	5.3	.23
		30	+	8.0	.64	5.8	7.1	.20

µmoles·ml^{-1} in both E. coli and S. typhimurium and 1 µmol·ml^{-1} for K. pneumoniae. Reduction product formation continued until NO_2^- in the medium was exhausted. When cells were grown in a medium containing NO_2^-, a lag period of about 2 hrs was observed before NH_4^+ and N_2O production began. This lag phase is probably due to the time needed to acclimate the cultures to anaerobic metabolism. As summarized in Table 1, the rate of N_2O produced by the three strains in the presence of yeast extract was comparable to the rate without yeast extract, whereas growth rates in the presence of yeast extract was considerably higher (compare lines 4 and 6 and lines 10 and 12 in Table 1). These results suggest that N_2O production is not controlled by NH_4^+.

In all three organisms, N_2O production accounts for about 5-18% of the total NO_2^- reduced, while NH_4^+ production accounts for the remainder. Therefore NH_4^+ production is the major pathway of NO_2^- removal from the culture medium.

Kinetics of the Aerobic Conversion of NO_3^- and NO_2^-

E. coli and S. typhimurium failed to grow aerobically with NO_3^- as the sole nitrogen source (Table 2). This is due to the fact that the dissimilatory nitrate reductase was not synthesized under aerobic conditions. On the other hand, NO_2^-, as the sole nitrogen source, supported growth in all three organisms. Growth rates for E. coli and S. typhimurium were relatively low compared to K. pneumoniae (Table 2, lines 3, 4, 7, 8, 11 and 12). N_2 was not detected in the gas phase under aerobic conditions. This result suggests that N_2O formation is controlled by O_2. Although production of NH_4^+ (for growth) occurred aerobically, no detectable levels of NH_4^+ accumulated in the medium.

N_2O Production by Mutant Strains of K. pneumoniae

A mutant approach has been used to study the mechanism of N_2O production by enteric bacteria. We have isolated two classes of NO_2^- reduction defective mutants of K. pneumoniae. The first class of mutants which are blocked in N_2O production carried defects in either the formate dehydrogenase or pyruvate-formate lyase complex; the latter strains require acetate for anaerobic growth. These mutant strains were resistant to growth inhibitory concentrations of chlorate and defective in anaerobic NO_2^- respiration. Although N_2O production from NO_2^- was blocked in these strains, NH_4^+ production from NO_2^- was unaffected.

The second class of NO_2^- reduction defective mutants of K. pneumoniae were blocked in NH_4^+ production from NO_2^- without any effect on N_2O production; one strain was studied in detail. The rate of N_2O production by these strains accounted for almost 100% of the rate at which NO_2^- was removed from the medium. Since the

Table 2. Growth of enteric bacteria in NO_3^- or NO_2^- containing media under aerobic conditions. E. coli strain K-10, S. typhimurium strain LT-2, and K. pneumoniae strain M5A1 were grown anaerobically as described in Materials and Methods. Samples were removed at time intervals for cell density (OD_{420}), NH_4^+ and N_2O determinations. The reported values were calculated from the exponential phase of growth.

Strain	Growth conditions				Rate of product formation (μmol·h^{-1}·mg cell $protein^{-1}$)		
	N-source	Temperature	Final cell yield	μ (h^{-1})	NH_4^+ in the medium	NH_4^+ in the cell	N_2O
E. coli	NO_3^-	30	0	0	UD*	UD	UD
		37	0	0	UD	UD	UD
	NO_2^-	30	2.2	.0073	UD	.083	UD
		37	2.2	.012	UD	.13	UD
S. typhimurium	NO_3^-	30	0	0	UD	UD	UD
		37	0	0	UD	UD	UD
	NO_2^-	30	4.0	.025	UD	.26	UD
		37	3.4	.036	UD	.36	UD
K. pneumoniae	NO_3^-	30	5.0	.33	UD	3.5	UD
		37	5.0	.61	UD	6.2	UD
	NO_2^-	30	5.5	.89	UD	8.1	UD
		37	5.6	1.38	UD	10.1	UD

* = undetectable

rate of NO_2^- utilization was less in this mutant strain than its parent, this strain was found to be more sensitive to NO_2^- toxicity. This sensitivity to NO_2^- was used as a selection for mutants of this phenotype. Starting with these strains, we have isolated mutant strains which were blocked in N_2O production. Several classes of these mutants are currently under investigation to elucidate the biochemical and genetic lesions in the NO_2^- reduction and N_2O production pathways.

DISCUSSION

The results presented in Tables 1 and 2 suggest that there are phenotypically at least three kinds of NO_2^- reduction systems in enteric bacteria; assimilatory NO_2^- reduction to NH_4^+ (regulated by NH_4^+), dissimilatory NO_2^- reduction to NH_4^+ (regulated by O_2), and dissimilatory NO_2^- reduction to N_2O (controlled by O_2). However, the possibility exists that N_2O production by enteric bacteria is a side reaction catalyzed by the nitrite reductase to NH_4^+ system and as such does not represent a true dentrification reaction. Results discussed below suggest that N_2O production in these organisms is indeed through a separate denitrification reaction:

1) Although the rate of NH_4^+ production from NO_2^- at 30 C is stimulated by the presence of yeast extract in all three organisms, the rate of N_2O production is not affected by addition of yeast extract (Table 1, lines 4, 6, 10, 12, 16 and 18).

2) Mutant strains of *K. pneumoniae* have been isolated that are defective in the reduction of NO_2^- to NH_4^+, but which produce N_2O at rates comparable to the parental strain.

3) Anaerobic pyruvate degradation mutant strains of *K. pneumoniae*, which are defective in nitrate-respiration and either pyruvate-formate lyase or formate dehydrogenase complex produce NH_4^+ at normal rates. However, these mutants are defective in N_2O production.

These results suggest that *K. pneumoniae* utilizes formate as electron donor for N_2O production and NADH for NH_4^+ production (as suggested by Cole and Brown, 1980, for *E. coli*). Abou-Jaoudé and co-workers (1977) showed that *E. coli* (whole cells) is capable of utilizing formate as electron donor for nitrite reduction. It is difficult to visualize an enzyme system (NO_2^- reductase to NH_4^+) capable of utilizing two independent electron donors for production of different products from the same substrate. The simplest interpretation of all these results is that there are two separate NO_2^- reductases under anaerobic conditions, NO_2^- reductase whose catalytic product is NH_4^+ and a separate NO_2^- reductase producing N_2O

as the product. It is not known at this time whether the aerobic NO_2^- reduction system to NH_4^+ is the same as the anaerobic one.

Demonstration of N_2O production by enteric bacteria raises new questions about the role of these organisms in the removal of NO_3^- and NO_2^- from the environment, their significance in the production of atmospheric N_2O, and their impact on the geochemical nitrogen cycle. Also, the capacity of intestinal microorganisms such as E. coli to reduce nitrite has significant implications in the area of public health where more must be learned about how nitrites as potential mutagens and carcinogens are removed in nature. Finally, the door is now open to biochemical genetic studies of denitrification using E. coli and other enteric bacteria with their wealth of molecular biological information.

ACKNOWLEDGMENTS

We are indebted to Dr. R. C. Valentine for encouragement, advice and support, and Dr. J. Ingraham for helpful discussions. We also thank Drs. C. C. Delwiche and B. A. Bryan for the mass spectromic analyses. This work was supported by grants from NSF (PFR 77-07301) and SERI (XR-9-0836-1).

REFERENCES

Abou-Jaoudé, A., Chippaux, M. C., Pascal, M. C., and Casse, F., 1977, Formate: a new electron donor for nitrite reduction in Escherichia coli K-12, Biochem. Biophys. Res. Commun., 78:579.

Cole, J. A., 1978, The rapid accumulation of large quantities of ammonia during nitrite reduction by Escherichia coli, FEMS Microbiol. Lett., 4:327.

Cole, J. A., and Brown, C. M., 1980, Nitrite reduction to ammonia by fermentative bacteria: a short circuit in the biological nitrogen cycle, FEMS Microbiol. Lett., 7:65.

Delwiche, C. C., and Bryan, B. A., 1976, Denitrification, Ann. Rev. Microbiol., 30:241.

Hom, S. S. M., Hennecke, H., and Shanmugam, K. T., 1980, Regulation of nitrogenase biosynthesis in Klebsiella pneumoniae: effect of nitrate, J. Gen. Microbiol., 117:169.

Knowles, R., 1978, Common intermediates of nitrification and denitrification and the metabolism of nitrous oxide, in: "Microbiology-1978," D. Schlessinger, ed., Amer. Soc. Microbiol., Washington.

Payne, W. J., 1973, Reduction of nitrogenous oxides by microorganisms, Bacteriol. Rev., 37:409.

Tiedje, J. M., Caskey, N. V., Smith, M. S., Bleakley, B. H., and Firestone, R. B., 1979, Nitrous oxide production by bacteria that reduce nitrate to ammonium, Agron. Abstracts.

USE OF NITROGEN-13 AND NITROGEN-15 IN STUDIES ON THE DISSIMILATORY FATE OF NITRATE

James M. Tiedje

Departments of Crop and Soil Sciences and of
Microbiology and Public Health
Michigan State University
East Lansing, MI 48824

INTRODUCTION

Nitrate is a central nitrogen species in the biogeochemical nitrogen cycle because it is mobile, a biological oxidant, and a nitrogen source for growth for many organisms. As a result it has many competing fates as illustrated in Figure 1.

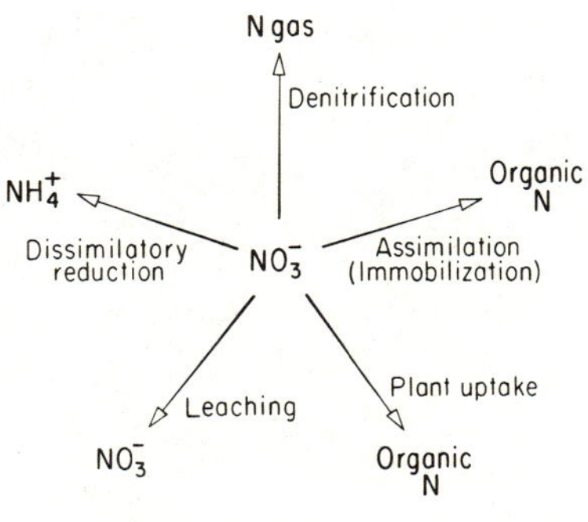

Fig. 1. Competing fates of nitrate in soil.

The top three processes are microbial and occur in soil. The subsequent fate of the metabolized N depends upon the chemical properties of the product: nitrogen gas diffuses out of the soil, ammonium is retained on clays via cation exchange, and organic-N becomes immobilized as part of soil organic matter though portions of it are slowly mineralized. The bottom two fates transport N out of the soil: plant uptake moves nitrogen to the surface where it is grazed or harvested though a portion is returned to the soil as organic nitrogen, and leaching carries nitrate below the rooting zone where it is often lost to groundwater.

In agriculture, people attempt to manage these competing fates for a more favorable economic product while in natural ecosystems a quasi steady state has evolved in which stability is the optimized product. People use the imput of energy to alter the natural steady state. Agricultural practices of the past -- plant variety development, cultivation, water management, weed control, fertilization -- all affect this competition. The goals of the future are basically no different; they are to continue to alter the fate of nitrogen so that products are favored for human use. Obviously, one wants to increase plant recovery of nitrogen at the expense of the permanent losses, especially denitrification.

The four biological processes in Figure 1 use two basic pathways which are illustrated below (Fig. 2).

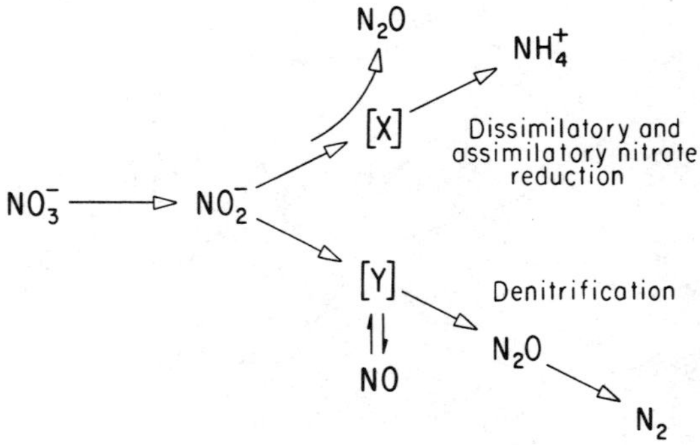

Fig. 2. The two basic pathways of nitrate reduction and the free intermediate products in each.

The first reaction, catalyzed by nitrate reductase, is common to all pathways though there seem to be a number of different enzymes in the many different organisms accomplishing this step. The second reaction is the key step since it is the branch point between pathways which conserve nitrogen and the one that loses it, denitrification. The reaction producing ammonia can be either dissimilatory or assimilatory. The former is under O_2 regulation and is insensitive to ammonium, while the latter is regulated by ammonium and is insensitive to O_2 (Cole and Brown, 1980). The dissimilatory reduction produces free ammonium because the rate of reduction is in excess of the assimilatory needs of the organism. This is because the dissimilatory process is linked to the central oxidative metabolism of the cell while the assimilatory process is balanced with growth.

The gaseous intermediates produced by the two pathways are also summarized in Figure 2. The denitrification pathway produces NO, N_2O, and N (Payne, 1973). It is now generally accepted that N_2O is a freely diffusable, obligate intermediate in denitrification (John and Hollocker, 1977; Firestone et al., 1980). The role of NO is less clear (John and Hollocher, 1977; Zumft and Vega, 1979) though in our recent work we found lines of evidence that support its involvement as an intermediate. First, ^{13}NO being reduced was exchanged out of cells in porportion to the ^{14}NO that was added (Firestone et al., 1979), and second, NO was found in steady state concentrations in all denitrifier cultures examined as long as nitrate or nitrite were being reduced (Betlach, 1979). In the diagram, NO is shown in equilibrium with an enzyme-bound intermediate (Y) which is the more conservative interpretation of our data (Firestone et al., 1979).

Nitrous oxide also seems to be commonly associated with the ammonia producing pathway. We have examined N_2O production by species of several genera known as "nitrate respirers" (Tiedje et al., 1979a) (which means they produce nitrite from nitrate). The proportion of nitrate-N that accumulated as N_2O is shown in Table 1. We have also confirmed by use of $^{13}NO_3^-$ that no ^{13}N-(N_2) is produced and that $^{13}NH_4^+$ was the only ^{13}N ion found when the organisms were grown in a medium containing high organic-N and ammonium (Tiedje et al., 1979). Thus these organisms have the critical characteristics of dissimilatory nitrate reduction to ammonium and do not possess the typical characteristics of denitrifiers (i.e., reduction of N_2O to N_2 and conversion of almost all of the NO_3^--N to N gasses).

While we have studied dissimilatory ammonia producers, others have reported N_2O production by organisms which assimilate nitrate (Yoshida and Alexander, 1970; Bollag and Tung, 1972). Since N_2O seems to be produced by many and perhaps all pathways producing ammonia from nitrate, it could lead to incorrect interpretations

Table 1. Production of nitrous oxide by bacteria which dissimilate nitrate to ammonia. Data of Bleakley and Tiedje (unpublished).

Organism	Percent NO_3^--N found as N_2O-N[a]
Escherichia	36
Klebsiella	30
Erwinia	19
Serratia	12
Enterobacter	6
Bacillus	3

[a]Grown on 3% tryptic soy broth without glucose and with 3.5 mM KNO_3. Incubation was for 2½ days.

about denitrification since this gas was formerly thought to be a specific indicator of this bacterial process.

The competition between the two dissimilatory pathways is particularly interesting since both occur under the same basic condition -- limiting oxygen -- yet one conserves nitrogen while the other converts nitrogen to gaseous forms unavailable to most biological species. The basic question is what controls this competition, how can it be evaluated, and can it be managed to the advantage of humankind?

The basic energetics of the two pathways is presented in Table 2. When compared on the basis of electron donor (H_2 used for illustrative purposes) denitrification yields the most energy. However, when compared on a nitrate basis, the nitrate to ammonium pathway provides more energy. In the latter case, however, the reduction of nitrite is not coupled to phosphorylation, but it is in denitrification. Thus the actual energy that is conserved and available for biosynthesis in the ammonium producing pathway is much less than theoretically possible. Though the ammonium producing pathway would seem to be less favorable energetically, in all cases it can accept more electrons per nitrate reduced than denitrification (Table 2). In anaerobic habitats the feature that

Table 2. Theoretical energy yield and electron accepting capacity of dissimilatory nitrate reduction to ammonium and denitrification.

Reaction	$\Delta G^{o'}$ (kcal/mole)		Electrons accommodated per NO_3^-
	H_2	NO_3^-	
Dissimilatory $NO_3^- \rightarrow NH_4^+$			
$NO_3^- + 4H_2 + 2H^+ \rightarrow NH_4^+ + 3H_2O$	-35.8	-143.3	8
Denitrification			
$2NO_3^- + 5H_2 + 2H^+ \rightarrow N_2 + 6H_2O$	-53.6	-133.9	5

Calculated from Gibbs free energies of formation from the elements as tabulated by Thauer et al. (1977).

is most often limiting to growth is the availability of electron acceptors. Therefore, the eight electron reduction of nitrate to ammonium represents an ideal electron sink. Summarizing the above reasoning, one would predict that in carbon limited habitats denitrification should be favored since the energy is more efficiently utilized for growth. However, in nitrate limited habitats nitrate reduction to ammonium should be favored since more electrons can be accommodated. This interpretation is supported by studies in carbon and nitrate limited chemostats which have shown that the principles elucidated above hold (Cole and Brown, 1980; Dunn et al., 1979).

The above arguments provide an indication of where one pathway might be naturally favored over the other. The actual parameters that determine the outcome of competition for the two physiologic types in the same habitat have not yet been adequately investigated. The competition should be controlled by two principle components: the kinetics of uptake of nitrate, and the suitability of the populations for the habitat which should determine their respective densities. The kinetics of competition are often evaluated by comparing Km's -- the lower the Km the greater the affinity and thus the winner of the competition. However, Km is not a dynamic term; rate obviously plays a role in competition which is not adequately evaluated by comparing Km's. This can be a particular problem with bacteria since many high affinity-low nutrient strains seem to also have low maximum growth rates. Healey (1980) has also noted the overemphasis on Km and has suggested that the ratio of maximum rates (Rmax) over Km, which is the slope at the lowest substrate concentrations be used to evaluate competition. This is a simple form that emphasizes both factors. He also points out how this expression is equally applicable for uptake kinetics (Michaelis-Menten) and growth kinetics (Monod). Though this evaluation has not yet been completed for the nitrate competition I have discussed, I feel it is a more useful approach than comparing Km's, which already exist in the literature and have not been particularly illuminating.

The second aspect that may control the competition is the density of organisms of each type which reflects their overall fitness for the habitat. This is probably more indicative of their success at competing for and efficiently utilizing the carbon of the habitat. It is not yet known how successful these two physiologic types are in different habitats and why, but this must be known if one is to better manage the competition.

METHODS

The study of the dissimilatory pathways of nitrate has long been limited by the lack of direct and sensitive methodology to measure the processes particularly in natural samples. To overcome these deficiencies we have used the three basic analytic approaches noted below, each with important advantages for particular experimental questions.

^{13}N

We have used the longest lived radioactive isotope of nitrogen, ^{13}N, because of the extreme sensitivity this provides and because one can measure the denitrified N_2 directly in any atmosphere. The sensitivity and basic features of this isotope compared to ^{15}N and ^{14}C are given in Table 3. We produced this isotope with a cyclotron accelerated proton beam using the nuclear reaction of $^{16}O(p,\alpha)^{13}N$ with oxygen in water as the target. The yields of $^{13}NO_3^-$ are greater than or equal to 10 mCi per bombardment (Tiedje et al., 1979b). This is enough to allow measurement of denitrification activities for several hours though it is rarely necessary to carry on an experiment this long because of the high sensitivity.

We have used three basic ^{13}N analytical systems which are completely described elsewhere (Tiedje et al., 1979b). Briefly these are 1) a gas stripping system which continuously removes, traps and counts ^{13}N-N_2O and ^{13}N-N_2 and is most useful for measurement of denitrification rates, 2) a gas chromatograph-proportional

Table 3. Methods of detection and detection limits of the nitrogen isotopes.

Isotope	Half-life	Max. spec. act. possible (mCi/matom)	Method of detection	Minimum detectable amt.[a] (moles)
^{15}N	Stable	> 99 atom %	Mass spectrometry Emission spectrometry	7×10^{-9}
^{13}N	9.96 min	1.4×10^{10}	β^+ 1.19 MeV γ 0.511 MeV	3×10^{-20}
^{14}C	5730 yr.	62.4	β^- 0.156 MeV	7×10^{-12}

[a]Assuming 1000 dpm minimum detectable radiation. For ^{15}N, assuming 0.001 atom % excess of ^{15}N and 1 mg N gas produced in 10 mg N gas sample. However, the sensitivity can be increased about two orders of magnitude by using the new quadrupole gas chromatograph-mass spectrometers because of their greater absolute sensitivity (D. D. Focht, personal communication).

counter which provides separation and quantitation of ^{13}N in any gas peak, and 3) a high pressure liquid chromatography coupled to a NaI crystal which serves as a gamma detector to provide separation and quantitation of ^{13}N in nitrate, nitrite, and ammonium. We also use NaI crystal well counters and a liquid scintillation counter to quantify total label in any sample, e.g., ^{13}N assimilated by cells.

The basic features of the ^{13}N method are compared in Table 4 to the other two most prominent methods for studying nitrate fate. The major advantages of ^{13}N are the first three characters -- sensitivity, direct N_2 quantitation and, because of the first, the ability to add isotope without altering the natural nitrate concentration. The major limitation of the ^{13}N method is the technical complexity of the production, detection, and data reduction aspects even if a suitable accelerator is available.

Table 4. A summary comparison of three methods for measuring dissimilatory fates of nitrate.

Character	^{13}N	^{15}N	Acetylene-CG-ECD
Sensitivity	V. sensitive	Insensitive	Sensitive
N_2 analysis	Direct	Difficult	Indirect
(NO_3^-)	Will not alter pool	Often increases pool	No change
Additions	$^{13}NO_3^-$	$^{15}NO_3^-$, remove $^{14}N_2$	Acetylene
No. of samples	Few	Many	Many
N mass balance	Difficult	Possible except for N_2	Incomplete
$NO_3^- \rightarrow NH_4^+$ studies	Less versatile	Preferred	Indirect
Location of expt.	Adjacent to accelator	V. good in field	Field possible

^{15}N

The use of stable isotopes (enriched or depleted) are somewhat insensitive, especially for distinguishing N_2 from denitrification but it is the most suitable approach for field nitrogen budget studies and for lengthy studies in which one needs to integrate processes over time. We have also found that study of the dissimilatory nitrate reduction to ammonia is most convenient by the ^{15}N method since the analysis of the $^{15}NH_4^+$ is particularly easy.

Acetylene Inhibition and Measurement of N_2O by Electron Capture Detector

This method for denitrification rate measurements relies on acetylene blocking further reduction of N_2O^{15} which then can be very sensitively measured by a high temperature electron capture detector (Kaspan and Tiedje, 1980). The advantages of the method are the sensitivity, the fact that no alteration of the indigenous nitrate pool is required, that many samples can be analyzed, and that the necessary equipment is not costly. The method is, however, indirect and acetylene also inhibits nitrification (Walter et la., 1979) which replenishes the nitrate pool for denitrification. The method can be reliably used for comparison of denitrification activities among samples, but it is not yet clear whether the measured rates can be extrapolated with any reliability to arrive at field losses of nitrogen. It is assumed that acetylene does not affect the other processes competing for nitrate; this needs further investigation. If acetylene does not affect the dissimilatory nitrate reduction to ammonium, it may then be possible to estimate this process indirectly by assuming that the amount of nitrate-N not recovered as either nitrate or nitrous oxide was reduced to ammonium.

COMPETITION FOR NITRATE BY PLANTS

It has been a commonly held axiom that denitrification is stimulated in the rhizosphere. This is based on the logic that there is more available carbon and less oxygen in this environment. We have shown that the denitrifying enzyme concentration decreases with increasing distance from roots, both in laboratory and field evaluations (Smith and Tiedje, 1979). One cannot conclude from this, however, that denitrification losses are actually greater in the rhizosphere.

To directly answer this question we developed an assay system based on the acetylene method and using an intact soil-plant system in an aerobic environment (Smith and Tiedje, 1979). The results are summarized in Table 5. In both soils, the high denitrification rates occurred only when the nitrate concentration was relatively high. When the nitrate concentration was low, the planted soils had less denitrification than the unplanted soils. This suggests that the plants were effectively competing for the nitrate and thereby reducing denitrification. If this interpretation is generally true, it would suggest that nitrogen lost to denitrification should be higher in agricultural ecosystems than in natural ecosystems. It remains to be determined, however, whether the proportion lost to denitrification varies with nitrate concentration in the field.

COMPETITION FOR NITRATE BETWEEN DENITRIFICATION AND DISSIMILATORY NITRATE REDUCTION TO AMMONIA

Studies With Bacterial Cultures

The energetic benefits which accrue to denitrifiers from growth on nitrate is well known. Virtually all denitrifiers are aerobes and incapable of anaerobic growth except in the presence of nitrate.

Table 5. Denitrification rates as affected by plants and soil nitrate concentrations. Adapted from Smith and Tiedje (1979).

Soil & Plant	Soil NO_3^-	Denitrification rate	
		Planted	Unplanted
		($p\ mol\ N_2O \cdot g\ soil^{-1} \cdot min^{-1}$)	
Brookston corn	High[a]	18.6	2.6
	Low	1.5	9.1
Miami orchardgrass	High	87.1	19.4
	Low	1.8	9.4

[a]Soil nitrate concentrations in $\mu g\ NO_3^- \text{-} N \cdot ml\ solution^{-1}$: high > 25; low < 6.

In contrast, the organisms dissimilating nitrate to ammonia are all capable of fermentation and can grow in the absence of nitrate; they include both obligate and facultative anaerobes. The prevalent denitrifiers include species of Pseudomonas, Alcaligenes, Flavobacterium, Paracoccus, and Bacillus (Payne, 1973; Gamble et al., 1977). The most studied dissimilatory ammonia producing organisms include several of the rumen genera such as Vibrio and Selenomonas, many of the Enterobacteriaceae, and the obligate anaerobe Clostridium (Payne, 1973; Tiedje et al., 1979a; Caskey and Tiedje, 1979). The energy benefit of nitrate reduction to the ammonium producing group is of two types: some of the organisms have electron transport phosphorylation linked to the nitrate to nitrite step which markedly enhances their growth yield. Others like Clostridium have no electron transport phosphorylation, however, they as well as the previous group reducing nitrite, can benefit from increased substrate level phosphorylation allowed by the nitrate or nitrite reduction (Hason and Hall, 1977; Caskey and Tiedje, 1980). This is because the reductive pathway can serve as an electron sink thereby allowing acetyl-SCoA to be used for ATP production (via acetyl-P) rather than as an electron acceptor.

Caskey and Tiedje (1979) have examined soil for predominant organisms which have the dissimilatory nitrate to ammonium pathway and isolated a number of Clostridia and Bacilli. The energy benefit in terms of molar growth yield from the shift in electron acceptor allowed by nitrate was examined for one of the soil Clostridia (Caskey and Tiedje, 1980). The data are summarized in Table 6. These organisms require organic nitrogen in the medium so it is necessary to use $^{15}NO_3^-$ in order to distinguish how much of the nitrate was reduced and to confirm that NH_4^+ was the only product. It is clear that there is an increase in growth yield of 15.7 percent from the nitrate reduction which is about the amount one might expect from the metabolic shift predicted. But this is not enough to suggest this type of organism could compete well with denitrifiers under carbon limited conditions. We found that this reduction in <u>Clostridium</u> was not affected by ammonium, glutamate, or methionine sulphoximine which is consistent with a dissimilatory type of regulation. One of the more interesting observations in this study (Caskey and Tiedje, 1980) was that growth on nitrate greatly enhanced the organism's ability to reduce sulfite to sulfide; a five-fold increase in activity observed for growth on sulfate and a ten-fold increase over growth without either anion. The cause of this enhancement is not known but it bears further investigation since it may lead to approaches to regulate the dissimilatory nitrate to ammonium pathway.

<u>Studies With Soils</u>

The ability of the dissimilatory reduction of nitrate to ammonium to compete with denitrification for nitrate was examined with ^{15}N techniques in anaerobically incubated soils by Caskey and me (Caskey and Tiedje, 1979). Results of several experiments are summarized in Table 7. It is assumed the remainder of the ^{15}N was lost to denitrification since little or no nitrate or nitrite was left at the end of the experiment. As has been noted many times before, $^{15}NH_4^+$ production from NO_3^- in unammended soils is meager, about 2 percent. Glucose which is a preferred substrate for Clostridia greatly stimulated this reduction, but acetate which is not used by this genus, did not. Buresh and Patrick (1978) showed that the positive response to glucose is further enhanced by preincubation, and that methanol did not stimulate the process. Apparently, there is a greater carbon limitation for the dissimilatory ammonium producing organisms than for the denitrifying organisms in soils.

Regulators of the assimilatory nitrogen pathways were added to soil on the chance that they might enhance the nitrate to ammonium pathway and thereby provide an approach to favor ammonium production. As shown in Table 7, they did not greatly affect the pathway. The response to methionine sulphoximine (MSX) may be significant; however, since the general evidence indicates that the pathway is not assimilatory, it is not clear how the MSX result should be interpreted.

Table 6. The effect of NO_3^- reduction to NH_4^+ on the molar growth yield of Clostridium KDHS2 isolated from soil. Adapted from Caskey and Tiedje (1980).

Measurement	No NO_3^-	With $^{15}NO_3^-$
$^{15}NO_3^-$ reduced	--	0.38 μmol ^{15}N/ml
$^{15}NH_4^+$ produced	--	0.37
Glucose used	8.0	7.2 μmol/ml
Cell yield	0.24	0.25 mg dry wt/ml
Molar growth yield	30	34.7 g dry wt/μmol glucose
Percent Increase	15.7%	

Table 7. Influence of various treatments on the dissimilatory reduction of nitrate to ammonium in anaerobically incubated soil. Adapted from Caskey and Tiedje (1979).

Percent of $^{15}NO_3^-$ recovered as $^{15}NH_4^+$ + organic ^{15}N					
Effect of carbon source[a]		Effect of assimilatory inhibitors[b]		Effect of inoculation of Clostridial spores[c]	
No addition	2	None	32	None	22
Glucose	40	NH_4Cl	38	Heat-Shock	27
Acetate	7	Glutamine	30	Inoculation	83
		MSX	46		

[a] Measured after 5 days in Conover soil.

[b] Measured after 1 day in glucose amended and heat shocked Kranzburg soil.

[c] Measured after 1 day in glucose amended Conover soil.

One factor which obviously can affect competition for nitrate is the density of cells of each physiological type. In the soil we studied, denitrifiers exceeded the spore forming ammoinum producing organisms by 100-fold (Caskey and Tiedje, 1979). If the population of Clostridia was increased to ten-fold greater than the denitrifiers (Table 7), then most of the nitrate (83%) was recovered as ammonium. This shows that the Clostridia could compete in soils if their density were greater. This experiment is not meant to suggest that soil should be inoculated with Clostridium or any other dissimilatory ammonium producers since I do not believe this approach is realistic. However, it does point out one of the problems accounting for the poor competition of this process with denitrification in the agricultural soil habitat.

Studies in Other Habitats

Kaspar and I have studied the partitioning of nitrate between denitrification and dissimilatory nitrate reduction to ammonium in samples from the rumen, anaerobic sewage sludge digestor, and lake sediments (Kaspar and Tiedje, unpublished information). These results are compared to results from other habitats in Table 8. Essentially the entire range of fates of nitrate-N from total denitrification to total ammonium production was found. Because of the energetic advantage of denitrification, it is perhaps surprising that there are several habitats where it fares so poorly. But as explained earlier, the dissimilatory ammonium producing pathway accepts more electrons per nitrate and could be more favorable in habitats where nitrate is limited. In lake sediments and sewage sludge there is virtually no nitrate. Denitrification potential is present in both habitats which verifies that this physiologic process is present. One of the most interesting results of the comparative study is the finding of no denitrification in the rumen. Why? There is certainly adequate metabolizable carbon and nitrate entering this environment. I believe that further resolution of the features which control nitrate partitioning in these different habitats could be useful in identifying approaches for better management of the nitrogen cycle.

One of the features that Kaspar and I (unpublished information) observed in these studies was the production of nitric oxide, albeit very low, in sludge and lake sediments but not in rumen samples. It was produced only during active nitrate reduction and not by sterilized samples. When NO was added to sterilized sludge samples, its abiotic consumption was substantial. This suggests that the net production we observed reflects an even greater total production of NO. We now typically find detectable NO in all denitrification studies, whether with pure cultures or natural samples. We did not observe this product before. The difference is that we now have a better analytical scheme (Kaspar and Tiedje, 1980). The sample is sealed in a system which recycles the headspace of

Table 8. Partitioning of nitrate between dissimilatory nitrate reduction to ammonium and denitrification, and the production of nitric oxide.

Habitat	Percent of NO_3 dissimilated to NH_4^+ [a]	Detection of nitric oxide	Investigator
Rumen	100	0	Kaspar and Tiedje (unpublished)
Sewage sludge	60-70	+	Kaspar and Tiedje (unpublished)
Marine sediment	20-70	+	Sørensen (1978), Koike and Hattori (1978)
Lake sediment	7-30	+	Kaspar et al. (unpublished), Chen et al. (1972)
Rice paddy soils	5-40	-[b]	MacRae et al. (1968)
Well-drained soils	2	0?[c]	Caskey and Tiedje (1979), many others

[a] Most of the remaining nitrate-N was lost to denitrification.

[b] Not measured.

[c] Not generally detected but in our recent work with the better analytical system we usually can detect NO from soil columns flushed with argon (Parkin and Tiedje, unpublished data).

the sample through the sampling loop of the gas chromatograph. Oxygen can be completely excluded and no syringes are used. I now believe that some NO is produced wherever denitrification occurs.

SUMMARY

Control of denitrification losses in agriculture will not be easy. The population capable of this process is diverse and widespread in nature. The only realistic approaches are management of the soil environment to reduce this process (e.g., water control, fertilization) or by use of chemical inhibitors. An inhibitor will, however, need to be specific so that assimilatory nitrate reduction is not altered. The aspect I dealt with in this text is the competition for nitrate from the viewpoint that if the basic characters of competition are understood, additional management or chemical approaches might be feasible which would be expressed through altering of competition. I think the dissimilatory nitrate to ammonium pathway, which is less well known, deserves further elucidation. This process has a high capacity for nitrate turnover and it is carried out by organisms which occupy the same habitats and respond to anoxia in the same manner as denitrification. If any solutions are to come for the problem of reducing denitrification losses, research in microbial ecology is going to have to provide the breakthroughs since denitrifying organism are so well entrenched in the soils of our planet.

ACKNOWLEDGMENT

The author's work is supported by the National Science Foundation, the Michigan Agricultural Experiment Station and USDA Regional Research Project NE-39.

REFERENCES

Balderston, W. L., Sherr, B., and Payne, W. J., 1976, Blockage by acetylene of nitrous oxide reduction in Pseudomonas perfectomarinus, Appl. Environ. Microbial., 31:504.

Betlach, M. R., 1979, Accumulation of intermediates during denitrification -- kinetic mechanisms and regulation of assimilatory nitrate uptake, Ph.D. Thesis, Michigan State University.

Bollag, J.-M., and Tung, G., 1972, Nitrous oxide release by soil fungi, Soil Biol. Biochem., 4:271.

Buresh, R. J., and Patrick, Jr., W. H., 1978, Nitrate reduction to ammonium in anaerobic soil, Soil Sci. Soc. Am. J., 42:913.

Caskey, W. H., and Tiedje, J. M., 1979, Evidence for Clostridia as agents of dissimilatory reduction of nitrate to ammonium in soils, Soil Sci. Soc. Am. J., 43:931.

Caskey, W. H., and Tiedje, J. M., 1980, The reduction of nitrate to ammonium by a Clostridium sp. isolated from soil, J. Gen. Microbiol., 119:217.

Chen, R. L., Keeney, D. R., Konrad, J. G., Holding, A. J., and Graetz, D. A., 1972, Gas production in sediments of Lake Mendota, Wisconsin, J. Environ. Qual., 1:155.

Cole, J. A., and Brown, C. M., 1980, Nitrite reduction to ammonia by fermentative bacteria: a short circuit in the biological nitrogen cycle, FEMS Microbiol. Lett., 7:65.

Dunn, G. M., Herbert, R. A., and Brown, C. M., 1979, Influence of oxygen tension on nitrate reduction by a Klebsiella sp. growing in chemostat culture, J. Gen. Microbiol., 112:379.

Firestone, M. K., Firestone, R. B., and Tiedje, J. M., 1979, Nitric oxide as an intermediate in denitrification: evidence from nitrogen-13 isotope exchange, Biochem. Biophys. Res. Comm., 91:10.

Firestone, M. K., Firestone, R. B., and Tiedje, R. M., 1980, Nitrous oxide from soil denitrification: factors controlling its biological production, Science, 208:749.

Gamble, T. N., Betlach, M. R., and Tiedje, J. M., 1977, Numerically dominant denitrifying bacteria from world soils, Appl. Environ. Microbiol., 33:926.

Hasan, S. M., and Hall, J. B., 1977, Dissimilatory nitrate reduction in Clostridium tertium, Zeit. Allg. Mikrobiol., 17:501.

Healey, F. P., 1980, Slope of the Monod equation as an indicator of advantage in nutrient competition, Microb. Ecol., 5:281.

Kaspar, H. F., and Tiedje, J. M., 1980, Response of electron-capture detector to H_2, O_2, N_2, CO_2, NO and N_2O, J. Chromatogr., 193:142.

Koike, J., and Hattori, A., 1978, Denitrification and ammonia formation in anaerobic coastal sediments, Appl. Environ. Microbiol., 35:278.

MacRae, I. D., Ancajar, R. R., and Salandanan, S., 1968, The fate of nitrate nitrogen in some tropical soils following submergence, Soil Sci., 105:327.

Payne, W. J., 1973, Reduction of nitrogenous oxide by microorganisms, Bacteriol. Rev., 37:409.

St. John, R. T., and Hollocher, T. C., 1977, Nitrogen 15 tracer studies on the pathway of denitrification in Pseudomonas aeruginosa, J. Biol. Chem., 252:212.

Smith, M. S., and Tiedje, J. M., 1979, The effect of roots on soil denitrification, Soil Sci. Soc. Am. J., 43:951.

Sørensen, J., 1978, Capacity for denitrification and reduction of nitrate to ammonia in a coastal marine sediment, Appl. Environ. Microbiol., 35:301.

Thauer, R. K., Jungermann, K., and Decker, K., 1979, Energy conservation in chemotrophic anaerobic bacteria, Bacteriol. Rev., 41:100.

Tiedje, J. M., Caskey, N. V., Smith, M. S., Bleakley, B. H., and Firestone, R. B., 1979a, Nitrous oxide production by bacteria that reduce nitrate to nitrite, Agron. Abstr.

Tiedje, J. M., Firestone, R. B., Firestone, M. K., Betlach, M. R., Smith, M. S., and Caskey, W. H., 1979b, Methods for the production and use of nitrogen-13 in studies of denitrification, Soil Sci. Soc. Am. J., 43:709.

Yoshida, T., and Alexander, M., 1970, Nitrous oxide formation by Nitrosomonas europaea and heterotrophic organisms, Soil Sci. Soc. Am. Proc., 34:880.

Walter, H. M., Keeney, D. R., and Fillery, I. R., 1979, Inhibition of nitrification by acetylene, Soil Sci. Soc. Am. J., 43:195.

Zumft, W. G., and Vega, J. M., 1979, Reduction of nitrite to nitrous oxide by a cytoplasmic membrane fraction from the marine denitrifier Pseudomonas perfectomarinus, Biochem. Biophys. Acta, 548:484.

SOIL DENITRIFICATION

Dennis D. Focht

Department of Soil & Environmental Sciences
University of California
Riverside, CA 92521

HISTORICAL ASPECTS

Scarcely a century ago, Gayon and Dupetit (1886) observed that upon addition of nitrate and nitrite to soil, both anions were rapidly converted to gaseous forms of nitrogen, namely N_2 and N_2O. The biological nature of the process was noted because soil exposed to boiling or chloroform vapors was inactive. Considerable fear was raised by Wagner (1895) that the nitrogen fertility status of the soil would be depleted by the addition of manure. He assumed incorrectly that manure was the sole source of denitrifying bacteria, and he suggested that manure be treated with sulfuric acid to destroy those potentially harmful bacteria. Deherain's (1897) astute observation that denitrification effected by indigenous soil bacteria was a normal part of the nitrogen cycle led him to recommend that nitrates and manure not be added at the same time because manure provided an available energy source for reducing both O_2 and NO_3^-, with the latter being converted to N_2. Excellent management recommendations for minimizing N losses were thus advanced by Deherain's awareness of basic microbiological principles.

Waksman (1927) concluded that denitrification was of no significance in well aerated agricultural soils. Nevertheless, Allison (1955, 1966) in two classical reviews of studies encompassing several decades of fieldwork, concluded that roughly 10-30% of the nitrogen added to soil could not be accounted for in crop removal, leachate, or increase in soil organic N. He attributed this loss to denitrification. Upon conclusion of a ten year field study, Pratt et al. (1972) also observed a nitrogen deficit despite the well-drained, open porous, sandy loam texture of the soils. They, too, attributed this loss to denitrification.

FIELD MEASUREMENT OF DENITRIFICATION

By Difference

Determination of denitrification by the difference method used by Allison (1955, 1966) and many others would appear to be logical if one can accurately measure the various N components of the system. For example:

$$N_D = N_i - N_C - N_L$$

where N_D is the ammount denitrified, N_i is the input (normally fertilizer), and N_C and N_L are the respective losses of N by crop removal and leaching. Though this equation seems reasonably straight-forward, there are assumptions and uncertainties that are not always resolved. Unquestionably, the most accurate figure is the input of fertilizer if one assumes the contribution from asymbiotic N fixation to be insignificant. This is probably valid in instances involving high N-consuming crops. Crop removal data are subjected to the normal distributional errors, while data from nitrate leaching is subjected to log-normal or highly variable distributions (Nielsen et al., 1973). Despite the use of the NO_3^-/Cl^- ratio, whereby the Cl^- ion serves as an internal standard because it is biologically inert, accurate measurement of the N losses through leaching are not possible over a growing season because of tremendous spatial variability in the field (Biggar and Nielson, 1976). Moreover, the assumption that the soil organic N pool remains constant because of its large size is not applicable to short term field studies because a net increase above the normal sampling variability may not be detected. Even in instances where ^{15}N tracers are used, there can be no assurance that a one-year balance study will yield meaningful data with regards to N losses through denitrification. Much of the spatial variability in soil organic N appears to be normalized or approaches some quasi-steady state over several years of the same continuous management, so that longer term N balance studies lead to more reliable estimates of denitrification by difference. There has been considerably interest recently in measuring denitrification directly, not necessarily to obtain absolute measurements of N fluxes from soil, but to verify if estimates of denitrification made by the difference method are realistic and are not badly distorted by sampling techinques or spatial variability.

Direct Measurement

Because the atmosphere contains 78% N_2 by volume, measurements of the evolution of N_2 are difficult. Changes of N_2 in the soil atmosphere are not necessarily a direct result of denitrification; they may be caused by partial pressure changes during the consumption of oxygen in blocked pore spaces. The N_2/Ar ratio used by oceanographers to calculate fluxes is generally unapplicable to soil because of the large uncertainty in estimating the diffusion coefficients, the concentration gradients, the depth of the denitrification zone and the inherent spatial variability of the soil profile. These uncertainties have been illustrated by Rolston et al. (1976, 1979), who attempted to measure denitrification directly with $^{15}NO_3^-$; in some instances the estimates coincided with denitrification as calculated indirectly by difference; in others the two methods gave comparable estimates. Other investigators have chosen simply to measure the amount of N_2O that evolves from the soil surface. This concept obviates the uncertainty in establishing concentration gradients and diffusion coefficients, particularly since the presence of "blocked pores" invalidates the use of gaseous diffusion coefficients (Stolzy and Flühler, 1978). Experimentally, it is easier to use the "box approach" (Fig. 1) because one does not have to sample the soil profile extensively for gaseous concentrations in order to establish a gradient.

Since the discovery that acetylene blocks reduction of N_2O to N_2 (Fedorova et al., 1973; Yoshinari and Knowles, 1976; Balderston et al., 1976), it is possible to measure denitrification directly in the field by the methods developed by Ryden et al. (1978, 1979 a,b). The method is easily adapted to the field and involves only a short (3-4 hr) exposure time. One set of boxes is placed over the soil and is subjected to a slow continuous flow of acetylene to maintain at least a 1% concentration throughout the profile. Another set of boxes placed away from the acetylene treatment is left untreated with acetylene. Nitrous oxide is collected on an adsorbent (Fig. 2) as the air is swept through slowly by a portable vacuum pump. An additional advantage of this method is that the adsorbent can be removed and stored for several months at room or field temperature prior to analysis in the laboratory. The acetylene-treated profiles represent the total gaseous N production ($N_2O + N_2$), while the untreated represents the actual net production of N_2O. Thus, N_2 production can be calculated by the difference between the two treatments.

Since acetylene is known to nonselectively inhibit many bacteria -- specifically the nitrifiers (Hynes and Knowles, 1978; Walter et al., 1979) -- a long incubation period is not recommended. Furthermore, the same site should not be used more than once a month. Thus, by sampling different sites in the field, the

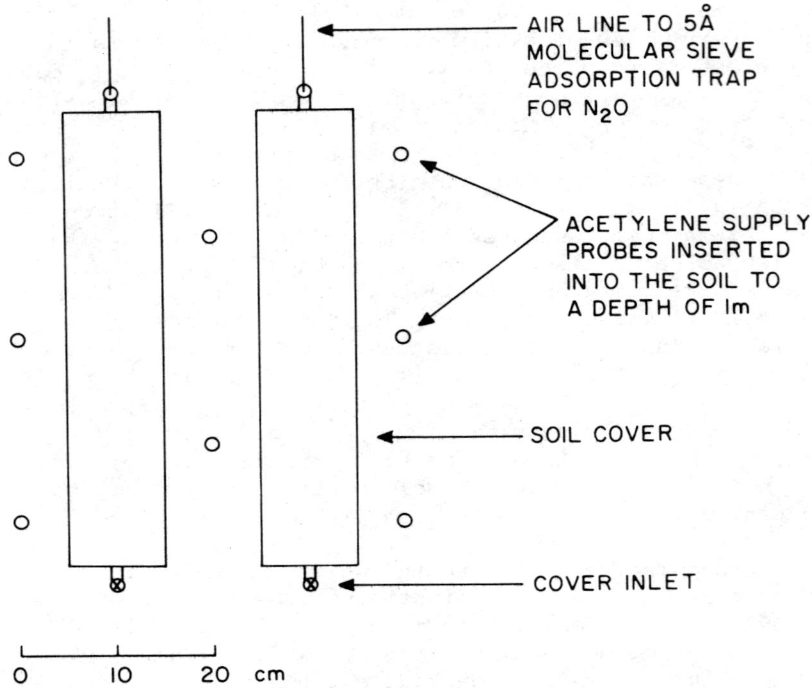

Fig. 1. Field schematic for direct measurement of denitrification (from Ryden et al., 1979b).

investigator will not only obtain a better random sample, but will minimize the injurious effects of acetylene to pertinent microbial processes.

When Rydan et al. (1979b) measured denitrification losses directly by the acetylene inhibition method, they found agreement with what they had calculated by the indirect difference method.

Other variations of the soil cover method (Denmead, 1979) allow for continuous accumulation of N_2O in the box; the concentration is measured continuously by infrared spectroscopy.

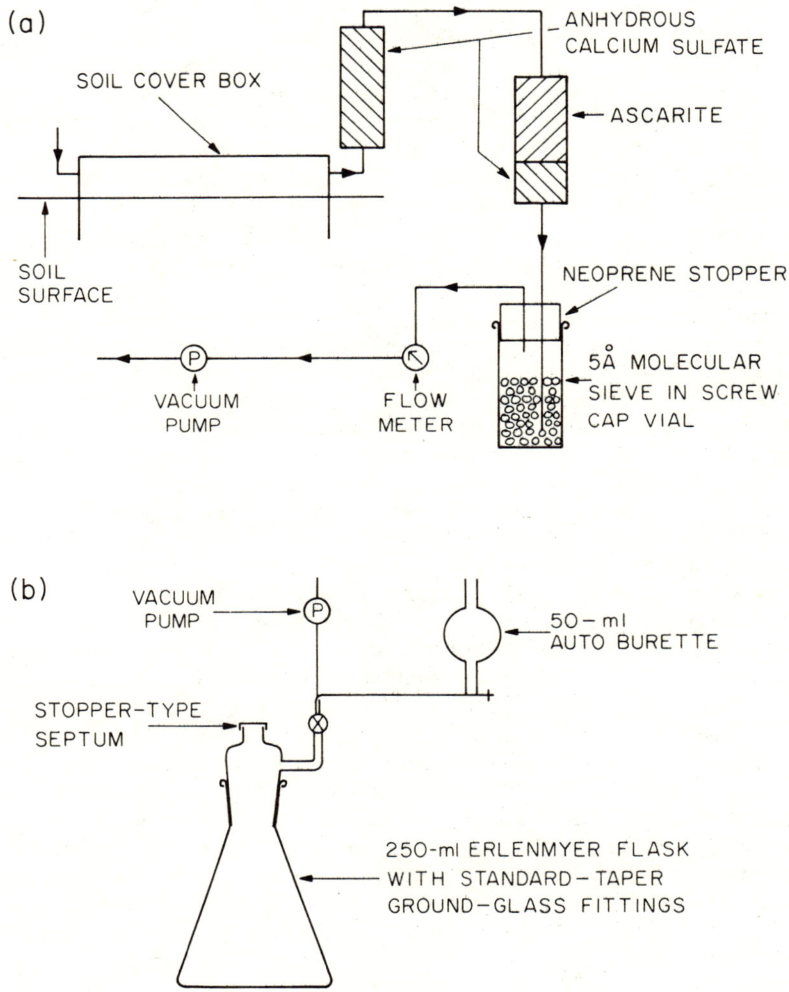

Fig. 2. Schematic representation of field collection and adsorption (a) and laboratory removal (b) of nitrous oxide (from Ryden et al., 1978).

THE SOIL ENVIRONMENT

Aeration

More attention has been focused upon soil aeration than any other factor affecting denitrification for good reason. Firstly, denitrification is a form of anaerobic respiration whereby nitrogen oxides are used in lieu of oxygen by the normal aerobic bacterial flora. Secondly, aeration is perhaps the most transient physical parameter in soil and is drastically affected by irrigation or rainfall. The addition of water to dry soil has the dual effect of decreasing oxygen availability by impeding gaseous diffusion and by enhancing microbial respiration rates. Thirdly, the extent of oxygen depletion will determine how rapidly and completely the nitrogen oxides are converted to N_2.

Denitrification in soil can be envisioned to occur in three possible ways. (1) Nitrogen oxides can be reduced continually as they leach to anoxic zones in the soil profile. Examples of this are flooded soils (e.g., rice paddies), sediments, and soils with shallow impermeable hard pans. (2) Denitrification can occur at a slow but continuous rate in anoxic microsites of moist soils. Such soils generally have high organic matter content and good structure which effect a high water holding capacity and formation of stable aggregates. (3) Denitrification may be rapid and transient, namely after a rainfall or irrigation. This seems to be more characteristic of well-drained thermic arid and semi-arid soils in which the organic matter content is not only lower than soils of temperate humid regions but is more closely confined to the surface. In all three cases, denitrification occurs only in the absence of oxygen.

Failure to appreciate the slow rate at which oxygen diffuses through water (about 100,000 times slower than through air) and the high rate of biological oxygen consumption has on occasion led to some confusion regarding whether or not denitrification is truly an anaerobic process. It is important therefore to recall that the soluble, not the gaseous oxygen concentration is what governs the state of anoxia. Generally when the pore volume of soil contains less than 10% air, gaseous diffusion becomes insignificant as a result of a large volume of "blocked and discontinuous air pores" surrounded by water (Wesseling and von Wijk, 1957; Stolzy and Flühler, 1978).

Concurrent mineralization and denitrification can occur in soil in all three cases mentioned previously as long as aerobic and anaerobic zones exist. Case 1 is represented by instances in which ammonium is nitrified in the aerobic zone near the surface, whereupon nitrate diffuses to the anoxic zone below in which it is denitrified. In fact, the aerobic zone in wet-land paddy soils serves as a sink for ammonium such that the cation will actually diffuse upwards. Losses of 10-30% of the applied $(^{15}NH_4)_2SO_4$ have been attributed to the concurrence of nitrification-denitrification in flooded soils (Patrick and Reddy, 1976; Broadbent and Tusneem, 1971). Though the losses are small by comparison to the complete losses that would be expected from addition of nitrate, they are nevertheless within the same range noted for arable soils (Allison, 1955; 1966). Minimizing N losses through denitrification should thus be focused on the primary factor controlling nitrification, namely the thickness of the aerobic zone (see Focht, 1979).

The occurrence of anoxic microsites in soil (Greenwood, 1961; Currie, 1961) is dependent upon good soil structure and high biological oxygen demand. If an aggregate is saturated with water, it is conceivable that at some point near the center, the soluble oxygen concentration will become depleted as a result of microbial oxygen consumption. Denitrification would thus occur in these anoxic microsites. Although no one has measured the soluble oxygen concentration gradient in an aggregate, the concept is nevertheless valid on the basis of the equations derived by Greenwood (1961) and Currie (1961). Using a spherical diffusion model it is easy to show that the critical radius at which the soluble oxygen concentration will drop to zero is less than 1 mm with the normal respiration rates reported for soil. Molar growth logistics coupled with a cylindrical diffusion model (similar to rice paddies) indicates that about 10^8 bacteria/cm^3, will remove all the oxygen from a 1 mm thickness (Focht, 1979). Thus, the occurrence of anoxic microsites are not difficult to envision in soil, and readily explain the concurrence of mineralization and denitrification as shown in Table 1. A relatively constant concentration of nitrate during a several week incubation period accompanied by a reduction in the proportion of $^{15}N_3^-$ can be explained only by dilution with $^{14}NO_3^-$, which originates from organic N. As expected, more mineralization is evident as the aerated pore content increases, while the reverse is true for denitrification.

The transient nature of denitrification in the field is illustrated in Figure 3, whereupon increases in N fluxes occur immediately as the soil moisture content increases. Wetting and drying cycles have several affects upon N metabolism by soil bacteria which are related to aeration, and available organic matter. Immediately upon wetting of soil, there is a rapid proliferation of microorganisms, which results in three major factors conductive to denitrification: 1) the rapid growth rates cause a considerable increase in the biological oxygen demand, which when coupled with the reduced diffusion of oxygen by water causes rapid depletion of oxygen; 2) the labile sources of organic substrate is also an available reductant which accelerates the rate of denitrification; 3) the mineralization of labile organic compounds

Table 1. Mineralization and denitrification in soil (from Focht et al., 1980).

	Air-filled porosity		
	0	5	10
Initial NO_3 (N_o), μg/g	183	167	151
Final NO_3 (N), μg/g	140	138	138
Initial $^{15}NO_3$ fraction (f_o)	0.873	0.861	0.848
Final $^{15}NO_3$ fraction (f)	0.824	0.788	0.748
Mineralized (M), μg/g	9.6	13.6	18.1
Denitrified (D), μg/g	52.6	42.6	31.1
Organic ^{15}N gain, μg/g	4.6	4.2	3.6
$^{15}NO_3$ loss, μg/g	44.8	35.1	24.8
Gaseous -^{15}N loss, μg/g[1]	14.3	12.1	9.7

[1]Calculated from cumulative emissions during the incubation.

liberates ammonia, which can be nitrified, and therefore, provide the oxidation product necessary for denitrification.

The degree of anoxia is also important in determining the extent of reduction of nitrogen oxide to N_2. As the soil becomes more anoxic, less N_2O is liberated, and the primary gaseous product becomes N_2 (Focht, 1974; Focht et al., 1979) as shown in Figure 4. The exact reasons are not known for certain, but it appears that N_2O does not yield as much energy per electron transfer as NO_3^- in pure cultures of Pseudomonas (Koike and Hattori, 1975a,b). Thus, Focht and Verstraete (1977) concluded that NO_3^- would be the preferred electron acceptor to N_2O. Nitrate also appears to inhibit further reduction of N_2O in soil (Blackmer and Bremner, 1978), and concentrations of nitrate and N_2O in soil are inversely correlated (Focht et al., 1979). Thus, the anomoly of peak N_2O concentrations (Fig. 4) with respect to soil aeration (or related indirect measurement of aeration) is explained as follows. Low N_2O concentrations in the presence of greater

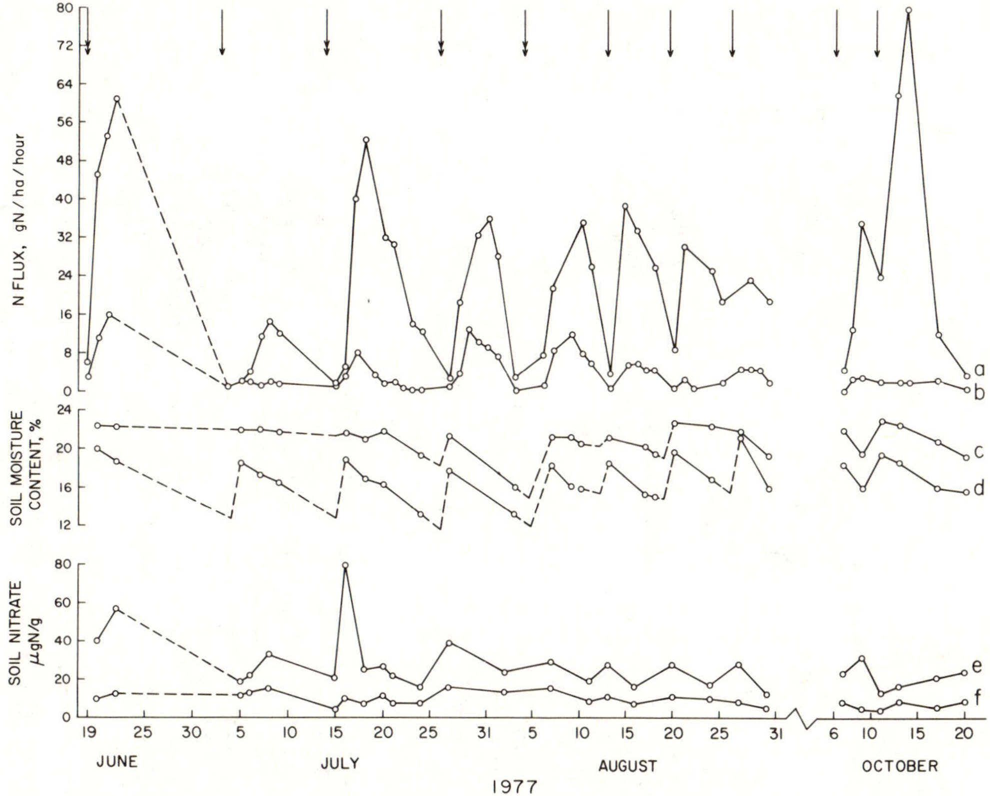

Fig. 3. Gaseous N fluxes, with (a) and without acetylene (b). Single arrows indicate irrigation; double arrows, fertilization and irrigation. Soil moisture and nitrate are for the 0-30 (d, e) and 30-60 cm (c, f) depths (from Ryden et al., 1979b).

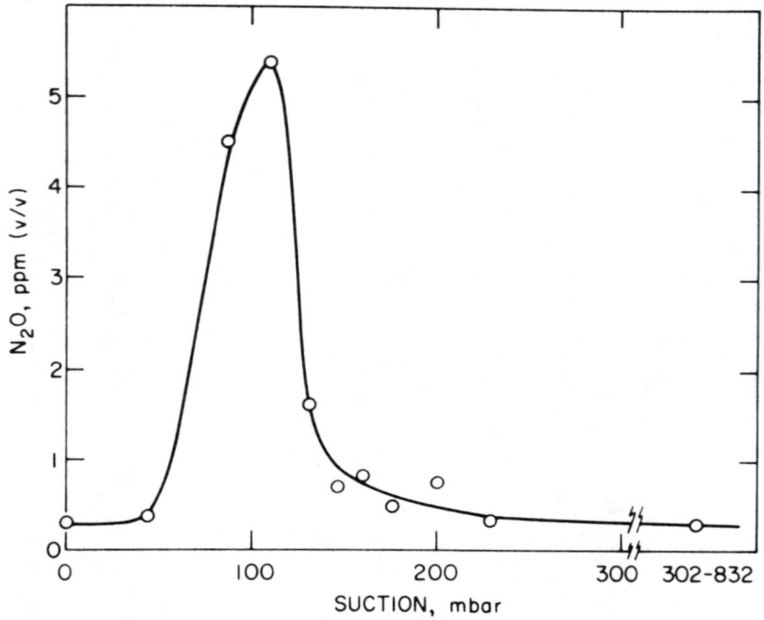

Fig. 4. Median nitrous oxide concentrations (p = 0.50) with respect to soil matter organic content (expressed as mbar suction). Each point represents the median N concentration of 41 or 42 gas samples collected within each class interval of suction readings.

aeration due to little or no denitrification, while low N_2O concentrations under poor aeration are the result of more vigorous and complete denitrification. Thus the anomolous N_2O peak represents incomplete denitrification and presumably is associated with high nitrate concentrations. Although there is some indication that N_2O is generated during nitrification, the available evidence thus far (Freney, 1979) indicates that it does not appear to constitute a significant amount by comparison to that normally produced during denitrification.

An indirect qualitative indicator of soil aeration and denitrification is the N_2/O_2 ratio used by Focht et al. (1979). In a field study, they found that the frequency of ratios significantly above ambient was 45% during the 59-84 day period as opposed to 20% during the 0-28 day period. However, lower concentrations of N_2O (many significantly less than ambient) were found during the 59-84 day sampling period. Thus, the higher N_2/O_2 and lower N_2O concentrations during the later stage of the experiment were attributed to more vigorous and complete reduction of nitrogen oxides to dinitrogen.

Reductant

Since reduction of nitrate requires a reductant, it is not surprising that denitrification is accelerated by the addition of labile organic compounds. Confusion nevertheless has occurred concerning the order of the reaction, i.e., whether or not denitrification was dependent or independent of nitrate concentration without prior consideration of whether the system was reductant or oxidant limited. Apparent zero order (vis a vis nitrate) reactions have in fact been shown to shift to apparent first order reactions when available reductant was added (Bowman and Focht, 1974; Kohl et al., 1976).

Wetting and drying cycles of soil are known to increase the amount of nitrogen lost by denitrification in arid and semi-arid soils (MacGregor, 1972). The frequency of irrigation also appears to have a marked effect not only upon the rate of denitrification, but upon the conservation of soluble carbon and nitrogen in the system (Focht et al., 1979). The effect of wetting and drying cycles -- particularly in arid climates -- in liberating more available reductant is presumably due to the disruption of H-bonding which destorys part of the structurally refractile humus that would otherwise be unavailable to microbial attack (Birch, 1958; Focht and Martin, 1979). Thus, the highest losses of indigenous organic N occur through nitrification-denitrification where drying cycles are more severe (Paul and Myers, 1971).

The presence of labile organic compounds is presumably why denitrification is greater in the rhizosphere, than in non-rhizosphere soil (Woldendorf, 1962). This peculiar anomoly suggests that more N is thus lost from grass lands than barren land and more or less confirms Allison's (1966) conclusions that a 10-30% loss of N from arable soils is a natural and non-preventable occurrence. This raises some intriguing questions concerning N balances if, indeed, more N is lost from soils that have a greater density of plant roots, then from where does the nitrogen come if the soil N does not become depleted? Initially, it is easy to conclude that a greater root density means a greater "mining" of the soil organic N. Presumably this accounts for the observation by Broadbent and Carlton (1979)
addition of ^{15}N-depleted ammonium sulfate beyond the yield requirements resulted not only in an increase in ^{15}N-depleted nitrate, but in a parallel increase in nitrate produced by mineralization of soil organic N. However, in the absence of significant inputs from asymbiotic fixation or fertilizer, the soil N should continually decrease. Since there is no indication that this occurs, the anomoly relating to high rates of denitrification in the rhizosphere may be more a result of experimental artifacts, namely the induction of higher than normal denitrification rates brought about by abnormally high nitrate concentrations.

Temperature

Optimal rates of denitrification have been reported to occur at 65 C by Nommik (1956) who attributed this high optimum to the predominance of thermophilic species of Bacillus. However, Keeney et al. (1979) found that gaseous nitrogen evolution exceeded the amount of nitrate added at temperatures of 50 C or higher. They concluded that thermophilic denitrification was a combination of biological and chemical reactions in which nitrate-respiring bacteria (Clostridium or Bacillus) generated nitrite which reacted chemically and biologically to form gaseous products.

Inasmuch as high temperatures (> 50 C) are found only in composts or hot springs, greater concern should be addressed to those temperatures normally observed in agricultural soils. Not surprisingly the denitrification steps can be shown to conform to the normal Arrhenius relationship between temperature and biochemical rate processes, between the 12-35 C range. As with all biological processes, the rate is more drastically effected with lower temperatures, such that the usual Q_{10} value of 2 for the mesophilic range may be increased to 20 in the 2-12 C range (Focht and Chang, 1975).

It is not clear what effect, if any, temperature has upon the relative rates of nitrogenous oxide reduction though it appears to be similar with respect to both the production and reduction of N_2O (Focht and Verstraete, 1977; Bailey and Beauchamp, 1973; Nommik, 1956). However, Bailey and Beauchamp (1973) found that the rates of nitrate reduction were more drastically effected than nitrite reduction by low temperatures. No reduction of nitrate occurred at 5 C during a 22-day incubation period, whereas considerable NO production was observed from the addition of nitrite. Chemodenitrification was offered as one possible alternative, though this does not appear to be a common occurrence in neutral soils at moderate temperatures. The biological explanation that nitrite represses nitric oxide reduction (Payne, 1973) would appear more feasible. Nevertheless, chemodenitrification cannot be ignored as inconsequential under conditions (e.g., low temperature) that favor nitrite accumulation since nitrite is a far more chemically reactive ion than nitrate.

Other factors

There are, of course, many other factors which affect denitrification, though for the most part they are not unique. The lack of nutrients (e.g., phosphorus, iron, molybdenum) will, of course, affect denitrifiers in the same way that other bacteria are affected. Similarly, toxicants will have the same non-selective effect. There has been renewed interest in semi-arid agriculture on the effect of salinity on aspects of the nitrogen cycle.

Since there are many genera of marine denitrifying bacteria that are similar to soil denitrifiers, interest in salinity effects upon denitrification would appear to be largely academic.

Denitrification occurs over a wide pH range from 3.4-11.2, though the optimum is between 7.0-8.0 (Focht and Verstraete, 1977). Proportionally more N_2O is evolved under acidic conditions (Nommik, 1956). Nitrification is also favored in the neutral range between 6.5-7.5 (Focht and Verstraete, 1977), and this factor once again illustrates the important link between the two seemingly different processes.

Chemodenitrification

In general, non-biological reduction of nitrogen oxides to gas occurs primarily under acidic conditions with nitrous acid (pK = 4.2):

$$3\ HNO_2 \dashrightarrow 2NO + HNO_3 + H_2O$$

Nitric oxide reacts rapidly with O_2 to form NO_2, which, in the presence of water, yields nitric acid. Thus, Broadbent and Clark (1965) conclude that there is very little chance that much NO would be lost even from acid soils. Though it is true that NO and O_2 react spontaneously at high concentrations of each, it is not true for low concentrations of either gas since the reaction kinetics are third order. An NO concentration of 10 ppm represents 10% per hour conversion to NO_2 under ambient O_2 concentrations. In the absence of oxygen, much higher concentrations of NO would be stable, and this NO might react with organics. Keeney et al. (1979) found that more gaseous N was evolved than was added in the form of nitrate during high temperature denitrification studies. They also observed significant generation of NO from sterilized NO_2^- amended soil at a pH of 6.2. They postulated that either nitrite or nitric oxide could be involved in nitrosation reactions of organic N, to result in liberation of N_2 from organic matter by a modified "Van Slyke type" reaction.

$$RNH_2 + HNO_2 \dashrightarrow ROH + H_2O + N_2$$

It has been shown by Cho and Sakdinan (1978) that a major portion of N_2 gas was produced from NO_2^- at pH 6.2 in the presence of high organic matter. It is not always easy to separate the contribution of gaseous losses by chemodenitrification from biological denitrification in soil though one unifying concept can be assured: in instances where nitrite accumulates, there are more opportunities for chemical reaction to occur by virtue of its far greater re- reactivity than nitrate. Thus, where nitrate is rapidly reduced to N_2O or N_2 (i.e., where NO_2^- or NO do not accumulate), it seems

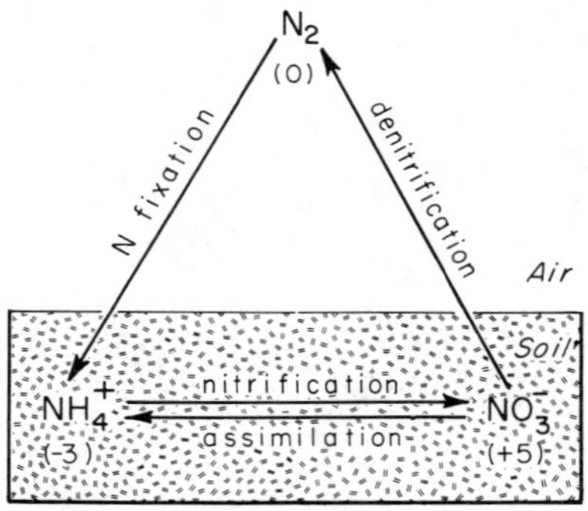

Fig. 5. The nitrogen triangle (from Focht and Martin, 1979).

unlikely that the contribution from chemodenitrification is significant.

<u>One Leg of the Nitrogen Triangle</u>

When stripped of secondary and tertiary reactions, the nitrogen cycle can be envisioned in the schematic shown in Figure 5, and the organic N envisioned as a reversible branch-point from ammonium. Two important aspects deserve mention. 1) The importance of procaryotic organisms in the cycling of nitrogen: denitrification, N-fixation and chemolithotrophic nitrification are effected only by bacteria. 2) The dependence of the denitrification process upon the nitrification process. Inasmuch as the denitrifying bacteria are ubiquitous in soil and respire either oxygen or nitrate, it would appear as germaine to focus upon factors governing the production and consumption of nitrate by organisms other than denitrifiers as a means of conserving soil nitrogen as it would to manage soil in a manner less conductive to denitrification. In this regard, the nitrogen dynamics of the rhizosphere <u>in situ</u> should yield more realistic findings than soil incubation studies. Finally, the realization that some N-fixing bacteria are denitrifiers (Neyra et al., 1977; Eskew et al., 1977; Zablotowicz et al., 1978), and that anaerobic N-fixation (acetylene reduction) is enhanced and coupled to denitrification in crushed-nodule preparations (Zablotowicz and Focht, 1979) should not go unnoticed as this may effect the interpretation of ^{15}N isotope data.

REFERENCES

Allison, F. E., 1955, The enigma of soil nitrogen balance sheets, Adv. Agr., 7:213.

Allison, F. E., 1966, The fate of nitrogen applied to soils, Adv. Agr., 18:219.

Bailey, L. D., and Beauchamp, E. G., 1973, Effects of temperature on NO_3^- and NO_2^- reduction, nitrogenous gas production, and redox potential in a saturated soil, Can. J. Soil Sci., 53: 213.

Balderston, W. L., Sherr, B., and Payne, W. J., 1976, Blockage by acetylene of nitrous oxide reduction in *Pseudomonas perfectomarinus*, Appl. Environ. Microbiol., 31:504.

Biggar, J. W., and Nielsen, D. R., 1976, Spatial variability of the leaching characteristics of a field soil, Water Resour. Res., 12:78.

Birch, H. F., 1958, The effect of soil drying on humus decomposition and nitrogen availability, Plant Soil, 10:9.

Blackmer, A. M., and Bremner, J. M., 1978, Inhibitory effect of nitrate on reduction of N_2O to N_2 by soil microorganisms, Soil Biol. Biochem., 10:187.

Bowman, R. A., and Focht, D. D., 1974, The influence of glucose and nitrate concentrations upon denitrification rates in sandy soils, Soil Biol. Biochem., 6:297.

Broadbent, F. E., and Carlton, A. B., 1979, Methodology for field trials with ^{15}N-depleted fertilizer, in: "Nitrate in Effluents From Irrigated Lands," P. F. Pratt (Principal Investigator), Final Report to the National Science Foundation, University of California, Davis.

Broadbent, F. E., and Clark, F. E., 1965, Denitrification, in: "Nitrogen," W. V. Bartholomew, and F. E. Clark, eds., American Society of Agronomy, Madison, WI.

Broadbent, F. E., and Tusneem, M. E., 1971, Losses of nitrogen from some flooded soils in tracer experiments, Soil Sci. Soc. Amer. Proc., 35:922.

Cho, C. M., and Sakdinan, L., 1978, Mass spectrometric investigation on denitrification, Can. J. Soil. Sci., 58:443.

Currie, J. A., 1961, Gaseous diffusion in the aeration of aggregated soils, Soil Sci., 92:40.

Deherain, P. P., 1897, La reduction des nitrates dans la terre arable, C. R. Acad. Sci., Paris, 124:269.

Denmead, O. T., 1979, Chamber systems for measuring nitrous oxide emission from soils in the field, Soil Sci. Soc. Amer. J., 43:89.

Eskew, D. L., Focht, D. D., and Ting, I. P., 1977, Nitrogen fixation, denitrification, and pleomorphic growth in a highly pigmented Spirillum lipoferum, Appl. Environ. Microbiol., 34:582.

Fedorova, R. T., Milekhina, E. I., and Ilkukhina, N. I., 1973, Possibility of using the "gas-exchange" method to detect extraterrestrial life: Identification of nitrogen-fixing organisms, Izv. Akad. Nauk SSSR, Ser. Biol., 6:797.

Focht, D. D., 1974, The effect of temperature, pH, and aeration on the production of nitrous oxide and gaseous nitrogen -- A zero-order kinetic model, Soil Sci., 118:173.

Focht, D. D., 1979, Microbial kinetics of nitrogen losses in flooded soils, in: "Nitrogen and Rice," N. C. Brady, ed., The Intern. Rice Res. Inst., Los Baños, Philippines.

Focht, D. D., and Chang, A. C., 1975, Nitrification and denitrification processes as related to waste water treatment, Adv. Appl. Microbiol., 19:153.

Focht, D. D., and Martin, J. P., 1979, Microbiological and biochemical aspects of semi-arid agricultural soils, in: "Agriculture in Semi-Arid Environments," G. Cannell and A. E. Hall, eds., Springer-Verlag, New York.

Focht, D. D., Stolzy, L. H., and Meek, B. D., 1979, Sequential reduction of nitrate and nitrous oxide under field conditions as brought about by organic amendments and irrigation management, Soil Biol. Biochem., 11:37.

Focht, D. D., Valoras, N., and Letey, J., 1980, Use of interfaced gas chromatography-mass spectrometry for detection of concurrent mineralization and denitrification in soil, J. Environ. Qual., 9:218.

Focht, D. D., and Verstraete, W., 1977, Biochemical ecology of nitrification and denitrification, in: "Advanced Microbial Ecology, Vol. 1," M. Alexander, ed., Plenum Press, New York.

Freney, J. R., Denmead, O. T., and Simpson, J. R., 1979, Nitrous oxide emission from soils at low moisture contents, Soil Biol. Biochem., 11:167.

Gayon, F., and Dupetit, G., 1886, Reduction des nitrates par les infiniments petis, Mem. Soc. Bordeaux, Ser. 3, 2:201.

Greenwood, D. J., 1961, The effect of oxygen concentration on the decomposition of organic materials in soil, Plant Soil, 14:360.

Hynes, R. K., and Knowles, R., 1978, Inhibition by acetylene of ammonia oxidation in Nitrosomonas europaea, FEMS Microbiol. Lett., 4:319.

Keeney, D. R., Fillery, I. R., and Marx, G. P., 1979, Effect of temperature on the gaseous nitrogen products of denitrification in a silt loam soil, Soil Sci. Soc. Amer. J., 43:1124.

Kohl, D. H., Vitayathill, F., Whitlow, P., Shearer, G., and Chien, S. H., 1976, Denitrification kinetics in soil systems: The significance of good fits to mathematical forms, Soil Sci. Soc. Amer. Proc., 40:249.

Koike, I., and Hattori, A., 1975a, Growth yield of a denitrifying bacterium, Pseudomonas denitrificans, under aerobic and denitrifying conditions, J. Gen. Microbiol., 88:1.

Koike, I., and Hattori, A., 1975b, Energy yield of denitrification: An estimate from growth yield in continuous cultures of Pseudomonas denitrificans, under nitrate-nitrite, and nitrous oxide-limited conditions, J. Gen. Microbiol., 88:11.

MacGregor, A. N., 1972, Gaseous losses of nitrogen from freshly wetted desert soils, Soil Sci. Soc. Amer. Proc., 36:594.

Neyra, C. A., Dobereiner, J., Lalande, R., and Knowles, R., 1977, Denitrification by N_2-fixing Spirillum lipoferum, Can. J. Microbiol., 23:300.

Nielsen, D. R., Biggar, J. W., and Erh, K. T., 1973, Spatial variability of field-measured soil-water properties, Hilgardia, 42:215.

Nommik, H., 1956, Investigations on denitrification in soil, Acta Agric. Scand., 6:195.

Patrick, W. H., and Reddy, K. R., 1976, Nitrification-denitrification reactions in flooded soils and sediments: dependence on oxygen sypply and ammonium diffusion, J. Environ. Qual., 5:469.

Paul, E. A., and Myers, R. J. K., 1971, Effect of soil moisture stress on uptake and recovery of tagged nitrogen by wheat, Can. J. Soil Sci., 51:37.

Payne, W. J., 1973, Reduction of nitrogenous oxides by microorganisms, Bacteriol. Rev., 37:409.

Pratt, P. F., Jones, W. W., and Hunsaker, V. E., 1972, Nitrate in deep soil profiles in relation to fertilizer rates and leaching volume, J. Environ. Qual., 1:97.

Rolston, D. E., Fried, M., and Goldhamer, D. A., 1976, Denitrification measured directly from nitrogen and nitrous oxide gas fluxes, Soil Sci. Soc. Amer. Proc., 40:259.

Rolston, D. E., Broadbent, F. E., and Goldhamer, D. A., 1979, Field measurement of denitrification: II. Mass balance and sampling uncertainty, Soil Sci. Soc. Amer. J., 43:703.

Ryden, J. C., Lund, L. J., and Focht, D. D., 1978, Direct in-field measurement of nitrous oxide flux from soils, Soil Sci. Soc. Amer. J., 42:731.

Ryden, J. C., Lund, L. J., and Focht, D. D., 1979a, Direct measurement of denitrification loss from soils: I. Laboratory evaluation of acetylene inhibition of nitrous oxide reduction, Soil Sci. Soc. Amer. J., 43:104.

Ryden, J. C., Lund, L. J., Letey, J., and Focht, D. D., 1979b, Direct measurement of denitrification loss from soils: II. Development and application of field methods, Soil Sci. Soc. Amer. J., 43:110.

Stolzy, L. H., and Flühler, 1978, Measurement and prediction of anaerobiosis in soils, in: "Nitrogen in the Environment," D. R. Nielsen and J. G. MacDonald, eds., Academic Press, New York.

Walter, H. M., Keeney, D. R., and Fillery, I. R., 1979, Inhibition of nitrification by acetylene, Soil Sci. Soc. Amer. J., 43:

195.
Wagner, P., 1895, Die geringe Ausnutzung des Stallmistickstoffs und ihre Ursachen, Deut. Landw. Presse., 22:98.
Waksman, S. A., 1927, Principles of Soil Microbiology, Williams and Wilkins, Baltimore, Maryland.
Wesseling, J., and von Wijk, W. R., 1957, Land drainage in relation to soil and crops. I. Soil physical conditions in relation to drain depth, in: "Drainage of Agricultural Lands," L. N. Luthin, ed., Amer. Soc. Agron., Madison, Wisconsin.
Woldendorf, J. W., 1962, The quantitative influence of the rhizosphere on denitrification, Plant Soil, 17:267.
Yoshinari, T., and Knowles, R., 1976, Acetylene inhibition of nitrous oxide reduction by denitrifying bacteria, Biochem. Biophys. Res. Commun, 69:705.
Zablotowicz, R. M., Eskew, D. L., and Focht, D. D., 1978, Denitrification in Rhizobium, Can. J. Microbiol., 24:757.
Zablotowicz, R. M., and Focht, D. D., 1979, Denitrification and anaerobic, nitrate-dependent acetylene reduction in cowpea rhizobium, J. Gen. Microbiol., 111:445.

NITRATE TRANSPORT PROCESSES AND COMPARTMENTATION IN ROOT SYSTEMS*

W. A. Jackson and R. J. Volk

Professors of Soil Science
North Carolina State University
Raleigh, NC 27650

BACKGROUND INFORMATION

During normal growing conditions, most higher plants in the vegetative stage tend to accumulate nitrate in the unreduced form in their root tissues and in their above-ground conducting tissues. Under certain stress conditions, this tendency is accentuated (Maynard et al., 1976). Accumulated nitrate thus serves as a storage deposit to sustain growth processes during subsequent periods when stresses are relieved, when the ambient supply becomes limited, or when reproductive growth creates demands for reduced nitrogen which exceed the nitrate uptake rate.

The manner in which accumulated nitrate becomes available for utilization is not well understood. Measurements of *in vivo* nitrate reductase activity (NRA) in the absence of added nitrate (Ferrari et al., 1973; Aslam and Oaks, 1975; Aslam et al., 1976) indicate nitrate is partitioned between a relatively small metabolic pool and a considerably larger storage pool. Within a homogeneous tissue, considerations of cell structure suggest these two pools correspond to cytoplasm and vacuole, respectively, although experiments using dimethylsulfoxide to preferentially degrade the plasmalemma relative to the tonoplast, raise some doubt

*Paper No. 6473 of the Journal Series of the North Carolina Agricultural Research Service, Raleigh. Supported in part by National Science Foundation Grant No. PCM 77-03152. The technical assistance of Penny Windsor and Peggy Longmire is appreciated.

(Delmar, 1979). Although, evidence from in vivo NRA studies indicates only limited movement from the storage pool to the metabolic pool, experiments with intact plants clearly show that much of the previously accumulated nitrate can be utilized. An example is an experiment with corn in which comparisons were made between plants whose nitrate supply was eliminated at silking and those whose external supply was maintained until physiological maturity (Friedrich et al., 1979). Nitrate concentrations in the roots and stems declined rapidly when the ambient supply was eliminated. At silking, 41% and 56% of the total nitrogen in roots and lower stems, respectively, was present as nitrate. Within a week after removal of the external source, the nitrate concentration in the roots had decreased by about 70% and that in the lower stems by 50%. In contrast, when the nitrate supply was continued until physiological maturity, the root tissue concentration remained as high as at silking although in all other tissues it did decline.

Simultaneous accumulation and depletion of nitrate in separate roots of the same plant is shown by a divided root experiment conducted with dark-grown corn seedlings. Five days after germination, all except two seminal roots were excised from each seedling. One root (A) was placed in nutrient solutions containing 15 mM nitrate, while the other (B) was placed in nitrate-free media. After 20 hours, both roots were rinsed thoroughly and exposed to 0.5 mM nitrate. Analysis of the ambient solution during the next ten hours revealed significant differences between the two roots (Fig. 1) although the methodology did not permit separate measurement of the nitrate consumed in reduction and translocation. Root A had accumulated more than 60 μmol nitrate per g during the 20-hour pretreatment in 15 mM nitrate. Upon transfer to 0.5 mM, more than one-third was consumed during the next six hours. There was an initial net efflux to the ambient solution, but after the first hour nitrate was absorbed at a relatively steady rate of 8-9 μmol hr^{-1} g^{-1} FW. Thus, until about six hours after transfer, reduction of nitrate within the root or translocation from it exceeded the quantity taken up. Thereafter a steady state concentration was established at about 40 μmol per g with net uptake equalling consumption. Root B, however, exhibited quite a different pattern; it steadily accumulated nitrate. During the two to six hour period after transfer, it reduced and/or translocated less than root A even though the net uptake rates were similar. After the sixth hour, however, similar rates of consumption by reduction and/or translocation occurred in the two roots, although net uptake from B exceeded that from A. The nature of the regulation of these events is not delineated, but a reasonably rapid responsiveness of the various components of the nitrate assimilation pathway in root systems is indicated.

Among the uncertainties about the nitrate storage and retrieval processes are a) the effectiveness with which previously

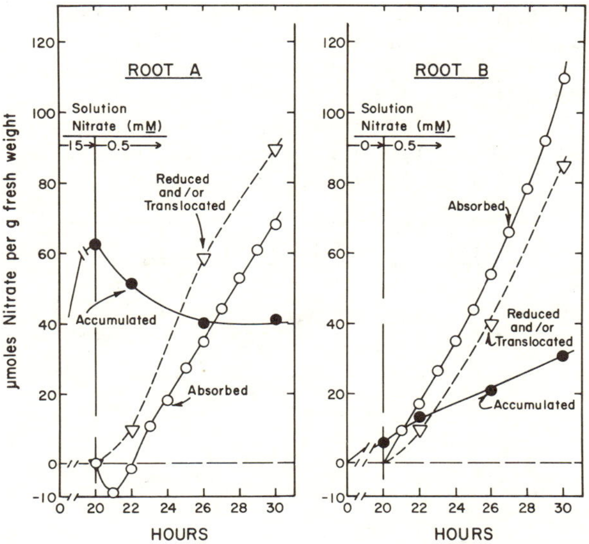

Fig. 1. Influence of 20-hour exposure to 15 mM nitrate to one-half of a dark-grown corn seedling divided-root system on subsequent uptake and accumulation from 0.5 mM nitrate (J. Toledo and W. A. Jackson, unpublished data).

accumulated nitrate is utilized in sustaining dry matter accumulation in the absence of exogenous nitrate, b) the partitioning between translocation to the xylem and reduction within the root system, c) the relative effectiveness with which previously accumulated nitrate and concurrently absorbed nitrate are utilized, d) the possible consequences of restricted influx and the relatively large efflux component of net uptake observed under some conditions of high prior nitrate loading (Jackson et al., 1976; c.f., Loeppert and Kronberger, 1979), and e) the nature of the other solute interchanges between cytoplasm and vacuole required to sustain turgor and charge balance when large quantities of nitrate are deposited or removed. Compartmentation of other inorganic ions (Cl^-, Na^+, K^+) has been examined in some detail by experiments involving compartmental efflux analysis and comparative rates of transfer of exogenous and endogenous ions to the xylem exudate (c.f., Hodges and Vaadia, 1964; Macklon, 1975, Jeschke, 1978). There is no comparable body of literature for nitrate compartmentation (Jackson, 1978).

Table 1. Translocation to the xylem exudate of exogenous (^{15}N-) and endogenous (^{14}N-) nitrate by 10-day-Old corn plants grown under 16/8-hour photoperiods and 25 ± 1C in ½-strength Hoagland's solution (7.5 mM ^{14}N-nitrate). Three hours after the onset of the photoperiod, plants were decapitated and placed in an identical solution containing 99 A% ^{15}N-nitrate. At this time the roots contained 63 μmol ^{14}N-nitrate per plant. Exudate was collected during each of the next two hours during which the ^{15}N-influx was 7 μmol hr^{-1} $plant^{-1}$.

Period (hr)	Nitrate translocated to xylem			Atom % ^{15}N
	$^{15}NO_3$	$^{14}NO_3$	Σ	%
	(μmol hr^{-1} $plant^{-1}$)			
0-1	3.46	1.96	5.42	63.7
1-2	3.35	1.38	4.73	70.8

The following sections describe experiments in which some of these matters are addressed. Included are examinations of utilization of ^{15}N-nitrate following a short pulse, utilization of previously accumulated nitrate in the absence of ambient nitrate, and concomitant utilization of both previously accumulated and concurrently absorbed nitrate. Except for the first, for which intact wheat seedlings were used, and for the data in Table 1, the experiments were conducted with dark-grown corn seedlings (see also Fig. 1). Leaving the endosperm of such seedlings attached results in a continual energy supply to the root system maintaining soluble carbohydrate concentrations in excess of 50 μmol (glucose-equivalent) g^{-1} FW and sustaining high rates of the nitrate assimilation pathway processes. Excision of the shoot results in a sizable exudation of xylem fluid which continues over many hours, providing data on translocation from the root system. The translocation data, together with analysis of the ambient solution and tissue permits measurements of nitrate uptake, accumulation, translocation and reduction. Two characteristics limit direct extrapolation of the observations with this model system to roots of intact autotrophic plants (Jackson, 1978). One is that the carbohydrate status is two- to three-fold higher than that of roots of intact plants. This may result in a greater proportion of the entering nitrate being reduced, and less translocated, in the decapitated seedlings. The second, which may tend in the same direction, is

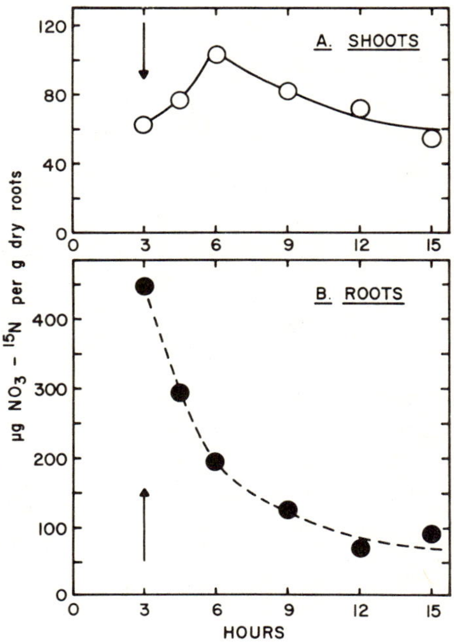

Fig. 2. Depletion of ^{15}N-nitrate in root tissue, and changes in ^{15}N-nitrate of shoots, by nitrogen-depleted wheat plants in Ca $(^{14}NO_3)_2$ following a three-hour exposure to Ca $(^{15}NO_3)_2$ (from Ashley et al., 1975).

that transpirational water flow is eliminated. If nitrate translocation is fostered by transpiration in intact plants, the shorter residence time of nitrate in the root system could decrease reduction. Nevertheless, the system provides a convenient means for examining each of the nitrate assimilation processes involved.

UTILIZATION OF ^{15}N-NITRATE AFTER A SHORT PULSE

Nitrogen-depleted wheat seedlings (Ashley et al., 1975) were exposed to 0.2 mM Ca $(^{15}NO_3)_2$ with 97.5 atom % ^{15}N for three hours followed by a further 12 hours in 0.2 mM Ca $(^{14}NO_3)_2$. Sequential harvests and analysis of both roots and shoots revealed a marked decline in the previously accumulated ^{15}N-nitrate of the roots (Fig. 2B). During the first three hours after transfer, ^{15}N-nitrate in the shoots increased moderately following which it declined (Fig. 2A) indicating translocation of the ^{15}N-nitrate and its subsequent reduction. Translocation of ^{15}N essentially

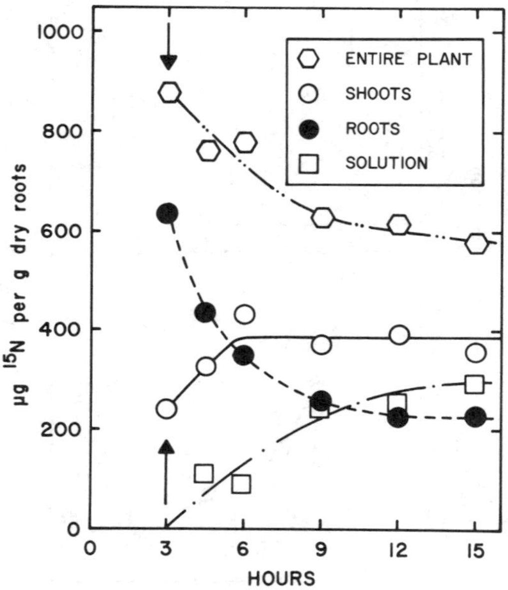

Fig. 3. Distribution of total ^{15}N in nitrogen-depleted wheat plants in Ca $(^{14}NO_3)_2$ following a three-hour exposure to Ca $(^{15}NO_3)_2$ (from Ashley et al., 1975).

ceased by the third hour after transfer, even though signigicant quantities of ^{15}N-nitrate and reduced-^{15}N remained in the root tissue at that time (Fig. 3). The evidence indicates some sequestering of the ^{15}N components as ^{14}N-nitrate continually entered the roots. There was also a sizable net efflux of ^{15}N-nitrate to the ambient solution (Fig. 3).

UTILIZATION OF ENDOGENOUS NITRATE IN THE ABSENCE OF NITRATE UPTAKE

The effectiveness with which previously accumulated nitrate was utilized for reduction and translocation was examined by exposing five-day-old dark-grown corn seedlings to complete solutions containing KNO_3 at 0.5, 5 and 50 mM for 19.5 hours at 30 C. They were then decapitated and transferred to nitrate-free media containing 0.5 mM KCl, labeled with Cl-36. At this time the tissue contained 31, 47 and 84 μmol nitrate per g root fresh weight. Analysis of the xylem exudate collected during the subsequent eight hours revealed a marked decrease in the nitrate concentration (Fig. 4A) indicating a limited capability of the previously accumulated nitrate to sustain the translocation process in the absence

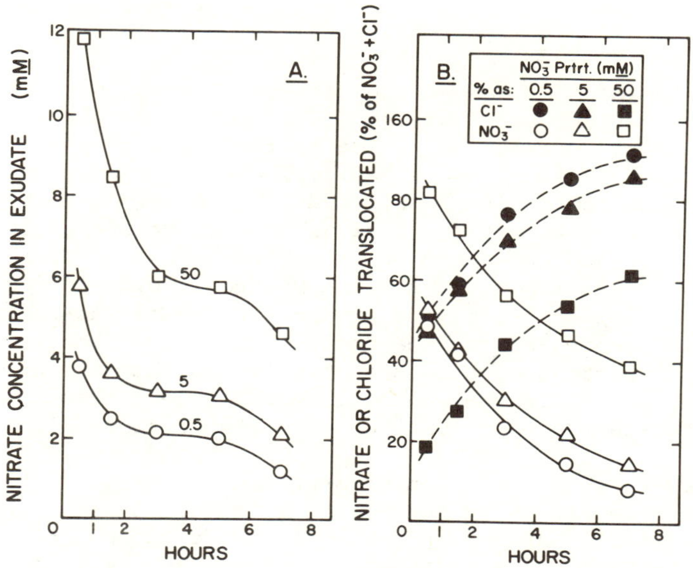

Fig. 4. Nitrate concentrations (A) and the proportions present as nitrate or chloride (B) in the xylem of decapitated dark-grown corn seedlings during an eight-hour exposure to 0.5 mM KCl following a 19.5-hour pretreatment in 0.5, 5, and 50 mM KNO_3 (W. A. Jackson and R. J. Volk, unpublished data).

of an ambient supply, even when the tissues contained high nitrate concentrations. It should be noted that decapitation by itself does not cause a restriction in nitrate translocation by these roots (Jackson et al., 1973; Ezeta and Jackson, 1975). A decrease in nitrate reductase activity of corn roots upon removal of ambient nitrate has also been observed (Oaks et al., 1972; Jackson et al., 1973). Thus the previously accumulated nitrate does not sustain translocation and nitrate reductase activity as effectively as does nitrate entering the tissues. But the data (Fig. 4A) also show that significant quantities of the previously accumulated nitrate continued to be translocated, albeit at slower rates, after the transfer (c.f., Hodges and Vaadia, 1964; Anderson et al., 1974). Again, there is some parallelism with the decline in nitrate reductase activity; significant activity remains some hours after transfer to nitrate-free media (Oaks et al., 1972; Aslam and Oaks, 1975). The data imply that movement out of storage deposits can occur with reasonable facility, although it is not as effective as transport across the plasmalemma from the ambient solution.

Estimates of the rates of deposition of exogenous chloride in the xylem exudate were obtained by dividing the Cl-36 activity recovered in each collection period by the ambient specific activity. Figure 4B shows the steadily changing proportions of the two ions deposited in the xylem during the eight-hour period. During the first hour, nearly equivalent proportions of the exogenous chloride and endogenous nitrate were transferred to the xylem in those seedlings previously exposed to 0.5 and 5 mM nitrate. A steady decline in the proportion present as nitrate occurred with all three pretreatments. Thus, translocation of endogenous nitrate was restricted relative to that of the exogenously supplied chloride, although high concentrations of endogenous nitrate tended to counter this restriction (Fig. 4B).

For all three pretreatments, approximately 20% of the initial nitrate was translocated to the xylem during the eight-hour exposure (Fig. 5). Analysis of the tissue at the end of the experiment showed a depletion about three-fold as great at that which could be accounted for by transport to the xylem. No nitrate accumulation was detectable in the solution at the end of the experiment. Hence, reduction of the previously accumulated nitrate was approximately twice that of translocation. It should be emphasized that the data show only what resulted between initiation and eight hours; they do not permit a conclusion that the approximately 2:1 ratio of reduction to translocation occurred throughout the experimental period. Differential partitioning between the two processes could have occurred during this time when the translocation rates changed appreciably (Fig. 4A). Nor do the data exclude the possibility that a significant component of reduction could have occurred in association with influx following prior efflux to external unstirred layers during the eight-hour period (Morgan et al., 1973). What the results do indicate is that about 60% of the previously accumulated nitrate was able to be reduced within the roots or translocated from them in the absence of an external nitrate supply.

Because of the evidence indicating compartmentation of other inorganic ions in root tissue (e.g., Hodges and Vaadia, 1964; Davis and Higinbotham, 1976; Jescke, 1978), it seems reasonable to conclude that nitrate would behave qualitatively in the same way. Accordingly, it would be anticipated that concurrently absorbed nitrate would be utilized in preference to previously accumulated nitrate for reduction and translocation. To our knowledge, there is little experimental evidence which bears directly on this point. Results of our preliminary experimentation described below do, however, support this expectation, at least for the translocation process.

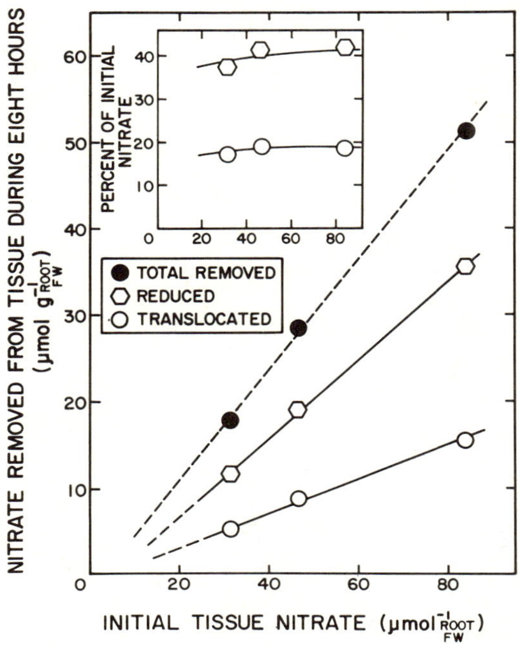

Fig. 5. Reduction and translocation of nitrate previously accumulated in roots of decapitated dark-grown corn seedlings during an eight-hour exposure to 0.5 mM KCl following a 19.5 h pretreatment in 0.5, 5 and 50 mM KNO_3 (W. A. Jackson and R. J. Volk, unpublished data).

CONCURRENT UTILIZATION OF ENDOGENOUS AND EXOGENOUS NITRATE

An indication of the extent of nitrate compartmentation within root systems can be obtained from relative rates of appearance of the nitrate isotopic species in the xylem exudate when roots are placed in ^{15}N-nitrate following prior exposure to ^{14}N-nitrate. Experiments of this sort with both light-grown and dark-grown seedlings clearly reveal preferential translocation of the exogenous ^{15}N-nitrate, although significant translocation of the endogenous ^{14}N-nitrate did occur. An example with light-grown corn plants exposed to 7.5 mM ^{14}N-nitrate for 10 days is shown in Table 1. Three hours after the onset of the photoperiod, plants were decapitated and placed in an identical solution containing 99 A% ^{15}N-nitrate. Analysis of the exuding xylem fluid revealed that 5.4 and 4.7 μmoles of nitrate per plant were translocated

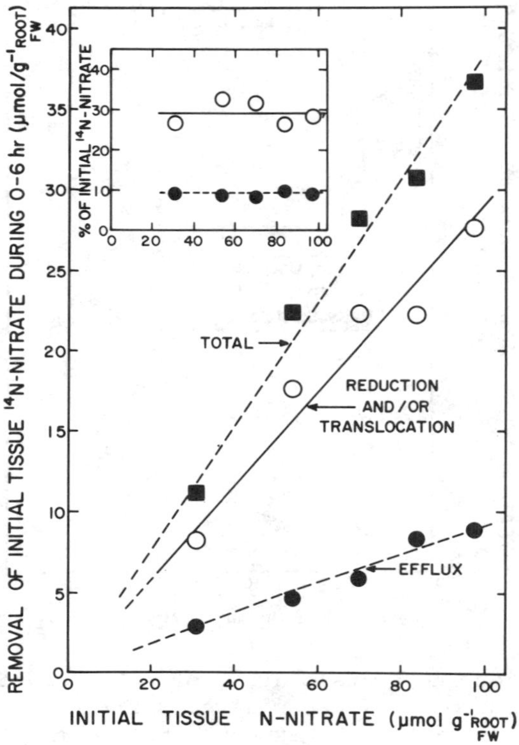

Fig. 6. Utilization of endogenous ^{14}N-nitrate (previously accumulated in root tissue of decapitated dark-grown corn seedlings during 18 hours exposure to $K^{14}NO_3$ from 0.5 to 50 mM) during a six-hour exposure to $K^{15}NO_3$ (R. J. Volk and W. A. Jackson, unpublished data).

during the following two hourly periods (Table 1). The root tissue initially contained 63 μmol ^{14}N-nitrate per plant and only 7 μmol ^{15}N-nitrate were absorbed during each hour. Nevertheless, 64 and 71% of the translocated nitrate originated from the exogenous ^{15}N-nitrate.

Similar preferential utilization of exogenous nitrate was observed with dark-grown corn seedlings preloaded with nitrate for 18 hours by exposure to $K^{14}NO_3$ from 0.5 to 50 mM. They then were transferred to 0.5 mM $K^{15}NO_3$ (99 A% ^{15}N) for a further six hours.

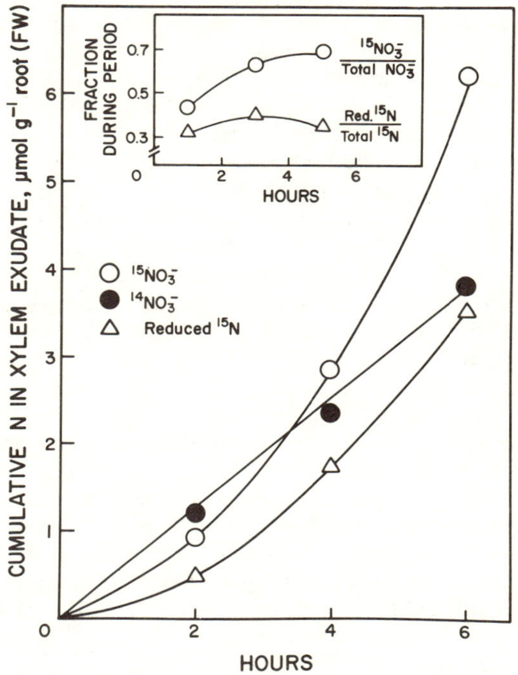

Fig. 7. Translocation of ^{15}N-nitrate, ^{14}N-nitrate and reduced-^{15}N to the xylem during a six-hour period in 0.5 mM $K^{15}NO_3$ following pretreatment in 0.5 mM $K^{14}NO_3$. The insert shows the proportion of ^{15}N-nitrate relative to total nitrate and the proportion of reduced-^{15}N relative to total ^{15}N in the xylem in each two-hour measurement period (c.f., Fig. 6).

Enrichment of the nitrate in the tissue and exudate, and the decrease in enrichment in the ambient solution, were determined (Volk et al., 1979) as was enrichment in the tissue reduced-nitrogen. During the six-hour period, about 30% of the initial tissue ^{14}N-nitrate was utilized for reduction and/or translocation, and about 10% was recovered in the ambient medium (Fig. 6). Due to the use of different sets of plants and a difference in time of exposure, the results shown in Figures 5 and 6 are not directly comparable. Nonetheless, the suggestion is that a smaller proportion of the endogenous nitrate was utilized in reduction and translocation when ambient nitrate was concurrently being absorbed.

The relative rates of appearance of exogenous (^{15}N-) and endogenous (^{14}N-) nitrate in the xylem (Fig. 7) provides some evidence for compartmentation although it appears not to be as marked as in the roots of light-grown plants (c.f., Table 1). During the first two hours after transfer, ^{15}N-nitrate was 43% of the total nitrate translocated in the seedlings initially having 31 μmol ^{14}N-nitrate g^{-1} (Fig. 7). The rate of ^{15}N-nitrate translocation increased steadily. During the four to six hour period, 69% of the exudate nitrate was present as ^{15}N-nitrate (Fig. 7, insert) while at six hours the total tissue nitrate contained only 54% ^{15}N. Preferential translocation of exogenous nitrate is clearly indicated although considerable turnover of the endogenous nitrate did occur. Reduced-^{15}N constituted about 35% of the total ^{15}N in the exudate with this proportion being relatively constant as the rate of deposition of the entering ^{15}N increased progressively (Fig. 7, insert). The xylem exudate of these seedlings contains substantial quantities of reduced-N (Ezeta and Jackson, 1975) with a sizable component likely being derived from the endosperm. The proportion of reduced-N in the xylem derived from the exogenous ^{15}N-nitrate increased steadily from 6% at the second hour to 14% at the sixth hour. The data indicate relatively effective reduction of the entering nitrate and subsequent translocation of the products of that reduction. Partitioning of the entering ^{15}N-nitrate over the six-hour period is shown in Table 2. Approximately half remained unreduced in the roots whereas about one-third was reduced and one-fifth was translocated as nitrate. Of that which was reduced, about one-third was translocated. Hence, effective entry of the ambient ^{15}N-nitrate into each of the nitrate assimilation pathways occurred. Especially significant is the observation that about 10% of the entering ^{15}N-nitrate was translocated in reduced form, constituting slightly more than a third of the total ^{15}N translocated. The partitioning shown in Table 2 differs from that of dwarf bean (Breteler and ten Cate, 1980) primarily by a greater proportion of the ^{15}N-nitrate being reduced in the root system and a larger translocation of the reduced-^{15}N.

As the initial ^{14}N-nitrate tissue concentrations increased, there was a progressive increase in the ^{14}N-nitrate translocated and a corresponding decrease (from 43% to 20%) in the atom % ^{15}N of the nitrate entering the xylem during the first two hours after transfer to 0.5 mM $K^{15}NO_3$ (Fig. 8). The more rapid translocation of nitrate observed with increasing nitrate concentration during pretreatment was largely a consequence of an enhanced translocation of the ^{14}N-nitrate; the previously accumulated ^{14}N-nitrate exerted relatively little impact on translocation of the entering ^{15}N-nitrate, implying a translocation capacity not saturated by endogenous ^{14}N-nitrate. There was, however, a tendency for translocation of products of ^{15}N-nitrate reduction to be restricted during the initial two hours, suggesting an impact on reduction of the ^{15}N-nitrate.

Table 2. Influx of ^{15}N-nitrate and its partitioning among reduction, accumulation and translocation by dark-grown corn seedlings during a six-hour period in 0.5 mM $K^{15}NO_3$* following 18 hours exposure to 0.5 mM $K^{14}NO_3$*.

	μmol g^{-1} root FW	% of Influx	% of Translocated
^{15}N-nitrate influx	33.7		
^{15}N-nitrate accumulation in roots	16.6	49	
Reduced-^{15}N accumulation in roots	7.3	22	
^{15}N retained in roots	23.9	71	
^{15}N-nitrate translocated to xylem	6.2	18	63
Reduced-^{15}N translocated to xylem	3.6	11	37
^{15}N translocated	9.8	29	
^{15}N-nitrate reduced	10.9	32	33

*In complete nutrition solution.

CONCLUDING STATEMENTS

Present information does not permit many definitive conclusions regarding the regulation of nitrate movement into and out of storage pools in roots (or other plant tissue). The data presented here does indicate that previously accumulated ^{14}N-nitrate was not in a static pool. In absence of ambient nitrate it was utilized effectively in reduction and translocation (Fig. 5) and, even in presence of ambient nitrate, sizable utilization occurred (Fig. 6). Quantities reasonably estimated to be present initially in the cytoplasm and vacuole of corn roots can be calculated assuming the tissue to be 90% H_2O, the cytoplasmic volume to be $\simeq$ 5% of the total root volume and the cytoplasmic concentration to be $\simeq$ two-fold that of the vacuole (c.f., Davis and Higinbotham,

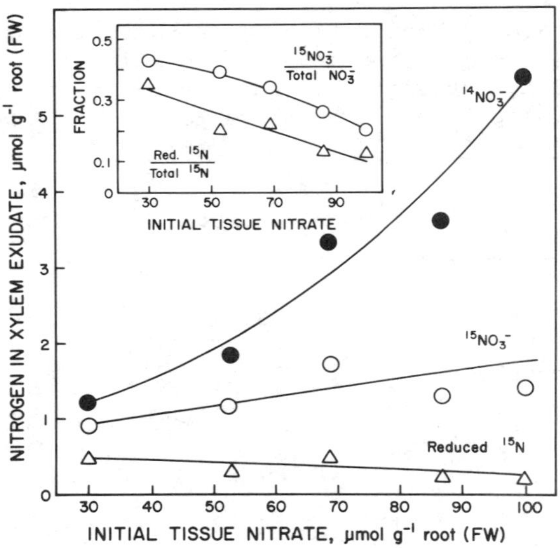

Fig. 8. The effect of increasing initial ^{14}N-nitrate concentrations on translocation of ^{14}N-nitrate, ^{15}N-nitrate, and reduced-^{15}N to the xylem during the first two hours exposure to $K^{15}NO_3$ (c.f., Fig. 6).

1976). On the assumption that metabolic and storage pools correspond to cytoplasm and vacuoles, respectively, it can be calculated that efflux, translocation and/or reduction in the dark-grown seedlings (Fig. 6) consumed considerably more previously-accumulated ^{14}N-nitrate during the six-hour period than could have been present initially in the metabolic pool, implying substantial removal from storage. It should also be noted that this removal occurred even though there was a sizable resupply of the metabolic pool from the ambient solution. It also seems reasonable to suggest that the relative utilization of stored nitrate in reduction and translocation may not be identical in all parts of the root system. Differences in nitrate reductase activities and stability between meristematic and basal regions (Aslam and Oaks, 1975; Oaks et al., 1977) indicate the strong possibility of such vertical compartmentation.

There is therefore a clear need to define the metabolic and storage pool sizes more accurately in different parts of the root system and to determine what regulates the fluxes between the pools. Studies with protoplasts and isolated vacuoles may permit direct comparisons of vacuolar and extravacuolar concentrations (e.g., Wagner, 1979) as well as direct measurement of fluxes across the tonoplast (e.g., Doll et al., 1979), and will no doubt help resolve some of the current uncertainties about nitrate compartmentation. But the involvement of reduction, translocation, and efflux in affecting metabolic nitrate pool sizes dictates further examination insofar as possible of the intact, functioning root system, and of different regions within it.

REFERENCES

Ashley, D. A., Jackson, W. A., and Volk, R. J., 1975, Nitrate uptake and assimilation by wheat seedlings during initial exposure to nitrate, Plant Physiol., 55:1102.

Anderson, W. P., Goodwin, L., and May, R. K., 1974, Evidence for vacuole involvement in xylem ion supply in the excised primary roots of two species, Zea mays and Allium cepa, in: "Structure and Function of Primary Root Tissue," J. Kolek, ed., Veda, Bratislava, Czechoslovakia.

Aslam, M., and Oaks, A., 1975, Effect of glucose on the induction of nitrate reductase in corn roots, Plant Physiol., 56:634.

Aslam, M., Oaks, A., and Huffaker, R. C., 1976, Effect of light and glucose on the induction of nitrate reductase and on the distribution of nitrate in etiolated barley leaves, Plant Physiol., 58:588.

Breteler, H., and ten Cate, C. H., 1980, Fate of nitrate during initial nitrate utilization by nitrogen-depleted dwarf bean, Physiol. Plant., 48:292.

Davis, R. F., and Higinbotham, N., 1976, Electrochemical gradients and K^+ and Cl^- fluxes in excised corn roots, Plant Physiol., 57:129.

Delmar, D. P., 1979, Dimethylsulfoxide as a potential tool for analysis of compartmentation in living plants, Plant Physiol., 64:623.

Doll, S., Rodier, F., and Willenbrink, J., 1979, Accumulation of sucrose in vacuoles isolated from red beet tissue, Planta, 144:407.

Ezeta, F. N., and Jackson, W. A., 1975, Nitrate translocation by detopped corn seedlings, Plant Physiol., 56:148.

Ferrari, T. O., Yoder, O. C., and Filner, P., 1973, Anaerobic nitrite production by plant cells and tissues: evidence for two nitrate pools, Plant Physiol., 51:423.

Friedrich, J. W., Schrader, L. E., and Nordheim, E. V., 1979, N deprivation in maize during grain-filling. I. Accumulation of dry matter, nitrate-N and sulfate-S, Agron. J., 71:461.

Hodges, T. K., and Vaadia, Y., 1964, Uptake and transport of radiochloride and tritiated water by various zones of onion roots of different chloride status, Plant Physiol., 39:104.

Jackson, W. A., 1978, Nitrate acquisition and assimilation by higher plants: processes in the root system, in: "Nitrogen

in the Environment," D. R. Nielsen and J. G. MacDonald, eds., Vol. 2, Academic Press, New York.
Jackson, W. A., Flesher, D., and Hageman, R. H., 1973, Nitrate uptake of dark-grown corn seedlings: some characteristics of apparent induction, Plant Physiol., 51:120.
Jackson, W. A., Kwik, K. D., and Volk, R. J., 1976, Nitrate influx and efflux by intact wheat seedlings: Effects of prior nitrate nutrition, Planta, 132:149.
Jeschke, W. D., 1978, Univalent cation selectivity and compartmentation in cereals, in: "Recent Advances in the Biochemistry of Cereals," D. L. Laidman, and R. G. Wyn Jones, eds., Academic Press, New York.
Loeppert, H., and Kronberger, W., 1979, Control of nitrate uptake by photosynthesis in Lemna paucicostata 6746, in: "Photosynthesis and Plant Development," Proc. Conf., R. Marcelle, H. Clijsters, and M. van Poucke, eds., Junk, The Hague.
Macklon, A. E. S., 1975, Cortical cell fluxes and transport to the stele in excised root segments of Allium cepa L. I. Potassium, sodium and chloride, Planta, 122:109.
Maynard, D. N., Barker, A. V., Minotti, P. L., and Peck, N. H., 1976, Nitrate accumulation in vegetables, Advances in Agronomy, 28:71.
Morgan, M. A., Volk, R. J., and Jackson, W. A., 1973, Simultaneous influx and efflux of nitrate during uptake by perennial ryegrass, Plant Physiol., 51:267.
Oaks, A., Wallace, W., and Stevens, D., 1972, Synthesis and turnover of nitrate reductase in corn roots, Plant Physiol., 50:649.
Oaks, A., Aslam, M., and Boesel, I., 1977, Ammonium and amino acids as regulators of nitrate reductase in corn roots, Plant Physiol., 59:391.
Volk, R. J., Pearson, C. J., and Jackson, W. A., 1979, Reduction of plant tissue nitrate to nitric oxide for mass spectrometric ^{15}N analysis, Anal. Biochem., 97:131.
Wagner, G., 1979, Content and vacuole/extravacuole distribution of neutral sugars, free amino acids, and anthocyanin in protoplasts, Plant Physiol., 64:88.

REDUCTION OF NITRATE AND NITRITE IN BARLEY LEAVES IN DARKNESS

M. Aslam, R. C. Huffaker*, and C. C. Delwiche**

*Plant Growth Laboratory and the Department of Agronomy and Range Science, and **Department of Land, Air, and Water Resources, University of California, Davis, CA 95616

INTRODUCTION

Reports are contradictory as to whether nitrate and nitrite are reduced in darkness in green leaves. Several workers report that nitrate is not assimilated in darkness in green leaves (Canvin and Atkins, 1974; Sawhney et al., 1978). Taken as evidence for a light requirement also for nitrite reduction is the coupling of nitrite reduction to photosynthetic electron flow at the level of ferredoxin (Miflin, 1974; Neyra and Hageman, 1974). Sadler and Scott (1974) observed that nitrite reduction is almost completely light-dependent and that the process does not require CO_2 fixation in barley leaves. Sawhney et al. (1978) reported that nitrate reduction occurs only in light, so accumulation of toxic levels of nitrite in the dark is avoided. It has been proposed that nitrite is not found in darkness under aerobic conditions because O_2 inhibits nitrate reduction (Sawhney et al., 1978; Atkins and Canvin, 1975).

Jones and Sheard (1978), however, reported the reduction of nitrate and nitrite in leaf slices in darkness. They also showed evidence that O_2 is required for nitrite assimilation. We recently observed that green barley leaves reduced nitrate at steady-state rates for several hours in both light and darkness, but the rate was twice as fast in light (Aslam et al., 1979). We further found that nitrite also was assimilated in both light and dark in green barley leaves (Aslam et al., 1979). Because nitrite can be lost to the atmosphere through the transpiration stream (Stutte and Weiland, 1978), it is important to follow the products of nitrite reduction into ammonium and organic N. Using $^{15}NO_2^-$, we now show that nitrite is reduced to the level of ammonium and organic nitrogen in green leaves in darkness, and that the process utilizes the available energy reserves very efficiently.

MATERIALS AND METHODS

Barley seedlings were grown in a growth chamber in nitrogen-free Hoagland's solution under continuous light as described (Aslam et al., 1979). After 7 days, leaf tips 10 cm long (10 leaves for each treatment) were excised and placed base down in small glass vials containing 5 mM KNO_3 or $NaNO_2$. Chloramphenicol at 50 μg/ml was added to uptake solutions to prevent bacterial growth.

To produce carbohydrate deficiency in the leaves, one set of seedlings was placed in darkness and another set was left in light. Thirty-six hours later, 10 leaf tips 10 cm long were excised, placed base down in small glass vials containing 10 ml of 5 mM $Na^{15}NO_2$ (95% ^{15}N), and incubated in light or dark for 24 hr.

Nitrate and nitrite uptake was measured as that disappearing from solution. Reduction was determined by subtracting the total amount accumulated by the leaves from the total uptake, as described (Chantarotwong et al., 1976; Aslam et al., 1979). Incorporation of $^{15}NO_2$-N into reduced nitrogen compounds was calculated from total reduced nitrogen and atom % excess of ^{15}N in reduced nitrogen according to the following formula: $^{15}NO_2$-N incorporation (μmoles g^{-1} dry wt) = (total reduced nitrogen (μg g^{-1} dry wt)) (1/100 of atom % excess ^{15}N) ÷ (molecular wt of N). Total nitrogen was determined after Kjeldahl digestion in sulfuric acid. The digest was steam-distilled in the presence of concentrated NaOH, and the distillate was collected in boric acid containing methyl-red indicator. The distillate was then titrated against standard 0.05 M H_2SO_4, and the nitrogen content was calculated on a dry-weight basis. After titration, the distillate was dried at 60-65 C and the isotopic concentration of ^{15}N was determined by mass spectrometry.

RESULTS AND DISCUSSION

Although leaves reduced only 50% of the nitrate absorbed, all of the nitrite absorbed was reduced over a 24-hr period in darkness (Fig. 1). A slight amount of nitrite accumulated in the leaves during the early hours, but that too was reduced with time. Not only was nitrite reduced in darkness, it was reduced more efficiently than nitrate. Carbohydrate-starved green leaves reduced only 20% of the nitrate but reduced 85% of the nitrite absorbed over a 24-hr period (Fig. 2). Nitrate reduction in carbohydrate-starved leaves stopped completely after 12 hr, whereas nitrite reduction continued at a constant rate up to 24 hr. When glucose was supplied to the

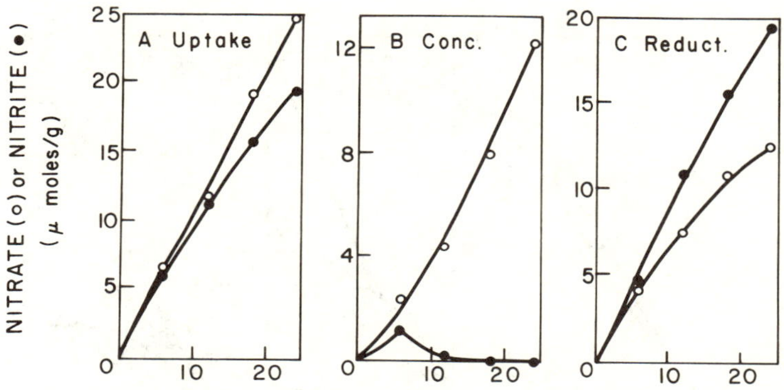

Fig. 1. Nitrate and nitrite assimilation is excised light-grown barley leaves in dark. Seedlings were grown for 7 days in continuous light. Tips 10 cm long of 10 leaves, were then excised, placed base down in small glass vials containing 10 ml of 5 mM KNO_3 or $NaNO_2$, and incubated in dark. Uptake was measured as the nitrate or nitrite disappearing from the solution. Nitrate was determined after conversion to nitrite by Klebsiella respiratory nitrate reductase (Thayer and Huffaker, 1980). Nitrite was determined as discribed by Aslam et al. (1979).

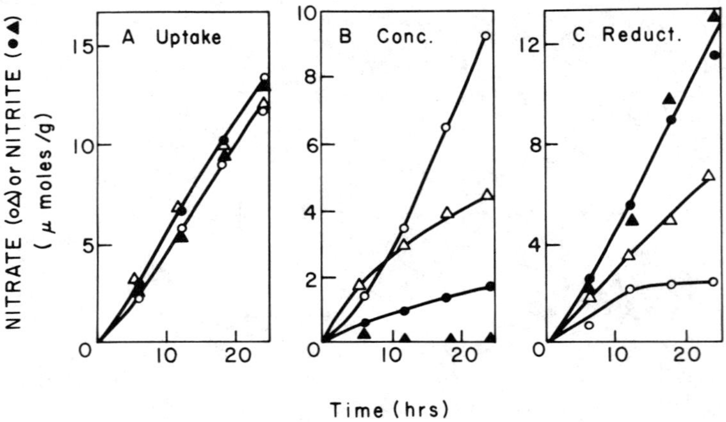

Fig. 2. Nitrate and nitrite assimilation in carbohydrate-deficient barley leaves in darkness in the absence (0, ●) and presence (Δ, ▲) of 0.1 M glucose. The experimental procedure was as in Figure 1, except that the seedlings after 6 days of light were placed in darkness for 36 hrs to deplete carbohydrates.

leaves, nitrate reduction also continued at a constant rate. About 50% of the nitrate absorbed was reduced in a 24-hr period. In contrast, all of the nitrite absorbed was reduced in the presence of glucose (Fig. 2). The results suggest that nitrate reduction was limited by energy supply.

Stutte and Weiland (1978) observed that inorganic nitrogen in several crop and weed species was lost as gaseous products to the atmosphere via the transpiration stream. We therefore employed two methods to follow the fate of the NO_2^--N in detached leaves. In the one method, NO_2^- reduction was determined by subtracting the internal concentration of NO_2^- from its uptake (Table 1, middle column). In the other method, $^{15}NO_2^-$ was followed into reduced product. All of the $^{15}NO_2^-$-N absorbed in light and darkness in carbohydrate-rich leaves was recovered as ammonium and organic nitrogen (Table 1). In carbohydrate-starved leaves in light, all $^{15}NO_2^-$-N was incorporated into reduced nitrogen, and in darkness 85 to 90% of $^{15}NO_2^-$-N went into reduced N; 10 to 15% remained as $^{15}NO_2^-$. Total recovery of $^{15}NO_2^-$-N indicates that in barley leaves NO_2^- was not lost to the atmosphere.

The reduction of nitrite appears to occur in the chloroplast (Atkins and Canvin, 1975; Jones and Sheard, 1978), utilizing reduced ferredoxin as the electron donor. It now seems that cellular metabolites may move into the chloroplast and reduce NADP. NADPH reduction could then reduce ferredoxin, supplying electrons for nitrite reduction. The much greater reduction of nitrite than nitrate in darkness in carbohydrate-starved leaves, which have very low levels of soluble sugars (2.8%, compared with 14.5% in

Table 1. Assimilation of $^{15}NO_2^-$-into reduced nitrogen in carbohydrate-rich and carbohydrate-deficient excised leaves in light and dark.

Treatment	Uptake	Reduction	Incorporation into reduced N
	μmoles (g dry wt x 24 hr)$^{-1}$		
a) Carbohydrate-rich			
Light	155	155	164
Dark	125	125	129
b) Carbohydrate-deficient			
Light	171	171	182
Dark	124	105	110

normal leaves), is interesting. Since nitrite is toxic at low levels, whatever energy is available seems to be used primarily for the reduction of nitrite instead of nitrate. In either case, both nitrate and nitrite are reduced in darkness in carbohydrate-rich leaves, but the available energy in carbohydrate-depleted leaves can be directed preferentially toward nitrite reduction.

SUMMARY

Light-grown leaves, rich in carbohydrates, have the ability to reduce both nitrate and nitrite in darkness. Carbohydrate-starved leaves are little able to reduce nitrate but reduce 85% of nitrite absorbed. This suggests that nitrite reduction is much more efficient than nitrate reduction and can utilize electrons from many sources.

ACKNOWLEDGEMENT

This work was supported in part by a grant from the National Science Foundation (PRF 77-07301).

REFERENCES

Aslam, M., Huffaker, R. C., Rains, D. W., and Rao, K. P., 1979, Influence of light and ambient carbon dioxide concentration on nitrate assimilation by intact barley seedlings, *Plant Physiol*., 63:1205.

Atkins, C. A., and Canvin, D. T., 1975, Nitrate, nitrite and ammonia assimilation by leaves: effect of inhibitors, *Planta*, 123:41.

Canvin, D. T., and Atkins, C. A., 1974, Nitrate, nitrite and ammonia assimilation by leaves: effect of light, carbon dioxide and oxygen, *Planta*, 116:207.

Chantarotwong, W., Huffaker, R. C., Miller, B. L., and Granstedt, 1976, *In vivo* nitrate reduction in relation to nitrate uptake, nitrate content and *in vitro* nitrate reductase activity in intact barley seedlings, *Plant Physiol*., 57:519.

Jones, R. W., and Sheard, R. W., 1978, Accumulation and stability of nitrite in intact aerial leaves, *Plant Sci. Lett*., 11:285.

Miflin, B. J., 1974, Nitrite reduction in leaves: studies on isolated chloroplasts, *Planta*, 116:187.

Neyra, C. A., and Hageman, R. H., 1974, Dependence of nitrite reduction on electron transport in chloroplasts, *Plant Physiol*., 54:480.

Sadler, R., and Scott, K. J., 1974, *in*: "Mechanisms of Regulation of Plant Growth," R. L. Bieleski, A. R. Ferguson, M. M. Cresswell, eds., Bulletin 12, The Royal Society of New Zealand, Wellington.

Sawhney, S. K., Naik, M. S., and Nicholas, D. J. D., 1978, Regulation of nitrate reduction by light, ATP and mitochondrial respiration in wheat leaves, Nature, 272:647.

Stutte, C. A., and Wieland, R. T., 1978, Gaseous nitrogen loss and transpiration of several crop and seed species, Crop Sci., 18:887.

Thayer, J. R., and Huffaker, R. C., 1980, Determination of nitrate and nitrite by high-pressure liquid chromatography: comparison with other methods for nitrate determination, Anal. Biochem., 102:110.

INFLUENCE OF LIGHT AND CO_2 ON NITRATE ASSIMILATION BY BARLEY SEEDLINGS

M. Aslam, R. C. Huffaker, D. W. Rains, and K. P. Rao

Plant Growth Laboratory and the Department of Agronomy
and Range Science, University of California
Davis, CA 95616

Although the role of light in nitrate assimilation has been studied extensively (Beevers and Hageman, 1969, 1972), controversy continues as to whether nitrate reduction normally occurs in darkness. Nitrate reduction under anaerobic conditions has been reported in tissue slices in darkness; it was detected as nitrite accumulation in the medium (Atkins and Canvin, 1975; Klepper et al., 1971; Radin, 1973; Sawhney et al., 1978a,b). Under aerobic conditions, however, nitrite does not accumulate in the dark (Aslam et al., 1979). That observation has been ascribed to two contradictory processes: 1) that O_2 inhibited nitrate reduction (Atkins and Canvin, 1975; Sawhney et al., 1978a,b); or 2) that both nitrate and nitrite reduction occurred in the presence of O_2. This contradiction can be resolved by following each of the processes involved: nitrate uptake, accumulation, and reduction or at least the disappearance of the product (nitrite). When the appearance of nitrite is the only assay, it is not possible to determine whether nitrate reduction was inhibited by O_2 or whether nitrite was further reduced. Either case would account for the inability to find nitrite in the incubation medium. Others did not find proof that $^{15}NO_3^-$ converts into amino acids in barley leaves in darkness during short-term experiments (Canvin and Atkins, 1974). The purpose of this report was to describe nitrate and nitrite assimilation by barley seedlings in light and darkness.

MATERIALS AND METHODS

Barley seeds were surface-sterilized in a 1% Clorox solution for 15 min, rinsed, germinated for 24 hr in distilled water, and grown for 6 days on a layer of cheesecloth supported by a stainless steel screen suspended over a 0.2 mM $CaSO_4$ solution (Aslam et al., 1979; Chantarotwong et al., 1976).

Uptake of nitrate was determined as that disappearing from the external solution over time. In vivo reduction of nitrate was determined by subtracting the total amount of nitrate in the seedlings from the total uptake.

The effect of CO_2 on nitrate assimilation was done by placing seedlings in a 15 l plexiglass chamber. Normal or CO_2-free air was passed through the chamber at 2 l/min.

RESULTS AND DISCUSSION

Our results show clearly that nitrate reduction in intact barley seedlings under aerobic conditions occurs at linear rates in both light and darkness (Fig. 1). In darkness, storage reserves apparently drive nitrate reduction, whereas in light, energy

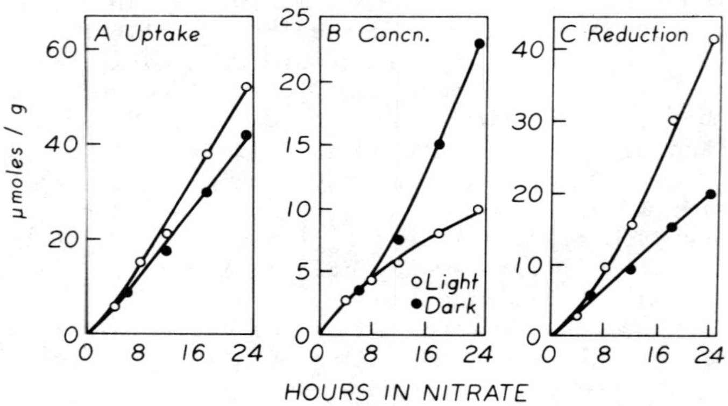

Fig. 1. Time course of nitrate uptake (A), accumulation (B), and in vivo reduction (c) in light (500 $\mu E/m^2 \times sec$) (O) and darkness (●) in light-grown seedlings. Seedlings were grown 5 days in darkness and 3 days in light, then placed in an uptake medium containing 1.0 mM KNO_3 and 5 mM $CaSO_4$ in one-fourth-strength Hoagland solution. Uptake was determined as the nitrate disappearing from the solution and reported on a fresh weight basis. Solutions were changed every 12 h. In vivo reduction was determined by subtracting the total amount of nitrate in both roots and shoots from the total uptake of nitrate at each assay period. Rates were calculated from the slopes after linearity occurred.

for nitrate reduction might be derived from both stored reserves and recently fixed photosynthate. Nitrate reduction in darkness occurred in both roots and leaves.

Since uptake supplies the nitrate flux through the metabolic pool (Aslam et al., 1976; Chantarotwong et al., 1976; Shaner and Boyer, 1976), decreases in uptake could be reflected by corresponding decreases in nitrate reduction. In the present studies, however, the slightly decreased uptake in darkness probably did not account for the decreased reduction rate, because the steady-state concentration of internal nitrate was doubled.

The slower nitrate reduction in darkness might be due partially to the activation of inhibitors in darkness, as reported by Jolly and Tolbert (1978) for soybeans. Our recent research (unpublished) shows that when excised barley leaves in darkness are supplied with glucose, they reduce more than 90% of the absorbed nitrate. Hence, reserve substrate supply seems to be more important than inhibitors in barley leaves.

CO_2 on Nitrate Assimilation

Part of the faster reduction in light is probably a function of recent products of CO_2 fixation. In the intact barley plants, CO_2 affected nitrate reduction almost exclusively, and not uptake (Fig. 2). A concomitant increase in tissue nitrate resulted, and there was less reduction in the absence of CO_2. The necessity for photosynthetic CO_2 fixation was shown in carbohydrate-deficient seedlings. Light did not increase nitrate reduction unless CO_2 was present (Fig. 3). It appears that recently fixed photosynthate could, if necessary, supply all the energy required for nitrate reduction. In the presence of CO_2 and light, rates of nitrate reduction were similar for both carbohydrate-"sufficient" and -"deficient" plants. Recently fixed photosynthate could supply energy for nitrate reduction by chloroplast shuttle systems. CO_2 and light increased nitrate reduction in corn also (Neyra and Hageman, 1976).

It has been reported that CO_2 affected nitrate reduction in corn seedlings secondarily by limiting uptake or movement of nitrate from the roots or stalk into the leaves (Neyra and Hageman, 1976). Others did not find proof that CO_2 influences conversion of nitrate to amino acids in short-term studies using barley leaves (Canvin and Atkins, 1974).

In summary, nitrate assimilation occurs in both light and darkness in barley seedlings, the light rate being two times that in the dark. Stored carbohydrate and recent products of photosynthetic CO_2 fixation can apparently supply the energy requirement for nitrate and nitrite reduction.

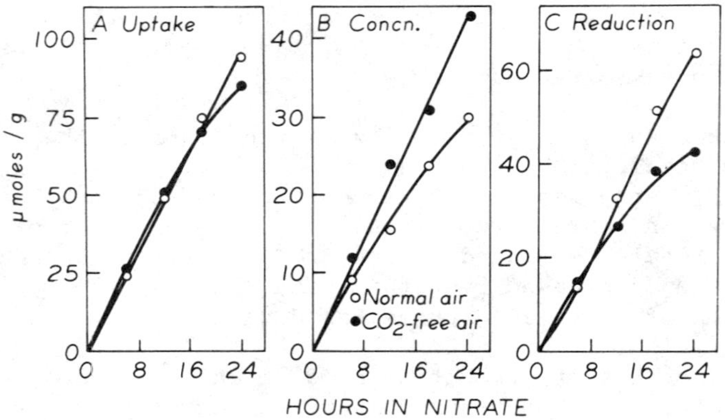

Fig. 2. Effect of CO_2-free (●) and normal (O) air on time course of nitrate uptake (A), accumulation (B), and reduction (C) in carbohydrate-sufficient seedlings in light. Seedlings were grown as in Figure 1. Ten seedlings per treatment were placed in the uptake medium, which contained 1 mM KNO_3 and 5 mM $CaSO_4$ in one-fourth-strength Hoagland solution. Seedlings were then transferred to a Plexiglas chamber through which CO_2-free or normal air was passed at 2 l/min. Uptake solutions were aerated with the same air.

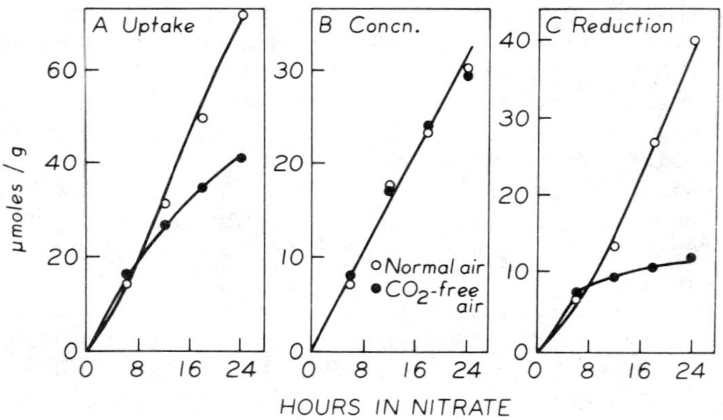

Fig. 3. Effects of normal (O) and CO_2-free (●) air on time course of nitrate uptake (A), accumulation (B), and reduction (C) in carbohydrate-deficient seedlings in light. Seedlings were grown as in Figure 1. After light treatment, seedlings were placed in darkness for 24 h to deplete carbohydrate. Other conditions were the same as in Figure 2.

ACKNOWLEDGMENT

This work was supported in part by a grant from the National Science Foundation (PFR 77-07301).

REFERENCES

Aslam, M., Huffaker, R. C., and Travis, R. L., 1973, The interaction of respiration and photosynthesis in induction of nitrate reductase activity, Plant Physiol., 52:137.

Aslam, M., Oaks, A., and Huffaker, R. C., 1976, Effect of light and glucose on the induction of nitrate reductase and on the distribution of nitrate in etiolated barley leaves, Plant Physiol., 58:588.

Aslam, M., Huffaker, R. C., Rains, D. W., and Rao, K. P., 1979, Influence of light and ambient carbon dioxide concentration on nitrate assimilation by intact barley seedlings, Plant Physiol., 63:1205.

Atkins, C. A., and Canvin, D. T., 1975, Nitrate, nitrite and ammonia assimilation by leaves. Effect of inhibitors, Planta, 123:41.

Beevers, L., and Hageman, R. H., 1969, Nitrate reduction in higher plants, Ann. Rev. Plant Physiol., 20:495.

Beevers, L., and Hageman, R. H., 1972, The role of light in nitrate metabolism in higher plants, in: "Photophysiology," Vol. 7, A. C. Giese, ed., Academic Press, New York.

Canvin, D. T., and Atkins, C. A., 1974, Nitrate, nitrite and ammonia assimilation by leaves: effect of light, carbon dioxide and oxygen, Planta, 116:207.

Chantarotwong, W., Huffaker, R. C., Miller, B. L., and Granstedt, R. C., 1976, In vivo nitrate reduction in relation to nitrate uptake, nitrate content and in vitro nitrate reductase activity in intact barley seedlings, Plant Physiol., 57:519.

Jolly, S. O., and Tolbert, N. E., 1978, NADH-nitrate reductase inhibitor from soybean leaves, Plant Physiol., 62:197.

Klepper, L., Flesher, D., and Hageman, R. H., 1971, Generation of reduced nicotinamide adenine dinucleotide for nitrate reduction in green leaves, Plant Physiol., 48:580.

Neyra, C. A., and Hageman, R. H., 1976, Relationships between carbon dioxide, malate, and nitrate accumulation and reduction in corn (Zea mays L.) seedlings, Plant Physiol., 58:726.

Radin, J. W., 1973, In vivo assay of nitrate reductase in cotton leaf discs. Effect of oxygen and ammonium, Plant Physiol., 51:332.

Sawhney, S. K., Naik, M. S., and Nicholas, D. J. D., 1978a, Regulation of NADH supply for nitrate reduction in green plants via photosynthesis and mitochondrial respiration, Biochem. Biophys. Res. Commun., 81:1209.

Sawhney, S. K., Naik, M. S., and Nicholas, D. J. D., 1978b, Regulation of nitrate reduction by light. ATP and mitochondrial

respiration in wheat leaves, Nature, 272:647.
Shaner, D. L., and Boyer, J. S., 1976, Nitrate reductase activity in maize (Zea mays L.) leaves. I. Regulation by nitrate flux, Plant Physiol., 58:499.

LIGHT INTERACTION WITH NITRATE REDUCTION

Steven L. Morrison, R. C. Huffaker, and C. Rick Guidara

Plant Growth Laboratory and the Department of
Agronomy & Range Science, University of California,
Davis, CA 95616

Light affects assimilation of nitrate via nitrate reductase in many ways. Two prominent examples are: an effect on uptake and movement of nitrate between storage and metabolic pools; and production and distribution of reducing power and generation of carbon skeletons via photosynthesis. Nitrate reductase activity (NRA) correlates more closely with nitrate flux into leaf tissue than leaf nitrate concentration. The possibility that nitrate "flux" through a pool may regulate NRA fits well with the postulation that NO_3^- exists in "metabolic" (active) pools, and "storage" (inactive) pools, where only the active pool is available for induction of NRA and for use as a substrate. The existence of "storage" and "metabolic" pools has been demonstrated in tobacco cells (Heimer and Filner, 1971; Ferrari et al., 1973), corn roots (Aslam and Oaks, 1975) and etiolated barley leaves (Aslam et al., 1976).

The mechanisms through which production and partitioning of reducing power for nitrate reduction and carbon skeletons for amino acid biosynthesis are controlled remain uncertain. An analysis suggests that reduction of nitrate by active nitrate reductase would continue to occur if: 1) nitrate were present in a "metabolic" pool; 2) production of reducing power occurred; and 3) intracellular compartmentation remained functionally intact to insure regulated movement of nitrate and reductant to the site of the enzyme. Evaluation of nitrate reductase activity and nitrate assimilation under varied environmental conditions therefore depends heavily on the localization of nitrate reductase and on the rates of generation and mobilization of reductant to the site of enzyme activity via intracellular shuttle systems. The presumed cytoplasmic location of nitrate reductase (Ritenour et al., 1969; Grant et al., 1970; Swader and Stocking, 1971; Dalling et al., 1972) and results of Beevers and Hageman (1969) suggested NADH, rather than NADPH, was the preferred donor for nitrate reductase in a wide variety of species. Stocking and Larson (1969) and Walker and Crofts (1970) suggested that reducing power generated in the light in the chloroplast was transferred to the cytoplasm via shuttle systems for use in nitrate reduction.

Several other types of information also suggest that carbohydrate supply may play a role in mobilization of nitrate and in cytoplasmic generation of reductant. Leaf disks infiltrated with nitrate and sugars (fructose-1,6-diphosphate, glyceraldehyde phosphate, or glucose) enhanced production of nitrate under dark-anaerobic conditions (Klepper and Hageman, 1969). Exogenously added glucose increased uptake and mobilization of nitrate in barley roots (Aslam and Oaks, 1975). The fact that infiltration of sugars increased reduction of nitrate under dark-anaerobic conditions suggested a cytoplasmic origin for the generation of reductant. Klepper et al. (1971) concluded that products of photosynthesis (glyceraldehyde-3-phosphate, hexoses, or phosphorylated hexoses) migrated to the cytoplasm and were metabolized by glycolytic enzymes to cytoplasmically generate NADH as reductant. Permeability of the plastid membrane to phosphorylated sugars has been documented by Heber and Willenbrink (1964), Urbach et al. (1965), Walker and Crofts (1970), and Stocking et al. (1963).

Although the role of light in providing reductant for nitrate and nitrite reduction seems understood, the integration of associated metabolic processes is not as clear. This study has addressed the following questions. How does light affect transport and mobilization of nitrate between various pools in the plant? Is reductant available for nitrate and nitrite reduction in the dark? How is net nitrate reduction related to nitrate flux through the system in light and dark?

MATERIALS AND METHODS

Plant Material

Barley (*Hordeum vulgare* var. Numar) was planted, 40 seeds/4 inch pot, in medium vermiculite and watered. After 24 hr imbibition in water, a stock nutrient solution was supplied which contained in mmole/l: $MgSO_4$, 2; KH_2PO_4, 1; $CaSO_4$, 2.5; and in µmol/l: $MnSO_4$, 18.3; H_3BO_3, 8.0; $ZnSO_4$, 3.8; $CuSO_4$, 1.5; $(NH_4)_6Mo_7O_{24}$, 0.1; NaCl, 28.2; and Fe as Fe-ethylenediamine di-(O-hydroxyphenylacetate), 110.4. Nitrate was supplied as 0.5 mM $Ca(NO_3)_2$ + 1.25 mM K_2SO_4, 2.5 mM $Ca(NO_3)_2$ + 1.25 mM K_2SO_4, or 2.5 mM $Ca(NO_3)_2$+ 5 mM KNO_3 + 1.25 mM K_2SO_4, to yield final nitrate concentrations of 1, 5, and 10 mM.

Plants were grown in growth chambers with a 16-hr day, 8-hr night cycle. Day temperature was maintained at 25 C and night temperature at 15 C. Relative humidity was maintained at 70%. Light was supplied from a mixture of warm white fluorescent and incandescent bulbs and maintained at 550 $\mu E \cdot m^{-2} \cdot sec^{-1}$ at the top of the canopy.

Experimental Procedures

Detached Leaf Experiments. The top 10 cm of 7-day-old primary leaves were excised under water, weighed, and placed in treatment solutions, 10 leaves per vial. For detached leaf experiments, these solutions contained 10 mM NO_3^- as 3.33 mM $Ca(NO_3)_2$ and 6.67 mM KNO_3. Light-air experiments were conducted in a growth chamber. Light-nitrogen treatments were conducted in clear temperature regulated-humidity controlled Lucite boxes linked to an IRGA. Dark-air and dark-nitrogen treatments were carried out in black Lucite boxes at 25 C temperature and 70% relative humidity.

Uptake of nitrate in this system remained linear for seven hours, and all experiments were conducted within this time span.

Leaf Slice Experiments. The 2-4 cm section (measured from tip) of primary barley leaves from 50 seven-day-old plants were clamped in a hand microtome and sectioned into 0.5 mm slices cut across the leaf blade (adapted from Smith and Epstein, 1964). One hundred leaf sections were floated in each treatment vial upon the treatment solutions. These solutions were minus NO_3^- or contained 1, 5, or 10 mM NO_3^- (as a mixture of Ca^{2+} and K^+ salts) in a 1 mM Tris or morpholinopropane sulfonic acid (MOPS) buffer (pH 7.2). Chloramphenicol was added at 25 μg/ml. Uptake of nitrate by these leaf slices remained linear for 6 hr. Treatments were normally continued until 20% of available nitrate was taken up (approximately 3-5 hr). At the end of the treatment period the solution was removed, leaf slices washed 5 sec with 1 mm Tris or MOPS buffer (pH 7.0) minus nitrate and frozen until analysis of tissue NO_3^- content was made.

Leaf slices were illuminated from the bottom by a Westinghouse MG 400 Bu/4 400-watt metal halide lamp. Neutral density filters were used to reduce the light intensity to 550 $\mu E \cdot m^{2} \cdot sec^{-1}$. Plastic screening was used to modulate light intensity in experiments to determine light saturation curves for nitrate reduction and photosynthesis.

Nitrite Analysis

Sample aliquots (0.1-0.2 ml) were diluted to 1 ml with distilled water. Two ml of a 1:1 mixture of 1% sulfanilamide (w/v) in 1.5 M HCl and 0.02% N-1-napthylethylenediamine-dihydrochloride was added and shaken. Absorbance at 540 nm was read after 15 min.

Nitrate Analysis

Nitrate was assayed spectrophotometrically at 210 nm following separation by HPLC on a Whatman Partisil-SAX anion exchange column (Thayer and Huffaker, 1980). The eluting buffer was 35 mM phosphate (pH 3.0) run at a flow rate of 1.5 ml/min. Frozen tissue samples were thawed and extracted in 80% ethanol until the slices were translucent and free of chlorophyll. Samples were diluted 1:10 with H_2O and assayed. Nitrate uptake was analyzed as NO_3^- loss from treatment solutions following filtration through a 0.45 µ Millipore filter.

RESULTS AND DISCUSSION

Figure 1 shows a hypothetical pathway for the uptake and reduction of nitrate. The "free space" NO_3^- ($NO_3^-{}_{F.S.}$) along with a presumed vacuolar NO_3^- pool ($NO_3^-{}_{vac}$), is considered a storage pool. Nitrate reduction is considered to occur in the cytoplasm in this model, hence $NO_3^-{}_{cyto}$ constitutes the "metabolic" NO_3^- pool. Nitrite reduction occurs in the plastid, producing NH_4^+ and under some conditions NO_x. Nitrate accumulates in stems, petioles, and leaves of many species. The localization of NO_3^- pools in the leaf is significant because determination of NO_3^- concentration and movement through the "metabolic NO_3^- pool" may depend on both intra- and intercellular movement of nitrate.

The results in Figure 2 suggest that nitrate is not uniformly distributed in primary leaves of barley plants, highest nitrate concentrations being associated with the midrib of the leaf. NO_3^- found in a 1 mm midrib strip was 8-fold more concentrated than in a similar strip cut halfway between the midrib and the edge of the blade. As much nitrate was contained in a 1-mm strip cut along the midrib as was contained in the remainder of the leaf.

Experiments with Detached Leaves

Rates of NO_3^- uptake and reduction by detached leaves of barley grown on 5 mM NO_3^- are shown in Table 1A and 1B. The ratio of nitrate uptake to water loss remained constant for all treatments (data not shown) suggesting NO_3^- uptake in detached leaves was dependent on transpiration as expected.

Primary leaves of plants grown on 5 mM NO_3^- are low in nitrate but actively assimilate nitrate. Leaf NO_3^- more than tripled during the light treatments. Nitrate reduction was 3-fold greater in light and air than for any other treatment. This was due to both greater uptake and more efficient reduction. Uptake of NO_3^- during light-N_2 and dark-air treatments was depressed 60-73% due to decreased transpiration, and the percent NO_3^- reduced declined from 71% to 44-47%. This indicated that a larger fraction of NO_3^- remained

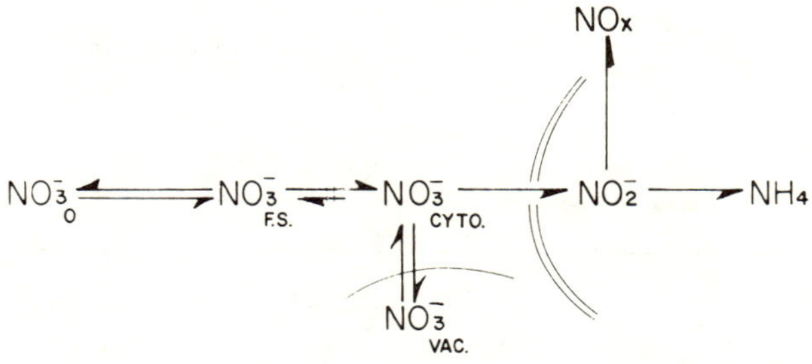

Fig. 1. A schematic flow diagram for the path of nitrate assimilation in barley leaf cell. $NO_3^-{}_o$ = external nitrate; $NO_3^-{}_{F.S.}$ = free space nitrate; $NO_3^-{}_{cyto}$ = cytoplasmic nitrate; $NO_3^-{}_{vac}$ = vacuolar nitrate; NO_x^- = oxides of nitrogen.

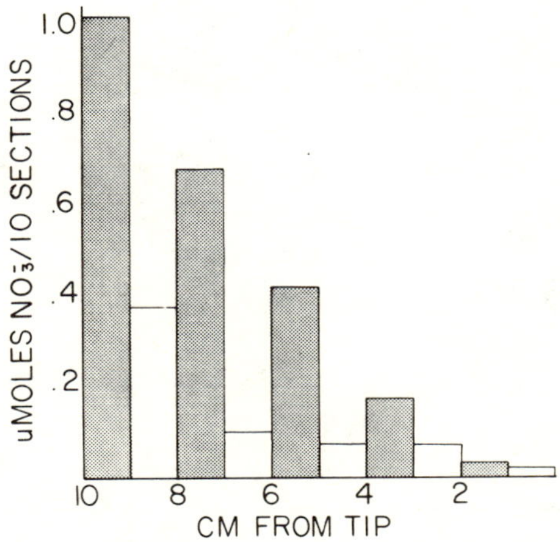

Fig. 2. µmoles NO_3^- contained in a 2 cm x 1 mm strip removed from the midrib of a 10-cm barley leaf (shaded bars) and from a 2-cm x 1-mm strip excised halfway between the midrib and edge of the leaf (open bars). Sections were obtained from the primary leaf of a 7-day-old barley plant grown on nutrient solution containing 5 mM nitrate.

Table 1. Rate of nitrate uptake and reduction by detached leaves of barley plants.

A) Assimilation of nitrate by detached, base fed, 10 cm barley leaves.*

	NO_3^- (μmoles/gfw x 6 hr)				
	Original	Uptake	Available	Final	Reduction
LT-Air	2.2	23.5±2.0	25.6	7.4±.4	18.2±1.9
LT-N_2	2.2	9.5±1.5	11.7	6.5±.8	5.2±1.1
DK-Air	2.2	6.9±1.4	9.1	4.8±.5	4.3±1.3
DK-N_2	2.2	6.3±0.9	8.5	2.9±.3	5.6±0.7

B) Percent NO_3^- and NO_2^- reduced as determined from data in 1A.

	NO_3^- assimilation to NO_2^-			NO_2^- reduced to RNH_2	
Sample	Available	Reduced	%		%
LT-air	25.6	18.2	71	18.2	100
LT-N_2	11.7	5.2	44	5.2	100
DK-air	9.1	4.3	47	4.3	100
DK-N_2	8.5	5.5	65	3.3	60

* Analysis of nitrate reduction is made by subtraction where: Original nitrate + nitrate uptake = total nitrate available; Total nitrate - final nitrate = nitrate reduced; LT = light; DK = dark; N_2 = nitrogen treatment.

and was partitioned into storage pools, or that reductant was not available for reduction of nitrate in an increasingly large metabolic pool. Although NO_3^- uptake and total NO_3^- available under dark-air and dark-N_2 treatments were not significantly different, the percent of total nitrate assimilated was increased in dark-N_2 treatments from 47-65%. Appearance of nitrite was observed in this system only during dark-N_2 treatments. On the basis of the balance sheet, 60% of the NO_2^- formed was further assimilated or lost from the system.

Experiments with Leaf Slices

Tissue slices have been used in a number of experiments to evaluate uptake of ions (Smith and Epstein, 1964; Rains, 1968). Use of leaf slices to evaluate nitrate assimilation has two initial advantages. First, it avoids problems related to long distance transport and transpiration in detached leaves. Second, it maintains cellular integrity necessary for ion uptake. NO_3^- assimilation by such a leaf slice system is shown in Figure 3. Uptake of nitrate by 0.5 mm leaf slices of barley from a 1 mm external NO_3^- solution was linear and not significantly different in light and dark. This pattern was similar to results obtained in whole plants (Chantarotwong et al., 1976; Aslam et al., 1979), but differed from that in detached leaves (Table 1A-1B). This discrepancy probably resulted because NO_3^- was efficiently loaded into the xylem in roots of intact plants (Shaner and Boyer, 1976a,b), while uptake in detached leaves depended on relative rates of transpiration. Reduction in the dark in the leaf slice system occurred at

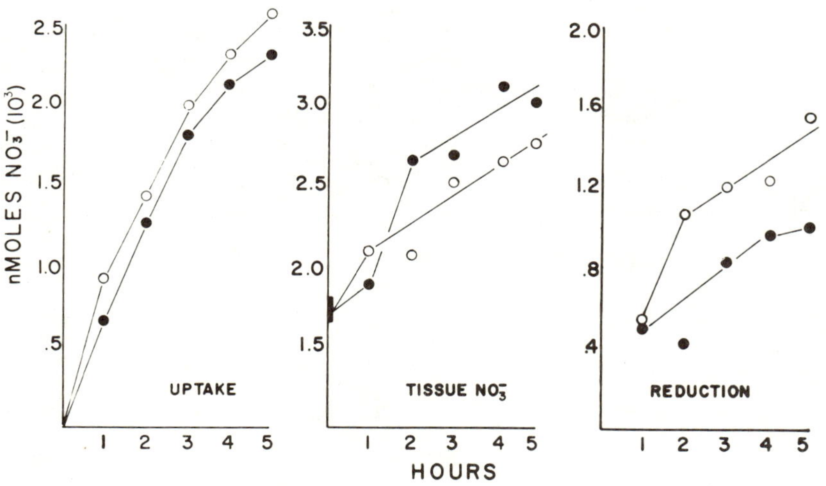

Fig. 3. Uptake and reduction of nitrate from a 1 mM nitrate solution by 100 0.5 mm leaf slices from primary leaves of plants grown 7 days on nutrient solution containing 5 mm nitrate.

50-70% the rate observed in the light. Leaf slices were incubated in dark under aerobic conditions versus "bubbled-N_2" at 0.2 $l \cdot min^{-1}$ (Table 2). Anaerobiosis accelerated NO_3^- uptake, lowered original tissue nitrate levels, doubled the rate of nitrate reduction, and produced nitrite, of which 61% was reduced further or otherwise lost from the system. Thus nitrate and nitrite reduction occurred in darkness as well as light, and nitrate was reduced under anaerobic conditions.

It is possible by varying nutritional conditions during the growth of the seedlings to "preset" a number of physiological parameters in primary barley leaves such as tissue nitrate, rates of nitrate uptake and reduction, and photosynthetic rate. The "preset" physiological parameters can be perturbed by changing treatments and the effects examined using leaf tissue slices. Results of an experiment which examined the effect of varying NO_3^- during both growth and treatment of leaf slices on rates of nitrate reduction are shown in Table 3. Rates of nitrate reduction responded to tissue nitrate concentration, preset rates of enzyme activity, and the rate of nitrate uptake as caused by varying concentrations of external nitrate. Although the rate of uptake of NO_3^- from a 1 mM NO_3^- solution was more rapid in slices from plants grown on 5 mM rather than 1 mM nitrate, plants grown on 10 mM nitrate took up less nitrate than plants grown on 1 mM NO_3^- (data not presented. This resulted from tissue free-space nitrate concentrations high enough to generate NO_3^- unloading (negative uptake);

Table 2. Nitrate assimilation by 0.5 mm leaf slices grown 7 days on nutrient solution containing 5 mM nitrate. Data are presented as nmoles NO_3^-/5 hr/100 leaf slices. Numbers in parenthesis are % of nitrate and nitrite reduced. T_o = nmoles tissue nitrate at beginning of the incubation. T_F = nmoles tissue nitrate at end of incubation.

	nmoles NO_3^- / 100 leaf slices				nmoles NO_2^-	% NO_2^-
Treatment	Uptake	T_o	T_F	Reduced	Produced	Reduced
DK-Air	69	1548	1308	309 (16)	4.8	(98)
DK-N_2	119	1548	776	891 (50)	345	(61)

Table 3. Effect of nitrate supply on nitrate reduction. Each 4 inch pan was supplied 100 ml of 1 or 4 mM nitrate during seven days of growth. Nitrate was then supplied to leaf slices as minus nitrate (.2 mM $CaSO_4$), 1 mM, or 10 mM nitrate. Results are given as nmoles NO_3^-/6 hr/50 leaf slices.

mM NO_3^- during treatment	mM NO_3^- during growth	
	1	5
	NO_3^- red	NO_3^- red
0	133	530
1	593	760
10	1217	1990
DK (1 mM KO_3^-)	--	471

hence, the higher rates of reduction in high nitrate plants were supported by tissue nitrate pools.

During investigation of the interaction of nitrite reduction and photosynthesis (Table 4), no significant inhibition of photosynthesis was observed at high light intensity even during high rates of nitrate reduction. Photosynthesis was inhibited 23% when leaf slices were incubated with 1 mM nitrate at low light intensity, but addition of nitrate also doubled the rate of dark $^{14}CO_2$ assimilation. Addition of $CaSO_4$ and K_2SO_4 to leaf slices stimulated rates of photosynthesis which suggested the possibility of a cation stimulation of $^{14}CO_2$ incorporation; hence we could not discriminate between cation and anion effects in this system.

Assimilation of nitrate occurs at approximately 1/20 the rate of photosynthesis, suggesting that low light intensity may inhibit nitrate reduction more than photosynthesis. Light saturation curves for nitrate uptake and reduction by plants grown on 1 mM nitrate were generated using the leaf slice system (Fig. 4). Negligible leaf nitrate accumulated in plants grown on 1 mM NO_3^- and nitrate uptake was high. Little effect of light intensity on nitrate uptake was observed. Tissue nitrate increased more rapidly at low light intensities than high, suggesting nitrate uptake was

Table 4. Effect of nitrate and light intensity on photosynthesis.

Treatment	Minus NO_3^- [a]	Minus NO_3^- [b]	1 mM NO_3^- [c]
1000 $\mu E/M^2\cdot sec$	2.32±1.16	--	2.29±.06
360 $\mu E/M^2\cdot sec$	1.94±.14	2.29±.18	1.45±.08
Dark	.06±.002	--	.11±.03

All units = µmoles ^{14}C/hr. 10 leaf slices.
[a]Tris-SO_4 = buffer (1 mM, pH 7.0)
[b]Tris-SO_4 = buffer (1 mM, pH 7.0) + .25 mM $CaSO_4$ + .25 mM K_2SO_4.
[c]Tris-SO_4 = buffer (1 mM, pH 7.0) + .25 mM $Ca(NO_3)_2$ + .5 mM KNO_3.

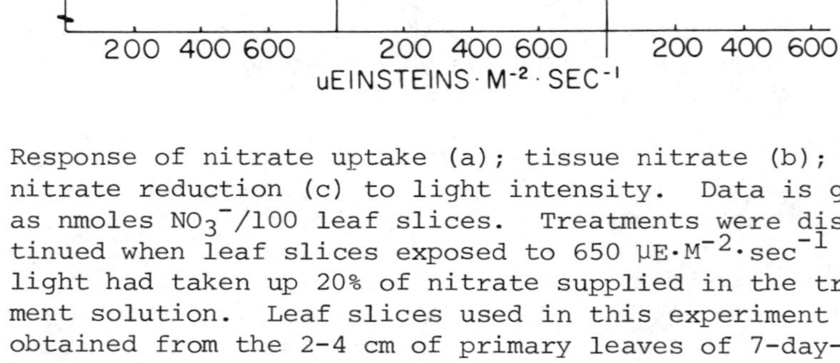

Fig. 4. Response of nitrate uptake (a); tissue nitrate (b); and nitrate reduction (c) to light intensity. Data is given as nmoles NO_3^-/100 leaf slices. Treatments were discontinued when leaf slices exposed to 650 $\mu E\cdot M^{-2}\cdot sec^{-1}$ of light had taken up 20% of nitrate supplied in the treatment solution. Leaf slices used in this experiment were obtained from the 2-4 cm of primary leaves of 7-day-old barley plants. Nitrate supplied during growth was 1 mM. 1 mM NO_3^- in 1 mM MOPS buffer (pH 7.2) was supplied during the treatment.

less responsive to light intensity than nitrate reduction. Nitrate reduction showed a saturation curve when measured as a function of light intensity, because more nitrate is accumulated at low than high light intensities.

CONCLUSIONS

The mechanisms controlling nitrate uptake, transport, and assimilation remain somewhat an enigma, although light intensity, CO_2 levels, temperature, water stress, and nitrate supply play a role (Beevers and Hageman, 1969). The synthesis and activity of nitrate reductase depends on a series of processes which maintain a flow of nitrate (Chantarotwong et al., 1976; Shaner and Boyer, 1976a,b) to a cellular metabolic pool (Ferrari, 1973; Aslam and Oaks, 1975; Aslam et al., 1976), and light may have multiple roles in maintaining these processes. Any evaluation of the effects of light on uptake and mobilization of nitrate between "storage" and "metabolic" pools must consider carbohydrate supply (Aslam and Oaks, 1975; Jackson and Volk, this volume), transpiration (Table 2) and nitrate distribution within and between different leaf tissues (Figure 2). Such studies should also be made with consideration of light effects on localized movement between vacuole and cytoplasm. Light, both directly and indirectly, plays a role in primary nitrogen assimilation (nitrate to glutamine) in that it affects production of reductant and ATP.

Based on our model (Fig. 1), nitrate reduction *per se* should continue as long as active enzyme, nitrate, and a source of reducing power are present. Generation of reducing power for nitrate and nitrite reduction in the dark or under anaerobic conditions may occur via one or a combination of reactions. These sequences are 1) oxidation of organic acids via the Krebs cycle reactions, 2) glycolysis, and 3) the pentose phosphate shunt.

Our results showed that nitrate reduction occurred in both light and dark (Table 1A-1B; Fig. 3; Table 4) and under anaerobic conditions (Table 1A-1B; Table 3). We conclude that failure to observe nitrite production under dark-aerobic conditions does not constitute evidence for the absence of nitrate reduction, since we observed disappearance of nitrate from the system. These conclusions are also supported by evidence obtained by Aslam et al. (this volume) from barley seedlings which showed $^{15}NO_3^-$ was incorporated in dark into a reduced fraction.

In contrast, Canvin and Atkins (1974) did not detect reduction of $^{15}NO_3^-$ into an amino acid fraction following a 15-min uptake period under dark or minus-O_2 conditions. Analysis of $^{15}NO_3^-$ incorporation into an amino acid fraction is complicated not only by nonuniform NO_3^- distribution in the leaf blade, but also by the possibility that transfer of the amido-nitrogen of glutamine by GOGAT occurs only in the light (Ito et al., 1978).

Our results may be interpreted in several ways. Differences in rates of nitrate reduction may result from changes in nitrate concentration in a small metabolic pool. This suggests nitrate reductase is substrate limited. Differences in rates of reduction may also result from changes in the amount of reducing power available for reduction of nitrate. Sherrard (1980) and Sawhney (1979) have suggested competition for reductant between nitrate reduction and respiration. This is a possible reason for the increase in nitrate uptake and reduction we observed under anaerobic conditions. A third possibility is that the energy status of the cell regulates not only the reductant supply but also nitrate concentration in the metabolic NO_3^- pool, thereby altering rates of nitrate reduction.

REFERENCES

Aslam, M., Huffaker, R. C., Rains, D. W., and Rao, K. P., 1979, Influence of light and ambient carbon dioxide concentration on nitrate assimilation by intact barley seedlings, Plant Physiol., 63:1205.

Aslam, M., and Oaks, A., 1975, Effect of glucose on the induction of nitrate reductase in corn roots, Plant Physiol., 56:634.

Aslam, M., Oaks, A., and Huffaker, R. C., 1976, Effect of light and glucose on the induction of nitrate reductase and on the distribution of nitrate in etiolated barley leaves, Plant Physiol., 58:588.

Beevers, L., and Hageman, R. H., 1969, Nitrate reduction in higher plants, Ann. Rev. Plant Physiol., 20:495.

Canvin, D. T., and Atkins, C. A., 1974, Nitrate, nitrite and ammonia assimilation by leaves: effect of light, carbon dioxide and oxygen, Planta (Berl.), 116:207.

Chantarotwong, W., Huffaker, R. C., Miller, B. L., and Granstedt, R. C., 1976, In vivo nitrate reduction in relation to nitrate uptake, nitrate content, and in vitro nitrate reductase activity in intact barley seedlings, Plant Physiol., 57:519.

Dalling, M. J., Tolbert, N. E., and Hageman, R. H., 1972, Intracellular location of nitrate reductase and nitrite reductase. I. Spinach and tobacco leaves, Biochim. Biophys. Acta, 238:505.

Ferrari, T. E., Yoder, O. C., and Filner, P., 1973, Anaerobic nitrite production by plant cells and tissues: evidence for two nitrate pools, Plant Physiol., 51:423.

Grant, B. R., Atkins, C. A., and Canvin, D. T., 1970, Intracellular location of nitrate reductase and nitrite reductase in spinach and sunflower leaves, Planta, 94:60.

Heimer, Y. M., and Filner, P., 1971, Regulation of the nitrate assimilation pathway in cultured tobacco cells. III. The nitrate uptake system, Biochim. Biophys. Acta, 230:362.

Heber, U., and Willenbrink, J., 1964, Sites of synthesis and transport of photosynthetic products within the leaf cell, Biochim. Biophys. Acta, 82:313.

Ito, O., Yoneyama, T., and Kumazawa, K., 1978, Amino acid metabolism in plant leaf. IV. The effect of light in ammonium assimilation and glutamine metabolism in the cells isolated from spinach leaves, Plant and Cell Physiol., 19:1109.

Klepper, L. A., Flesher, D., and Hageman, R. H., 1971, Generation of reduced NAD for nitrate reduction in green leaves, Plant Physiol., 48:580.

Klepper, L. A., and Hageman, R. H., 1969, Dependence of nitrate reduction on generation of reduced nicotinamide adenine dinucleotide by glycolytic metabolism of leaf tissue, Fed. Proc. Amer. Soc. Exp. Biol., 28:467.

Rains, D. W., 1968, Kinetics and energetics of light enhanced potassium absorption by corn leaf tissue, Plant Physiol., 43:394.

Ritenour, G. L., Joy, K. W., Bunning, J., and Hageman, R. H., 1967, Intracellular localization of nitrate reductase and glutamic acid dehydrogenase in green leaf tissue, Plant Physiol., 42:233.

Sawhney, S. K., Naik, M. S., and Nicholas, D. J. D., 1978, Regulation of nitrate reduction by light, ATP, and mitochondrial respiration in wheat leaves, Nature, 272:647.

Shaner, D. L., and Boyer, J. S., 1976a, Nitrate reductase activity in maize (Zea mays L.) leaves. I. Nitrate regulation by nitrate flux, Plant Physiol., 58:499.

Shaner, D. L., and Boyer, J. S., 1976b, Nitrate reductase activity in maize (Zea mays L.) leaves. II. Regulation by nitrate flux at low leaf water potential, Plant Physiol., 58:505.

Sherrard, J. H., and Hageman, R. H., 1980, The influence of oxygen on nitrate reductase in wheat leaves in relation to mitochondrial activity, Plant Physiol., S:289.

Smith, R. C., and Epstein, E., 1964, Ion absorption by shoot tissue: technique and first findings with excised leaf tissue of corn, Plant Physiol., 39:338.

Stocking, C. R., and Larson, S., 1969, A chloroplast cytoplasmic shuttle and the reduction of extraplastid NAD, Biochem. Biophys. Res. Commun., 37:278.

Stocking, C. R., Williams, G. R., and Ongun, A., 1963, Intracellular distribution of the early products of photosynthesis, Biochem. Biophys. Res. Commun., 10:416.

Swader, J. A., and Stocking, C. R., 1971, Nitrate and nitrite reduction by Wolffia arrhiza, Plant Physiol., 47:189.

Thayer, J. R., and Huffaker, R. C., 1980, Determination of nitrate and nitrite by high-pressure liquid chromatography: comparison with other methods for nitrate determination, Anal. Biochem., 102:110.

Urbach, W., Hudson, M. A., Ulrich, W., Santarius, K. A., and Heber, U., 1965, Verteilung und wanderung von phosphoglycerat zwischen den chloroplasts und dem zytoplasm während der photosynthese, Z. Naturforsch. B., 20:890.

Walker, D. A., and Crofts, A. R., 1970, Photosynthesis, Ann. Rev. Biochem., 39:389.

INTERACTION BETWEEN NO_3^-, NO_2^-, AND NH_4^+ DURING ASSIMILATION IN DETACHED BARLEY LEAVES

S. S. Goyal and R. C. Huffaker

Plant Growth Laboratory and the Department of Agronomy
and Range Science, University of California
Davis, CA 95616

INTRODUCTION

Numerous reports show that plants may utilize both NO_3^- and NH_4^+ ions as a source of nitrogen. In microorganisms, ammonium often feedback-inhibits NR activity and represses NR induction (Payne, 1973). Because NO_3^- reduction seems to occur in the cytoplasm (Beevers and Hageman, 1972; Neyra and Hageman, 1978) and NO_2^- and NH_4^+ are assimilated in the chloroplast, feedback-inhibition by products may not be as serious in green leaves. Little information is available concerning the effects of each of these nitrogenous compounds on the assimilation of the others. This paper deals with that problem.

MATERIALS AND METHODS

Plant Material

Barley seeds (*Hordeum vulgare* L. cv. Numar) were surface-sterilized by continuous stirring in 1% (v/v) chlorox solution for 15 min. The seeds were germinated in vermiculite and irrigated daily with 1/4 strength Hoagland solution (Hoagland and Arnon, 1950) lacking N. All plant material was grown in a growth chamber under respective day and night temperatures of 25 C and 16 C, a daylength of 16 hours, a relative humidity of 70%, and light intensity at leaf level of 550 $\mu E/m^{-2}.sec^{-1}$. The top 10 cm of leaf blades from 10-day-old seedlings were used in all experiments.

Nitrate, Nitrite, and Ammonium Uptake

The uptake of nitrogen was measured as the amount of ion

disappearing from substrate solutions. Ten leaf blades per treatment (each treatment repeated three times and each experiment repeated twice) were placed base down in a glass vial containing 5 and 10 ml, respectively, of experimental solution for a period of 12 and 8 hours for experiments conducted in darkness and light. The environmental conditions during the tests were the same as those under which plant materials were grown, except the temperature was 25 C in both light and darkness. At the end of uptake period, the leaves were removed from the solution, blotted dry, weighed, and frozen immediately. The tissue was then homogenized in 4 volumes of 50 mM $KHPO_4$, pH 7.4 (7 volumes and pH 6.5 whenever NH_4^+ was assayed), with a mortar and pestle and centrifuged at 30,000 x g for 15 min. The supernatant was used for analysis.

Uptake Solutions

Potassium nitrate, potassium nitrite, and ammonium sulphate were used as sources of NO_3^-, NO_2^- and NH_4^+, respectively. All uptake solutions contained 5 mM $CaSO_4$.

In vivo Assimilation of N

The assimilation of each ion was determined by subtracting the amount of ions recovered in leaves from the amounts taken up. Whenever NO_3^- and NO_2^- were used simultaneously, the amount of NO_3^- reduced was added to the amount of NO_2^- taken up from substrate solution. The sum of both was considered the amount of NO_2^- available in the leaves. Similarly, the amount of NO_2^- reduced was added to the amount of NH_4^+ taken up.

Nitrate, Nitrite, and Ammonium Assay

Nitrate and nitrite in solutions and nitrate in leaf extracts were assayed by HPLC method of Thayer and Huffaker (1980). Nitrite in leaves was assayed by adding 0.75 ml of 1% (w/v) sulfanilamide in 1.5 N HCl and 0.75 ml of 0.02% (w/v) N-(1-naphthyl)ethylenediamine dihydrochloride. After 30 min absorbance was recorded at 540 nm. Ammonium was analyzed by adding 0.1 ml of Nessler's reagent in a total volume of 2.1 ml (Baily, 1962) and recording the absorbance at 425 nm and by ammonia electrode (Orion Research Inc., Cambridge, MA, Model 95-10) in solutions and leaf extracts, respectively.

RESULTS

Nitrate Reduction as Affected by NH_4^+ and NO_2^-

The effects of NH_4^+ and NO_2^- on the reduction of NO_3^- are shown in Tables 1 and 2, respectively. Even though NH_4^+ decreased the reduction of NO_3^- slightly, it had no effect on the total N

Table 1. Effect of NH_4^+ on reduction of NO_3^-.

Treatment NO_3^- (mM)	NH_4^+	N assimilated, % of total available in leaves: Light NO_3^-	Light NH_4^+	Light Total N*	Darkness NO_3^-	Darkness NH_4^+	Darkness Total N*
5.0	0	85	87	75	61	64	44
5.0	2.5	77	90	77	59	80	62
5.0	5.0	71	89	75	50	73	54

* Nitrogen not detected as NO_3^- or NH_4^+.

Table 2. Effect of NO_2^- on reduction of NO_3^-.

Treatment NO_3^- (mM)	NO_2^-	N reduced, % of total available in leaves: Light NO_3^-	Light NO_2^-	Light Total*	Darkness NO_3^-	Darkness NO_2^-	Darkness Total*
5.0	0	69	100	69	37	100	37
5.0	2.5	57	99	75	38	100	76
5.0	5.0	52	99	80	--	--	--

* N not detected as NO_3^- or NO_2^-.

use efficiency in either light or dark. Nitrite also had a slight inhibitory effect on the reduction of NO_3^- in light, but promoted the percent total N assimilation in light as well as in darkness (Table 2).

Nitrite Reduction as Affected by NO_3^- and NH_4^+

Tables 3 and 4 show that almost all of the NO_2^- available in these leaves was reduced regardless of the presence of NO_3^- or NH_4^+,

Table 3. Effect of NH_4^+ on reduction of NO_2^-.

Treatment		N assimilated, % of total available in leaves					
NO_2^- (mM)	NH_4^+	Light			Darkness		
		NO_2^-	NH_4^+	Total N*	NO_2^-	NH_4^+	Total N*
5.0	0	99	87	86	100	79	79
5.0	2.5	99	90	90	100	76	76
5.0	5.0	98	89	88	100	84	84

* N not detected as NO_2^- or NH_4^+.

Table 4. Effect of NO_3^- on reduction of NO_2^-.

Treatment		N reduced, % of total available in leaves					
NO_2^- (mM)	NO_3^-	Light			Darkness		
		NO_2^-	NO_3^-	Total*	NO_2^-	NO_3^-	Total*
5.0	0	98	0	98	98	0	98
5.0	2.5	98	54	86	99	50	81
5.0	5.0	98	52	77	98	60	77

* N not detected as NO_3^- or NO_2^-.

in both light and darkness. Ammonium had little effect on the total N assimilation. However, NO_3^- decreased the total N assimilation efficiency by 21% when included at equimolar concentrations of 5 mM, in both light and darkness (Table 4).

Ammonium Utilization as Affected by NO_3^- and NO_2^-

Results in Table 5 indicate that NO_3^- promoted utilization of NH_4^+ in light but did not effect the total N use efficiency. In

Table 5. Effect of NO_3^- on assimilation of NH_4^+.

Treatment NH_4^+ (mM)	Treatment NO_3^- (mM)	N assimilated, % of total available in leaves: Light NH_4^+	Light NO_3^-	Light Total*	Darkness NH_4^+	Darkness NO_3^-	Darkness Total*
5.0	0	72	--	72	70	--	70
5.0	2.5	85	69	77	74	31	56
5.0	5.0	89	71	76	80	26	56

* N not detected as NH_4^+ or NO_3^-.

Table 6. Effect of NO_2^- on assimilation of NH_4^+.

Treatment NH_4^+ (mM)	Treatment NO_2^- (mM)	N assimilated, % of total available in leaves: Light NH_4^+	Light NO_2^-	Light Total*	Darkness NH_4^+	Darkness NO_2^-	Darkness Total*
5.0	0	69	0	69	67	0	67
5.0	2.5	83	99	82	76	100	76
5.0	5.0	89	99	88	84	100	84

* N not detected as NH_4^+ or NO_2^-.

the dark NO_3^- did not seem to have much beneficial effect on NH_4^+ utilization and decreased total N assimilation efficiency. Nitrite increased both NH_4^+ utilization and total N use efficiency in light as well as in dark (Table 6).

DISCUSSION

Ammonium and Nitrite on Nitrate

The effect of reduced N, mainly NH_4^+, or amino acids on the level or activity of NR has been examined extensively in higher plants (Schrader and Hageman, 1967; Schrader et al., 1972; Breteler and Smit, 1974; Oaks et al., 1979; Sihag et al., 1979); however, few reports deal with the actual *in vivo* reduction of NO_3^-. Results presented in this paper show very minor inhibition of *in vivo* NO_3^- reduction by NH_4^+ and NO_2^-. Total N use efficiency did not change with the addition of NH_4^+, but NO_2^- increased the total N use efficiency, especially in darkness (Tables 1 and 2). The probable reasons for the lack of feedback inhibition of NO_3^- reduction in this study could be: a) very low level of NO_2^- was maintained by these leaves (the steady-state level of NH_4^+, although higher than NO_2^-, was still low enough to have no major effect on NO_3^- reduction); b) separate intracellular sites of assimilation of NO_3^-, NO_2^-, and NH_4^+.

Nitrate and Ammonium on Nitrite

Little information is available on the effect of NO_3^- and NH_4^+ on the assimilation of NO_2^-, although many workers have studied the inhibition and promotion of NR by NH_4^+. Our results show that almost all the NO_2^- available in leaves was reduced whether or not NO_3^- or NH_4^+ was present. This agrees with the general concept that little or no NO_2^- accumulates in barley leaves (Aslam et al., 1979). Total N use efficiency remained essentially unchanged by NH_4^+ but was lowered by NO_3^-. This is primarily because NO_3^- assimilation efficiency of these leaves was lower than that of NO_2^-, which then resulted in lower total N use efficiency.

Nitrate and Nitrite on Ammonium

Ammonium as a sole source of N has been reported to be deleterious (Bennet et al., 1964; Goyal, 1974) as well as beneficial (Wahhab and Bhati, 1957; Schrader et al., 1972) for growth of many higher plants. Some reports have shown that a mixture of NO_3^- and NH_4^+ proved to be better than NO_3^- or NH_4^+ alone (Erickson, 1965; Gamborg et al., 1968). Essentiality of small amounts of NO_3^- for radish plants in order to use NH_4^+ was observed by Goyal (1974) but the reasons were not understood. Little information is available on the effect of NO_2^- on the assimilation of NH_4^+. We observed a small beneficial effect due to NO_3^- on the utilization of NH_4^+ in both light and darkness (Table 5). Total N use efficiency did not change in light, but NO_3^- lowered total N use efficiency in darkness. This was because NO_3^- assimilation in darkness is lower than in light and also lower than NH_4^+ assimilation. Ammonium assimilation and total N use efficiency were increased by NO_2^-, in light as well as in dark. Nitrate and nitrite aided in the utilization of NH_4^+, probably through the same mechanism, which is not understood.

In conclusion, we found little or no product feedback inhibition of NO_3^- and NO_2^- reduction. We think it is due to a low steady-state level of NO_2^- and NH_4^+ maintained by these leaves and to intracellular separation of these nitrogenous compounds during assimilation. The reasons for the small beneficial effects of NO_3^- and NO_2^- on NH_4^+ assimilation observed in this study are not clear.

ACKNOWLEDGMENT

This work was supported in part by the National Science Foundation (Grant No. PFR 77-07301) and by NASA (Grant NSG 2406).

REFERENCES

Aslam, M., Huffaker, R. C., Rains, D. W., and Rao, K. Prasad, 1979, Influence of light and ambient carbon dioxide concentration on nitrate assimilation by intact barley seedlings, Plant Physiol., 63:1025.

Baily, L. J., 1962, Techniques in Protein Chemistry, Elsevier, Amsterdam, 299.

Beevers, L., and Hageman, R. H., 1972, The role of light in nitrate metabolism in higher plants, in: "Photophysiology," Vol. 7, Chap. 4, A. C. Giese, ed., Academic Press, New York.

Bennet, W. F., Pesck, J., and Hanway, J. J., 1964, Effect of NO_3^- and NH_4^+ on growth of corn in nutrient solution sand culture, Agron. J., 56:342.

Breteler, H., and Smit, A. L., 1974, Effect of ammonium nutrition on uptake and metabolism of nitrate in wheat, Neth. J. Agric. Sci., 22:73.

Erickson, T., 1965, Studies on the growth requirements and growth measurements of cell cultures of *Haplopappus gracilis*, Physiol. Plant., 18:976.

Gamborg, O. L., Miller, R. A., and Ojima, K., 1968, Nutrient requirements of suspension cultures of soybean root cells, Exp. Cell Res., 50:151.

Goyal, S. S., 1974, Studies on the inhibitory effects of ammoniacal nitrogen on growth of radish plants (*Raphanus sativus*) and its reversal by nitrate, Ph.D. Dissertation, University of California, Davis.

Hoagland, D. R., and Arnon, D. I., 1950, The water culture method for growing plants without soil, Calif. Agric. Exp. Sta. Circ. No. 347.

Neyra, C. A., and Hageman, R. H., 1978, Pathway of nitrate assimilation in corn (*Zea mays* L.) leaves. Cellular distribution of enzymes and energy sources for nitrate reduction, Plant Physiol., 62:618.

Oaks, A., Stulen, I., and Boesel, I. L., 1979, Influence of amino acids and ammonium on nitrate reduction in corn seedlings, Can. J. Bot., 57:1824.

Payne, W. J., 1973, Reduction of nitrogenous oxides by microorganisms, Bacteriol. Rev., 37:429.

Schrader, L. E., and Hageman, R. H., 1967, Regulation of nitrate reductase in corn seedlings by endogenous metabolites, Plant Physiol., 42:1750.

Schrader, L. E., Domska, D., Jung, Jr., P. E., and Peterson, L. A., 1972, Uptake and assimilation of ammonium-N and nitrate-N and their influence on the growth of corn (*Zea mays* L.), Agron J., 64:690.

Sihag, R. K., Guha-Mukherjee, S., and Sopory, S. K., 1979, Effect of ammonium, sucrose, and light on the regulation of nitrate reductase level in *Pisum sativum*, Physiol. Plant., 45:281.

Thayer, J. R., and Huffaker, R. C., 1980, Determination of nitrate and nitrite by high-pressure liquid chromatography: comparison with other methods for nitrate determination, Anal. Biochem., 102:110.

Wahhab, A., and Bhati, H. M., 1957, Effect of various sources of nitrogen on rice paddy yields, Agron. J., 49:114.

VACUOLAR NITRATE AND THE ISOLATION OF VACUOLES FOR LOCALIZATION AND TRANSPORT STUDIES

Robert C. Granstedt and R. C. Huffaker

Plant Growth Laboratory and the Department of Agronomy & Range Science, University of California, Davis, CA 95616

INTRODUCTION

The vacuole is presumed to be the site of nitrate storage in plants. Various authors (Heimer and Filner, 1971; Ferrari et al., 1973; and Aslam et al., 1976) have demonstrated the existence of an intracellular pool from which nitrate can be made available for reduction. Washout kinetics obtained from tobacco suspension cells indicate that this pool may represent vacuolar nitrate (Ferrari et al., 1973). However, no direct evidence exists that the vacuole is indeed the main storage area of cellular nitrate.

Many vacuolar components have been identified by utilizing the analysis of isolated vacuoles. Components found to be residing in the vacuole include hydrolases (Boller and Kende, 1979), glucosinolates (Grob and Matile, 1979), nicotine (Saunders, 1979), acid invertase and sucrose (Leigh et al., 1979), dhurrin (Saunders and Conn, 1978), K^+, N^+, Mg^{++}, Ca^{++} and Cl^- (Lin et al., 1977), and glucose, fructose and some organic acids (Wagner et al., 1979). Although a tonoplast specific enzyme activity has not been identified as a vacuolar marker, Boller and Kende (1979) identified several soluble activities which could be useful in identifying the presence of vacuoles. In particular, α-mannosidase and acid phosphatase activities were identified as largely or entirely in the vacuoles of tobacco cells, tulip petals, and pineapple leaves. In addition they utilized α-mannosidase activity as a vacuolar marker to aid in the identification of other vacuolar enzyme activities.

The investigation of the vacuole as the site of nitrate storage requires a procedure producing large quantities of vacuoles uncontaminated by other organelles or plasma membrane. This paper presents a procedure developed for the isolation of large numbers of high quality vacuoles from barley, a plant whose nitrate utilization has been extensively studied. Since α-mannisidase and acid phosphatase activities could be useful vacuolar markers in barley, the activities of these two enzymes and their relationships to the vacuole were studied. The distribution of their activities was then related to that of nitrate.

MATERIALS AND METHODS

Tissue Digestion

The first leafes of 6-8 day-old, light grown *Hordeum vulgare* L. var. Numar were excised, the lower epidermis was stripped away, and the leaves were chopped into 1 cm pieces. About 3 grams of tissue were placed in a Petri dish containing 20 ml of 0.6 M mannitol, 20 mM arginine, modified Murashige and Skoog salts (9.4 mM K_2SO_4 was substituted for NH_4NO_3 and KNO_3), 2% Cellulysin, 1% Hemicellulase, 0.5% Macerase and 25 mM K_2HPO_4-citric acid adjusted with H_2SO_4 to pH 5.5. The tissue was incubated at 26 C for 2 h with gentle shaking.

Protoplast Recovery

After digestion, the resulting protoplast suspension was filtered to remove undigested material and layered on a sucrose pad consisting of 0.6 M sucrose and 25 mM K_2HPO_4-citric acid at pH 5.5. This gradient was centrifuged at 100 x g for 15 min and the protoplasts were recovered from a band on top of the sucrose pad. Protoplasts were resuspended in 0.6 M mannitol and 25 mM K_2HPO_4-citric acid at pH 5.5. They were then placed on top of a sucrose pad and centrifuged at 100 x g for 15 min. This washing was repeated twice, excluding a sucrose pad on the last centrifugation. The pelleted protoplasts were resuspended in 2.5 ml of 0.6 M mannitol. A sample of this resuspended pellet was used as the protoplast preparation from which comparisons were made.

Vacuole Recovery

To the purified protoplast suspension was added 12.5 ml of 0.06 M K_2HPO_4, containing 12% ficoll, 1 mM DTT and +/- 1 mg/ml BSA, adjusted to pH 8.0 with H_2SO_4. Above this was layered 5 ml of 0.6 M mannitol containing 1 mM (Na^+) EDTA and 50 mM K-phosphate buffer at pH 7.0. The vacuoles were recovered in the top 4 ml of the gradient, and samples from the gradient were saved for analysis.

Enzyme Assays

Acid phosphatase (substrate:PNP-phosphate) was determined by the method of Leigh et al. (1979) except that the reaction was run at pH 5.0 and 0.1 M Na_2CO_3 was substituted for 1 M NaOH. α-Mannosidase activity was assayed in the same manner as acid phosphatase except that the substrate was 1.8 mg/ml p-nitrophenyl-α-mannopyranoside. Chlorophyll was determined according to MacKinney (1939).

Nitrate

Nitrate was assayed according to Thayer and Huffaker (1980).

RESULTS AND DISCUSSION

Vacuoles were recovered from protoplasts by a combination of pH and osmotic shock in the presence of 10% ficoll. Upon centrifugation, many of the intact vacuoles floated up through the 10% ficoll layer into a layer of 0.6 M mannitol (0% ficoll) which provided osmotic stability. The fractionation of this flotation system is described in Figure 1 with the 0% and 10% fraction labels referring to the final ficoll concentration in each layer. High concentrations of free chloroplasts and/or intact protoplasts resulted in high chlorophyll concentrations (Table 1). Many unlysed protoplasts and vacuoles remained at the 0/10% ficoll interface. The pellet fraction consisted mostly of chloroplasts but also contained some intact protoplasts (Table 1).

Since only one intact vacuole emerged from a lysing protoplast, cellular components which are exclusively vacuolar should have the same activity or concentration on a per vacuole basis or on a per protoplast basis. In Table, α-mannosidase and acid phosphatase activities as well as nitrate concentration are calculated on this basis except that cross-contamination is accounted for by utilizating the concentration of both protoplasts and vacuoles found in each fraction. However, extensive protoplasts and vacuolar lysis in the 10% ficoll layer resulted in the release of protoplast and vacuolar contents. Consequently, when the activity or concentration of these components in the 10% fraction is calculated on a per protoplast plus vacuole basis, the resulting ratio is far in excess of that observed in the protoplasts prior to lysis (Table 2). This is the result of the large amount of free activity or high concentration in relation to the relatively small number of surviving protoplasts or vacuoles in the 10% fraction where lysis occurred.

In the 0% fraction 1.45×10^6 vacuoles/ml were recovered, a 37% yield of vacuoles from the total protoplasts subject to lysis. This vacuolar fraction was contaminated by less than 1% unlysed protoplasts (counted visually) and by very few free chloroplasts as indicated by the low chlorophyll concentration found in this fraction (0% fraction, Table 1). This procedure compares favorably with that of Boller and Kende (1979) who reported 10% proto-

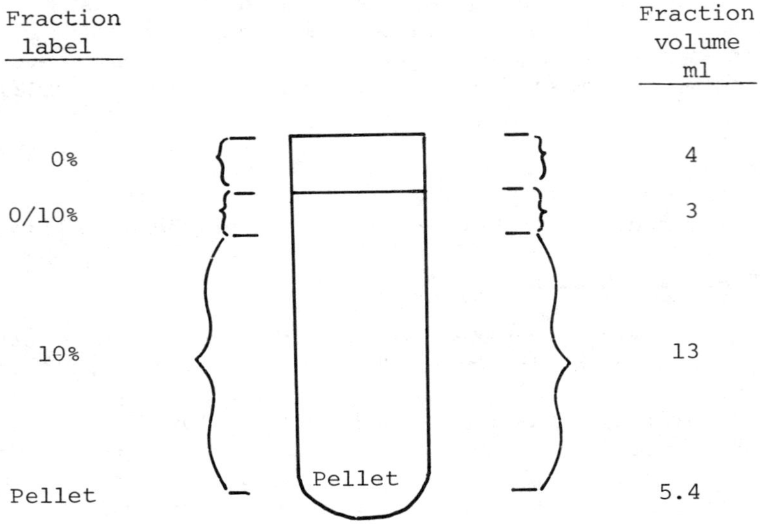

Fig. 1. Vacuole floatation system. The fraction labels refer to the final ficoll concentrations. Purified vacuoles were recovered in the top 4 ml (0% fraction label) and the pellet was resuspended to the indicated volume.

plast contamination of their preparations, Saunders and Conn (1978) who reported yields of 5-10% intact vacuoles from lysed protoplasts, and Wagner and Siegelman (1975) who reported a total of 10^6 vacuoles from their preparations.

High concentrations of α-mannosidase activity were associated with high concentrations of unlysed protoplasts and vacuoles (0/10% fraction, Table 1) and intact vacuoles (0% fraction, Table 1). Upon resuspension and centrifugation of the 0% fraction, the high concentration of α-mannosidase activity remained associated with the high concentration of intact vacuoles (plus contaminating protoplasts) (Table 3). A large amount of activity appeared not to associate with the vacuoles, but this can be clearly accounted for by the small recovery (i.e., extensive lysis) of the vacuoles subjected to this treatment (total % recovery, Table 3). The observations that the concentration of α-mannosidase activity was

Table 1. The recovery of unlysed protoplasts, vacuoles, α-mannosidase activity, acid phosphatase activity, nitrate and chlorophyll from the vacuole flotation system following protoplast lysis and centrifugation.

	Protoplasts	Vacuoles	α-Mannosidase activity	Acid phosphatase activity	Nitrate	Chlorophyll
	$\frac{x10^4}{ml}$	$\frac{x10^4}{ml}$	$\frac{nmoles}{ml \cdot min}$	$\frac{nmoles}{ml \cdot min}$	$\frac{nmoles}{ml}$	$\frac{\mu g}{ml}$
Protoplasts (prior to lysis)[1]	575.73	49.63	53.467	2213.9	1707.04	535.4
Fractions[2]						
0%	1.26	145.32	7.869	143.9	346.7	2.5
0/10%	100.98	49.06	9.363	238.4	379.0	102.5
10%	3.52	0.88	3.057	101.4	193.7	6.2
Pellet	32.56	0	2.288	132.9	19.4	337.8
Total % recovery	72%*		79%	57%	116%	166%

*Protoplasts and vacuoles recovered from lysed protoplasts.
[1]Total volume applied = 2.5 ml
[2]For fraction volumes, see Figure 1.

Table 2. α-Mannosidase and acid phosphatase activities and nitrate concentration expressed on a per protoplast and vacuole basis, and as a percentage of total protoplast activity or concentration.[a]

	nmoles α-mannoside / proto.+vac. / min ($\times 10^{-6}$)	% per protoplast activity[b]	nmoles P / proto.+vac / min ($\times 10^{-4}$)	% per protoplast activity[b]	nmoles nitrate / proto.+vac. / min ($\times 10^{-4}$)	% per protoplast concentration[b]
Protoplasts (prior to lysis)	8.55	100	3.54	100	2.73	100
Fractions						
0%	5.39	63	0.97	28	2.37	87
0/10%	6.24	73	1.59	45	2.53	93
10%	69.5	813	23.0	651	44.02	1,613
Pellet	7.03	82	4.08	115	0.60	22

[a] Derived from Table 1, i.e., (activity or concentration/ml)/(protoplasts + vacuoles/ml).

[b] This ratio examines the activity or concentration found in a fraction on a per vacuole and protoplast basis, and compares it to the appropriate activity or concentration found in the protoplast preparation (on a per vacuole and protoplast basis) prior to lysis. This gives an estimation of the % of the cellular component present in the vacuole, e.g., since the 0% gradient fraction is mostly vacuoles, (5.39/8.55)100 = 63%, therefore 63% of the cellular α-mannosidase activity is associated with the vacuoles.

2.5 times higher in the 0% fraction associated with the vacuoles than in the 10% fraction where the protoplasts were lysed (Table 1), and that this activity remained associated with the vacuoles upon subsequent resuspension and centrifugation provides evidence that much of the recovered activity was intimately associated with the isolated vacuoles. Analysis of the vacuolar fraction (0% fraction) revealed that each vacuole contained 63% of the protoplast activity observed prior to lysis when the recovered activity was calculated on a per protoplast plus vacuole basis. This seems a conservative estimate since only 79% of the total applied α-mannosidase activity was recovered following protoplast lysis (total % recovery, Table 1). Although all of the α-mannosidase activity was not vacuolar as was reported in other plants by Boller and Kende (1979), it is concluded from these results that the majority of activity is vacuolar.

Boller and Kende (1979) also reported that essentially all of the acid phosphatase activity of a protoplast was localized in the vacuole. Grob and Matile (1979), Saunders (1979) and Mettler and Leonard (1979) found this activity enriched in their vacuolar fractions. Like the α-mannosidase activity, enrichment of acid phosphatase activity in the flotation system following protoplast lysis was associated with unlysed protoplasts and vacuoles (0/10% fraction, Table 1), and vacuoles (0% fraction, Table 1). Upon subsequent resuspension and centrifugation, the highest concentration of activity remained associated with the high concentration of vacuoles in spite of extensive lysis of the vacuoles (Table 3). The 1.4 times concentration of activity observed in the flotation system in association with the vacuoles (0% fraction) as compared to the 10% fraction (Table 1), and the continued association upon resuspension and centrifugation (Table 3) indicates that much of the activity recovered in the 0% fraction was vacuolar. Comparing the vacuolar acid phosphatase activity with that of the activity per protoplast prior to lysis reveals that 28% of the protoplast activity was vacuolar (0% fraction, Table 2). An overall recovery of only 57% of the acid phosphatase activity applied prior to the lysis (Table 1) indicates that the vacuolar acid phosphatase activity estimate could be low, but most of the acid phosphatase activity does not appear to be associated with barley vacuoles in this study.

High nitrate concentrations were recovered in association with either high concentrations of unlysed protoplasts and vacuoles (0/10% fraction) or vacuoles (0% fraction, Table 1). Resuspension and centrifugation in 0.6 N KCl (Table 3) were unsuccessful because the high chloride concentration interferred with the assay of nitrate. However, the almost 1.8-fold concentrations of nitrate in the 0% layer over that of the 10% layer where lysis occurred indicates that nitrate was in fact associated with the vacuoles. The protoplasts prior to lysis contained 2.73×10^{-4} nmoles of nitrate per protoplast whereas the recovered vacuoles contained

Table 3. Two ml of the recovered vacuoles (0% fraction) were resuspended with 12.0 ml of 0.6 N KCl and 25 mM K-phosphate buffer pH 7.0 and centrifuged at 100 x g for 5 min. Protoplasts and vacuoles were counted, and α-mannosidase and acid phosphatase activities were determined in the top 12 ml and bottom 2 ml of the resultant distribution.

	Protoplasts	Vacuoles	α-mannosidase activity	Acid phosphatase
	numbers/ml	numbers/ml	nmoles/ml min	nmoles/ml min
Top	400	36,600	0.886	17.8
Bottom	9,000	204,200	3.187	67.3
Total% recovery	34%*		102%	92%

*Combined recovery of protoplasts and vacuoles applied.

2.73×10^{-4} nmoles of nitrate per vacuole (Table 2), when cross-contamination is taken into account. Thus it was concluded that the vacuole was indeed the major site of nitrate deposition in the cell, accounting for 87% of the cellular nitrate (0% fraction, Table 2). This conclusion is in agreement with that obtained indirectly from washout experiments performed on tobacco suspension cells (Ferrari et al., 1973).

This isolation procedure resulted in a high vacuole yield quite free of adhering plasma membrane and chloroplast or protoplast contamination. α-Mannosidase activity in this presentation was largely vacuolar and a portion of the acid phosphatase activity was also vacuolar. Nitrate was associated with both unlysed protoplasts and vacuoles and with α-mannosidase activity, indicating that it was in fact associated with the vacuoles. Further analysis indicated that most of the intracellular nitrate was present in vacuoles in barley, and that the vacuole represents the site of cellular nitrate storage.

ACKNOWLEDGMENT

Thanks to Christine Granstedt for much of the preliminary typing, and the Department of Agronomy & Range Science and NSF Grant PFR 77-07301 for their support of this project.

REFERENCES

Aslam, M., Oaks, A., and Huffaker, R. C., 1976, Effect of light and glucose on the induction of nitrate reductase and on the distribution of nitrate in etiolated barley leaves, Plant Physiol., 58:588.

Boller, T., and Kende, H., 1979, Hydrolytic enzymes in the central vacuole of plant cells, Plant Physiol., 63:1123.

Ferrari, T. E., Yoder, O. C., and Filner, P., 1973, Anaerobic nitrite production by plant cells and tissues: evidence for two nitrate pools, Plant Physiol., 51:423.

Grob, K., and Matile, P. H., 1979, Vacuolar location of glucosinolates in horseradish root cells, Pl. Sci. Lett., 14:327.

Guy, M., Reinhold, L., and Michaeli, D., 1979, Direct evidence for a sugar transport mechanism in isolated vacuoles, Plant Physiol., 64:61.

Heimer, Y., and Filner, P., 1971, Regulation of the nitrate assimilation pathway in cultured tobacco cells. III. The nitrate uptake system, Biochim. Biophys. Acta, 230:362.

Leigh, R. A., Rees, T., Fuller, W. A., and Banfield, J., 1979, The location of acid invertase activity and sucrose in the vacuoles of storage roots of beet root (Beta vulgaris), Biochem. J., 178:539.

Lin, W., Wagner, G. J., Siegelman, H. W., and Hind, G., 1977, Membrane bound ATPase of intact vacuoles and tonoplasts isolated from mature plant tissue, Biochim. Biophys. Acta, 465:110.

Mackinney, G., 1939, Criteria for purity of chlorophyll preparations, J. Biol. Chem., 132:91.

Mettler, I. J., and Leonard, R. T., 1979, Isolation and partial characterization of vacuoles from tobacco (Nicotiana glutinosa) protoplasts, Plant Physiol., 64:1114.

Saunders, J. A., and Conn, E. E., 1978, Presence of the cyanogenic glucoside dhurrin in isolated vacuoles from sorghum, Plant Physiol., 61:154.

Saunders, J. A., 1979, Investigations of vacuoles isolated from tobacco. I. Quantitation of nicotine, Plant Physiol., 64:74.

Thayer, J. R., and Huffaker, R. C., 1980, Determination of nitrate and nitrite by high-pressure liquid chromatography: comparison with other methods for nitrate determination, Anal. Biochem., 102:110.

Wagner, C. J., and Siegelman, H. W., 1975, Large scale isolation of intact vacuoles and isolation of chloroplasts from protoplasts of mature plant tissues, Science, 190:1298.

Wagner, G. C., 1979, Content and vacuole/extravacuole distribution of neutral sugars, free amino acids, and anthocyanin in protoplasts, Plant Physiol., 64:88.

THE EFFECT OF PH, TEMPERATURE, AND NO_3^- CONCENTRATION ON NH_4^+ ABSORPTION AND $N_2(C_2H_2)$-FIXATION BY SOYBEANS

Lynn Marcus-Wyner and D. W. Rains

Department of Agronomy & Range Science
University of California
Davis, CA 95616

ABSTRACT

Nodulated and nonnodulated soybeans (Glycine max (L) Merr., Amsoy 71) were supplied with complete nutrient solution either lacking or containing nitrogen. Intact plants, at the flowering stage were transferred from solution culture to a mist assay chamber. In this chamber plant roots were sprayed with an uptake solution ranging from 0.05 to 1.0 mM NH_4Cl and 0.5 mM $CaSO_4$. Concurrently root nodules were exposed to 0.12 ml/ml C_2H_2. Simultaneous measurements were made for NH_4^+ absorption and C_2H_2 reduction.

The short-term uptake pattern for NH_4^+ resembles a hyperbolic curve. Uptake was linear for over 6 hours, except for an initial rapid uptake period. Ammonia absorption capacity was consistently greater in uninoculated plants than in inoculated plants. Varying the pH range of experimental solutions from 4.0 to 6.8 did not significantly influence NH_4^+ uptake in nodulated plants, but had a stimulatory effect in nonnodulated plants. Short-term exposure to lower pH levels did not alter acetylene reduction rate.

When 0.1 to 2.0 mM KNO_3 was added to uptake solutions NH_4^+ absorption was not significantly influenced in either nodulated or nonnodulated plants. In addition, pretreatment of plants for 16 hours with 0.5 mM KNO_3 and 0.5 mM NH_4Cl did not change the absorption rate. Acetylene reduction rate decreased when nodulated plants were pretreated with nitrogen, but reduction rate did not change when plants were exposed to varying NO_3^- concentrations for short time periods. When the temperature of the uptake solution was decreased from 28 to 22 C acetylene reduction rate also decreased, but NH_4^+ absorption rate did not significantly change in either nodulated or nonodulated plants.

In general, it appears that neither process NH_4^+ absorption nor acetylene reduction is very sensitive to the short-term environmental changes in our mist system. Acetylene reduction is more sensitive to temperature changes and NH_4^+ is more sensitive to pH changes.

ACKNOWLEDGMENT

This work was supported in part by a grant from the National Science Foundation (PFR 77-07301).

RECURRENT DIVERGENT AND MASS SELECTIONS IN MAIZE WITH PHYSIOLOGICAL AND BIOCHEMICAL TRAITS: PRELIMINARY AND PROJECTED APPLICATION

R. H. Hageman and R. J. Lambert

Crop Physiologist and Maize Breeder, respectively
Agronomy Department
University of Illinois
Urbana, IL 61801

FOREWORD

This manuscript was prepared for presentation at a workshop. We have intentionally included controversial and speculative points with the hope that this would stimulate discussion. Our goal has been and is to identify physiological or biochemical traits that will be useful in the selection of superior cultivars.

INTRODUCTION

The importance of fertilizer N in the production of maize grain is illustrated by the parallelism between the amount of N applied and the increase in grain yield in Illinois (Hageman, 1979). Over the past three decades, maize yields (state-wide averages) have progressively increased from 3.1 to 7.9 MT/ha while the use of fertilizer N has progressively increased some eight-fold. Because grain N has remained relatively constant (ca 10% ± .5) during this period, the amount of grain N has increased only in proportion to increases in yield. This shows that fertilizer N has a direct effect on grain dry matter. It is our belief that the role of N in enhancing maize productivity is to maximize sink (grain) development and to develop and maintain an effective photosynthetic apparatus which should be conducive to vegetative development (canopy) and high and/or sustained grain filling.

Although corn hybrids used in commercial production have been improved over the past 40 years (Duvick, 1977), selection of the hybrids may have been dictated in part by their responsiveness to fertilizer N. With recently developed hybrids, Welch (1979) has shown that grain yields decreased (46%) when the annual application of fertilizer N (180 k/ha) to highly fertile soils of high organic content was omitted.

Nitrate is the predominant form of soil nitrogen available to the maize plant because in soils conducive to plant growth, soil organisms rapidly convert the ammonical form to nitrate. Upon absorption by the plant, nitrate is assimilated as follows:

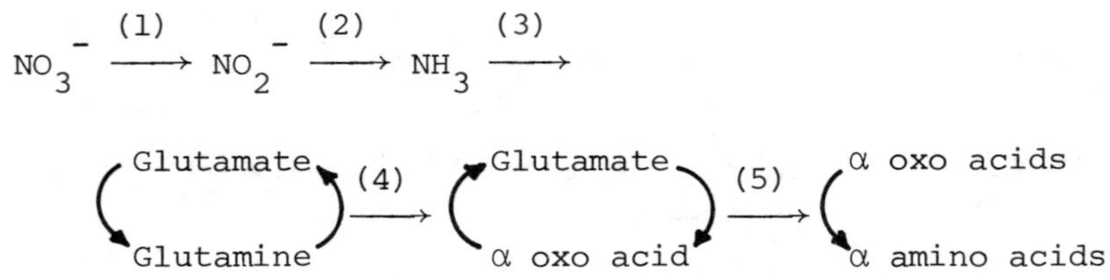

(1) nitrate reductase (NADH); (2) nitrate reductase (Fd, leaves); (3) glutamine synthetase (ATP); (4) glutamic:2-oxoglutarate transferase (Fd, leaves); (5) transaminases. Nitrate reductase is localized in the cytoplasm while nitrite reductase is found in the chloroplasts (Beevers and Hageman, 1969; Ritenour et al., 1972). Based on the proportion of nitrate to reduced N in the xylem exudate, of maize seedlings, Polisetty (1977) found that one-third of the nitrate is reduced in the root (range 25 to 50%).

Nitrate reductase (NR) is the logical point to affect regulation of nitrate assimilation because it is (a) the first and rate limiting step in the pathway based on relative activities of the component enzymes and the relatively high Km for nitrate; (b) relatively unstable *in vitro* and *in situ*, especially when the plant is subjected to high temperatures or water stress; and (c) substrate inducible (Beevers and Hageman, 1969; Hageman, 1979). Relative to nitrite and ammonia, nitrate can be accumulated (50 to 80 μmol g^{-1} fresh weight of maize stalk) without apparent injury (Reed and Hageman, 1980). Nitrite is seldom found in normal plant tissue and ammonia concentrations are usually low (Hewitt et al., 1976). For these reasons it was initially believed that measurements of leaf nitrate reductase activity (NRA) would reflect the presence of nitrate in the leaf and the availability of reduced N and thus would serve as a simple selection criterion.

Initial experiments in 1960 and 1961 with maize hybrid Hy2 x Oh7, which had in the past consistently outyielded hybrid WF9 x C103, showed that Hy2 x Oh7 with more (ca 30-40%) NRA throughout the growing season yielded no more grain than WF9 x C103 (Zeiserl et al., 1963). In a subsequent trial in 1964, Hy2 x Oh7 outyielded

WF9 x C103 by 100% (unpublished). With wheat when nitrate was not limiting, input of reduced N as estimated from leaf NRA of the entire canopy or entire plant and integrated over time was correlated ($r > +.9$) with the actual accumulation of reduced N (Eilrich and Hageman, 1973; Dalling et al., 1975; Brunetti and Hageman, 1976) With six maize hybrids, significant correlations were found between canopy NRA, integrated over the growing season, and grain yield and plant and grain reduced N content; however the correlation values were low, indicating that factors other than the level of leaf NRA were involved in grain yield (Deckard et al., 1973). Dalling et al. (1975) showed that transport of vegetative nitrogen to the wheat grain was one of these factors. Deckard et al. (1973) identified one maize genotype with relatively low leaf NRA, and high vegetative nitrate content that accumulated large amounts of reduced N. These observations and the work of Chantarotwong et al. (1976) indicates that the availability of nitrate as well as the level of enzyme may affect the rate of production of reduced N. This view is supported by the stimulation of the in vivo NRA by nitrate (Klepper et al., 1971) and the positive correlations between nitrate uptake and accumulation of reduced N by several maize genotypes (Chevalier and Schrader, 1977). Preliminary studies with the in vivo NRA assay with and without nitrate and malate plus glucose, indicates that availability of both substrate (NO_3^-) and energy (NADH) may limit nitrate reduction (Hageman, 1979). For a single maize genotype, Shaner and Boyer (1976) have shown that it is the flux of nitrate rather than the nitrate content of the leaf that regulates the level of NRA in maize leaves. Unfortunately among maize genotypes, a given flux rate will not result in the same amount of NRA (Reed and Hageman, 1980a,b). For these and other reasons, the measurement of leaf NRA, will not serve as a useful tool for the identification of superior cultivars. Rao et al. (1977) reached a similar conclusion from their studies with two wheat cultivars.

These observations indicate that we need a better understanding of when and how much reduced N should be available to permit development and maintenance of an effective photosynthetic apparatus, maximum sink (grain) initiation and high and/or sustained grain filling among genotypes grown under diverse environments.

DIVERGENT RECURRENT SELECTION

In 1974, a divergent recurrent selection program was initiated using the synthetic maize, super-stiff stalk synthetic as the parental material (Dunand, 1980). This synthetic was chosen because it has genes favorable for productivity and several important commercial inbreds including B73 have been derived from this material. NRA (in vivo + NO_3) of the leaf blade of the first and second leaf above the ear assayed at 10 and 20-days postanthesis was the sole selection criterion. NRA was chosen as the selection criterion because (a) the level of activity should reflect in a gross way the influx of nitrate into the cells as NRA is substrate inducible and there is no evidence to date of a maize genotype that maintains high NRA in the absence of nitrate; and (b) we hoped that high levels of leaf NRA during the postanthesis period would be associated with a leaf that maintained its photosynthetic apparatus (extended leaf duration). Other studies have indicated a close association between leaf area duration and yield (Evans et al., 1975). In the selection cycle all plants sampled were selfed. Using a 20% selection intensity the plants (20 or 40) with highest or lowest NRA were identified. An equal number of kernels from the 20 high NRA ears (cycle O selfed high) were bulked and this seed planted in the Hawaiian winter nursery. These plants were allowed to pollinate randomly and the grain harvested and the kernels (cycle 1 high) bulked. The low NRA material was handled separately in the same manner. The cycle 1 high and cycle 1 low kernels were planted on the Agronomy Farm at Urbana and the selection process repeated to generate the successive cycles.

A preliminary evaluation in 1978 of four cycles of material (high and low NRA selection) and the parental material (C_0) show that divergence in NRA has been significant and continuous. The increase in nitrate reductase activity was +.91 relative units cycle^{-1} for the high selections and a decrease of -1.55 relative units cycle^{-1} for the lows. Selection for NRA had no effect on grain yield. Selection for high but not low NRA caused an increase in grain N (+.12% cycle^{-1}) and stover dry weight (8.1 g cycle^{-1}). These increases were statistically significant at the 5% level. The increase in % grain N is of interest because most breeders have normally found a negative relationship between % grain N and yield. Selection for NRA had a negligible effect on maturity as measured by % grain moisture at harvest or by visual evaluation of time and degree of leaf senescence. In 1979, replicated tests at three sites on the farm were repeated with five cycles of the selected material. While these data have not been completely processed the effects of selection on grain yields and % grain N were similar to those obtained the previous year.

In 1977, comparisons of yield and grain protein were made with plants obtained from the third cycle selfed ears selected for high and low NRA populations. Fifty families (ears) were chosen out of a total of 200 from each of the high and low NRA selection populations. Kernels from each ear were planted in triplicate plots on highly fertile soil (280 kg N ha^{-1}). The mean grain yields of the 50 high and 50 low NRA entries did not differ, however there was four times more genetic diversity in the high than in the low NRA selections. Frequency plots of these data show that the grain yield of four of the high NRA entries exceeded the mean by 2-fold while five of the high NRA entries were half as productive as the mean population. Why some of the high NRA genotypes were superior in productivity while others were extremely inferior is not known. It is possible that some of the genotypes were unable to cope with a relatively high input of reduced N during grain fill or equally possible that NRA as measured is in no way related to yield. That NRA, as measured, reflects the input of reduced N is again indicated by the fact that the mean percentage grain protein was 0.11% higher (significant at the 1% level) in the high than in the low NRA selections. The genetic diversity in grain protein was 30% higher in the high than in the low NRA selections. Four classes (high NRA-high yield; high NRA-low yield; low NRA-high yield; and low NRA-low yield) were selected from this material. Each class contained from four to five entries. In 1980, these classes are being grown along with a standard tester line. The resultant hybrid seed will be evaluated for yield performance in the near future.

These results were most disappointing in that divergence in postanthesis NRA affected neither yield nor leaf area duration as judged visually by time of leaf senescence or grain moisture content at harvest. This study did establish that the level of leaf NRA was heritable, but does not define the cause of the difference in level of activity. Thus the actual trait that is heritable could be: (a) the ability of the plant to absorb, store and translocate nitrate to the leaf cells during the postanthesis period; (b) the ability of the plant to provide an adequate level of reductant for the enzyme; (c) differences in sensitivity of the NR induction site to nitrate; (d) differences in stability or inactivation of NR *in situ* among the genotypes or (e) other processes. This study shows that although a wide range of genetic diversity may exist in the maize synthetic used, the major problem is to identify a physiological or biochemical trait that is closely associated with grain yield.

MAIZE HYBRIDS WITH HIGH AND LOW POSTANTHESIS LEAF NRA

In 1978, a detailed study was initiated with four maize hybrids to identify physiological or biochemical traits that might be associated with grain yield. Previous work had shown that two of these hybrids B37 x B73 (A) and B37 x H96 (B) had higher leaf NRA during the postanthesis period and leaves that senesced later than did the other two hybrids C123 x B14A (C) and Mo17 x H95 (D). NRA (*in vivo* + NO_3^-) and protease activities (casein as substrate at pH 5.5 and 7.5) and chlorophyll were measured for the ear leaf and the fourth leaf above and below the ear leaf. Nitrate and reduced N were determined on the stalk (includes sheath) and the upper, middle and bottom leaves. These measurements were made at intervals during grain development. Stover and grain dry weights and harvest indices for dry weight and N were determined at maturity (black layer formation).

As previously observed, hybrids A and B had more (50 to 75%) NRA in all leaves throughout the grain filling period than hybrids C and D when activity was expressed per g fresh weight. When expressed on a per leaf basis, the differences were similar except for the fourth leaf above the ear where the activity was 5 to 10% higher in hybrids A and B than in C and D. The larger upper leaves of hybrids C and D raises the possibility that excessive availability of reduced N could be inversely related to leaf size (weight). Some evidence in support of this view was obtained in the divergent selection studies, however no consistent trends were observed. We wonder whether these changes in leaf weights are transitory, independent of genotype and caused by an imbalance of input of C and N. NRA decreased progressively in all leaves of all hybrids giving rise to a series of essentially parallel lines. Activity had essentially disappeared in the fourth leaf above and below the ear by 50 days postanthesis and in the ear leaf by 60 days postanthesis for hybrids C and D. Activity was present in all leaves of hybrids A and B 60 days postanthesis. There was no mathematical correlation between NRA and nitrate content of leaves or stalks.

Leaves of hybrids A and B, that initially had less chlorophyll at 20 days postanthesis, maintained and retained their chlorophyll longer than did hybrids C and D (low NRA). Leaf chlorophyll of hybrid D was negligible by 60 days postanthesis and hybrids A, B and C had 0.015, 0.15 and 0.23 relative units by 70 days postanthesis (maximum chlorophyll content ranged from 0.75 to 1.0 units). Duration of functional leaves (judged visually) was from 7 to 14 days longer for hybrids A and B than for C and D. Unfortunately photosynthetic activity was not measured. In view of these results it is not clear why divergent selection for high postanthesis leaf NRA was not associated with delayed leaf senescence. Does this indicate that senescence of leaves is controlled by (a) some other mechanism (e.g., hormonal) and operates independently of nitrogen metabolism or (b) a complex interaction of both? (The maize inbreds B73, B37 and B14A used as parental lines in these four hybrids were derived directly or indirectly from super-stiff stalk synthetics.)

Nitrate content per plant (largely in stalk) was comparable for all four hybrids at or near anthesis, however the high NRA hybrids maintained a higher nitrate content during the latter stages of grain fill. At 70 days postanthesis the nitrate content was 15, 13, 4 and 0.05 µmol plant^{-1} for hybrids A, B, C and D, respectively. Based on the work of Shaner and Boyer (1976) it can be assumed that the flux of nitrate from the soil, stalks and midribs to the cytoplasm of the leaf blade cells is higher in hybrids A and B than in C and D during this period. This could be the cause of the differences in level of NRA.

Patterns of nitrate content per plant differed during grain fill: hybrid B had a progressive increase (50%) until 40 days postanthesis; hybrid D a 25% increase by 20 days; hybrid C a 7% increase by 30 days; and hybrid A a 12% increase by 20 days. After reaching these maxima, the nitrate content of all hybrids except A, decreased to the previously mentioned levels at 70 days postanthesis. Hybrid A maintained a relatively constant nitrate content throughout. Because nitrate content of the whole plant is the result of nitrate uptake from the soil less assimilation and loss, these data permit some deductions about the rates and duration of nitrate uptake from the root system. These data indicate that for hybrid B, the uptake rate exceeds the assimilation rate between 10 and 40 days postanthesis. After 40 days, the uptake rate was less than the assimilation rate and total plant nitrate decreased. Hybrids C and D had reduced or possibly negligible rates of nitrate uptake 30 and 20 days postanthesis, respectively. This is supported by the near depletion of the plant nitrate and loss of NRA by these two hybrids. With hybrid A, it appears that nitrate absorption from the soil continues throughout grain development. These deductions are based on the assumption that the *in vivo* NRA is a reasonable reflection of *in situ* nitrate assimilation. The concepts that corn genotypes could vary in duration of nitrate uptake, uptake rates, partitioning and remobilization are worthy of consideration.

Proteolytic activities (casein as substrate at pH 5.5 and 7.5) increased in all leaves of all hybrids during grain fill. The rate of sustained increase in both activities was initiated earlier and were higher in the low than in the high NR hybrids. For all hybrids the pH 5.5 activity was higher than the pH 7.5 activity throughout grain fill and the development of the pH 5.5 activity preceded the development of the pH 7.5 activity. The high NRA hybrids developed only limited amounts of the pH 7.5 activity.

With respect to above-ground vegetation of all hybrids the leaves (sheaths excluded) lost the greatest amount of reduced-N during grain development. The middle leaves which had the most reduced N at anthesis lost the most N during grain development although roughly proportional amounts were lost from the upper and lower leaves. The fastest rate of loss of reduced-N from the leaves occurred 30 days postanthesis in most instances. The loss of reduced N from the ear leaf of hybrid C between 10 and 20 days postanthesis was a major exception. This early loss of N is consistent with the initially high level of pH 5.5 protease activity of hybrid C.

In general, the pattern of loss of reduced-N from the leaves is coincident with the onset and development of the pH 5.5 proteolytic activity for all hybrids and for the pH 7.5 activity with hybrids C and D. Leaves of hybrids A and B that failed to develop appreciable levels of pH 7.5 activity and had lower levels of pH 5.5 activity retained more (nearly double) reduced N per leaf at 70 days postanthesis than did hybrids C and D that had relatively higher levels of both proteolytic activities. The coincidence of these events indicates a causal relationship between the proteolytic activities and the loss of reduced N from the leaves. This relationship is probably complex and may be more apparent than real as no numerical relationship can be established between the proteolytic activities and the loss of N from the leaves or gain of N by the grain.

The stalks of hybrids C and D lost a large amount (more than half) of their reduced N during the latter stages of grain fill. In contrast, stalks of hybrids A and B lost no more than one-fourth of their reduced N and none in the last phases of grain fill. Since these stalks included leaf sheaths, the actual amount of N derived from the stalks per se was not determined. Subsequent studies with other hybrids indicate that the loss of reduced N from sheath would account for one-third of the total N loss from stalks plus sheaths. We were unable to extract proteolytic enzymes from the stalk and thus are unable to explain the mode of N mobilization from the stalk. Extractable free amino-N compounds from the stalk were low indicating that most of the N in the stalk was protein.

The inverse relationship between NRA and proteolytic activities as well as the delay and lower amounts of proteolytic activities of the two high NRA hybrids, indicates a possible cause and effect relationship between the two processes. Some support for this view is afforded by the work of Martin and Thimann (1972). They observed that serine or cysteine enhanced senescence (proteolytic activity) while arginine acted as an antagonist of the enhancement in excised oat leaves. However preliminary examination of 1979 data obtained with 10 maize genotypes indicated no obligatory relationship between NRA and proteolytic activities.

Several observations are apparent from the decreases of reduced N from the above-ground vegetation (stover) that occurred more or less concurrently with the increases in reduced N in the ear. (a) For all hybrids there was little loss of reduced-N from the stover until 32 days after anthesis. During this initial period the ear accumulated approximately half of its total reduced N at maturity. This indicates that nitrate assimilation is directly or indirectly the source of the early accumulation of N by the ear. Since cob development is complete within 12 days after anthesis (Hanway, 1963), the reduced N accumulated by the ear must be predominantely in the grain. (b) Although hybrids A and B had higher levels of NRA than hybrids C and D, no significant differences were observed in the rates of reduced N accumulation in the ears of the four hybrids during the first 32 days of development. However hybrids C and D lost more reduced N from the stover (especially the ear leaf) than hybrids A and B during this initial period. Would this early loss of N from the leaves adversely affect photosynthesis over the grain filling period? Is this early loss of N due to decrease in in situ nitrate reduction or increased in situ hydrolysis of protein, or both? Factors other than the level of NRA appear to control the initial accumulation of N by the ears. (c) Higher rates of accumulation of reduced N by ears of hybrids C and D occurred coincident with the appearance of higher proteolytic activities in hybrids C and D. (d) Decreases in reduced N of the stover between 32 and 72 days postanthesis accounted for only a part of the increases in reduced N of the ears over the same time period, except for hybrid D. During this period, hybrid D had the highest level of proteolytic activity and the lowest NRA and nitrate content of all hybrids. Current assimilation of nitrate would seem the logical source for a portion of the reduced N accumulated by the ears of the three hybrids during the latter portion of grain fill. Conclusions (b) and (c) are based on the premise that reduced N is not redistributed from root to shoot during grain development according to the findings of Friedrich and Schrader (1979). (e) Losses of reduced N from the stover of hybrids C and D (high proteolytic activities) was greater than for hybrids A and B (low proteolytic activities). The coincidences of these processes indicate a cause and effect relationship, however current work in our laboratory indicates that the system is more complex. When comparable leaves were harvested at intervals from a population of hybrid maize during grain development, the main increase in proteolytic activities (pH 5.5 and 7.5) occurred after much of the N was lost. Based on this finding and the work of Miller and Huffaker (1979) and Storey and Beevers (1977) it is possible that the level of proteolytic activities present in the leaves at anthesis is adequate to completely mobilize the maize leaf proteins. One can speculate that in the absence of a sink the amino acids produced are resynthesized into new proteins, as the sink develops, the amino acids are diverted and the pool(s) of amino acids in the leaf may be depleted. The depletion of amino acids, hormonal effects or both could serve as regulatory factors of protease synthesis. Alternatively, other unmeasured proteases that show enhanced activity concurrent with the loss of N from the leaf or compartmentalization may be involved. Why the plant develops the extremely high levels of proteolytic activities after the bulk of the N has been mobilized remains unclear. This could be a "survival trait" that plants acquired when soil N supplies were not as plentiful as under current agricultural practices.

There was no consistent relationship between the "high NRA, low protease" hybrids A and B or "low NRA, high protease" hybrids C and D and grain yield, grain N plant^{-1}, dry weight harvest index and dry weight or N content of the total plant. Hybrids C and D had grain of higher percentage N and harvest index for N and retained less N in the stover at maturity than did hybrids A and B.

These results indicate that high proteolytic activity in the leaves and efficient remobilization of N are more directly related to the accumulation of grain N than the level of leaf NRA. Although the leaves of the high NRA hybrids did not senesce as rapidly as those of the low NRA hybrids, this longer duration of apparently functional leaves did not enhance yields. This was surprising, as the major portion of the grain dry weight is derived from current photosynthesis (Bidinger et al., 1977). Either photosynthate supply did not limit grain yield or photosynthesis was low in the green leaves of genotypes A and B during the later stages of grain yield.

Table 1. Three year grain yields for Urbana, IL.

Hybrid	Grain yield			
	1975	1976	1978	$\overline{X}$
		bu/A		
A	187 (11.7)*	129 (8.1)	102 (6.4)	139 (8.7)
B	176 (11.0)	174 (10.9)	119 (7.4)	156 (9.8)
C	148 (9.3)	142 (8.9)	119 (7.4)	136 (8.8)
D	176 (11.0)	136 (8.5)	120 (7.5)	140 (8.8)

*Values in parenthesis are in MT/ha.

While this study shows that postanthesis levels of NRA were not related to grain yield of these four hybrids when grown on the Agronomy South Farm, University of Illinois, in the spring and summer of 1978, a different conclusion would have been drawn from the 1975 data. In 1975-1976, hybrids A and B had higher leaf NRA (10 and 20 days postanthesis) than hybrids C and D and these differences between the hybrid pairs were comparable to those recorded in 1978. Based on the three-year average grain yields, hybrid B would be the superior cultivar. Whether the high leaf NRA, higher plant nitrate content, delayed leaf senescence or other traits of hybrid B are associated with its higher yield or whether any trait will be a more useful selection tool than yield remains to be established. It seems plausible that identification of traits associated with yield will be useful. For example, it might be possible to divergently select for a plant with high nitrate content and a high harvest index for N.

The yield by year interaction illustrates one of the problems encountered when attempting to utilize physiological or biochemical traits in the identification of superior cultivars. The successful hybrid maize companies utilize multiple year and location evaluation for yield in selecting those genotypes that are to be released for commercial production. Trait evaluation may have to follow the same "numbers game" as has been so successfully employed by the plant breeders.

Auxiliary Studies

The same hybrids used in the preceding field project were grown under growth chamber conditions in nutrient medium (7.5 mM nitrate). The objectives were to relate uptake, flux and reduction of nitrate to accumulation of reduced N by the maize seedlings and to compare these results with those obtained in the field study.

Hybrid D took up more nitrate than the other three hybrids (based on plant age and accumulation of total N per plant) primarily because of a larger root system. Hybrid A took up the least nitrate of any of the hybrids. The correlations between total N (nitrate plus reduced N $plant^{-1}$) accumulated by harvest and root dry weight or shoot:root ratios were $r = +.97$ and $-.90$, respectively. Correlation with shoot dry weight was low. Although the larger root system indicates enhanced partitioning of photosynthate to the root of hybrid D, our data do not elucidate the role of photosynthate in increasing nitrate uptake. Because of the limited number of observations, the validity of these correlations may be questioned. These correlations are consistent with the observations with tobacco, cotton and soybean plants that "the rate of nitrate uptake is coequal with the rate of root growth" (Raper et al., 1977a,b). There were no genetic differences in partitioning of nitrate (% of total) among plant parts, however, the hybrids differed in the amounts of nitrate stored in stalks and midribs. As seedlings, hybrids D and B accumulated more nitrate $plant^{-1}$ than A and C. In the field at maturity hybrids A and B had appreciable levels of nitrate while C and D had negligible amounts.

Although two of the hybrids (A and C) with the highest NRA (measured at harvest only) had the lowest concentrations of nitrate in all plant parts, NRA was not correlated with the accumulation of nitrate or reduced N for the four hybrids. Comparison of data for hybrid pair A and D (one high and one low NRA based on previous postanthesis assay) indicate that accumulation of reduced N was more dependent on nitrate uptake and flux than the amount of NRA per plant. For the other matched pair, hybrid C with more NRA, accumulated more reduced N and stored less nitrate than hybrid B, although both hybrids had comparable uptake and flux of nitrate. For hybrids C and B the equality in uptake is associated with equality in total N (nitrate plus reduced N) $plant^{-1}$. These data show that both flux and NRA affect the accumulation of reduced N.

Comparisons of data of all hybrids (disregarding harvest age) show that nitrate uptake, flux or reduction (as measured) are not numerically related to the accumulation of reduced N. Differences in plant age (23 v 25 days after planting for hybrids A and D and C and D, respectively) may be one factor that precluded correlation among these parameters. For example, for each pair the rates of nitrate uptake and flux are consistent with the accumulation of total N (nitrate plus reduced N) $plant^{-1}$. Relative rates of plant growth, differences in diurnal patterns of uptake and flux and differences in requirement for nitrate for induction, assimilation and other factors could affect these relations.

Of all possible comparisons of the five parameters measured, only nitrate uptake versus flux was correlated ($r = +.99$). The occurrence of such a relationship would be useful, however additional measurements with more genotypes and variations in uptake rates are needed to establish validity. Among the four genotypes, NRA was not correlated with nitrate flux when the data was expressed on a per plant or per unit of weight basis. That nitrate flux and NRA were closely associated for a given maize genotype (Shaner and Boyer, 1976) but not among genotypes is not surprising as it seems probable that genetic differences may exist in systems that affect induction, stability and activity of NR.

Under field conditions, hybrid C was consistently classified as a "low NRA" type when assays were made during the postanthesis period, however during vegetative development under growth chamber conditions, hybrid C was a "high NRA" type. Under growth chamber conditions the reduced N content of hybrids A, D, C and B were 6.1, 8.4, 5.9 and 5.3 mmol $plant^{-1}$, respectively, while under field conditions they contained 265, 241, 290 and 291 mmol $plant^{-1}$ (grain plus stover at maturity). Hybrid D had the most reduced N at the seedling stage and the least at maturity. The initial large root mass and associated high levels of uptake, flux and reduction observed for hybrid D under growth chamber conditions were apparently not achieved or maintained under field conditions.

These observations illustrate the problems encountered in attempting to develop a simple physiological and biochemical screening criterion useful in identifying superior cultivars at the seedling stage.

In another study, hybrids C and D were grown under growth chamber conditions on solution or vermiculite medium that provided 2.5, 7.5 and 15 mM nitrate. The objectives were as previously given.

Increases in nutrient nitrate concentrations caused increases in (a) shoot but not root dry weight and (b) increases in nitrate uptake, flux and reduction and accumulation of nitrate and reduced

N by the aerial plant parts of both hybrids. Increases in nitrate supply resulted in decreases in NRA and negligible increases in reduced N in the roots of both hybrids.

At 2.5 and 15 but not at 7.5 mM external nitrate, hybrid B had higher rates of nitrate uptake and flux. The differences were much greater at 2.5 than at 15 mM nitrate. Hybrid B had lower NRA at all levels of external nitrate and accumulated less reduced N than C except when the plants were grown at 2.5 mM nitrate. The observation that hybrid B is apparently more efficient than hybrid C in absorbing nitrate when the nitrate supply is low (2.5 mM) provides a possible explanation for the ability of hybrid B to accumulate nitrate during the initial phases of grain fill under field conditions. This explanation is based on the premise that soil nitrate levels and availability are low during the postanthesis period. Alternatively it is equally possible that the patterns of root growth and senescence are different.

Correlations were noted among the various parameters regardless of whether the values were expressed on a per plant or unit weight basis. The r values (weight basis) between: (a) nitrate flux and shoot reduced N were: +.999** hybrid B, +1.0** hybrid C and +.91* both hybrids; (b) NRA of leaf blade and shoot reduced N were +.95 hybrid B, +.997** hybrid C and +.87 both hybrids; and (c) product of nitrate flux times NRA and shoot reduced N were +.98 hybrid B, +.997** hybrid C and +.993** both hybrids. The regression equation of reduced N (y axis) versus NRA was y = 9.32X + 8.81. This indicates either that the *in vivo* leaf NRA is underestimating nitrate reduction or that the root is supplying reduced N to the shoot. We favor the latter view because the intercept value of the correlation equation was 8.81 mg reduced N and this would account for 37, 31 and 27% of the shoot reduced N at 2.5, 7.5 and 15 mM nitrate. The decreasing proportion of reduced N supplied by the root with increasing external nitrate supply is consistent with the increase in shoot to root ratio and the increase in shoot and decrease in root NRA. Assays of xylem exudate of several maize genotypes (comparable plants) indicated that from 25 to 50% of the nitrate is reduced in the root (Polisetty, 1977). While interesting, these and all subsequent correlations presented need to be viewed with caution as the number of observations were limited and all parameters are directly affected by the variable nitrate treatment. The correlation values between nutrient nitrate concentration and shoot reduced N were +.988 hybrid B, +.892 hybrid C and +.903* for both hybrids.

Symbols ** and * indicate significance at the 1 and 5% level, respectively.

The roots of the two hybrids did not differ in reduced N concentration. In response to increased nitrate supply, the increase in reduced N of the roots was linear and limited relative to that of the shoots. The accumulation of reduced N by the root was negatively related to root NRA. This could indicate either that the *in vivo* assay was not representative of the *in situ* nitrate reduction or that as the external nitrate supply increased, less photosynthate was available to the roots. The latter view is supported by the accelerated growth and nitrate reduction by the shoots and negligible root growth. If the *in vivo* NRA is a reasonable reflection of the *in situ* reduction, the decrease in NRA could indicate that less reduced N is exported to the shoots with increase in nitrate supply. With cotton, Radin (1977) found that reduced N was transported from shoot to root as nutrient nitrate level was increased.

When values were expressed on a whole plant basis, the regression equation between reduced N (whole plant) and NRA (root plus shoot) was y (red N) = .267X - 1.63. The negative intercept would indicate that the *in vivo* activities of root and shoot were overestimating the *in situ* reduction as measured by accumulation of reduced N by the whole plant. The correlation value (r) between reduced N (whole plant) and shoot NRA was +.926** (both hybrids) and between reduced N (whole plant) and root NRA was -.678 (both hybrids). Our tentative interpretation of these values is that the shoot assumes an increasing and the roots a decreasing role in supplying reduced N to the plant as nutrient nitrate is increased. At the onset of the growing season, 15 mM nitrate would be an average level of nitrate in fertile fields used for corn production.

If the flux of nitrate is the only factor that affects the accumulation of reduced N by the plant, hybrid B should have accumulated more reduced N than hybrid C. However, hybrid B had more reduced N than C only at 2.5 mM nitrate. At 7.5 mM nitrate where the nitrate flux rates of the two hybrids was comparable, hybrid C with its higher level of leaf NRA accumulated significantly more reduced N than C. Based on these and preceding data, we conclude that both flux and NRA levels are factors affecting accumulation of reduced N and that flux and NRA levels are not correlated among genotypes.

The vermiculite studies were used to show that NRA measured in leaf samples representative of the entire canopy and integrated over time were correlated with the increase in reduced N in the shoot over the same time period. For hybrid C the correlation values (r) were +.987, +.990 and +.913 for the time intervals 12 to 17 and 17 to 24 days after planting and over both periods, respectively. For hybrid B the corresponding values were +.997*, +.972 and +.934*. For both hybrids over both time periods (12 observations) the r value was +.921**. We conclude that the *in vivo* NRA measurements of leaf NRA provided a reasonable estimate of the ability of these hybrids to accumulate reduced N under these conditions. It should be noted that these two hybrids had similar growth characteristics and rates of nitrate uptake and flux at 7.5 and 15 mM nitrate.

PROJECTED APPLICATION

From these and other studies, it was concluded that more useful information could be obtained with another approach. This approach, mass selection using half-sib families, was initiated in May 1980 on the Agronomy South Farm. One hundred families (ears) were used as the starting material. These ears had been taken from third cycle plants that had been selected for high grain yield and high percentage grain N by the half-sib procedure from the maize synthetic super-stiff stalk. Although the primary selection trait will be grain yield, various physiological and biochemical traits dictated by our proposed "ideotype" plant will also be measured. These traits will be correlated with grain yield among the 100 families. A main advantage of this procedure is that some of the plants of each family can be destructively sampled and others reserved for grain yield.

Our proposed "ideotype" plant has the following characteristics:

a) High dry matter and reduced N accumulated by or near anthesis.

b) Low nitrate reduction by anthesis plus 15 days.

c) High photosynthetic rates during grain fill.

d) Prolonged grain filling period.

e) High harvest indices for dry weight and N.

f) High grain yield (primary trait).

Measurements of leaf photosynthesis, nitrate reductase and protease activities will be made and correlated with the "primary standards" of changes of dry weight and reduced N content of the various plant parts. Initially, plants will be harvested at 40 days after planting, ear emergence, 40 days after anthesis and at maturity (black layer formation). The harvested plants will be subdivided, dried, weighted and assayed for nitrate and reduced N. It is visualized that consistent positive or negative correlations of these various parameters with the actual grain yields will provide information on the role of N in productivity in maize. It should also indicate when and how much reduced N should be available for best plant performance.

With the half-sib family procedure mathematical estimates of genetic variability (heritability) of all measured parameters can be computed.

REFERENCES

Beevers, L., and Hageman, R. H., 1969, Nitrate reduction in higher plants, Annu. Rev. Plant Physiol., 20:495.

Bidinger, F., Musgrave, R. B., and Fischer, R. A., 1977, Contribution of stored preanthesis assimilate to grain yield in wheat and barley, Nature, 270:431.

Brunetti, N., and Hageman, R. H., 1976, Comparisons of *in vivo* and *in vitro* assays of nitrate reductase in wheat (*Triticum aestivum* L.) seedlings, Plant Physiol., 58:583.

Chantarotwong, W., Huffaker, R. C., Miller, B. L., and Granstedt, R. C., 1976, *In vivo* nitrate reduction in relation to nitrate uptake, nitrate content and *in vitro* nitrate reductase activity in intact barley seedlings, Plant Physiol., 57:519.

Chevalier, P., and Schrader, L. E., 1977, Genotypic differences in nitrate absorption and partitioning of N among plant parts in maize, Crop Sci., 17:897.

Dalling, M. J., Tolbert, N. E., and Hageman, R. H., 1972, Intercellular location of nitrate reductase and nitrite reductase. II. Wheat roots, Biochim. Biophys. Acta, 283:513.

Dalling, M. J., Halloran, T. M., and Wilson, H. J., 1975, The relation between nitrate reductase activity and grain nitrogen productivity in wheat, Aust. J. Agr. Res., 26:1.

Deckard, E. J., Lambert, R. J., and Hageman, R. H., 1973, Nitrate reductase activity in corn leaves as related to yield of grain and grain protein, Crop Sci., 13:343.

Dunand, R. T., 1980, Divergent phenotypic recurrent selection for nitrate reductase activity and correlated responses in maize, Ph.D. Thesis, University of Illinois, Urbana-Champaign.

Duvick, D. V., 1977, Genetic rates of gain in hybrid maize yields during the past 40 years, Maydica, 22:187.

Eilrich, G. L., and Hageman, R. H., 1973, Nitrate reductase activity and its relationship to accumulation of vegetative and grain nitrogen in wheat (*Triticum aestivum* L.), Crop Sci., 13:59.

Evans, L. T., Wardlow, I. F., and Fischer, R. A., 1975, Wheat, *in*: "Crop Physiology," L. T. Evans, ed., Cambridge University Press, London.

Friedrich, J. W., and Schrader, L. E., 1979, N deprivation in maize during grain filling. II. Remobilization of ^{15}N and ^{35}S and the relationship between N and S accumulation, Agron. J., 71:466.

Hageman, R. H., 1979, Integration of nitrogen assimilation in relation ot yield, in: "Nitrogen Assimilation of Plants," E. J. Hewitt and C. V. Cutting, eds., Academic Press, New York.

Hanway, J. J., 1963, Growth stages of corn (Zea mays L.), Agron. J. 55:487.

Hewitt, E. J., Hucklesby, D. P., and Notton, B. A., 1976, Nitrate metabolism, in: "Plant Biochemistry," J. Bonner and J. E. Varner, eds., Academic Press, New York.

Klepper, L. A., Flesher, D., and Hageman, R. H., 1971, Generation of reduced nicotamide adenine dinucleotide for nitrate reduction in green leaves, Plant Physiol., 48:580.

Martin, C., and Thimann, K. V., 1972, The role of protein synthesis in the senescence of leaves. II. The influence of amino acids on senescence, Plant Physiol., 50:432.

Miller, B. L., and Huffaker, R. C., 1979, Changes in endopeptidases, protein and chlorophyll in senescing barley leaves, Agron. Abstracts.

Polisetty, R., 1977, Characteristics of nitrate uptake from nutrient solution and root nitrate reductase activity among corn genotypes, Ph.D. Thesis, University of Illinois, Urbana-Chamgaign.

Radin, J. W., 1977, Contribution of the root system to nitrate assimilation in whole cotton plants, Aust. J. Plant Physiol., 4:811.

Rao, K. P., Rains, D. W., Qualset, C. O., and Huffaker, R. C., 1977, Nitrogen nutrition and grain protein in two spring wheat genotypes differing in nitrate reductase activity, Crop Sci., 17:283.

Raper, C. D., Patterson, D. T., Parsons, L. R., and Kramer, P. J., 1977a, Relative growth and nutrient accumulation rates for tobacco, Plant Soil, 46:743.

Raper, C. D., Patterson, D. T., Parsons, L. R., and Kramer, P. J., 1977b, Relationship between growth and nitrogen accumulation for vegetative cotton and soybean plants, Bot. Gaz., 138:129.

Reed, A. J., and Hageman, R. H., 1980a, The relationship between nitrate uptake, flux and reduction and the accumulation of reduced N in maize (Zea mays L.). I. Genotypic variation, Plant Physiol., in press.

Reed, A. J., and Hageman, R. H., 1980b, The relationship between nitrate uptake, flux and reduction and the accumulation of reduced N in maize (Zea mays L.). II. The effect of nutrient nitrate concentration, Plant Physiol., in press.

Ritenour, G. L., Joy, K. W., Bunning, J., and Hageman, R. H., 1967, Intercellular localization of nitrate reductase, nitrite reductase and glutamic dehydrogenase in green leaf tissue, Plant Physiol., 42:233.

Shaner, D. L., and Boyer, J. S., 1976, Nitrate reductase activity in maixe (Zea mays L.) leaves. I. Regulation by nitrate flux, Plant Physiol., 58:499.

Storey, R. L., and Beevers, L., 1977, Proteolytic activity in relationship to senescence and cotyledonary development in *Pisum sativum* L., *Planta*, 137:37.

Welch, L. F., 1979, Nitrogen use and behavior in crop production, Bull., 761, Agric. Expt. Station, College of Agric., Urbana, IL.

Zeiserl, J. F., Rivenbark, W. L., and Hageman, R. H., 1963, Nitrate reductase activity, protein content and yield of four maize hybrids at varying plant populations, *Crop Sci.*, 3:27.

INHERITANCE OF NITRITE REDUCTASE

S. Heath-Pagliuso and R. C. Huffaker

Plant Growth Laboratory and the Department of Agronomy
and Range Science, University of California
Davis, CA 95616

INTRODUCTION

Although nitrite reductase (NiR) resides primarily in the chloroplast (Dalling et al., 1972; Lips et al., 1972; Miflin, 1974; Ritenour et al., 1967; Washitani et al., 1977) it has not been determined whether this enzyme is coded for by chloroplast DNA or by nuclear DNA.

Electrophoretic variants can be used to determine inheritance patterns for enzymes from appropriate crosses of individuals exhibiting mobility differences. The resulting F_1 and F_2 progeny phenotypic ratios are compared to the expected ratios. If the enzyme exhibits a maternal inheritance pattern, it is coded for by chloroplast DNA. If the enzyme exhibits nuclear (Mendelian) inheritance, it is coded for by nuclear DNA.

In this paper we show that NiR is inherited as an enzyme whose mRNA is coded for by nuclear DNA.

MATERIALS AND METHODS

Plant Material

NiR mobility variants of *Avena barbata* brot. were identified by electrophoretic survey. Sampling for electrophoresis requires only a small portion of leaf material so the plant is not destroyed. Individuals of differing mobility classes were maintained in the greenhouse. The parental types were crossed and F_1 progeny seed was obtained. The F_1 progeny of this cross were analyzed on starch gels. The five F_1 individuals sampled were allowed to grow to maturity and selfed seed was collected. Twenty individuals from a single F_2 family were analyzed electrophoretically.

Electrophoresis

Plants were grown in vermiculite and watered with full-strength Hoagland's nutrient solution whenever necessary over a period of 12 days. On day 13, plants were given 100 mM KNO_3 in quarter-strength Hoagland's solution. On day 14, one leaf from each plant was excised and crushed. The exudate obtained was soaked onto a paper wick and used for electrophoresis. The wicks were inserted into the gel at a point 4 cm from the bottom.

The gel and tray buffer systems were modified after Brewer (1970) as follows: the pH of the electrode buffer was changed from 8.5 to 8.6, and gel buffer pH is adjusted to pH 8.0. The constituents of the gel buffer were also altered to contain the following: 15 mM tris-Sigma 7-9, 22 mM boric acid, 3 mM EDTA, 1 mM cysteine, and 0.04 g/l FAD. Gels are run at 30-35 mA in a 4 C chamber with ice packs until the marker dye (Bromophenol blue) reached a point 11 cm from the origin of the wicks.

The staining procedure was modified after that presented by Hucklesby and Hageman (1973) in order to accommodate horizontal starch gels. NiR accepts electrons from reduced methyl viologen; therefore, the bands appear as white regions against a red background.

RESULTS

Figure 1 is a diagrammatic representation of the two types of inheritance patterns as they would appear on starch gels. The appropriate corsses are shown and the expected phenotypic ratios for a single locus with two alleles. _A. barbata_ is a tetraploid organism; hence, the banding patterns observed on the gels is actually more complicated than diagrammed in Figure 1. Each parental type has two NiR loci which appear to be different from one another. In addition, one locus appears to have a mutant NiR (see Fig. 2). The NiR isozyme designated by B in the diagram is identical in both parents. In the other genome, the parents differ electrophoretically for NiR (A and A'). Each parent has two NiR isozymes. The hybrids of these two parental types all exhibited the triple banded phenotype, AA'BB. The F_2 family appeared to be segregating, that is, three phenotypes appeared in the family array. The Chi-square value calculated on the expected 1:2:1 ratio for nuclear inheritance is .474 ($.9<P<.75$).

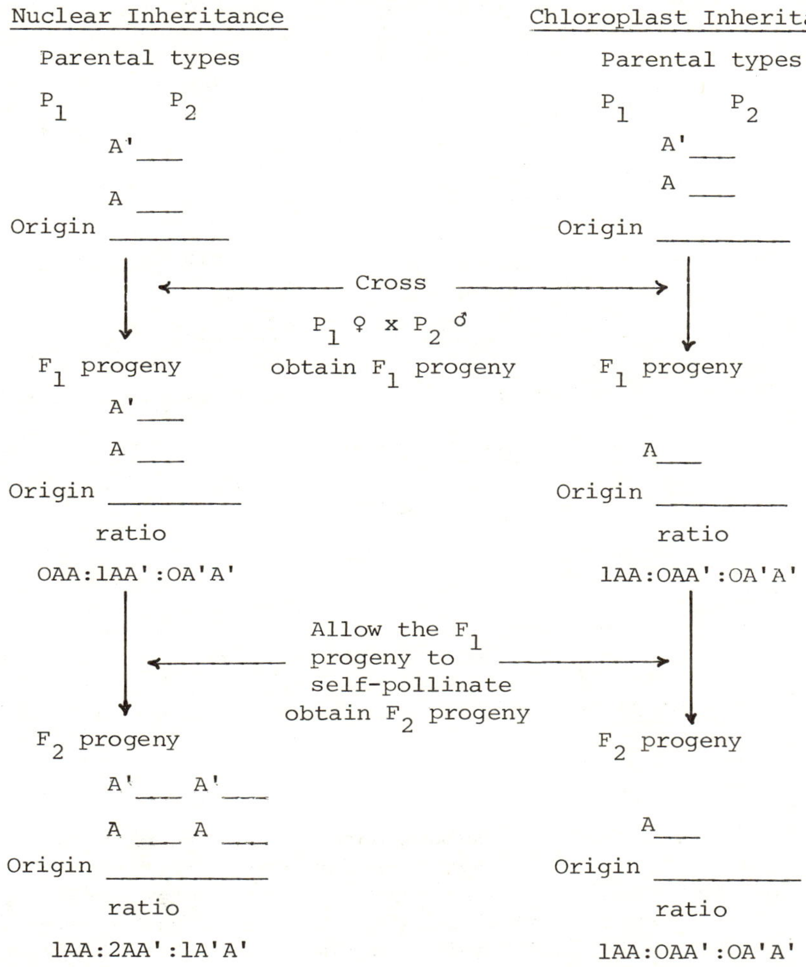

Fig. 1. A diagrammatic representation of the two possible inheritance patterns and the expected progeny phenotypic ratios for an enzyme exhibiting two mobility classes upon electrophoresis. A and A' are the relative distances the two NiR enzymes travel from the origin. Individuals of type A were used as the females in the first cross to obtain F_1 seed. The F_1 seed were allowed to self-pollinate and F_2 seed was collected. Following the scheme on the left gives the expected F_1 and F_2 gel banding patterns and ratios for nuclear inheritance. The scheme on the right gives the expected F_1 and F_2 gel banding patterns and ratios for chloroplastic inheritance.

NiR Phenotypes for Avena barbata

Parent 1		Parent 2	
B ___	4.3 cm	B ___	4.3 cm
A ___	3.7 cm	A' ___	4.0 cm
Origin ________		Origin ________	

Expected F_2 Phenotypes and Ratios

Nuclear Inheritance		Chloroplastic Inheritance	
B___B___B___	4.3 cm	B___	4.3 cm
A'___A'___	4.0 cm		
A___A___	3.7 cm	A___	3.7 cm
Origin ________		Origin ________	
ratio		ratio	
1AABB:2AA'BB:1A'A'BB		1AABB:OAA'BB:OA'A'BB	

Fig. 2. Figure 2 shows the actual banding pattern phenotypes on gels of parents of Avena barbata Brot. different electrophoretically at a single locus and the phenotypes among the F_2 progeny resulting from a cross between them if inheritance at the locus is nuclear and if it is chloroplastic. A and B refer to two separate NiR loci. A and A' refer to mobility variants at locus A. The numbers at the right indicate the distance each enzyme traveled from the origin in cm. The crosses are as diagrammed in Figure 1.

CONCLUSION

The Chi-square value obtained for the F_2 family fits the hypothesis that NiR inheritance is under nuclear control not chloroplastic. It appears that the messenger RNA for NiR is made on nuclear DNA although the subcellular location of NiR seems to be the chloroplast. This implies that 1) NiR is made in the cytoplasm and is transported to the chloroplast, 2) or that the mRNA is transported to the chloroplast and NiR is made inside the chloroplast. The genetic analysis does not distinguish between the two alternatives. However, these data strongly suggest that NiR is a chloroplast protein whose genetic information is of nuclear origin.

ACKNOWLEDGMENT

We thank Dr. Ladizinsky for making the *Avena barbata* crosses. We also thank Dr. R. W. Allard for making the hybrid seed available to use. A special thanks goes to Elmarie Hutchinson for proof-reading the early drafts of this paper. This work was partially supported by grants from NSF (PFR 77-07301) and USDA (AGR 516-15-119).

REFERENCES

Brewer, G., 1970, An Introduction to Isozyme Techniques, Academic Press.

Dalling, M. J., Tolbert, N. E., and Hageman, R. H., 1972, Intracellular location of nitrate reductase and nitrite reductase, *Biochim. Biophys. Acta*, 283:505.

Hucklesby, D. P., and Hageman, R. H:, 1973, A staining method for nitrite reductase on polyacrylamide gels after electrophoresis, *Anal. Biochem.*, 56:591.

Lips, S. H., and Avissar, Y., 1972, Plant-leaf microbodies as the intracellular site of nitrate reductase and nitrite reductase, *Eur. J. Biochem.*, 29:20.

Miflin, B. J., 1974, The location of nitrite reductase and other enzymes related to amino acid biosynthesis in the plastids of root and leaves, *Plant Physiol.*, 54:550.

Ritenour, G. L., Joy, K. W., Bunning, J., and Hageman, R. H., 1967, Intracellular localization of nitrate reductase, nitrite reductase and glutamic acid dehydrogenase in green leaf tissue, *Plant Physiol.*, 42:233.

Washitani, I., and Sato, S., 1977, Studies on the function of proplastids in the metabolism of *in vitro* cultured tobacco cells. I. Localization of nitrite reductase and NADP-dependent glutamate dehydrogenase, *Plant and Cell Physiol.*, 18:117.

MODELLING DYNAMIC ASPECTS OF NITROGEN IN SOILS AND PLANTS

H. van Keulen

Centre for Agrobiological Research (CABO)
Bornsesteeg 65, 6708 PD
Wageningen, The Netherlands

INTRODUCTION

It is well over a century since the disagreements between Von Liebig and Gilbert/Lawes on the mineral nutrition of crop plants surfaced. Most of the questions that occupied these scientists at the time seem to have come to a state where reasonable agreement among scientists exists. However one can hardly avoid the conclusion that a hundred years of agricultural and plant physiological research have not advanced the understanding of the basic principles involved in plant nutrition sufficiently to permit the formulation of an unequivocally accepted quantitative description. True enough, the processes involved are complex and for a long time, even up to today may not have been amenable to direct measurements. It also appears however, that the interest in the nature of the processes has been only moderate, probably because a ceiling yield could be achieved anyway by increasing the amount of applied mineral fertilizer. Changing conditions and attitudes in recent years, which led on the one hand to a growing awareness of the (relatively) high costs of fertilizer manufacturing and on the other hand to a growing concern over the possible environmental effects of dumping high doses of plant nutrients on agricultural crops, have stimulated research aimed at a more efficient utilization of the nutrient resource. The basis of any improvement however, is a thorough understanding of the system, the way it operates and the constraints to which it is exposed.

In this presentation a model will be outlined in which an attempt is made to integrate the existing knowledge on nitrogen cycling in the soil-plant system. Because of the nature of the research project in which it was developed, it is especially suitable for annual species, growing under conditions of limited moisture supply. Certain modules could have a wider applicability, but since in many cases descriptive rather than explanatory formulations have been used, such an extrapolation may be dangerous at the moment.

THE SIMULATION MODEL PAPRAN

The model has been developed in the framework of a research project, concentrating on actual and potential production under semi-arid conditions. It therefore concentrates on the effects of moisture availability and of nitrogen supply on dry matter production. The description given here refers to the version applied to the natural pasture, an abandoned cropland vegetation, consisting of annual grasses and herbs. Basically the same formulation is being used for a parallel model, describing the production of a wheat crop under these conditions. Since the validation studies with that version are still in a preliminary stage it will not be discussed here.

The main elements of the model are given in the schematized relational diagram of Figure 1. The mixture of annual plants is considered as a homogeneous stand with defined physiological properties. In the real situation the appearance of the canopy is rather heterogeneous, especially at the onset of the growing season. That heterogeneity is especially important when the vegetation is being exploited by grazing. The consequences of this phenomenon for the performance of the model are being studied at the moment. The model describes the growth of an annual crop from the moisture balance and the nitrogen balance in the soil below it.

Plant Growth

Plant growth starts in the soil with germination of part of the seed store. The process is initiated when the average moisture content in the upper 10 cm of the soil is above field capacity and proceeds until a temperature sum of 150 days by degrees C above 0 C has accumulated. When the soil dries out before that due to evaporation, the germinating seeds are supposedly killed and a new wave of germination starts only after additional precipitation.

The initial amount of biomass present at establishment depends on the seed stock in the soil in a given season and the conditions during germination. The description used in the present version does not allow for such a detailed treatment, hence the initial value is estimated independently for each season.

After establishment, the gross rate of CO_2-assimilation is obtained from the leaf area index and the radiation intensity for

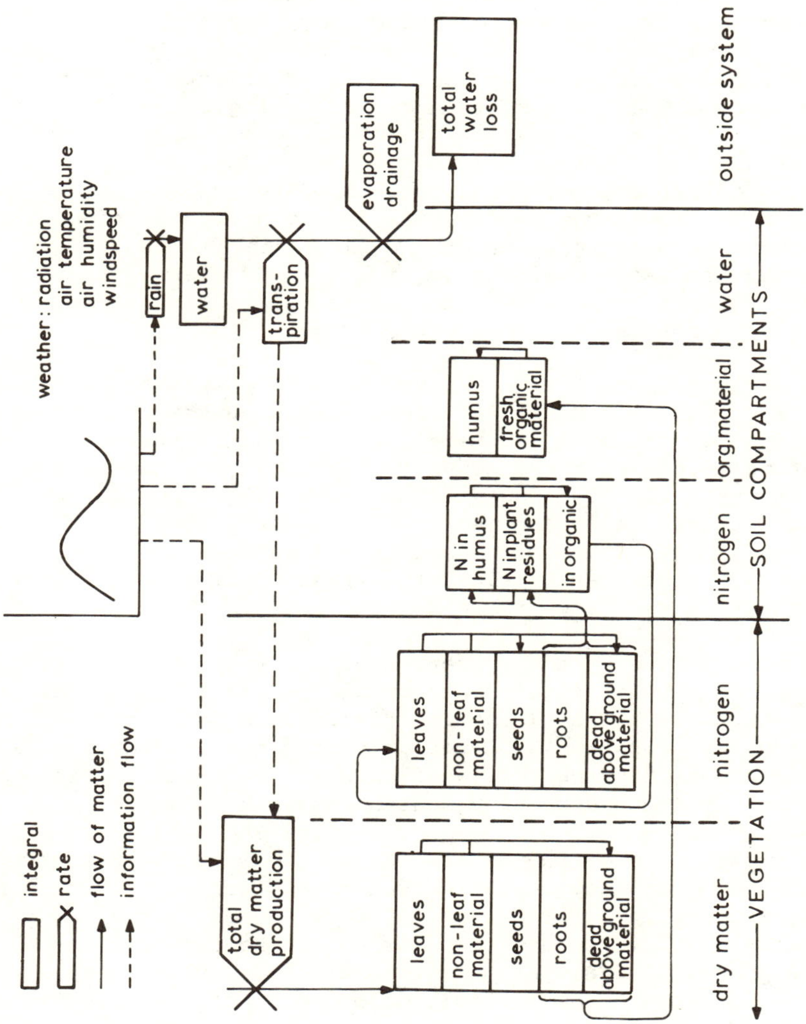

Fig. 1. Simplified relational diagram of the simulation model PAPRAN.

a given photosynthesis light response curve of the individual leaves (Goudriaan and van Laar, 1978). The rate of maintenance respiration, that is energy required to maintain the existing cell structures (Penning de Vries, 1974) is calculated from the dry weight present and the protein content of the structural material. The potential rate of dry weight increase then follows from the balance between gross assimilation and maintenance respiration, by multiplying with a conversion efficiency, expressing growth respiration, the energy lost in the conversion of primary photosynthates into structural plant material. That efficiency also depends on the composition of the material being formed (Penning de Vries, 1974).

The potential rate of transpiration of the vegetation is obtained from the leaf area index and environmental conditions influencing the evaporative demand: level of irradiance, vapor pressure and temperature of the ambient air and windspeed. The values are integrated in a descriptive formula, using average daily weather parameters (van Keulen, 1975).

The ratio of potential daily growth rate and potential transpiration rate, the water use efficiency, is a central parameter in the model. This ratio is assumed to be constant, irrespective of the moisture status of the canopy. Theoretical considerations (van Keulen et al., 1980b) as well as experimental results (Lof, 1976) suggest that its value may change slightly when plants are under water stress. The actual amounts of water transpired during such periods are however so small that these differences hardly affect the dry matter accumulation.

The actual rate of transpiration of the vegetation is calculated from the potential rate accounting for the influence of moisture status in the soil and the distribution and activity of the root system. In each soil compartment water is freely available until about 25 percent of available water is left, after which uptake is reduced until cessation at wilting point.

The growth rate of the vegetation is obtained by multiplying the water use efficiency by the actual rate of transpiration. Nitrogen deficiency in the tissue undoubtedly leads to reduced growth rates. Its influence on water use is however far less clear. A substantial amount of literature data suggest improved water use efficiencies as a result of optimum nitrogen supply (de Wit, 1958; Viets, 1962). This is confirmed by the experimental results obtained in the northern Negev desert of Israel, where application of nitrogenous fertilizer resulted in appreciable increases in total dry matter production, without affecting the moisture balance in the soil (van Keulen, 1975). In the present version of the model transpiration is therefore unaffected by the nitrogen status of the tissue. The point will be treated again in the discussion.

Allocation of Dry Matter. The total daily increment in dry matter is divided between leaves, non-leaf vegetative above-ground material (stems, leaf sheaths), roots and seeds. The allocation between the various organs is in first instance governed by the phenological state of the vegetation, represented by its development stage. First, a part of the available photosynthates is allocated to the roots. For this annual vegetation, the fraction varies between 0.5 at establishment and 0.025 toward maturity under optimum growth conditions. Insufficient moisture supply may modify this fraction, according to the functional balance principle: water stress favors root growth relative to that of above-ground material. A similar approach could be followed to account for the influence of nitrogen deficiency, but insufficient experimental data are available for a sound quantitative description.

A constant fraction of the material available for above-ground growth is diverted to the developing seeds after the flowering stage. The remainder is divided between leaves and non-leaf material, the actual proportions being governed by development stage (progressively less material is invested in new leaves with increasing age) and nitrogen status of the vegetation (nitrogen deficiency shifts growth in the direction of non-leaf material).

Death of the Vegetation. Vegetative plant parts may die as a result of stress, due to water or nitrogen shortage or due to senescence when the plants life cycle is completed.

The rate of dying of the material due to water shortage is governed by the balance between the rate of water loss by the vegetation and the rate of water uptake by the root system. Under severe moisture stress, the stomata are closed and water loss by the vegetation is restricted to cuticular transpiration. The rate of moisture loss through the cuticle is a function of the atmospheric demand, a constant conductivity being assumed. The rate of moisture uptake by the root system depends on the moisture status in the soil. Dying of the tissue is thus a function of both states of the system, whereas the buffering capacity of the vegetation is taken into account by applying a time constant of five days for dying. The death rate acts on both leaf and non-leaf material. The present description of tissue dying simulates actual situations during moisture stress reasonably well, but it has to be verified experimentally (van Keulen et al., 1980a).

Dying of the tissue due to nitrogen deficiency is a function of the nitrogen content of the vegetation. Dying starts when the nitrogen concentration drops below the threshold value for unrestricted growth. The relative death rate increases gradually from there on, to reach a maximum value of 0.3 day^{-1} at the absolute minimum nitrogen concentration.

Dying of vegetative material due to senescence proceeds in reality from early stages of plant development. Leaves have a limited life span and the first ones may disappear relatively early. Since, however, in the present version of the model the leaves are not distinguished in age classes, senescence is disregarded before flowering. After that event senescence is accelerated as a result of translocation of especially plant nutrients to the developing seeds. At the end of the plant's life cycle, the relative death rate assumes a maximum value of 0.1 day^{-1}. This value causes almost complete drying of the standing vegetation in a fortnight, whigh is in good agreement with field observations.

Nitrogen in the Vegetation

Inorganic nitrogen present in the soil is taken up by the vegetation. No distinction is being made at present between NO_3^- and NH_4^+, since neither denitrification nor leaching play a role under the semi-arid conditions, and the vegetation may take up both forms of the element. It is assumed that the density and activity of the root system is sufficient to explore the total rooted volume. All the inorganic nitrogen in that volume is therefore available for uptake within a very short time, either by mass flow with the transpiration stream or by diffusion along developing concentration gradients resulting from low N-concentrations at the root surface (van Keulen et al., 1975).

The demand for nitrogen in the vegetation is created by the difference between the current concentration in the tissue and the maximum concentration possible. The latter value is a function of the development stage of the vegetation and varies between the different plant organs. For each of the components the value is high in the early stages and declines toward maturity.

When the supply of nitrogen is limited, that is when the total demand is not met by the actual uptake, the amount taken up by the vegetation is distributed between shoot and root in proportion to their relative demands. Neither experimental evidence (van Dobben, 1961) nor simulation results support the hypothesis that the closeness to the source gives the root system special advantage. Nitrogen transferred to the aerial plant parts is again distributed between leaves and non-leaf material in proportion to the relative deficiencies in the organs. The nitrogen demand created by the developing seeds after flowering is met by translocation from the vegetative tissue. All nitrogenous compounds accumulated in the seed are thus assumed to have passed through the vegetative tissue first. This translocation process results in nitrogen depletion of the vegetative tissue, when the supply from the soil does not meet the seed requirements. In annual vegetations, both natural pastures (Penning de Vries et al., 1980) and crop plants (Spiertz, 1978) it is often observed that accumulation of nitrogen ceases after flowering. It is not clear, whether reduced root activity, or exhaustion of the soil nitrogen store, or spatial separation of moisture and nitrogen or a combination of these factors is responsible for this phenomenon. When the vegetative tissue becomes increasingly deficient in nitrogen, translocation to the seeds is hampered, resulting in lower protein contents in the grain. A lower limit for the nitrogen content has been introduced in the model, forcing cessation of the carbohydrate flow to the seeds when this limit is approached. This description is based on the observation that under limited N-supply the protein content of seeds is species-dependent only, irrespective of growing conditions (van Keulen, 1977). Lack of understanding of the underlying processes necessitated the adaptation of this descriptive formulation.

Death of vegetative tissue leads to loss of nitrogen from the vegetation. The nitrogen concentration in the dying tissue depends on the cause of death. Tissue dying from water shortage or senescence disappears with a concentration equal to that of the live material. Tissue, dying from nitrogen deficiency contains only the irreversibly incorporated nitrogen, the remainder being translocated to the remaining plant parts. This description mimicks the real-world situation where continuous breakdown and transfer of nitrogenous compounds permits the growth of new tissue at the expense of older ones.

Nitrogen in the Soil

The complexity and importance of nitrogen transformations in the soil is illustrated by the voluminous literature on the subject (Bartholomew and Clark, 1965; Tandon, 1974; van Veen, 1977). Nevertheless our understanding of the processes that play a role and their quantitative consequences is still very limited as is demonstrated by the attempts to use the available information for the development of detailed models of the soil nitrogen system (Beek and Frissel, 1973; Hagin and Amberger, 1974; van Veen, 1977). The approach followed by van Veen, which is microbiologically based is theoretically sound, since the transformations in the soil are governed by microbial activities. In practice however, the distinction of a separate pool of microbial biomass or proteins, is difficult (if not impossible) because of the available experimental techniques. Initialization of such a model for a specific site, as well as validation of the results is therefore hardly feasible. In the present approach nitrogen in the soil system is divided into three states: inorganic nitrogen, nitrogen in 'fresh' organic material (including plant residues which have not yet passed through the microbial pool and the microbial tissue itself) and nitrogen in 'stable' organic material (soil humus, which has at least once undergone a transformation through the microbes).

Organic Matter Transformations. The rate of decomposition of the two organic fractions in the soil is based on first-order kinetics: under optimum conditions a constant relative rate of decomposition is assumed. The specific rate is different for various compounds (Hagin and Amberger, 1974): it is of the order of 1 day^{-1} for easily decomposable proteins and sugars, ± 0.05 day^{-1} for cellulose and hemi-cellulose and ± 0.01 day^{-1} for lignin. In the model these different rates are introduced in a step-wise manner, as the original amount of added fresh organic material reduces. Different compositions of the added material can be accounted for by changing the switch-values for the rate constants. The rates of decomposition may be modified by environmental conditions of soil moisture, temperature and C:N ratio of the decomposing material. When the overall C:N ratio, including the mineral nitrogen present is below 25, decomposition proceeds at the potential rate (Parnas, 1975).

In this description it is implicitly assumed, that the activity of decomposing bacteria is never limiting the rate of decomposition, while instantaneous adaptation to different substrates is assumed.

The stable organic material, which is assumed to have a constant C:N ratio of 10, decomposes at a much lower rate (Harpaz, 1975). Again soil moisture conditions and temperature may influence its rate. Accretion of stable organic material results from stable compounds of the fresh organic material. It is assumed that 'humus-formation' takes place when the overall C:N ratio of the decomposing material is below 25. At that stage mineral nitrogen is released, of which 20% is assumed to be incorporated in the stable fraction. The constant C:N ratio applied, also yields the rate of humus accretion.

Soil Nitrogen Processes. During decomposition of the fresh organic material, mineral nitrogen is being released, when the carbon of the substrate is used for build-up of microbial tissue and supply of energy for the functioning of the microbial population. The rate of release of mineral N is therefore a direct function of the rate of decomposition of the substrate and its nitrogen content. At the same time, however, the build-up of the microbial biomass requires nitrogen for the build-up of proteins and amino acids. The rate at which nitrogen is incorporated in the tissue is again dictated by the rate of decomposition of the substrate which provides energy and structural material. The basic assumption is, that all carbon released during breakdown of the organic compounds is used for build-up of microbial biomass, with an average biosynthesis efficiency (Sörenson, 1975). The required protein content of the microbial tissue thus determines the N-requirement during decomposition. The composition of the decomposing material then dictates whether mineral nitrogen is immobilized in the tissue. Gross release of nitrogen is thus dependent on the nitrogen content of the originally added fresh material. Net release, i.e., the balance between gross release and immobilization must in this description be considered as originating from the microbial component of fresh organic material. The formulation leads to realistic simulations of the switch from net immobilization to net release at total C:N ratio's between 25-30.

The rate of change of nitrogen in the stable organic compound is the balance between its rate of mineralization governed by the decomposition rate and the rate of incorporation dictated by the rate of release from the fresh organic material.

The amount of mineral nitrogen in each soil compartment changes by release from or immobilization into the organic components and by uptake by the plants. Furthermore transport between compartments is taken into account. The time interval of one day employed in the model is much larger than the time constant for solute transport (de Wit and van Keulen, 1972; de Wit and Goudriaan, 1974). A mimicking procedure is therefore again necessary. Transport is assumed to take place only with movement of the water, that is during infiltration. The concentration of mineral nitrogen transported over the lower boundary of a compartment is obtained by 'mixing' the solutes present in the compartment and those transported into it, with water in the layer and all water flowing through. Such a description takes into account mass transport and (part of) the effects of mathematical dispersion, inherently present in such compartmentalized models (Goudriaan, 1973). Upward transport of solutes with the soil evaporative flow or diffusion along developing concentration gradients is not taken into account. The present description is satisfactory for our purpose, where the main interest is in the availability of nitrogen to the vegetation, rather than in its exact distribution within the profile.

Ammonium fertilizers, especially when applied to soils, with a high pH may lead to considerable losses of nitrogen through volatilization. Again the time constant of the chemical processes involved are lower than the time resolution of the model. Ammonium when present in the top soil compartment is assumed to volatilize at a constant relative rate, whereas at the same time nitrification proceeds. The relative rates for these processes are of the order of 5-10 day^{-1}, so that appreciable losses only occur during the first week after application when no precipitation occurs during that period. This is a rather crude description and a more realistic one would increase the generality of the model.

Processes Not Considered. Some of the processes taking place in the soil-plant system have not been included in the model:

1) Denitrification is disregarded for the moment, because anaerobic conditions are unlikely to occur in the semi-arid regions for which the model is mainly developed. The possibility that through intensive biological activity anaerobic pockets may occur around plant roots or inside soil structural elements is recognized, but simulation of such situations is extremely complex (Leffelaar, 1979).

2) Adsorption of NH_4^+ onto the exchange complex or fixation into the lattice of clay minerals has not been taken into account. In most situations these processes play a minor role in the annual nitrogen balance. In specific cases however, especially when nitrification is hampered, they could be of importance. Appropriate process models would have to be developed then.

Soil Water Balance

The description of the soil moisture balance in this model is essentially identical to that used in the model ARID CROP (van Keulen, 1975; van Keulen et al., 1980a) and will be treated here only briefly. For the description of the water balance, the soil depth is divided into homogeneous compartments. Both their number and their thickness can be easily adapted.

Infiltration. Infiltration into the soil either from rain or irrigation is obtained from the rate of moisture supply, taking into account possible run-off effects. The latter has been treated rather rudimentary. A more detailed treatment has been elaborated (Rietveld, 1978) but as yet not been applied in lower resolution models.

When water is added to the soil, the change of moisture content in each compartment is set equal to the difference between the moisture content at field capacity and the current moisture content. Instantaneous 'equilibrium' is thus assumed. The compartments are filled in this way from the top one downwards.

Soil Surface Evaporation. Under semi-arid conditions evaporation from the soil surface is the most important source of non-productive water loss. In the model potential evaporation follows from a Penman-equation. Subsequently that is modified by taking into account the effects of shading by the vegetation and of drying of the topsoil. The total water loss is proportioned over the different compartments by means of a mimicking procedure (van Keulen, 1975). This procedure yields reasonable results in winter rainfall conditions, but seems not to be applicable without modification under summer rainfall conditions, probably because of the steep temperature gradients that may develop in such situations.

Root Water Uptake. From the potential rate of transpiration and the total rooted depth, the required rate of water uptake per unit root length is calculated. In each compartment that rate may be reduced due to low moisture contents or due to low soil temperatures. The relation between soil moisture content and root water uptake is of the Viehmeyer-type: water remains freely available until about 70% of the available water has been used. After that a sharp reduction follows until the permanent wilting point. Partial compensation is accounted for, when a portion of the root system is in dry soil layers. Temperature effects on root water uptake take into account both the change in viscosity of the water and in activity of the root system.

RESULTS AND DISCUSSION

The performance of the model was studied by analyzing its behavior under conditions prevailing in the northern Negev desert of Israel. In this semi-arid region (average annual rainfall 250 mm) long-term experiments on the productivity of natural pastures have been carried out (Tadmor et al., 1974; van Keulen, 1975) both with and without the application of nitrogenous fertilizers. For the simulation runs actual weather data were used, collected at a nearby station of the Israeli Meterological Service, with the exception of rainfall which was recorded at the site.

The initial conditions at the onset of the rainy season were assumed identical for all years: a total amount of 3000 kg of fresh organic material, consisting of roots, stubble, etc., of last year with a nitrogen content of 0.01 kg N kg^{-1} (dry matter) present in the upper 60 cm of the soil profile. When fertilizer was applied, this was assumed to be evenly distributed in the upper 10 cm of the profile in ammoniacal form.

The results of the simulations (Figs. 2, 3, 4) do not show a consistent picture:

1) For the 1966/1967 growing season only application-yield data are available for testing. The yield without fertilizer application is reasonably well estimated, but the simulated fertilizer response curve deviates considerably from the measured one. Since no nitrogen uptake data are available it is impossible to determine

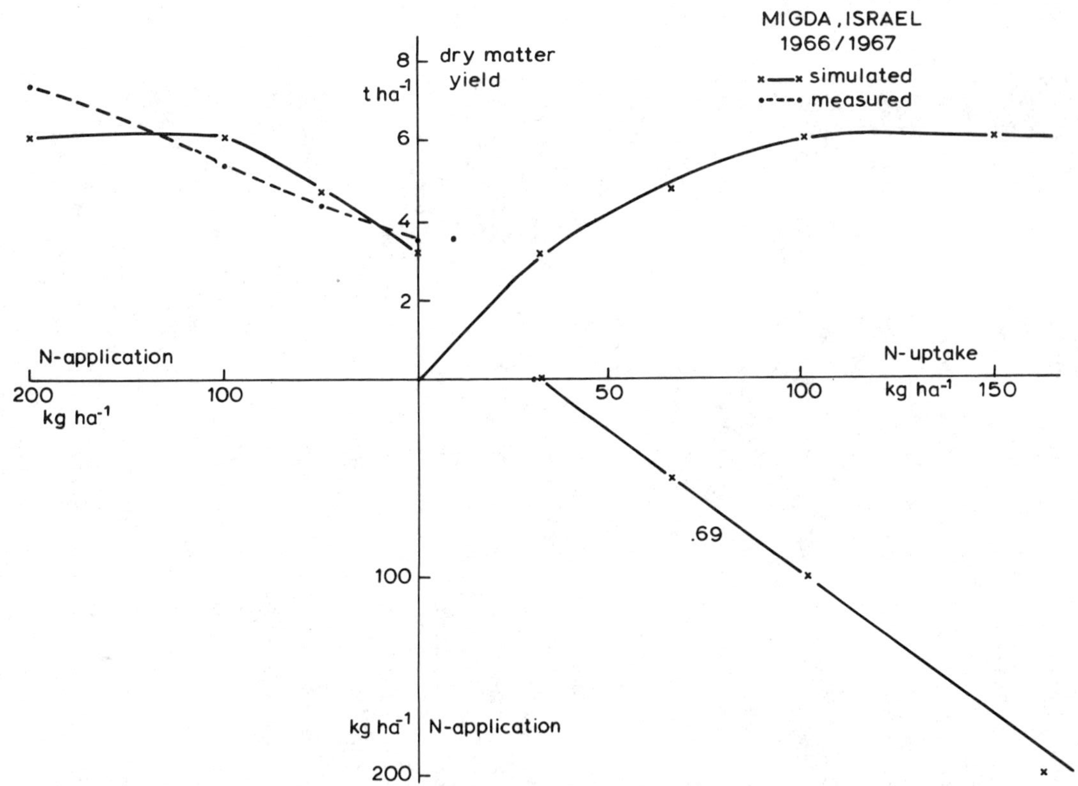

Fig. 2. Comparison between simulated and measured results of fertilizer experiments on natural vegetation in Migda, Israel in the 1966/1967 growing season (right hand quadrants only simulated).

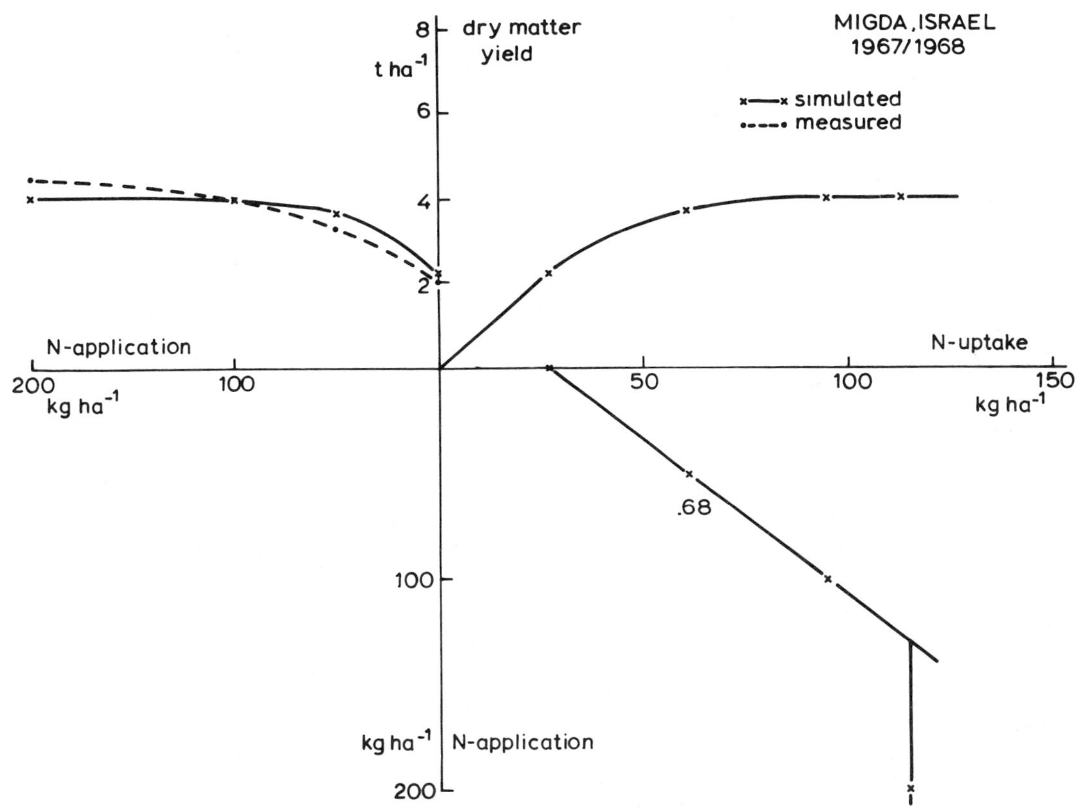

Fig. 3. Comparison between simulated and measured results of fertilizer experiments on natural vegetation in Migda, Israel in the 1967/1968 growing season (right hand quadrants only simulated).

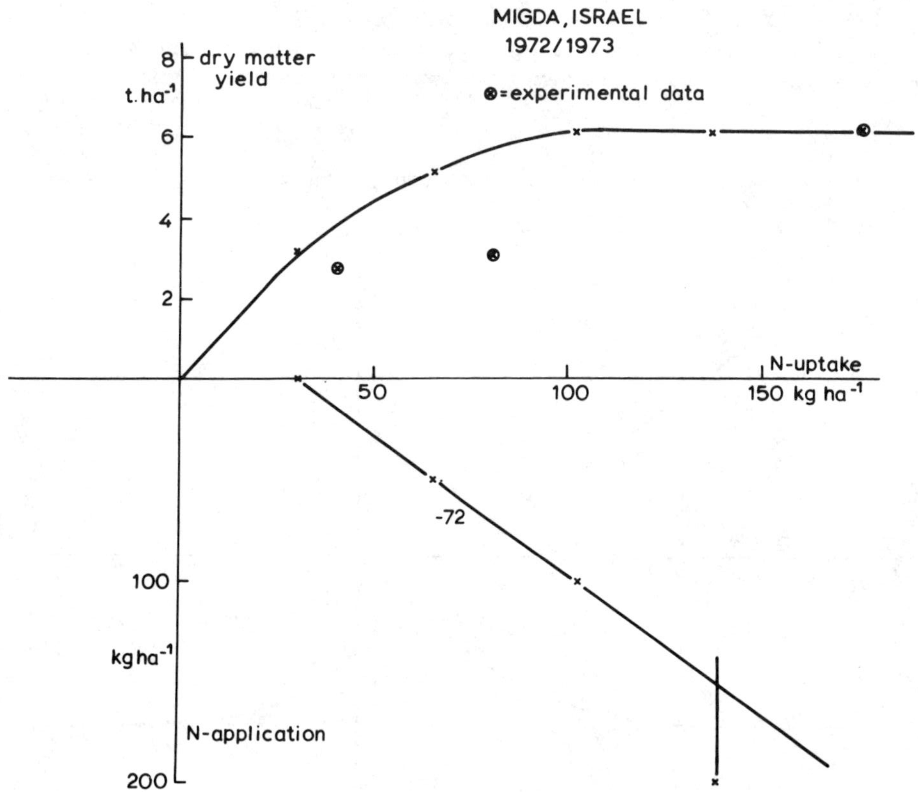

Fig. 4. Comparison between simulated and measured influence of nitrogen uptake on dry matter production of natural pasture in Migda, Israel in the 1972/1973 growing season.

whether the discrepancies in the fertilizer response are the result of inaccuracies in the description of the response of the vegetation to increased nitrogen availability or of a misrepresentation of the application-uptake relation.

2) For 1967/1968, a drier year than the previous one (mainly as a result of a much more unfavorable distribution of the rainfall), the simulated yield-application curve is fairly close to the measured one over the full range of applications. In such years there is only limited benefit of the application of nitrogenous fertilizer. However under these semi-arid conditions where neither leaching nor denitrification play an important role, the applied nitrogen will largely remain in the soil and become available in subsequent years. When the soils are sufficiently deep and fertilizer is applied in nitrate form, all applied nitrogen should eventually be recovered in the vegetation. Whether this is also true in the short run is questionable, since inputs of fertilizer N will result in the creation of a non-equilibrium, where part of the applied N may temporarily be tied up in the soil organic matter.

3) The simulated yield-uptake curve for 1972/1973 deviates considerably from the measured one. Part of the explanation must be the leguminous component, which, at peak biomass, comprised about 25% of the total dry matter yield. Even when that is taken into account, the measured yield is low in relation to the amount of N absorbed. The reason for this lack of dilution is not clear.

The influence of nitrogen shortage on the transpiration/assimilation ratio and hence the water use efficiency is subject to debate. There are, on the one hand, many reports showing improvements in water use efficiency at optimum N-levels (c.f., Viets, 1965). More recent experimental evidence suggests however a constant ratio irrespective of the nutritional level of the plant (Goudriaan and van Keulen, 1979). The results of the model (Fig. 5) strongly indicate that total water loss under the Negev conditions is virtually identical with and without N-application. Results obtained under Sahelian conditions (PPS, 1980) however show that under N-limiting conditions more moisture remains in the soil at the end of the growing season. This aspect of plant physiology certainly warrants more attention.

The results presented here indicate that the performance of the model when compared to the real-wold is rather variable. One of the major drawbacks is, that for a model of this complexity

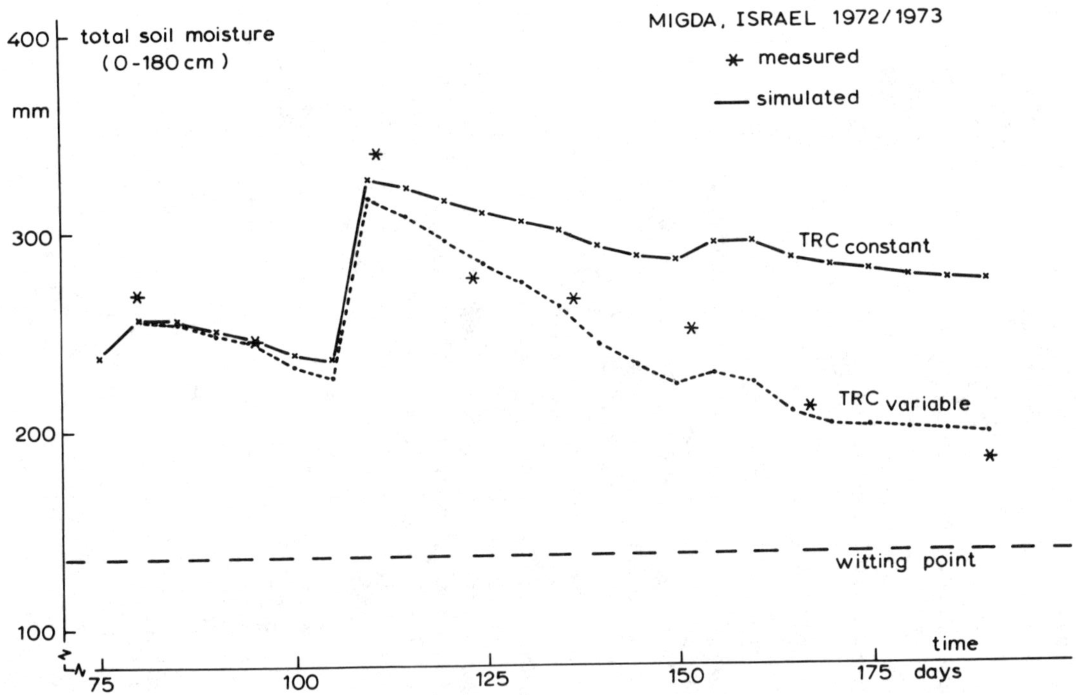

Fig. 5. The influence of two alternative hypotheses with respect to transpiration/photosynthesis ratio, on the simulated moisture balance below the natural vegetation in Migda, Israel in the 1972/1973 growing season, in comparison to the measured data.

the use of only gross output data for validation is hardly sufficient. It contains so many relations and parameters that almost any result can be obtained by changing their value. On the other hand, however, data sets, to validate separate elements of the model are extremely scarce. This again stresses the point that systems analysis, model development and simulation can never be a substitute for experimental work. It should be used as a framework by which problems can be more clearly recognized and analyzed, and hence may lead to the design of more relevant experiments.

REFERENCES

Bartholomew, W. V., and Clark, F. W., 1965, Soil nitrogen, Agronomy 10, American Society of Agronomy, Madison, Wisconsin.

Beek, J., and Frissel, M. J., 1973, Simulation of nitrogen behavior in soils, Simulation Monographs, PUDOC, Wageningen.

Dobben, W. H., van, 1961, Nitrogen uptake of spring wheat and poppies in relation to growth and development, Jaarboek I.B.S. (Dutch with English summary), 1961:54.

Goudriaan, J., 1973, Dispersion in simulation models of population growth and salt movement in the soil, Neth. J. Agric. Sci., 21:269.

Goudriaan, J., and van Laar, H. H., 1978, Relations between leaf resistance, CO_2-concentration and CO_2-assimilation in maize, beans, lalang grass and sunflower, Photosynthetica, 12:241.

Hagin, J., and Amberger, A., 1974, Contribution of fertilizer and manures to the N- and P-load of waters. A computer simulation, Deutsche Forschungsgmeinschaft, Bonn, Technion Research and Development Foundation Ltd., Haifa.

Harpaz, Y., 1975, Simulation of the nitrogen balance in semi-arid regions, Ph.D. Thesis, Hebrew University, Jerusalem.

Keulen, H. van, 1977, Nitrogen requirements of rice, with special emphasis on Java, Contr. Centr. Res. Inst. Agric. Bogor, no. 30.

Keulen, H. van, 1975, Simluation of water use and herbage growth in arid regions, Simulation Monographs, PUDOC, Wageningen.

Keulen, H. van, Seligman, N. G., and Goudriaan, J., 1975, Availability of anions in the growth medium to roots of an actively growing plant, Neth. J. Agric. Sci., 23:131.

Keulen, H. van, Seligman, N. G., and Benjamin, R. W., 1980a, Simulation of water use and herbage growth in arid regions: re-evaluation and further development of the model ARID CROP, Agricultural Systems, in press.

Keulen, H. van, van Laar, H. H., Louwerse, W., and Goudriaan, J., 1980b, Physiological aspects of increased CO_2-concentration, Experientia, 36:286.

Lof, H., 1976, Water use efficiency and competition between arid zone annuals, especially the grasses Phalaris minor and Hordeum mirinum, Agr. Res. Rep. 853, PUDOC, Wageningen.

Leffelaar, P. A., 1979, Simulation of partial anaerobiosis in a model soil in respect to denitrification, Soil Sci., 128:110.

Parnas, H., 1975, Model for decomposition of organic material by micro-organisms, Soil Biol. and Biochem., 7:161.

Penman, H. L., 1948, Natural evaporation from open water, bare soil and grass, Proceedings Royal Society London, A 193:120.

Penning de Vries, F. W. T., 1974, Substrate utilization and respiration in relation to growth and maintenance in higher plants, Neth. J. Agric. Sci., 22:40.

Penning de Vries, F. W. T., Krul, J. M., and van Keulen, H., 1980, Productivity of Sahelian rangelands in relation to availability of nitrogen and phosphorus from the soil, Proc. Workshop "Nitrogen Cycling in West-African Ecosystems," Ibadan, Dec. 1978.

Rietveld, J., 1978, Soil non wettability and its relevance as a contributing factor to surface runoff on sandy soils in Mali, Rep. Dept. of Theor. Prod. Ecol. Agric. University, Wageningen.

Sörenson, L. H., 1975, The influence of clay on the rate of decay of ammonia acid metabolites synthesized in soils during decomposition of cellulose, Soil Biol. and Biochem., 7:171.

Tandon, H. L. S., 1974, Dynamics of fertilizer nitrogen in Indian soils. I. Usage, transformation and crop removal of N, Fertilizer News, 19:3.

Veen, J. van, 1977, The behaviour of nitrogen in soil. A computer simulation model, Ph.D. Thesis, University of Amsterdam.

Viets, F. G., Jr., 1962, Fertilizers and the efficient use of water, Adv. Agron., 14:223.

Wit, C. T., 1958, Transpiration and crop yields, Verslagen landbouwkundig Onderzoek (Agricultural Research Reports), 64:6, PUDOC, Wageningen.

Wit, C. T. de, and van Keulen, H., 1972, Simulation of transport processes in soils, Simulation Monographs, PUDOC, Wageningen.

Wit, C. T. de, and Goudriaan, J., 1974, Simulation of ecological processes, Simulation Monographs, PUDOC, Wageningen.

INTEGRATION OF NITRATE AND AMMONIUM ASSIMILATION IN HIGHER PLANTS

R. Novoa[1], R. S. Loomis, and D. K. McDermitt[2]

Department of Agronomy and Range Science
University of California
Davis, CA 95616

INTRODUCTION

Because of its role in proteins and nucleic acids, nitrogen can be viewed as the central element around which carbon metabolism is organized in living material. Nitrogen accumulates to high levels in soils and vegetation of natural systems but it is strongly limiting to the productivity of most agricultural systems. One reason is that more nitrogen is removed in exploitation than returns from sources such as rainfall, dust, seed and fixation by free-living or symbiotic bacteria. Another is that the annual plants as employed in agriculture are capable of placing greater demands on nutrient cycling than occur in natural systems. The result is that nitrogen economy is a critical issue in crop management and thus also in agricultural research.

The requirements, pathways, controls and efficiency of nitrogen assimilation by higher plants then becomes a central issue. The nitrate source dominates in the nutrition of agricultural plants because of its rapid formation from ammonium in aerated soils. However, once in the plant it must once again be reduced to the amino level. The eight-electron cost of reduction of nitrate to the amino level and resulting problems in pH balance represent major costs in plant growth. That gives emphasis to a need

---

[1]Present address: Instituto Investigaciones Agropecuarias, Santiago, Chile.
[2]Present address: Monsanto Agricultural Products Co., St. Louis, MO.

for an understanding of the biochemistry of the cell. But nitrogen assimilation encompasses also the larger dimensions of the organismal and systems levels of organization. At the organismal level, for example, we find dynamic and integrative aspects which transcend cell physiology behavior. The early growth of the plant is strongly autocatalytic -- new nitrogen flows to leaf area and leaf protein and this leads to more photosynthesis. But the increased leaf area requires more water and the increased photosynthate supply requires a larger supply of nitrogen for its use in growth of stems and reproductive organs. So feedback from a limiting root system leads to adjustments in growth patterns. And later in the season, remobilization of nitrogen from senescing tissues may become a major activity.

Thus, a full understanding of nitrogen assimilation requires an integrated view of cellular and organismal events in a changing external milieu. The current balance of tissue types and growth centers with carbon and nitrogen supplies reflects past conditions and at the same time determines the possible course of future behavior.

The only effective research approach we have for adaptive, integrative issues of this sort is through simulation models (Pate, 1980). The number of time-varying elements in the system is far too large to be dealt with in any other way than by monitoring a few main variables such as total organic nitrogen within the plant. We are attempting a two-stage approach to such integration. The first stage involves the organization of cell physiology models of light (leaf) and dark (root) cells of a higher plant. Our progress in that area is the subject of this report. These models represent the carriers, enzymes and compartments involved in the uptake of nitrate and ammonium ions, and the synthesis and export or use of amino acids. It was found that internal regulation of those models was rather weak. It seems that regulation must come from outside the metabolizing cells through factors such as substrate supply (nitrogen, carbohydrate) and product use which are controlled at a higher level, in the organism. The second stage of the research centers on that higher level. We will show the integration of the cell models into a very simple whole plant and outline several issues which are revealed for cell physiology and for more sophisticated treatment of the whole plant.

THE MODELS

The models described below follow the dynamic, state-variable approach (Loomis et al., 1979). They were presented in detail and with extensive documentation by Novoa (1979). The key steps in conceptualizing a systems model include description of the physical system (what state variables and what process) and the mathematical formulation of their rates of change. Information about morphology, feedback, external weather and other factors can be introduced through auxillary variables or subroutines. The state variables generally represent real properties of the system such as the volume of the vacuole and the concentration of nitrate in the cytosol. Collectively, they describe the current condition (state) of the system as we might observe it at some point in time. That is also the condition which controls current metabolism. Those states change over time through the action of rate variables representing, in the cell model, net activities of enzymes and carriers. Thus, for a root cell, the change in state variable N6 representing $[NH_4^+]$ in cytosol is given by:

$$CN6 = R1 + R2 + \underline{R3} - R4 - R5 - R6 - \underline{R7} - \underline{R8}$$

where

R1 and R2 are passive and active uptake of NH_4^+ from free space to cytosol,

R4 and R5 are passive and active uptake of NH_4^+ from cytosol to vacuole,

R6 is NH_4^+ export to xylem,

R3 is NH_4^+ synthesis by nitrate reduction,

R7 is NH_4^+ use by glutamine synthetase, and

R8 is NH_4^+ use by glutamic dehydrogenase.

The rate variable CN6 is calculated in each iteration of the model and is then used to update the N6 state variable to a new point in time using, in this case, the Adams' integration routine.

The contributing enzyme and carrier processes, R1 through R8, are each formulated in separate differential equations in which the rates are expressed as functions of appropriate state and auxillary variables. To illustrate the calculation of an enzyme process, consider R7, the rate of NH_4^+ use by glutamine synthetase:

$$\underline{R7} = (VMXR7*ATP*N14*N6)/((ATPKI*GLUKM1*N6)+(AMMKM*ATP*N14)+ (GLUKM1*N6)+(ATP*N14*N6)+(ATPKM*N14*N6))$$

where N6 is $[NH_4^+]$ in cytosol, N14 is [glutamate] in cytosol, ATP is [ATP] in cytosol, ATPKI is the ATP inhibition constant, VMXR7 is Vmax of the enzyme and ATPKM, AMMKM and GLUKM1 are the Kms of the enzyme to ATP, NH_4^+ and glutamate. For simplicity, morphological details are omitted from this equation. ATP, N6 and N14 obviously are state variables. VMXR7 and the Km values may be introduced as parameters or as auxillary variable functions, for example, of temperature or some other condition. Analogous equations are written for each of the other enzyme and carrier processes in the cell.

In practice, one calculates all rates then updates all states in an iterative fashion. The time step for updating, DELT, is chosen short enough to simulate a smooth change in the state variable. Thus, if the real system assumes a new steady state in 10 min following a perturbation, the simulation might advance in steps of 1 to 2 min.

In the cell models, our conceptualization includes:

1) Transport functions representing rates of membrane crossing. Those are viewed as having passive components fit with Fick's law following electrochemical gradients. Membrane permeability in the key parameter. Active components are viewed as carrier- and energy-dependent following Michaelis-Menten kinetics. Feedback is introduced where appropriate.

2) Enzyme rates estimated with equations fit to the individual mechanisms. Where the mechanism of an enzyme is unknown, a "nearest neighbor" mechanism is used. For example, glutamine synthetase is taken as a bi uni uni bi, ping pong, ter ter mechanism (following the nearest neighbor concept), and is fit by the corresponding equation from Segal (1975). ATP, NADH, reduced ferredoxin and carbon skeleton sources (pyruvate, α-ketoglutarate and oxaloacetate) are taken as nonlimiting unless reduced linearly in proportion to the supply of available glucose. The central enzyme systems are illustrated in Figure 1.

3) Morphological factors include wall, vacuolar, cytoplasmic and chloroplastic compartments. The standard fractions for our cell volumes are 10% wall, 25% cytosol and 4% nucleus and mitochondria with 5% for chloroplasts and 56% for vacuoles in light cells, and 61% for vacuoles (no chloroplasts) in dark cells.

A diagram of our nonphotosynthetic, "root" cell is presented in Figure 2 and a flow chart of its nitrogen metabolism is shown in Figure 3. The soil solution is viewed as penetrating the free space with uptake into the cytoplasm by both active and passive mechanisms. The active process for nitrate follows a Michaelis-Menten formulation and is made sensitive to temperature following Ezeta and Jackson (1975), pH (van den Honert and Hooymans, 1955; Rao and Rains, 1976), and noncompetitive inhibition by ammonium

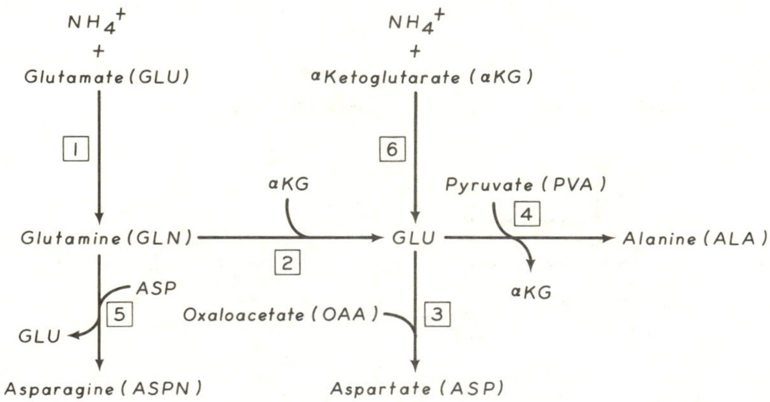

Fig. 1. Central enzymes in the synthesis of amino acids from ammonium ions: 1) glutamine synthetase; 2) glutamate synthase; 3) aspartate aminotransferase; 4) alanine aminotransferase; 5) asparagine synthetase; and 6) glutamic dehydrogenase. These enzyme systems are simulated in the nitrogen model. Other amino acids are generated as needed from the five basic amino compounds shown here according to a set of transamination rules.

(D. W. Rains, personal communication). That process is also subject to feedback due to nitrate in the cytosol (Cram, 1973; Smith, 1973). Ammonium uptake is dependent upon pH and temperature (van den Honert and Hooymans, 1961). Nitrate and ammonium ions and various amino acids move into and out of the vacuole by similar rules. Export from these dark cells occurs by active transport to an imaginary xylem tissue following Michaelis-Menten kinetics.

The Vmax of the nitrate reductase activity is allowed to vary linearly with cytosol nitrate level thus mimicking the inducible nature of that system (Aslam et al., 1976; Radin, 1978). Temperature dependence (Ezeta and Jackson, 1975) and glutamine feedback (Steward and Rhodes, 1977) are also included. Nitrite reduction is very much faster than nitrate reduction and is not included in the computer program so ammonium production equals nitrate reduction. The ammonium can flow into amino acids via glutamine synthetase and glutamic dehydrogenase. The main activity goes through glutamine synthetase due to its lower Km but that may be altered in the simulation by other conditions. From glutamine, nitrogen is transferred to glutamate and asparagine and then by transamination to asparate and alanine.

In all, rates of more than 50 enzymes and carriers are calculated in each iteration.

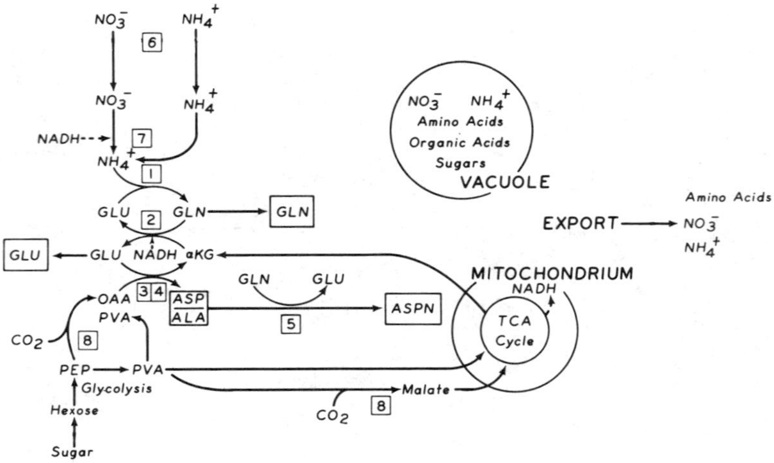

Fig. 2. Physical model of a nonphotosynthetic (root) cell, its enzyme systems and compartments. Glutamic dehydrogenase (6; NH_4^+ + αKG → glutamate) is omitted for simplicity. Enzymes are numbered as in Figure 1; nitrate reduction (7) and carboxylation reactions which help maintain the supply of organic acids are added.

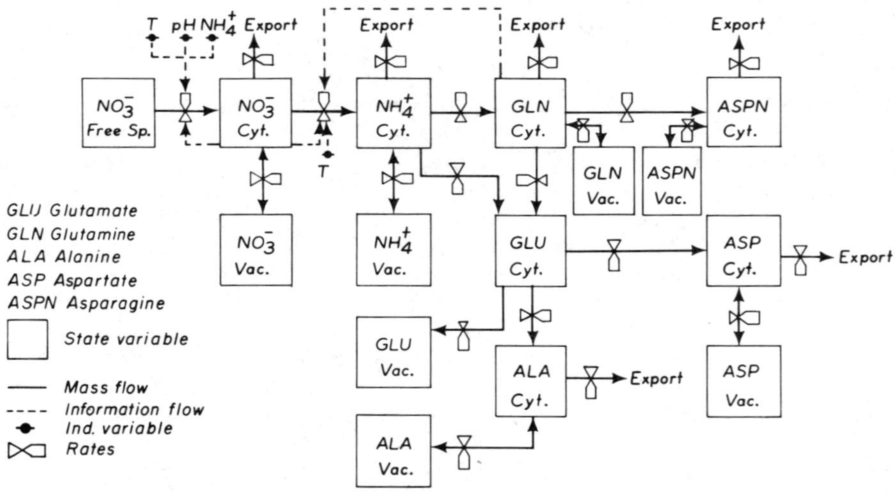

Fig. 3. Flow diagram of nitrogen metabolism in a nonphotosynthetic cell. State-variable notation is used. Nitrite is omitted here as an intermediate since it is not modeled in the computer program.

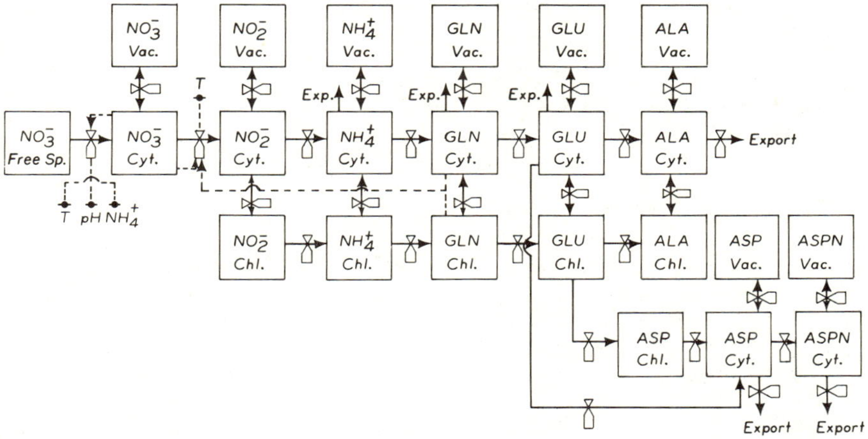

Fig. 4. Flow diagram of nitrogen metabolism in a photosynthetic cell. Notation as in Figure 3; glutamate dehydrogenase is omitted from the diagram but is included in the program while nitrite is included here but not in the program.

The light (leaf) cell model is formulated in a similar way with a central role for the chloroplast in nitrite reduction and the generation of reductant, carbon skeletons and sugar. This system is more complex than that of the dark cell, involving interactions of more organelles and cytosol and thus more membrane crossings. Import is from wall-xylem free space. Export of sugars and amino acids is to an imaginary phloem tissue. The nitrogen flow chart is shown in Figure 4.

Nitrate uptake to the cytosol and its reduction there to nitrite is treated as in root cells. In real plants, nitrite reduction to ammonium is generally viewed as occurring in the chloroplast and dependent upon reduced ferredoxin. But since nitrite is not included specifically in this prelminiary program, chloroplast synthesis and the dependence on ferredoxin and light is not a part of the simulation. Rather, in the light, the chloroplast membrane is given a high permeability to ammonium and the energy and reductant cost of ammonium synthesis is sharply reduced. Amino acid formation occurs in both compartments. The ammonium supply in the cytosol comes by release from the chloroplast and by uptake from the leaf free space (i.e., the xylem). Several enzyme processes are made subject to effects of light: glutamine synthetase in the chloroplast is given a higher Vmax in light (Mitchell and Stocking, 1975) and glutamate synthase is made dependent upon reduced ferredoxin supply.

Transport is somewhat different than in the root cells. In the dark, permeability coefficients for tonoplast and plasmalemma are smaller than for the root cells. In the light, those permeabilities are increased by a factor of 4. The main effect of that is to increase the supply of vacuolar nitrate to the cytosol during the light period (Jones and Sheard, 1976; Aslam et al., 1976).

Early tests with the separate root and leaf cell models revealed a very high capacity for nitrogen assimilation. Using enzyme parameters (Vmax and Km) common to the literature, there was very little regulation in the system as long as nitrate was in moderate supply (100 to 500 μM) in the free space and carbon sources were not limiting. In essence, nitrate was very rapidly assimilated to amino acids which then accumulated in the cells. Feedback control (e.g., glutamine on nitrate reductase; Stewart and Rhodes, 1977) had little effect mainly because the inhibition levels suggested in the literature require very high concentrations of products (near 10 mM, for example, for ammonium feedback on nitrate reductase) relative to those being simulated. Reaction reversibility has not been included in the program, except for passive transport, and that would serve to slow the net assimilation rate. Many other explanations are also possible. The most likely, however, is that supplies of carbon skeletons, energy and reductant are seldom nonlimiting in the real world.

Our attention thus was turned to introducing variations in carbon supply into the cell models. That requires some sort of organismal level integration to simulate the generation (photosynthesis) and use (growth, storage and respiration) of reduced carbon. This was done by incorporating the root- and leaf-cell models along with a "sink-cell" model into a very simple whole-plant model. The sink cells serve as variable users only of carbon and nitrogen assimilates. For example, with a large, rapidly growing sink of low protein composition, strain can be imposed on carbon supply. Leaves and roots were also endowed with capabilities for growth and thus serve as modest sinks in their own right.

Morphological characterizations include weights of root, leaf and sink tissues and leaf area per plant, and per unit area of land (i.e., as a plant community). The leaves serve as a light-dependent source of reduced carbon represented as glucose and of associated supplies of ATP and reductant. The leaves can export that carbon to a "reserve pool" which is accessed by root and sink tissues. The leaves also reduce nitrate and export amino acids. The root takes up nitrogen from the external medium and exports it on to the other tissues in inorganic and organic forms. Thus, the root is a consumer of glucose and exporter of nitrogen while the sink tissue is a consumer of both carbon and nitrogen. The realism of this simple plant model is enhanced by allowing root and leaf to grow dependent on their "meristematic" capacity and the supplies of reduced carbon and nitrogen. Photosynthesis is calculated with a dependence on leaf-area index and solar altitude (a function of day, hour and latitude) to provide an input of grams glucose m^{-2} h^{-1}. Respiration is dealt with in more detail. Respiration associated with biosynthesis is calculated for each amino acid and for assembling those in biomass following the general strategy of Penning de Vries et al. (1974). Maintenance respiration is taken as a linear function of accumulated biomass (Hunt and Loomis, 1979). In addition to respiration, glucose is also drained away for carbon skeletons of amino acids (calculated individually) and other biomass components (treated collectively).

BEHAVIOR OF THE COMMUNITY MODEL

This preliminary community model is very "stiff" since the events being simulated involves a very wide range of time constants. Certain of the intermediate pools in nitrogen assimilation (e.g., NH_4^+ and glutamate in the chloroplast) remain at low concentration with high turnover rates. For that reason, DELT must be kept near 4 s to avoid wide fluctuations in those pools and thus instability in the model output. That requires 21,600 iterations per simulated day and, with the large size of the model, becomes very expensive of computer time for simulating longer-term community events. We now believe that we can reduce that problem by introducing reversibility of enzyme reactions, fusing certain catenary sequences (as we have already done for nitrite) and reducing the detail of carbon costs for each amino acid. But it is also clear that best integration from molecules to community requires two models: one such as we have, characterizing cell behavior, to provide general rules on nitrogen assimilation for a second, more realistic community model.

The present model has not been validated. That requires comparison of output for the integrated levels (tissue, plant, community) with real systems. Our plant is too simple to give quantitative predictions of real systems at the higher levels. Alternatively, we can compare at the tissue level (e.g., with P. Filner's tissue culture results; not yet attempted), and examine the model for realistic qualitative behavior in sensitivity analyses. Our focus here will be on such sensitivity analyses. For orientation, the community had 7 plants m^{-2} with about 325 g m^{-2} dry weight leaves, 100 g sink and 50 g roots; leaf-area index was near 3.2 or enough for full cover. The weather is that of Davis, CA in midsummer.

In Figure 5, we compare simulated net, long-term (72 h) nitrate uptake with that observed by Lancaster (1977) in short-term (10 h) experiments with *Avena fatua* and *Bromus mollis*. The model was parameterized with the Vmax and Km of active uptake as obtained

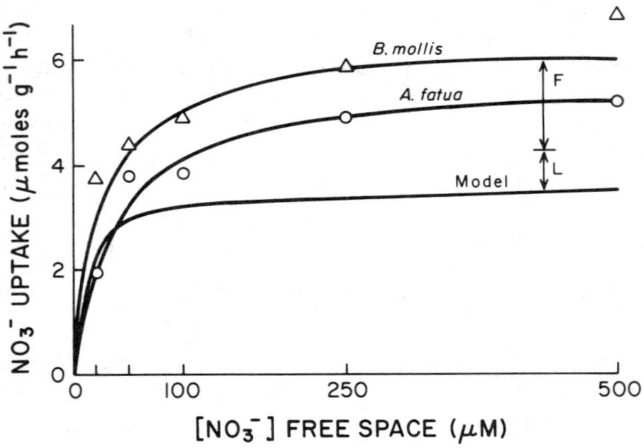

Fig. 5. Nitrate uptake per g fresh weight of roots simulated in the plant model and as observed by Lancaster (1977) for Avena fatua and Bromus mollis. F indicates the approximate reduction in uptake due to feedback from [NO_3] cyt and L indicates the passive loss.

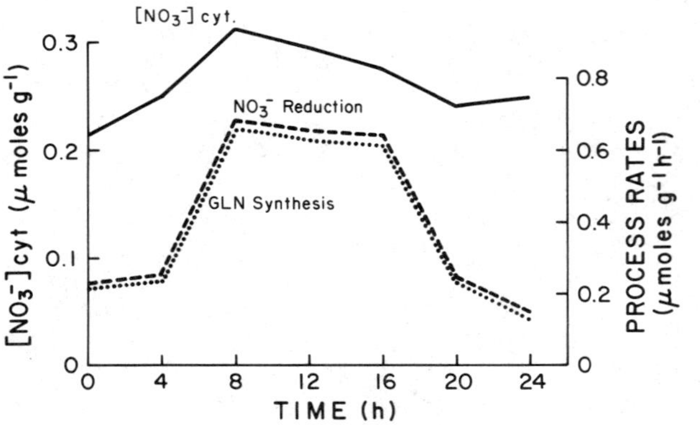

Fig. 6. Simulated diurnal course of nitrate concentration in leaf cytosol and rates of nitrate reduction and glutamine synthesis. (Fresh weight basis; 4 h observation interval.)

by Lancaster for Bromus. But after 72 h of simulation, a high concentration of nitrate has accumulated in the root cytosol (ca. 2.7 mM) and the lower uptake by the model reflects passive leakage (L) and feedback (F) on the active process. It appears that the combination of carrier and passive transport systems employed here gives a realistic behavior. With 50 g m^{-2} dry weight of roots at a dry/fresh ratio of 0.15, uptake (from Fig. 5) at 100 μM nitrate was 15 mg N m^{-2} h^{-1} or just sufficient to support a crop growth rate of 18 g dry matter m^{-2} day^{-1} at 2 percent nitrogen.

A strong diurnal pattern occurred with nitrate reduction due mainly to the role of light in its effects on tonoplast permeability and reduction rate in leaves. That is shown in Figure 6 with a sharp peak in the concentration of nitrate in leaf cytosol during the day. Part of the enhanced rate of reduction also was due to the increase in leaf temperature which occurs in the light. For the same reasons, the effects of nitrate starvation also develop mostly during the daylight hours (Fig. 7). In this simulation, the crop was set in an environment with zero external nitrate after a period of "growth" at high nitrate. The vacuoles provided a considerable internal pool which was sufficient to keep the crop growing for some time -- after 3 "days", the growth rate was still at 50% of the high nitrogen rate.

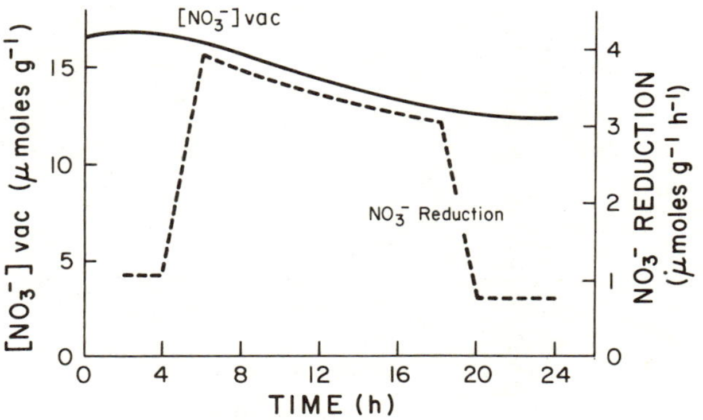

Fig. 7. Simulated diurnal course of nitrate concentration in leaf vacuoles and nitrate reduction rate in leaves during nitrate starvation. The plants were "grown" with 500 μM nitrate in the root environment and then switched to 0 nitrate at time 0. (Fresh weight basis, 1 h observation interval.)

The system proved sensitive to the light-dark permeability multiplier. A less dramatic increase in permeability (a multiplier of 2 rather than 4 in the light) lead to a deficiency in amino acids for growth. That could be overcome by increasing nitrate uptake and transport by roots, increased reduction in roots or alteration of enzyme parameters. But in each case, the required changes would fall outside the range for "normal plants" as established from the literature, suggesting that a large effect of light on membrane permeability is realistic.

The importance of membrane permeability was also evident in the chloroplast where glutamate, ammonium and glutamine concentrations were sometimes subject to oscillations. As the model is structured, high glutamate levels there lead to rapid synthesis of glutamine and exhaustion of the ammonium pool. In a sensitivity analysis, an attempt was made to buffer the chloroplast glutamate level by further increasing the permeability of the chloroplast membrane to glutamate. That would allow a buffering exchange with the much larger cytosol volume. Glutamate levels were indeed stabilized, but glutamine synthesis was then greater in dark than in light, an abnormal condition.

The predicted specific rate of glutamate synthesis via glutamic dehydrogenase was about 10 times greater in roots than in leaves (Table 1) due mainly to the high concentration of ammonium there relative to leaf chloroplasts. Glutamate synthase (GOGAT) was also important in roots but in leaves it was clearly dominant with about 85% of total glutamine synthesis being accomplished by the chloroplastic GOGAT system. A sensitivity analysis with glutamic dehydrogenase deleted in both roots and leaves reduced total glutamate synthesis by 80% for roots but only 5% per plant (Table 1).

Sensitivity to variations in other attributes of system were assessed in a similar way. Table 2 shows the results of variations in the Km of nitrate reductase. In a catenary enzyme system such as this, large changes in Km had little effect on throughput since compensatory increases in substrate concentrations occurred quickly. A similar result was found when the Km of glutamine synthetase to ammonium ion was varied.

Morphological changes can be handled as a change in the fractional volume of vacuole, cytosol or chloroplast with or without a similar proportional change in the amount of enzyme or carrier associated with that volume. Volume changes alone had only small effects but when the associated capacities were also changed, the system responded quickly in line with the new enzyme or carrier level. However, a similar test in which one-half of the leaf weight was removed, showed that the enzyme system had considerable reserve potential (Table 3). Substrate levels (nitrate,

Table 1. Simulated rates of glutamate synthesis per g fresh weight and per plant in various sites in the presence and absence of glutamic dehydrogenase. Nitrate supply is 500 μM in the root free space.

	Specific rate (μmoles $g^{-1} \cdot h^{-1}$)			Rate per plant (μmoles $plant^{-1} \cdot h^{-1}$)			
	Root cyt.	Leaf cyt.	Leaf chl.	Root cyt.	Leaf cyt.	Leaf chl.	Total
GLU synthase	0.70	0.02	1.32	35.0	5.3	318.0	358.3
GLU dehydrogenase	0.49	0.00	0.04	24.6	0.0	8.7	33.3
TOTAL	1.19	0.22	1.36	59.6	5.3	326.7	391.6
GLU synthase alone	0.21	0.04	1.46	10.4	9.0	352.0	371.4

Table 2. Effects of changes in the Km of nitrate reductase for nitrate on the simulation of nitrate assimilation. Nitrate supply is 500 μM in the root free space; 400 μM in the standard Km.

Km	NO_3^- reduction rate Leaf (μmoles $g^{-1}.h^{-1}$)	Root	$[NO_3^-]$ cyt. Leaf (μmoles g^{-1})	Root	$[NH_4^+]$ cyt. Leaf (μmoles g^{-1})	Root
200	0.56	1.06	0.09	2.53	0.15	0.018
400	0.52	1.01	0.12	2.57	0.15	0.017
800	0.47	0.94	0.19	2.65	0.14	0.012

Table 3. Simulation results of leaf behavior obtained with reduced leaf size. Nitrate supply is 500 μM in the root free space; the normal "plant" has 46 g (dry) leaves m^{-2}; results are given on a fresh basis.

Weight of leaves	Normal	0.5
Rates (μmoles $g^{-1}\cdot h^{-1}$)		
NO_3^- reduction in leaf cytosol	0.60	1.16
GLN synthesis in chloroplast	0.55	1.15
GLU synthesis in chloroplast	1.09	2.20
GLU export from leaves	0.30	0.58
Levels (μmoles g^{-1})		
$[NH_4^+]$ leaf cytosol	0.17	0.64
$[NH_4^+]$ chloroplast (x 10^{-3})	0.75	0.37
[GLN] chloroplast (x 10^{-2})	0.48	1.60
[GLU] chloroplast	0.057	0.111
Glucose (mmoles $plant^{-1}$)	42.5	29.0

glutamate, etc.) generally increased dramatically so that nitrogen metabolism operated at higher rates per unit enzyme (g^{-1} fresh wt). But the example shown in Table 3 shows a sharp drop in glucose supply by 36 hours (2 photo- and 1 nyctoperiods) after defoliation. Clearly, photosynthate supply was becoming limiting in a way which would restrict nitrogen assimilation and the use of amino acids for growth.

This final point indicates an area for needed research. We hypothesize that a "functional balance" between carbon and nitrogen must exist (in line with Radin, 1978; Pate and Layzell, 1981; and van Keulen, 1977). And we have included in the model those feedback loops from final stages (amino acid pools) to early stages (nitrate uptake and reduction) which are well established by experiment. But those feedbacks are inadequate to control nitrogen assimilation. The simulation model tells us that the coupling between carbon and nitrogen metabolism must be much more direct -- through supplies of carbon skeletons, energy and reductant. Sensitivity analysis to those issues remain to be explored.

ACKNOWLEDGMENT

We are particularly indebted to S. Morrison, R. C. Huffaker and D. W. Rains for their valuable advice during the course of this work. Support for this work was provided by a grant from the National Science Foundation (PFR 77-07301).

REFERENCES

Aslam, M., Oaks, A., and Huffaker, R. C., 1976, Effect of light and glucose on the induction of nitrate reductase in corn roots, Plant Physiol., 58:588.

Cram, W. J., 1973, Internal factors regulating nitrate and chloride influx in plant cells, J. Exp. Bot., 24:328.

Ezeta, F. H., and Jackson, W. A., 1975, Nitrate translocation by detopped corn seedlings, Plant Physiol., 56:148.

Hunt, W. F., and Loomis, R. S., 1979, Respiration modelling and hypothesis testing with a dynamic model of sugar beet growth, Ann. Bot., 44:5.

Jones, R. W., and Sheard, R. W., 1975, Accumulation and stability of nitrite in intact aerial leaves, Plant Sci. Lett., 11:285.

Lancaster, D. L., 1977, Ecological significance of nitrate uptake and reduction by annual grass species, Ph.D. Thesis, University of California, Davis, CA.

Loomis, R. S., Rabbinge, R., and Ng, E., 1979, Explanatory models in crop physiology, Ann. Rev. Plant Physiol., 30:339.

Mitchell, C. A., and Stocking, C. R., 1975, Kinetics and energetics of light driven chloroplast glutamine synthesis, Plant Physiol., 55:59.

Novoa, R. A., 1979, A preliminary dynamic model of nitrogen metabolism in higher plants, Ph.D. Thesis, University of California, Davis, CA.

Pate, J. S., 1980, Transport and partitioning of nitrogenous solutes, Ann. Rev. Plant Physiol., 31:313.

Pate, J. S., and Layzell, D. B., 1981, Carbon and nitrogen partitioning in the whole plant -- a thesis based on empirical modelling, in: "Aspects of Nitrogen Metabolism," D. Bewley, ed., Spruijt, The Netherlands, in press.

Penning de Vries, F. W. T., Brunsting, A. H. M., and van Laar, H. H., 1974, Products, requirements and efficiency of biosynthesis: a quantitative approach, J. Theor. Biol., 45:339.

Radin, J. R., 1978, A physiological basis for the division of nitrate assimilation between roots and leaves, Plant Sci. Lett., 13:21.

Rao, K. P., and Rains, D. W., 1976, Nitrate absorption by barley. I. Kinetics and energetics, Plant Physiol., 57:55.

Segal, I. H., 1975, "Enzyme Kinetics," John Wiley, New York.

Stewart, G. R., and Rhodes, D., 1977, Control of enzyme levels in the regulation of nitrogen assimilation, in: "Regulation of Enzyme Synthesis and Activity in Higher Plants," H. Smith, ed., Academic Press, London.

van den Honert, T. H., and Hooymans, J. J. M., 1955, On the absorption of nitrate by maize in water culture, Acta Bot. Neerl., 4:376.

van den Honert, T. H., and Hooymans, J. J. M., 1961, Observations on the absorption of ammonium by cut discs of potato tuber as compared to maize roots, Acta Bot. Neerl., 10:261.

van Keulen, H., 1977, Nitrogen requirements of rice with special reference to Java, Contr. Centr. Res. Inst. Agric. Bogor. (Indonesia), No. 30.

A NEW APPROACH TO THE ANALYSIS OF REDUCTIVE AND DISSIPATIVE COSTS IN NITROGEN ASSIMILATION

D. K. McDermitt and R. S. Loomis

Monsanto Agricultural Products Co., St. Louis, MO 63133, and Department of Agronomy and Range Science, University of California, Davis, CA 95616

INTRODUCTION

The relative costs of assimilation of atmospheric nitrogen and nitrate nitrogen is an important question for the interpretation of higher plant productivity. On the basis of the energetics of N_2 reduction by bacterial nitrogenase (Burris and Winter, 1968) and thermodynamic and stoichiometric considerations (Bergersen, 1971; Hardy and Havelka, 1975), it would appear that the cost of N_2 fixation is comparable to that for NO_3^- reduction. However, those analyses did not account fully for losses associated with H_2 evolution (Schubert and Evans, 1976) or for nodule growth and maintenance in the case of N_2 fixation, nor for the cost of pH balance during NO_3^- assimilation (Dijkshoorn, 1962; Raven and Smith, 1976).

Physiological studies to determine the costs of N_2 fixation and NO_3^- reduction in higher plants have given conflicting results. On the basis of growth analysis of subterranean clover, Gibson (1966) concluded that the carbohydrate requirement of growth on NO_3^- was about the same as that on N_2. Minchin and Pate (1973) studied dry weight changes, nitrogen accumulation and respiration in pea roots, nodules and shoots. They found that total root respiration per unit N assimilated in the root was the same for nodulated and NO_3^- grown plants. They did not give complete data on the size and growth rate of NO_3^--fed plants, but indicated that the "increments of root dry weight and nitrogen" were greater in plants grown on NO_3^- than in nodulated plants. But nodule respiration accounted for only 25 to 40% of total root respiration, casting some doubt on the validity of comparing total root respiration per unit N assimilated in the two cases.

More recent workers have reached different conclusions. Ryle et al. (1978, 1979) made detailed studies of photosynthesis, nitrogen assimilation, respiration and growth in soybean, cowpea and white clover. They concluded that there was a substantially higher respiratory burden associated with nodulated roots fixing N_2 than with non-nodulated roots of plants grown on nitrate (e.g., 34% vs. 21% of gross daily photosynthesis in white clover). Silsbury (1977) reached a similar conclusion for swards of subterranean clover. From CO_2 exchange measurements and a model described in McCree and Silsbury (1978), he found that on a whole plant basis, plants fixing N_2 respired 38% of their current daily assimilate in growth related processes, while plants grown on nitrate respired only 27%. These data imply that there may be a higher carbohydrate demand for N_2 fixation than for NO_3^- reduction.

Here we propose a new method of determining the cost of growth on different nitrogen sources. It is based upon measurements of true growth yield (see below) and elemental composition, and allows a calculation of substrate use efficiency. It is valid even if the size and growth rates of plants grown on different nitrogen sources vary. The method could also be used to evaluate the whole plant performance of legumes infected with rhizobial strains varying in H_2 evolution relative to N_2 reduction (Evans and Barber, 1977).

GROWTH EFFICIENCY

Recently, we showed that the substrate utilization efficiency for growth (E_G) can be calculated from growth yield and composition data (McDermitt and Loomis, 1980). E_G is defined as the fraction of substrate electrons which are recovered in the biomass during dry matter accumulation. Substrate consumption due to maintenance is not included. If the biochemical pathway analyses of Penning de Vries et al. (1974) are correct, E_G has a value of 0.88 ± 0.01 (s.d.) when NO_3^- is the nitrogen source (McDermitt and Loomis, 1980), and it is nearly constant over a wide range of plant compositions. E_G is also closely related to energetic efficiency defined as the fraction of enthalpy in the substrate which is retained in the product. This is true because molar heats of combustion in organic compounds are nearly proportional to the number of valence (or available) electrons which they contain (Thornton, 1917; Minkevich and Eroshin, 1973).

Processes which reduce E_G are those which divert substrate electrons to electron acceptors not conserved in dry biomass. Quantitatively, the most important process leading to loss of efficiency is ATP formation with the reduction of O_2 via oxidative phosphorylation. Substrate level reductions of molecular oxygen by various oxidase and peroxidase enzymes must also be considered.

By this concept, N_2 fixation is extremely inefficient. There is a requirement of about 4 ATP per 2e passing through nitrogenase (Winter and Burris, 1968), and in legumes, as much as 40 to 60% of the electron flux through nitrogenase may go to the production of H_2 which is lost from the plant (Schubert and Evans, 1976). In contrast, nitrate reduction which requires no ATP and produces no H_2 occurs with 100% efficiency. All electrons used in the process are retained in the product (NH_3). Thus, E_G should be lower for nodulated plants grown on N_2 than for non-nodulated plants grown on NO_3^- even if overall carbohydrate costs are comparable for the two processes. This should be particularly true when photoreduction of NO_3^- occurs (see below and Appendix).

Two quantities must be known in order for E_G to be calculated. One is the yield of biomass per unit of substrate consumed for growth. This quantity is usually given in weight units (g dry biomass/g substrate) and is called the true growth yield, Y_G (Pirt, 1965; Thornley, 1976). Y_G of higher plants has been obtained from gas exchange measurements (e.g., McCree and Salisbury, 1978, and references therein). The other quantity is the weight yield which would occur in the absence of dissipative processes (i.e., weight yield assuming 100% efficiency). If this were known, then the ratio of the measured yield (Y_G) to the maximal yield would give efficiency.

The maximal weight yield can be calculated in a straightforward way from the elemental formula of biomass. We call this maximal weight yield the glucose value (GV) (McDermitt and Loomis, 1980). So:

$$(1) \quad E_G = \frac{Y_G}{GV} \quad \text{(dimensionless).}$$

CALCULATION AND PROPERTIES OF THE GLUCOSE VALUE

In deriving GV, it is easier to think in terms of the inverse molar quantity, the glucose equivalent (GE). GE is defined as the molar quantity of glucose which contains the same number of available or valence electrons as a mole of product. Glucose is chosen as standard because carbohydrate is the main product of photosynthesis and glucose is a universal starting point in metabolism. Any hexose could be used without changing the results.

GE is calculated from the elemental formula of the product, $C_cH_hO_xN_nS_s$. The "product" may be an individual organic compound, mixture of compounds or organic biomass. First, the quantity of glucose needed to supply carbon is calculated. This is c/6 moles. If the product is more reduced than carbohydrate, then additional glucose is required to supply additional electrons. The level of reduction of the product is calculated from the oxidation numbers of the elements and their stoichiometrics in the elemental formula. For oxygen and hydrogen, these are 2- and 1+, respectively. The oxidation numbers of nitrogen and sulphur vary with the species assimilated. For nitrogen, these are 5+ (NO_3^-), 0 (N_2) and 3- (NH_3); for sulphur, the oxidation numbers are 6+ ($SO_4^=$) and 2- (H_2S). Using these oxidation numbers the overall level of reduction of the product (r) can be calculated as r = 2x-h-kn-ms. x,h,n and s are the stoichiometries of O,H,N and S, respectively, in the elemental formula; k = the ground state oxidation number for nitrogen (5+, 0 or 3-), and m = the ground state oxidation number for sulphur (6+ or 2-). For example, if NO_3^- and $SO_4^=$ are assimilated, k = 5+ and m = 6+; if NH_3 and H_2S are assimilated, k = 3- and m = 2-. If r is negative, then the product is more reduced than carbohydrate (r = 0) and N and S in their ground states, and additional substrate must be supplied to provide reducing equivalents. One mole of glucose provides 24 equivalents of electrons so -r/24 moles of glucose are required. Then:

$$(2) \quad GE = \frac{c}{6} - \frac{r}{24} \quad \left(\frac{\text{moles glucose}}{\text{mole product}}\right).$$

Note that for relatively reduced compounds, -r/24 is positive. For organic acids, r is positive, -r/24 is negative and GE is decreased accordingly. Some sample calculations are shown in Table 1.

When the ground states for N and S are chosen to correspond to the standard states taken for combustion products (e.g., N_2, H_2SO_4), the heat of combustion of the glucose equivalent is a good estimator of the heat of combustion of the product (McDermitt and Loomis, 1980). GE has also been shown to estimate the cost of forming many individual biological compounds to within about ± 10% of values obtained from metabolic pathway analysis (McDermitt and Loomis, 1980). When NO_3^- or NH_3 are nitrogen sources, the discrepancies between the glucose equivalents and pathway values were entirely accounted for by ATP formation and substrate level reductions of O_2.

Returning to whole biomass, the glucose value is calculated from GE for biomass, which in turn, requires that the elemental formula per "mole" be known. A "mole" of biomass is taken as 100g including minerals. Note that the formula weight will usually be less than 100g; the remainder is minerals. In the elemental analysis only C,H,O and organic N must be measured. NO_3-N cannot be included because it is not yet reduced. Sulphur may be neglected with only small error (<1%) because it is present in very low quantity. Phosphate does not undergo oxidation-reduction so

Table 1. Sample calculations of the glucose equivalent (GE) for various materials. The elemental formula leads directly to the calculation of the reduction level (r). For the GE calculation in the last column, carbon skeleton requirement (c/6 where 6 is the number of carbons per glucose) is summed with the reducing requirement (-r/24) where 24 is the number of electrons made available from the oxidation of one glucose molecule.

Compound or material	Elemental formula	r-value[1]	Glucose Equivalent (moles glucose/ mole product)
Malic acid	$C_4H_6O_5$	2x5-6=4	4/6-4/24=0.500
Leucine	$C_6H_{13}O_2N$	2x2-13-5=-14	6/6-(-14/24)=1.583
Caffeine	$C_8H_{10}O_2N_4$	2x2-10-5x4=-26	8/6-(-26/24)=2.417
Fructose	$C_6H_{12}O_6$	2x6-12=0	6/6-0/24=1.000
Palmitic acid	$C_{16}H_{32}O_2$	2x2-32=-28	16/6-(-28/24)=3.833
Phaseolin[2]	$C_{4.38}H_{6.91}O_{1.52}N_{1.13}S_{0.01}$	2x1.52-6.91-5x1.13-6x0.01= -9.58	4.38/6-(-9.58/24)= 1.129
Phaseolin	(k=0, m=6+)	2x1.52-6.91-0-6x0.01=3.93	4.38/6-(-3.93/24)= 0.894
Phaseolin	(k=3-, m=6+)	2x1.52-6.91-(-3x1.13)-6x0.01= -0.54	4.38/6-(-0.54/24)= 0.752
Corn grain[3]	$C_{3.72}H_{6.90}O_{2.83}N_{0.153}$	2x2.83-6.90-(5x0.153)=-2.005	3.72/6-(-2.005/24)= 0.704

[1] k=5+ and m=6+ except as otherwise noted. r=2x-h-kn-ms. See text for definition of symbols.
[2] From Benedict and Osborn (1907).
[3] Derived from Latshaw and Miller (1924).

it is grouped with minerals and need not be measured. The glucose value is calculated from GE as:

$$(3) \quad GV = [GE \times \frac{180}{100}]^{-1} \quad (\frac{\text{g biomass}}{\text{g glucose}}).$$

GE is the minimal amount of glucose necessary to form a product, and its inverse (in weight units), GV, gives the maximal yield per gram of substrate.

RELATIVE COSTS OF GROWTH ON N_2 AND NO_3

Using equations 1, 2, and 3, it is possible to calculate the growth efficiencies of plants grown on different nitrogen sources. N_2 fixation is expected to be less efficient than NO_3^- reduction because of its ATP demand and H_2 production. We have previously shown that for plants grown on NO_3^-, E_G varies only a little with large compositional changes (McDermitt and Loomis, 1980). Thus, it seems likely that differences observed in E_G due to grown on N_2 or NO_3^- will largely result from the differing efficiencies of N_2 and NO_3^- assimilation.

Measurement of E_G will give insight into the energetics of nitrogen assimilation, but it does not include the entire cost. A portion of the respiration resulting from N_2 assimilation and all respiration from NO_3^- reduction is due to reductive processes which do not lead to decreases in E_G. Thus the respiration resulting from N_2-fixation is a sum of dissipative and reductive terms. Growth respiration (R_G) for the whole plant can also be expressed as a sum of dissipative and reductive terms and it includes nitrogen assimilation. McDermitt and Loomis (1980) showed that growth respiration can be calculated as:

$$(4) \quad RG = 2.64 \; [GE \; (\frac{1-E_G}{E_G}) - \frac{r}{24}] \quad (\frac{\text{g } CO_2}{\text{g new biomass}}).$$

The derivation can be outlined as follows. The total weight of substrate consumed to produce
the true yield (Y_G^{-1}). The fraction of this substrate which is used for dissipative processes such as the reduction of O_2 and H^+ is $1-E_G$. Multiplication of $1-E_G$ by total substrate used per g new biomass gives the total substrate consumed in dissipative processes. Equations 1 and 2 are then applied to give:

$$\frac{1-E_G}{Y_G} = \frac{1}{GV} (\frac{1-E_G}{E_G}) = \frac{180}{100} \, GE \, (\frac{1-E_G}{E_G}) \quad (\frac{\text{g glucose}}{\text{g biomass}}).$$

To convert units to g CO_2/g biomass, multiply by 6 x 44/180. The resulting coefficient is 2.64. The quantity of glucose required to supply reductant is -r/24 moles glucose/100 g new biomass. To obtain gCO_2 respired per g new biomass, multiply by 6 x 44/100 giving -2.64 r/24. Total respiration for growth is the sum of respiration for dissipative and reductive processes giving equation 4. r is usually negative for biomass so -r/24 is positive.

Two features of equation 4 stand out. First, total growth respiration can be analyzed as a sum of two terms, one due to reductive processes and the other due to dissipative processes. This is important because N_2 fixation and NO_3^- reduction differ significantly in these areas. Second, R_G depends only upon composition and nitrogen source, and not upon growth rate. This is important because size and growth rate may vary with different nitrogen sources (e.g., Ryle et al., 1978).

The simplest application of equation 4 is to calculate R_G for plants grown on N_2 and NO_3^- and compare the results. However, it was observed long ago that plants grown on combined nitrogen are different in color and composition from plants grown on N_2 (e.g., Bond, 1941). R_G can vary due to compositional differences alone even when E_G and nitrogen source are constant (see below for example). Thus, if the purpose of the study is to compare the biochemical costs of N_2 fixation and NO_3^- reduction per se, this simple approach is not entirely valid. On the other hand, if compositional changes are seen as part of a syndrome of effects due to different nitrogen sources, then such changes may pose no problem. However, it is important to remember that under these circumstances a difference in R_G cannot be attributed solely and directly to the cost of assimilating N_2 or NO_3^-.

An approach which has been used in the past is to measure the fraction of daily gross photosynthesis which is lost as total respiration (Ryle et al., 1978; 1979) or growth respiration (Silsbury, 1977) on different nitrogen sources. If daily growth rate (g dry weight/day) and gross photosynthesis (g CO_2/day) are known, then this fraction is easily calculated as R_G x growth rate/gross photosynthesis. This calculation is useful but it is still an integration of the metabolism of the entire plant including variations in composition and size of plant.

We would like a means to calculate R_G on different nitrogen sources which is either independent of composition or appropriately corrected for composition changes. Then differences in R_G would be due to differences in nitrogen assimilating pathways alone.

It may be possible to accomplish this as follows. Measure compositions and growth efficiencies for plants grown on N_2 and on NO_3^-, and calculate R_G for growth on N_2. Then use the composition from N_2-grown plants, and E_G and k (= 5+) appropriate for NO_3^--grown plants to calculate a second value for R_G. This second R_G is that which would result from an efficiency and level of reduction characteristic of growth on NO_3^- but of a composition characteristic of growth on N_2. If this approach is valid then

$$(5) \quad \Delta R_G = R_G\ (N_2) - R_G'\ (NO_3)$$

should primarily reflect differences in nitrogen assimilation costs because a constant composition is used. The prime serves as a reminder that the composition used to calculate R_G' (NO_3^-) is that for plants grown on N_2 not NO_3^-.

For equation 5 to be correct, E_G (NO_3) must be nearly constant with varying composition. We have previously given theoretical reasons (McDermitt and Loomis, 1980) why we expect E_G on NO_3^- to vary over a narrow range (0.872-0.901). This was true over widely varying compositions, but over the legumes examined, E_G (NO_3) showed even smaller variations (0.872-0.883). Thus, the error associated with assuming that E_G (NO_3^-) is the same for two closely related compositions (legumes on N_2 and NO_3^-) is likely to be small.

On the other hand, a significant difference between E_G (NO_3) and E_G (N_2) is expected. This arises because N_2 fixation has a high carbohydrate demand for dissipative processes such as ATP synthesis and H_2 production. By contrast, NO_3^- reduction is 100% efficient. Thus, it is reasonable to assume that any error which arises from applying E_G (NO_3) to the composition of N_2-grown plants will be small compared to the difference between E_G (N_2) and E_G (NO_3^-). The validity of this assumption will be at least partially apparent when the measurements are made. Further tests of the assumption that E_G (NO_3^-) is constant could be made on other legume varieties or species where compositions are variable.

Nitrate reduction involves no loss in efficiency, so it is not surprising that E_G (NO_3^-) was found not to vary strongly with nitrogen content. However, N_2 fixation heavily involves dissipative processes so E_G (N_2) will probably vary with nitrogen content. This means that it would not be well to calculate ΔR_G by applying E_G (N_2) and k = 0 to compositions of NO_3^--grown plants.

PHOTOREDUCTION OF NO_3^-

The second term in equation 4 (-r/24) depends upon the levels of oxidation of N_2 and NO_3^- in relation to NH_3. As long as carbohydrate is the source of reducing power, these costs are fixed and independent of conditions. However, if photoreduction of NO_3^-

occurs which is not competitive with CO_2 reduction, then taking $k = 5+$ will over-estimate the true carbohydrate demand of NO_3^- reduction. The value of $-r/24$ which estimates total reductive carbohydrate demand will be too large, and hence, GE will also be too large. These over-estimates however, will be exactly compensated by an apparent increase in E_G so that photoreduction will automatically be included in R_G.

This latter point can be further examined. With photoreduction, the respiratory load associated with NO_3^- reduction will be lower than expected if carbohydrate supplied all reductant; thus Y_G which is measured by CO_2 exchange, will be increased. But as noted above, GE will be calculated taking $k = 5+$ and will be too large. Hence, GV will be too small and $E_G = Y_G/GV$ will be an over-estimate. An example showing that the increase in apparent growth efficiency quantitatively compensates for photoreduction is given in the Appendix.

CONCLUSION

The analysis given here is expected to be used in conjunction with existing methods of analysis for growth yield and maintenance respiration (Thornley, 1970, 1976; McCree, 1970; McCree and Silsbury, 1978; Silsbury, 1977). However, it offers three extensions beyond existing methods: (1) growth efficiency as a measure of dissipative processes can be determined. (2) Growth respiration can be measured as a sum of dissipative and reductive processes. This is important in the biochemical analysis of the energetics of nitrogen assimilation (Bergersen, 1971) and metabolism in general (Penning de Vries et al., 1974). (3) A method has been proposed to assess the costs of N_2 and NO_3^- assimilation independent of variations in composition. However, the method assumes that E_G (NO_3^-) is constant over compositions, or at least that the variation in E_G (NO_3^-) is much less than the difference between E_G (N_2) and E_G (NO_3^-). These points remain to be confirmed experimentally.

A few comments should be made about measurement of composition. Since respiratory measurements used to calculate Y_G are on a whole plant basis, the composition should also be representative of the whole plant. In addition, GE and GV are calculated on the basis of hexose as starting material. This will hold for whole plants but in general it may not be true for individual organs. Thus, when composition is measured, it should be from a bulked sample including roots. In addition, the units of R_G are properly given as g CO_2/g new biomass. This means that the formula used to calculate GE should represent newly formed material. If the bulked composition is to represent new biosynthesis averaged over the whole plant, it follows that the plant must be in steady state growth with constant composition.

We shall conclude with a numerical estimate of the growth respiration associated with NO_3^- reduction for two compositions. The first composition if given in Table 1 for corn grain. E_G is taken as 0.88 and k = 5+. Then GE = 0.704 and R_G = 0.474 g CO_2/g dry weight. In g C/g dry weight, R_G x 12/44 = 0.129 g C/g dry weight. Corn contains 0.00153 g atoms N/g dry weight, and converting from g atoms N to g N we have R_G = 6.0 g C/g N. This is the same value obtained by Minchin and Pate (1973) for pea roots. Penning de Vries (1976) gives an example elemental formula for "biomass" equivalent to $C_{3.20}H_{7.52}O_{2.12}N_{0.329}S_{0.047}$. In carbon and nitrogen content, this formula is similar to the composition of subterreanean clover reported by Silsbury (1979) and it is much more highly reduced than corn grain. For this composition, R_G is 5.0 g C/g N, showing that respiratory measurements per unit N can vary even when efficiency and nitrogen source are the same. These calculations are not meant as predictions, but simply to demonstrate that equation 4 gives reasonable results.

APPENDIX

The Contribution of Photoreduction to Nitrate Assimilation

Assume photoreduction of NO_3^- occurs to the extent that 3 out of 8 reducing equivalents come directly from the photosystems. The carbohydrate demand for NO_3^- reduction will then be smaller and can be calculated by giving N an effective oxidation number of 2+ rather than 5+. For corn with the composition given in Table 1, GE = 0.6844 and -r/24 = 0.0644. GV = $[0.6844 \times 1.8]^{-1}$ = 0.8117. If E_G is taken as 0.88, then Y_G = 0.7143. So R_G is 2.64 [0.6844 (.12/.88) + .0644] = 0.2464 + 0.1700 = 0.4164 g CO_2/g. Notice that this value of R_G is less than 0.474 g CO_2/g calculated previously for the same composition on NO_3^- but without photoreduction.

In general, we will not know the extent of photoreduction, and in practice the oxidation number for N will be taken to be 5+ when NO_3^- is given. This corresponds to -r/24 = 0.0835, GE = 0.7035 and GV = 0.7897. But the measured Y_G will reflect photoreduction so the value estimated above using 2+ is expected to be correct. Hence, the apparent E_G will be 0.7143/0.7897 = 0.9045 which is larger than before. So the estimate of R_G is 2.64 [0.7035 (0.0955/0.9045) + 0.0835] = 0.1961 + 0.2204 = 0.4165 g CO_2/g which is the same as before. More significant digits are carried in the calculation than could be experimentally justified so that round-off errors are minimized.

The rationale for the first calculation was that we knew the quantitative extent of photoreduction and that all calculations were made accordingly. In the second, more realistic case, we accept a given Y_G with photoreduction built in, but assume the normal nitrogen oxidation state (N = 5+) when calculating GE and GV. Since the results are the same either way, we are encouraged to believe that the cost calculation will give correct results even with photoreduction.

One final point may be made from this illustration. Here, a value for E_G was assumed so that Y_G could be calculated from GV. But experimentally, GV is obtained from elemental composition and Y_G is measured by gas exchange methods. Thus, an independent estimate of E_G can be found. However, if Y_G is large due to photoreduction, then the apparent E_G will also be increased. This effect is farily small, but it might be useful in the analysis of photoreduction, and it should be remembered in any analysis of efficiency.

ACKNOWLEDGEMENT

This work was supported in part by a grant from the National Science Foundation (PFR 77-07301).

REFERENCES

Benedict, F. G., and Osborne, T. B., 1907, The heat of combustion of vegetable proteins, J. Biol. Chem., 3:119.
Bergersen, F. J., 1971, The central reactions of nitrogen fixation, Plant and Soil, Special Volume: 511.
Bond, G., 1941, Symbiosis of leguminous and nodule bacteria. I. Observations on respiration and on the extent of utilization of host carbohydrates by the nodule bacteria, Ann. Bot. 5:313.
Dijkshoorn, W., 1962, Metabolic regulation of the alkaline effect of nitrate utilization in plants, Nature, 194:165.
Evans, H. J., and Barber, L. E., 1977, Biological nitrogen fixation for food and fiber production, Science, 197:332.
Gibson, A. H., 1966, The carbohydrate requirements for symbiotic nitrogen fixation: A "whole-plant" growth analysis approach, Aust. J. Biol. Sci., 19:499.
Hardy, R. W. F., and Havelka, U. D., 1975, Nitrogen fixation research: A key to world food?, Science, 188:633.
Latshaw, W. L., and Miller, E. C., 1924, Elemental composition of the corn plant, J. Agric. Res., 27:845.
McCree, K. J., 1970, An equation for the rate of respiration of white clover plants grown under controlled conditions, in: "Prediction and Measurement of Photosynthetic Productivity," Proc. I.B.P./PP. Tech. Meet. Trebon, PUDOC, Wageningen.
McCree, K. J., and J. H. Silsbury, 1978, Growth and maintenance requirement of subterranean clover, Crop Sci., 18:13.
McDermitt, D. K., and Loomis, R. S., 1980, Elemental composition of biomass and its relation to energy content, growth efficiency and growth yield, Ann. Bot., in press.
Minchin, F. R., and Pate, J. S., 1973, The carbon balance of a

legume and the functional economy of its nodules, J. Expt. Bot., 24:259.

Minkevich, I. G., and Eroshin, V. K., 1973, Productivity and heat generation of fermentation under oxygen limitation, Folia Microbiol., 18:376.

Penning de Vries, F. W. T., 1976, Use of assimilates in higher plants, in: "Photosynthesis and Productivity in Different Environments," J. P. Cooper, ed., Cambridge Univ. Press, Cambridge.

Penning de Vries, F. W. T., Brunsting, A. H. M., and van Laar, H. H., 1974, Products, requirements and efficiency of biosynthesis: A quantitative approach, J. Theor. Biol., 45:339.

Pirt, S. J., 1965, The maintenance energy of bacteria in growing cultures, Proc. Roy. Soc. Ser. B., 163:224.

Raven, J. A., and Smith, F. A., 1976, Nitrogen assimilation and transport in vascular land plants in relation to intracellular pH regulation, New Phytol., 76:415.

Ryle, G. J. A., Powell, C. E., and Gordon, A. J., 1978, Effect of source of nitrogen on the growth of Fiskeby soyabean: The carbon economy of whole plants, Ann. Bot., 42:637.

Ryle, G. J. A., Powell, C. E., and Gordon, A. J., 1979, The respiratory costs of nitrogen fixation in soyabean, cowpea, and white clover. II. Comparisons of the cost of nitrogen fixation and the utilization of combined nitrogen, J. Exp. Bot., 30:145.

Schubert, K. R., and Evans, H. J., 1976, Hydrogen evolution: A major factor affecting the efficiency of nitrogen fixation in nodulated symbionts, Proc. Natl. Acad. Sci. USA, 73:1207.

Silsbury, J. H., 1977, Energy requirement for symbiotic nitrogen fixation, Nature, 267:149.

Silsbury, J. H., 1979, Growth, maintenance and nitrogen fixation of nodulated plants of subterranean clover (Trifolium subterraneum L.), Aust. J. Plant Physiol., 6:165.

Thornley, J. H. M., 1970, Respiration, growth and maintenance in plants, Nature, 227:304.

Thornley, J. H. M., 1976, "Mathematical Models in Plant Physiology," Academic Press, London and New York.

Thronton, W. M., 1917, The relation of oxygen to the heat of combustion of organic compounds, Phil. Mag., 33:196.

Winter, H. C., and Burris, R. H., 1968, Stiochiometry of the adenosine triphosphate requirement for N_2 fixation and H_2 evolution by a partially purified preparation of Clostridium pasteurianum, J. Biol. Chem., 243:940.

STRATEGIES FOR ACHIEVING SELF SUFFICIENCY IN NITROGEN ON A MIXED FARM IN EASTERN CANADA BASED ON USE OF THE FABA BEAN

D. G. Patriquin, D. Burton, and N. Hill

Biology Department
Dalhousie University
Halifax, Nova Scotia, Canada B3H 4JI

INTRODUCTION

There have been substantial advances in the past decade in our understanding of the biology of N_2 fixation in legumes, but these have not yet led to appreciably lowered reliance on industrial N_2 fixation.

This study is an investigation into the potential for enhancing N_2 fixation in locally grown faba beans (Vicia faba L. minor) through selection of efficient rhizobia-legume associations. In western Canada where this crop was introduced in 1971, inoculation was reported to be essential for nodulation and N_2 fixation (Candlish and Clark, 1975). Evaluation of growers practices suggested that increasing the use of biologically fixed N_2 locally depended more on making better use of existing N_2 fixation than on increasing the specific rates of N_2 fixation. Plants growing for the first time on dykeland soils that had received no inoculant were well nodulated and exhibited high C_2H_2-reducing activity. A number of practical problems required consideration including: is "starter" N required, how much N_2 is fixed, and how should inorganic fertilizers be managed for crops grown in rotation with the beans?

Faba beans do stimulate growth of grain crops following it in rotation (unpublished observations). Thus we suspected that there would be a positive N balance for the faba bean. However, our observations indicated that a net gain in soil-N could be achieved only by recycling some of the N removed in harvested seeds. This aspect of the utilization of recently fixed N has not received much attention because pulses are commonly grown for export. An opportunity to study the problems and potential of a system in which inputs of N were provided largely by faba bean N_2 fixation and in which the bean-N was recycled via manures was provided by an upland farmer who grew faba beans on one-third of his cultivated land. He had not applied any commercial fertilizer-N for four years.

STUDY SITES

Both of the farms included in this study are located in the western part of the Annapolis Valley, Nova Scotia, in a cool, humid temperate climatic zone. There are approximately 120 frost free days, 2800 degree days above 42 F, and the average annual rainfall is about 114 cm of which 45 cm occur between May 1 and September 30 (MacDougall et al., 1969).

At the dykeland farm (Farm A), cereals, corn and faba beans are grown in rotation on 135 ha of reclaimed saltmarsh. The soil is a deep silty loam, drained by a system of mole and tile drains. The combination of a shallow water table and effective surface drainage provides almost ideal soil water conditions. In 1978, faba beans were sampled in a field where they were grown for the first time. Barley was the previous year's crop. No inoculant was used. "Treflan" herbicide was incorporated in the soil before planting.

Thirty hectares of cultivated land on the upland farm (Farm B) encompass three soil series of capability classes 3 and 4 (moderately severe to severe limitations). Individual fields are level to rolling in slope, sand to silty clay loam in texture, and contain 2.9 to 5.8% organic matter. The principal limitations are described as "texture and imperfect drainage" on the silty loam soils, and droughtiness on sandy soils (MacDougall et al., 1969). 2100 laying hens are maintained (Rhode Island Red x Light Sussex) in a floor operation with deep litter. From 1954 to 1975 cereals were grown for poultry feed according to conventional recommendations. Faba beans were introduced in 1968. In 1975 oat yields (98 imperial bushels per acre) were the highest recorded in the Provincial Soils and Crops Department competition. However, with the rising costs of petroleum products, it was suggested that a chemically intensive system might not be economically viable in the future. In an attempt to determine the biological limitations to production in a recycling system, no commercial fertilizer or pesticides have been used since 1975.

Because of excessively wet field conditions in the spring, only one-third of the acreage on the upland farm was seeded in 1979, the year most of our data were obtained. Crop yields in 1979 were much below, slightly below and slightly above yields of the previous few years for faba beans, oats and wheat, respectively.

METHODS

Faba beans were taken throughout prescribed areas of about 0.5 ha chosen for their accessibility. An outbreak of Chocolate Spot disease occurred at the dykeland site (Farm A) in mid-August; sampling was then restricted to regions not severely affected. For C_2H_2 reduction assays of nitrogenase activity (Hardy et al., 1968) roots and nodules from 6 plants were placed individually, or in pairs, in 1000 ml jars with 8 kPa C_2H_2. The jars were incubated in the soil for 0.5 hr and gas samples taken for analysis of C_2H_2 using a Carle model 9500 gas chromatograph equipped with flame ionization detector and a 0.32 x 50 cm column containing 80-100 mesh Porapak T. Hourly rates of C_2H_2 reduction were converted to daily rates by multiplying by 18.6; this was the ratio of total C_2H_2 reduction in one day estimated by integration of rates measured at 4 to 5 hr intervals to the rate measured at 1100 hr, the time of routine sampling. The integrated seasonal values of C_2H_2 reduction were converted to values of N_2 fixation by use of a molar ratio of nodule C_2H_2-reducing activity to whole plant N_2 fixation of 1.8 (Hudd et al., 1980). The general magnitude of N_2 fixation was also estimated for the upland crop in 1979 by comparing the total N accumulated in faba beans and associated weeds on September 21 with the total N accumulated in weeds and in grasses planted in 6 microplots (25 cm diameter) in a separate study. Estimates of N_2 fixation based on this apparent molar ratio (2.4) are referred to as "conservative estimates of N_2 fixation."

Nitrate reductase activity of faba bean leaves was assayed by an _in vivo_ technique. Twenty or more 0.7 cm leaf discs, taken from all leaves, were treated as described by Patriquin et al. (1978). Fresh weights of discs and punched leaves were determined in order to calculate whole plant nitrate reductase activity.

For assay of denitrification by the C_2H_2 blockage technique (Yoshinari et al., 1977), duplicate samples, each consisting of 7 or more 15 cm depth x 1.8 cm diameter soil cores were placed vertically and compactly in 1000 ml jars. Acetylene (8 kPa) was added, jars were incubated in the soil and gas samples were taken at 3 and 24 hr for analysis of N_2O (Patriquin et al., 1978).

Fifteen faba bean plants, not including roots, were taken for biomass measurements. Plant densities were counted in 20-one m^2 quadrats. To sample other crops and weeds in 1979, six to ten 35 x 35 cm quadrats were placed randomly in each of 1) a 0.9 ha field of winter wheat, 2) a 3.5 ha field of oats, 3) a 6 ha field which was fallowed early in the summer and later seeded with winter wheat, and 4) one-half of a 3 ha field in which clover was undersown in winter wheat in 1978. Aboveground plant material was removed, sorted (crop, leguminous weeds and non-leguminous weeds), dried, weighed and subsampled to provide duplicate 100 mg samples for analysis of N by the standard Kjeldahl technique. For oats and wheat, the ratio of grain to straw was measured on two composite samples taken separately. The biomass values reported for clover and fallow fields are the highest values observed during the summer, which were on June 13. A weed survey in a bean field (7 ha) was conducted on August 21, but the final preharvest samples were taken on September 21. Belowground biomass was measured for a 30 x 30 cm sod of wheat taken on August 1; the ratio of root-N to aboveground-N was 0.21. For oats, clover and weeds, we have assumed that the belowground-N to aboveground-N ratio is 0.15. Combine yields were calculated from the area of the field measured by a counter on the grain drill, and from the number of boxes of seeds transported from the field. One level box (4 ft high) contained 100 imperial bushels.

Total (Kjeldahl) N was determined on seven 10 g (fresh weight) samples of manure taken from a manure pile or the barn, and either not exposed, or spread out on a sheet out of doors and exposed for 3 or 24 hr to simulate different periods of exposure before manure is turned into the soil. Volatilization of ammonium after incorporation of manure (5.6 metric tons/ha) in the soil was measured in 8 microplots created by enclosing soil for 3 days in 25 cm diameter bottomless buckets. The ammonia was absorbed in 100 ml of 0.1 N H_2SO_4 contained in 15 cm diameter petri dishes placed on stands at 5 cm above the soil, and was measured by a colorimetric technique (Strickland and Parsons, 1972).

For routine sampling of soils, 15 or more cores (15 cm depth x 1.8 cm diameter) were taken within the crop sampling areas. Nitrate in air dried soils from Farm A was extracted in 0.025 molar aluminum sulfate solution and measured using an ORION specific ion electrode. Nitrate in frozen soils from Farm B was measured on a 2:1 water extract by a colorimetric technique (Cataldo et al., 1975). An average value of soil-N for the upland farm was estimated by multiplying the Kjeldahl N content of soil samples from each of 10 fields by their bulk densities (determined by the excavation technique) (Blake, 1965) weighting the values according to the area of each field.

Compositions of weed populations in oat, bean and wheat fields were examined on 3 occasions between June 20 and August 15 on Farm B, and in a wheat field on a neighboring farm on July 11, 1979. At each of 15 sites in each field cover by each species in a 4 m^2 area was estimated using the Braun-Blanquet scale transformed according to Dagnelie (1960). Values for different dates were added to give overall values and the ranks of these values are reported in the results. For seed bank studies, soil samples were taken between May 30 and July 11, 1979. Nine samples were taken from the bean field, and three from each of the other fields. Each sample consisted of twenty 15 cm depth x 1.8 cm diameter cores split into top and bottom fractions. The soil was mixed with an equal volume of coarse silica sand and spread over a base on silica sand in 1290 cm^2 trays with separations between shallow and deep fractions. These were placed in a greenhouse and seedling emergence monitored until December 5.

RESULTS

Nitrogen Balance in Faba Beans

At site on Farm A in 1978 (Fig. 1) and at site on Farm A in 1979 (weekly or biweekly assays, data not shown), nitrogenase activity was initiated at about 2½ weeks after planting and reached a maximum during vegetative growth, while nitrate reductase activity was highest during reproductive growth. More than half of the total N in the crop on Farm A accumulated after vegetative growth (Fig. 1). Cumulative N_2 fixation, calculated from C_2H_2 reduction rates, kept pace with N accumulation during vegetative growth, but not thereafter. These data suggest that atmospheric N_2 and soil-N were major sources of N during vegetative and reproductive growth successively. A sequence of high nitrogenase activity followed by high nitrate reductase activity has also been reported for *Phaseolus vulgaris* (Franco et al., 1979).

Increase in total seed-N (Fig. 1) occurred partially at the expense of pod-N, which declined during August, but most of the decline in total leaf-N was associated with leaf fall rather than with transport of N out of the leaves. The N content of green leaves declined from 5.50 to 4.05% during August while the number of green leaves per plant declined from 127 to 31 (counts for 3 plants August 1 and 31). Old leaves about to fall contained 4.4 to 5.3% N. Loss of N by leaf fall is estimated as (4.05/5.50) x (total leaf-N on August 1 minus total leaf-N on August 31) = 159 mg N per plant or 56 kg N per hectare.

Comparison of estimated inputs of N_2 with outputs of N in harvested seeds (Table 1) suggests that N_2 fixation was not sufficient to provide a positive N balance at the sites on both Farm A and B in 1978. There would be very large negative balances if straw was harvested, as it is when the plants are used for sileage.

Rapid decline of soil N under continuous growth of pulses is referred to in older literature (Harmsen and van Schreven, 1955) and more recently has been documented for soybeans (Johnson et al., 1975). The short term stimulation of crops following pulses was believed to be due to the accumulation of crops following pulses was believed to be due to the accumulation of N in legume residues in a readily decomposable form. The large amounts of N returned to the soil in high N (Fig. 1) residues after harvest of faba beans

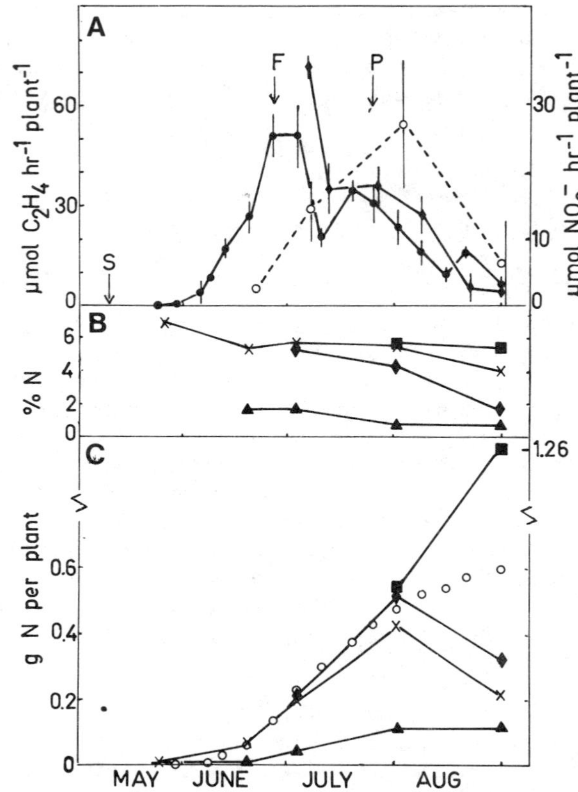

Fig. 1. Seasonal distribution in field grown faba beans of (a) whole plant nitrogenase and nitrate reductase activities and of (b) percent N and (c) total N in different plant parts. Calculated cumulative N_2 fixation is indicated in (c). (a): nitrogenase activity of Farm A plants (●) and Farm B (1978) plants (♦); nitrate reductase activity of Farm A plants (o); bars indicate standard errors; S = seeding, F = first flowers, P = first pods. (b & c): Farm A plants, stems (▲), leaves (x), pods (♦), seeds (■), cumulative N_2 fixation (o); roots (+ nodules) on August 1 contained 1.55% N, 0.061 g N/plant.

(Table 1) could likewise be expected to mineralize rapidly, and be lost by leaching and/or taken up by weeds and the succeeding crop.

Table 1. Crop and weed biomass and N data. Numbers in brackets are standard errors.

Crop	Biomass at harvest (metric ton/ha): Total crop	Biomass at harvest (metric ton/ha): Seeds	Biomass at harvest (metric ton/ha): Weeds	Biomass at harvest: Combine (bu/ac)	Inputs of N (kg N/ha): Seed	Inputs of N (kg N/ha): Man.	Inputs of N (kg N/ha): N_2 fix	Output: Seed-N (kg/ha)	Field residues (kg N/ha): Straw (+ roots)	Field residues (kg N/ha): Weeds (+ roots)	Field residues (kg N/ha): Oct. 28 weeds (- roots)
Farm A											
Beans 1978 (cv Minden)	16.2 (1.29)	7.16 (1.50)			9		217	303[a]	204[b]		
Farm B											
Beans (cv Akerp.)											
1978	9.57 (0.47)	4.83			9		207[c]	222	110		
1979	4.56 (1.27)	2.34 (0.44)	1.23 (0.18)	33	9		123	101	57.2	38.6	29.3[d]
Oats 1979 (cv Garry)	3.11 (0.52)	1.74	0.23 (0.04)	49	2.2	105[e]		39.9	22.9	4.4	3.3
Wheat 1979 (cv Lennox)	8.66 (1.81)	2.86	1.88 (0.64)	40	1.8	105	3.5[f]	41.3	28.6	18.2	23.0
Clover 1979 (cv Alsike)	3.73 (1.12)		1.65 (0.50)		0.3		51		117	41.0	
Fallow 1979			3.49 (0.06)				7.5			70.6	

Total inputs (+) and output (-) of N for Farm B (kg N/farm)[g]: 119+ 1470+ 1859+ 2188-

Denitrification, assayed biweekly at Farm A, was undetectable in most samples (sensitivity about 1 g N/ha per day); the maximum value for an individual sample was 3.4 g N/ha per day. Soil nitrate concentrations were low (3 to 7 ppm).

Nitrogen Budget for Farm B

Grain yields on Farm B declined by about 50% after commercial fertilizer use was terminated in 1976. Yields of beans were unaffected. General field crop ratings of soil samples taken from 6 of 10 fields in the fall of 1978 were in the range medium to high plus for Ca, Mg, K and P, and pH values were in the range 6.0 to 6.5. Thus we consider it likely that the declines in grain yields were due to low levels of available N. To gain some idea of whether the apparent N limitation was attributed to inadequate inputs of N to the farm through N_2 fixation, or to the nature of N cycling within the farm, a nitrogen budget (Fig. 2) was constructed from data obtained in 1978 and 1979 (Table 1 and footnotes to Fig. 2).

Notes to Table 1. (a) Output-N is corrected for harvest losses estimated as 8% for beans (seeds left in field after harvest of Farm B crop in 1978, counted in twenty 25 x 25 cm quadrats), and as 3% for grains (Kepner et al., 1978). (b) Straw-N for the beans includes the leaf loss before harvest (estimated as 11.8 mg N/g stem tissue from data for Farm A crop) x 0.75 assuming 25% of this N is recycled in season. Root-N for upland crop was estimated as 3.7 mg N/g stem tissue based on data from Farm A crop. (c) C_2H_2-reducing activity of Farm B crop in 1978 was measured only from July 7 onwards; to calculate seasonal activity we have assumed that the ratio of total activity before July 7 to that from July 7 onwards was the same as that of the Farm A crop (Fig. 1). (d) Biomass values in metric tons (m.t.) per ha for October 28 weeds were: bean field 1.34 (0.23), oat field 0.12 (0.016), wheat field after harvest 0.78 (0.20). Wheat planted in the fall of 1979 had a biomass on October 28 of 0.63 (0.086) m.t./ha and the weed biomass was 0.13 (0.08); total aboveground N was 20.6 and 4.6 kg N/ha for wheat and weeds, respectively. (e) Avg. manure-N was 1.88% on fresh weight basis (range 1.24 to 2.42) or 2.89% on dry weight basis. Field application rate was 5.6 m.t./ha. Losses of NH_3 from microplots after incorporation of manure were small (range 0.12 to 0.72 g N/ha). (f) N_2 fixation by leguminous weeds and clover was conservatively (Edmeades and Goh, 1978) estimated as 50% of the aboveground N. (g) Based on acreage of 10 ha in beans, 7 ha in oats, 7 ha in wheat, 3 ha in clover, and 3 ha under fallow, the approximate pre-1979 values. Data for faba beans were averaged.

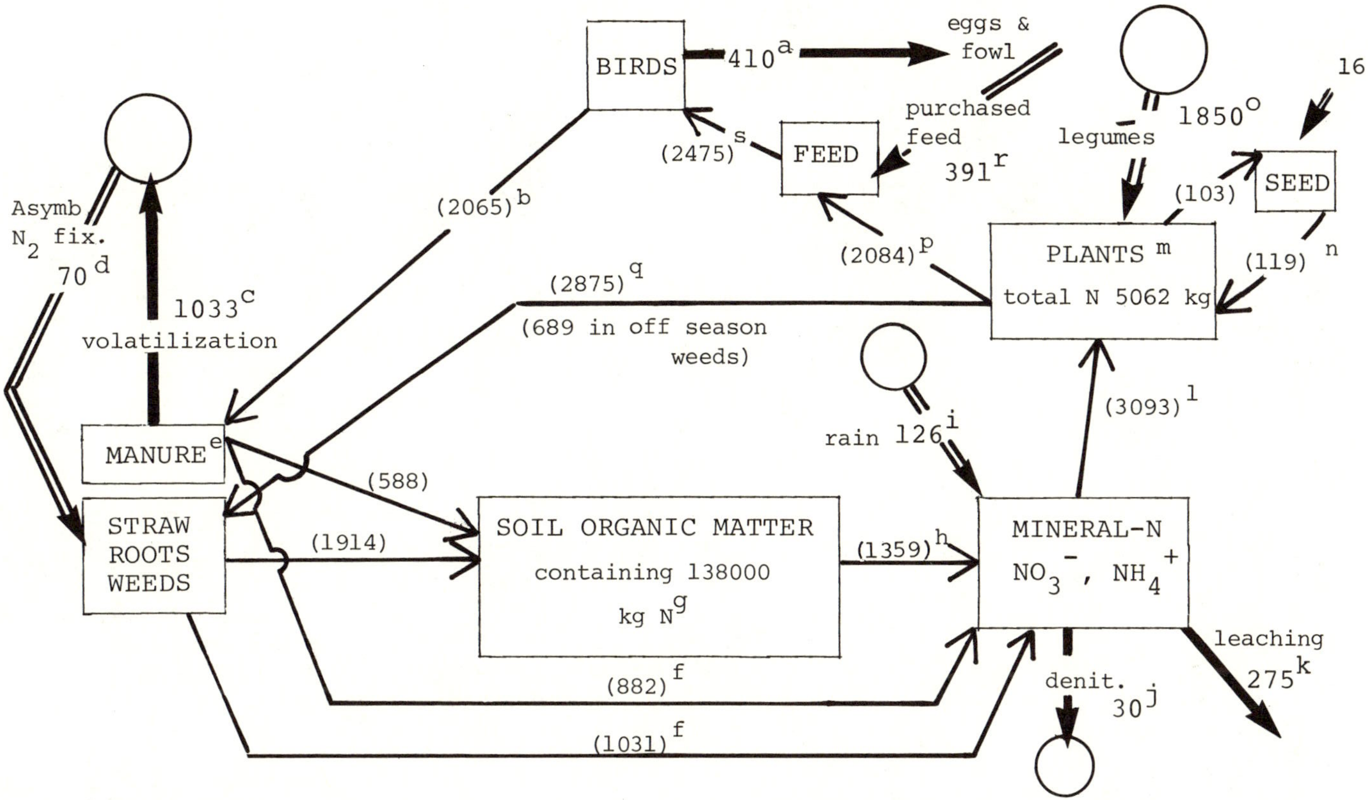

Fig. 2. Nitrogen budget for Farm B. Numbers in brackets are flows of N between compartments within the farm. Big arrows and accompanying numbers not in brackets are flows of N into and out of the farm. Circles represent the atmosphere. Units are kg N per farm per year. Calculations refer to 10 ha in beans, 7 in wheat, 7 in oats, 3 in clover and 3 fallow.

The results suggest: 1) For Farm B as a whole (Fig. 2), inputs of N through N_2 fixation are more than sufficient to balance losses associated with export of products, volatilization, leaching and denitrification, and the farm is therefore potentially self sufficient in N. 2) Losses due to leaching and denitrification appear to be low (footnotes (j) and (k) to Fig. 2) and therefore we conclude that most of the difference between inputs of legume N_2 fixation and manure-N to the fields (Table 1) and the outputs of harvested seed-N from the fields is accumulating in the fields, i.e., in the soil. This difference amounts to 37.7 kg N/ha, or to 24.0

Notes to Fig. 2. (a) Export of 384000 eggs x 1 g N/egg plus approx. 500 fowl at 1.8 kg, 18% protein. (b) Feed-N minus exported N. (c) Assumed 50% of manure-N was volatilized (National Academy of Sciences, 1978). (d) For 5200 kg/ha low N (0.26%) straw on wheat fields this is a conservative estimate of asymbiotic N_2 fixation (Brouzes et al., 1969). (e) The manure output is the field application rate. This is lower than inputs after volatilization losses. The latter could be reduced if the output is not sustainable, for example by application of gypsum (Roberts, 1897), available locally, to the roots. (f) It is assumed that 60% of manure-N and 35% of residue-N is mineralized in one year (Reddy et al., 1979; Mathers and Goss, 1979). (g) Soil-N in top 15 cm x 1.25 (Magdoff, 1978). (h) Calculated as (plant-N + groundwater-N + denitrification) minus (legume N_2 fixation + seed-N + manure & residue mineralization + rain-N). (i) Data from J. Underwood, Nova Scotia Dept. Environment. (j) Denitrification is assumed to be low because soil nitrate values (< 6 ppm dry soil except after manure application in fall when they rose to 14 ppm) were similar to those at Farm A where denitrification was very low. (k) Twenty samples of tile drain water were collected over one year from May 1979 in a field which was cultivated intermittently and later fertilized with manure and sown with winter wheat. Average nitrate-N in tile drain water from May to November was 2.0 ppm (range 0.6 to 3.1) and from December to April 1980, 5.5 ppm (3.6 to 7.9). The average values were multiplied by total rainfall for the 2 periods (84.1 and 36.1 cm, respectively) and by a percolation factor of 0.25 (provided by J. Kerekes, Canadian Wildlife Service as a maximum possible value) giving a value for leaching loss of 9.16 kg N/ha. (l) Calculated as plant-N minus (legume N_2 fixation + seed-N). (m) Table 1, total plant-N including weeds. (n) Oat and clover seed are purchased, other seed is produced internally. (o) Table 1. (p) Harvest-N (Table 1) minus seed-N. (q) Calculated as plant-N minus harvest-N. (r) Calculated as the difference between the supply and the requirement (see (s)). This value is 16%. About 25% of the bulk feed requirement is bought as grain. (s) Feed requirement is 91000 kg of 17% protein.

kg N/ha if a conservative estimate of N_2 fixation in faba bean is used. The current output in harvested seed-N is 73 kg N/ha implying that significant gains could be made if the excess N accumulating in the soil at present could be diverted to the grains. 3) Substantial amounts of N cycle through weeds.

Post-harvest weeds and weeds on uncultivated fields may play an important role by retaining N (Table 1) that would otherwise be lost by leaching in this high rainfall region. Lysimeter studies in New York state (Bizell and Lyon, 1927) indicated losses of 77.5, 8.8 and 2.8 kg N/ha under bare soil, rotation and grass, respectively. Present losses on Farm B, of the order of 10 kg N/ha, appear to be tolerable but losses of the order of 30 to 40 kg N/ha would negate most of the potential gains in yields.

Composition, Seed Banks and Biomass of Weeds

Factors affecting the distribution and abundance of weeds were studied initially in the context of their being considerations in crop rotation strategy, that in turn being a factor determining the nature of N cycling between crops.

The composition of weed populations (Table 2) was consistent with the basic tenet of agriculture that weeds whose life cycles are most closely synchronized to the crop make the best competitors (Bunting, 1959). Winter annuals and perennials were the dominant weeds amongst winter wheat, which has the longest growing season of any of the crops. Summer annuals were dominant amongst oats which have the shortest growing season, and summer annuals and perennials were dominant amongst faba beans, which have a long summer growing season.

Seed banks of Raphanus raphanistrum (wild radish), the most abundant summer annual amongst oats and beans, were much lower under winter wheat than under the summer annual crops (compare numbers for shallow horizons, Table 2). This is due to germination of the weed with wheat following tillage operations in late summer, and subsequent frost kill during the winter. Few seeds of Vicia tetrasperma, the most abundant weed amongst winter wheat were evident in the seed banks, but pods of V. tetrasperma of similar geometric form to that of wheat grains were found in the wheat seed, suggesting that this weed is distributed with the seed.

Within fields, the ratio of crop biomass to total (crop + weed) biomass increased with increasing total biomass, reaching values close to 1.0 at sites of highest total biomass (Fig. 3). When fertilizer-N was applied shortly after planting in the bean field, this relationship was upset in favor of the weeds (Fig. 3a).

Table 2. Dominant weeds and their seed banks in fields of oats, faba beans and winter wheat on Farm B, and in a field of winter wheat on Farm R about 5 km distant. The three most common species in each field are listed with those of the oat field first, followed by additional species, if they differ from the oat field species, from the faba bean and winter wheat fields, respectively. Nomenclature is after Roland and Smith (1969). The habit of each species is indicated by letters in brackets; (S) = summer annual, (W) = winter annual, (P) = perennial. cov = ranked cover values for each field. top = top, and bot = bottom halves of 15 cm deep soil cores. Seed bank numbers are tens per square meter of field surface sampled. Under cov, np = not present.

Species	Oats[a]			Faba beans[b]			Winter wheat[c]			Winter wheat[d] (R)		
	cov	top	bot	cov	top	bot	cov	top	bot	cov	top	bot
Raphanus raphanistrum (S)	1	20	33	2	81	15	21	13	85	17	0	13
Ambrosia artemisiifolia (S)	2	0[e]	0	20	0[f]	0	7	0	0	np	0	0
Plantago major (P)	3	79	380	4	308	59	22	92	79	9	7	20
Agropyron repens (P)	18	0	0	1	2	5	9	26	26	7	20	7
Solidago graminifolia (P)	14	20	0	3	111	22	np	0	0	np	0	0
Vicia tetrasperma (W)	np	0	0	np	0	2	1	3	13	1	0	7
Phleum pratense (P)	18	0	7	11	94	50	2	98	52	2	7	7
Galeopsis tetrahit (S)	7	0	0	6	33	51	3	46	52	13	0	0
Poa pratensis (P)	np	7	13	11	15	13	4	20	33	3	327	72
Total seed bank		242	603		1055	465		1447	1624		766	439
Standard error		17	46		112	42		114	162		234	93

[a] Faba beans were planted in this field in 1977 and oats in 1978; intense cultivation 1978-1979. [b] Faba beans on this field in 1978. [c] Hay field prior to 1979. [d] Continuous winter wheat for 5 years, no herbicide. [e] After a vernalization period 70 and [f] 20 plants/m^2 germinated in the trays.

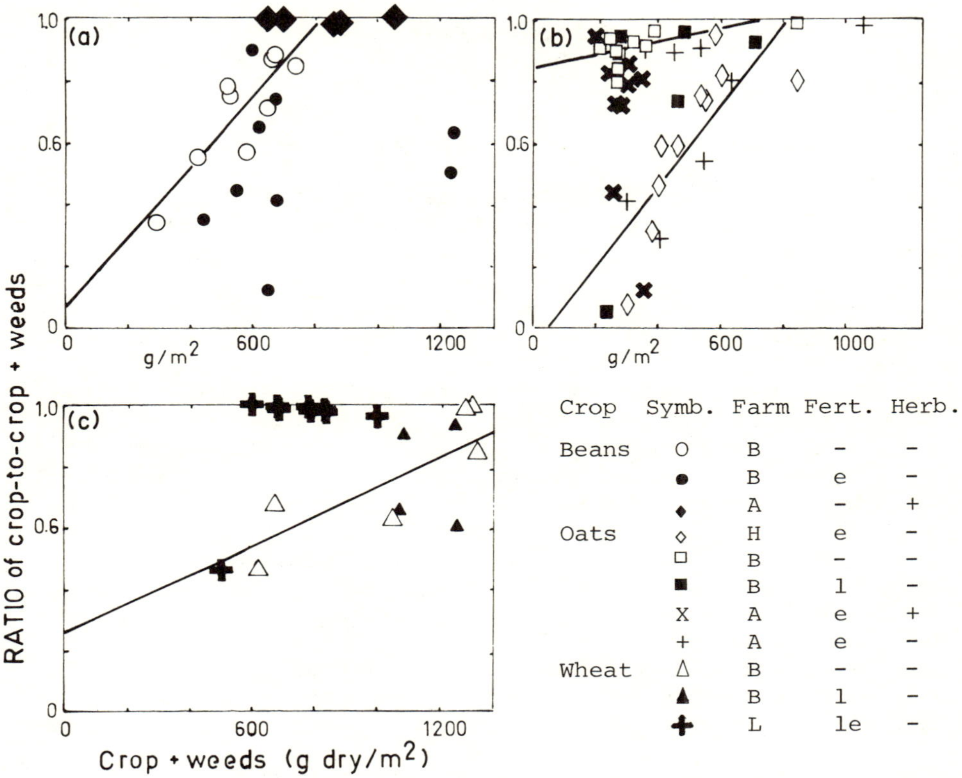

Fig. 3. Ratio of crop biomass to crop + weed biomass in relation to total (crop + weed biomass in 35 x 35 cm quadrants randomly placed in (a) fields of faba beans, (b) oat fields, and (c) fields of winter wheat. Sampling was on August 1 for winter wheat, August 21-24 for faba beans and oats, except for oats from Farm B which were sampled on September 9. Open symbols are used and regression lines are indicated for fields or treatments in which there were significant (α = 0.05) linear regressions between the two statistics. Farm B refers to the Aldhouse (upland) farm, Farm A to the Warren (dykeland) farm, and Farms H and L are other farms in Annapolis Co. Under FERT, e refers to application of fertilizer-N at about the time of seeding, and l to later application. On Farm B, 100 kg N as ammonium nitrate were broadcast on 2 x 2 m plots in wheat and bean fields on May 30, and in the oat field on August 7. HERB refers to the use of herbicide.

For beans and wheat on Farm B, and for oats on Farm H (see Fig. 3) (not treated with herbicide because of rain), regressions through the origin do not significantly ($\alpha = 0.05$) reduce the fits from those of regressions indicated in Figure 3. This was not true for oats on Farm A for which the ratio of crop to total biomass was high at low total biomass values. This field was also characterized by a significantly lower seed bank than other fields (Table 2) and by low post-harvest weed biomass (Table 2). Because of abundant Canada thistle on this field in 1978 (under oats), the field was deep ploughed in the fall of 1978 to break up perennial rhizomes, and harrowed following two rotovatings before planting oats in 1979.

Our interpretation of the regressions in Figure 3 is that increasing total biomass is related to increasing soil fertility, and provided that the crop has an initial advantage (given by seed bed preparation, optimal planting time, etc.), higher soil fertility enables the crop to compete more effectively with weeds for available nutrients. At low soil fertility when crop plants do not grow fast enough, or are not sufficiently dense to take up all available nutrients (or available N), weed growth acts as a negative feedback mechanism recycling nutrients that might otherwise be lost, and increasing soil fertility in the long run. When weed seed banks are low, as in the intensively cultivated oat field, post-harvest weed growth may be insufficient to take up a large fraction of the available N, which may then be lost through leaching.

Predicted Changes in Crop, Weed and Soil Nitrogen

As a first approximation, changes in soil-N can be described by the relation $dS/dt = I - kS$ where S = soil organic N, I = rate of addition of organic N to the soil and k = mineralization constant (Magdoff, 1978). Our data suggest that inputs of N to the fields exceed outputs, and thus we conclude that soil organic N, and accordingly the amount of N mineralized from this each year (kS) are increasing annually and will continue to increase until output of N is equal to input of N. The weed and crop biomass data (Fig. 3) suggest that crops will obtain a greater proportion of the available N as total available N increases, and hence that crop yields will increase faster than the increase in total available N. We have estimated changes in soil-N and yields with time as follows.

The field system is regarded as a black box into which net inputs of N are provided by manure-N (M) and faba bean N_2 fixation (F), and net outputs occur in the products removed from the field, i.e., cereal grain-N (G), and in faba bean seed-N (B). Transformations involving straw, roots and weeds are considered as recycling within the box, and denitrification, asymbiotic N_2 fixation, rain-N and (sowing) seed-N are ignored as they are individually small, and collectively add up to about zero (Fig. 2). Since the mineral-N pool is small, we assume that the difference between inputs and outputs is accumulating in soil organic N, and

$$dS/dt = F + M \ (G + B)$$

F and M are regarded as constants. Nitrogen in cereal grains comes from manure-N and from soil-N. For 1979 (Table 1), this is described as

$$G_{79} = M_G + k_G \times S_{79}$$

M_G is the grain-N derived from manure, and is given by

$$M_G = M \times A \times G/P \times P/PW$$

where A is the fraction of manure-N that is mineralized during the crop season, G/P is the ratio of seed-N removed at harvest to total plant-N (Table 1), and P/PW is the ratio of total crop-N to (weed + total crop)-N (as in Table 1 except that we use a P/PW value of 0.73 for oats, on the assumption that the oat fields are not normally intensively cultivated). N_2 fixation in the clover and fallow fields (Table 1) is included in M as we consider the forage legumes as green manure. Based upon the percentage of N in the manure, A is estimated as 0.50 for corn and sorghum in the southwest U.S. (Mathers and Goss, 1979). Allowing for the colder climate in Nova Scotia, and for differences in growing season between winter wheat and oats, we assume that the values of A are 0.45 and 0.4 for winter wheat and oats, respectively. k_G is the ratio of (grain-N minus M_G) to soil-N, where soil-N is the N in the top 15 cm multiplied by 1.25 to allow for use of N from deeper horizons (Magdoff, 1978). The terms M_G and k_G include explicitly or implicitly the ratio P/PW and we assume that this ratio will increase (to a maximum of 1) in direct proportion to increases in available N, given by

$$(M_s + k_{PW} \ S_t) \ / \ (M_s + k_{PW} \ S_{79})$$

where M_s is the manure-N mineralized during the crop season (M x 0.45 or 0.4) and k_{PW} is the ratio of (crop + weed)-N to total soil-N in 1979. Thus

$$G_t = (M_G + k_G \ S_t) \times (M_s + k_{PW} \ S_t) \ / \ (M_s + k_{PW} \ S_{79})$$

where the latter expression cannot exceed a value of 1/(P/PW).

Nitrogen in faba beans is derived partly from N_2 fixation, which is not available to weeds, and in part from soil-N which could also be used by weeds. Thus the harvested bean-N is described by

$$B = B_F + k_B S_t \times (F + k_{PW} S_t) / (F + k_{PW} S_{79})$$

B_F is the proportion of N in seeds derived from N_2 fixation and is given by F x ratio (harvested seed-N/total plant-N) obtained from Table 1. The second part of the expression is analagous to the previous expression describing soil N in cereal grains. Data from both 1978 (upland farm only) and 1979 were used in calculation of the various parameters for beans; it was assumed that P/PW in 1978 was 1.0 (cf. Table 1 and Fig. 3).

Grain-N, weed-N at harvest and soil-N were calculated according to these assumptions by an iterative procedure beginning with S = 1900 kg/ha, and up to a value of 6900 kg/ha (Fig. 4) which cover the range of soil-N values on Farm B. Data for oats and wheat were combined in Figure 4. This model is a considerable simplification of the real situation, and is used only to give an impression of the general magnitude of changes in yields and weeds that could be expected with time and with differences in soil-N. Calculations based on crop data obtained over a period of years and from more soil types might change the absolute values, but would not alter the relative changes greatly. If differences in the mineralization constant for different soils, and the carry-over of residue-N and manure-N for more than one year were considered, the slopes of the curves would probably be lower than in Figure 4. In any case, it is reasonable to conclude that (1) changes in yield are likely to take place very slowly; from an initial value of S of 4600 kg/ha, the average value of soil-N on Farm B in 1979, grain yields increase by 11% and bean yields by 2% after 20 years in Figure 4; and (2) considerable variation can be expected in the amount of weeds at harvest, with practically none in fields of highest fertility, and possibly severe weed problems in fields of lowest fertility. Both these conclusions are consistent with the experience of the grower on Farm B since 1976.

Introduction of a Regular Rotation

These various considerations suggest that management practices on Farm B are a step in the right direction toward achieving high yields and self sufficiency in N, but that the changes are likely to occur at a frustratingly slow pace. Nevertheless, we have attempted to be conservative in our evaluation of the potential for self sufficiency, and it seems fairly clear that the potential is there. Thus alternative strategies may be sought to increase the yields more quickly.

This is a matter of changing the internal structure of the system so that at the current level of accumulated soil fertility more of the manure-N and recently fixed N_2 is removed as product, and less accumulates in the soil or cycles through weeds.

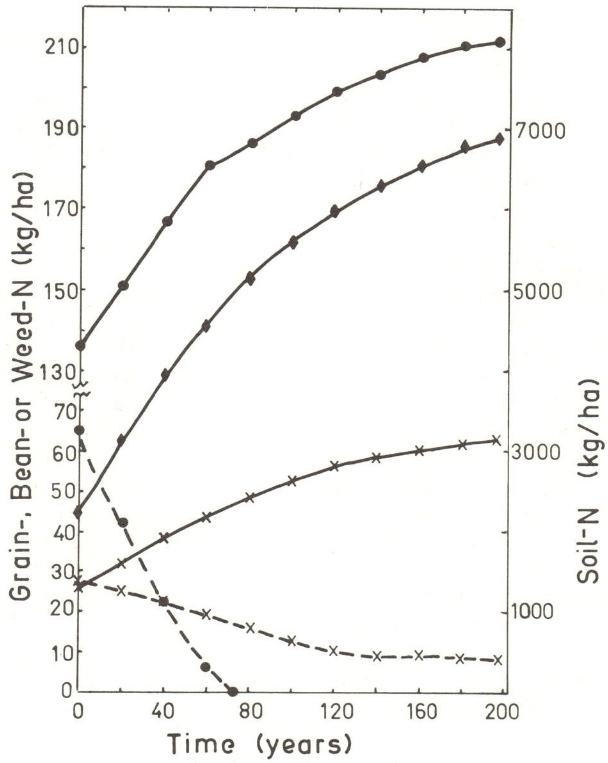

Fig. 4. Predicted changes in crop, weed and soil nitrogen with time. Present range of soil-N on Farm A is 2630 to 6230 kg/ha (N to 15 cm depth, x 1.25), and the mean value is 4600 kg. (—●—) faba bean seed-N (corrected for 8% harvest loss), (—x—) grain-N (3% harvest loss), (--●--) weed-N in bean fields at harvest, (--x--) weed-N in grain fields at harvest, (—♦—) soil-N.

Figuratively described, we need to change the interest rate rather than increase the bank account. The division of the land to various crops as it existed under a ferterilizer- and herbicide-intensive system is probably not the most appropriate division now.

Based in part on these considerations, in 1980, the grower on Farm B adopted a new management scheme involving a regular rotation of beans - oats - clover - winter wheat. The clover is seeded with oats in year 2, and is turned under in year 3 prior to planting winter wheat. Manure is applied to the oats and to the wheat.

Some of the considerations involved in formulating this scheme are as follows. 1) A regular rotation is required for weed control, facilitates planning, and the legume-cereal sequences provide for mopping up of N released from high N legume residues. 2) Faba beans are a long season crop; manure-N applied in the fall (after beans) may be immobilized long enough, or temperatures may be low enough to avoid substantial losses of N. Fall application of manure would facilitate field operations generally because of late wet springs in this area. 3) Underseeding of oats with clover provides an overwintering green manure crop prior to winter wheat and will help keep weeds under control after oats are harvested. 4) Wheat leaves behind a lot of low N straw (Table 1). This could be expected to immobilize mineral-N during vegetative growth of beans in the following year thereby stimulating N_2 fixation, and to release N during reproductive growth when NRA is high (Fig. 1). Production of CO_2 by decomposing straw may also stimulate N_2 fixation in the beans (Shrivashankar and Vlassak, 1978).

Maintenance of nearly continuous plant cover is a key consideration. Thus it is desired to control annual weeds during the crop season, but not to eliminate them so that they provide a self-seeding cover when crops are not present. The crop rotation keeps individual species off balance. Inclusion of a short season crop (oats) provides time for recurrent cultivation if necessary, but this is seen as a last resort because of the potential for increased losses of N by leaching. Introduction of biological controls for Canada thistle (e.g., Peschken and Harris, 1975) could reduce the need for intensive cultivation. Our observations suggest that weed problems are likely to be most severe on soils of low fertility. Denser seeding of crops and/or heavier application of low manure may be effective in controlling weeds on such soils. The latter would have to be accompanied by lower applications of manure on soils of highest fertility, where the response to manure is likely to be less, and overall would result in more N being cycled through crops than if manure was applied at a uniform rate throughout.

The total yields of N predicted from per hectare crop yields in Table 1 would be less than in the past (1824 versus 2188 kg) simply because 7½ rather than 10 ha are planted with faba beans. However, production of beans has generally exceeded requirements in the past, and with the addition input of clover-N before winter wheat, the balance may be redressed in favor of the grains. The difference between inputs of N from manure and N_2 fixation and outputs in harvested seed in this scheme is estimated as 34 (conservative estimate of N_2 fixation in faba beans) to 46 kg N/ha, so the potential increases in yields are high (present output 73 kg N/ha), and if realized would result in yields similar to those achieved with inputs of commercial fertilizer-N.

CONCLUSION

These studies suggest that for Farm B, there is a potential to be self sufficient in feed production and in N on a sustainable basis without having to increase the land base. A large input of N from N_2 fixation, and efficient recycling of N in manure and plant residues are required. The resources required for N_2 fixation are available now and are being utilized. The major part of the N_2 fixation is provided by the faba bean which also provides the high protein component of the poultry feed. The limiting factors to achieving self sufficiency are ones concerned with management of the N once it is fixed, rather than with the amounts of N_2 fixed. A regular rotation of crops appears to be essential.

Traditional crop rotation systems are up to 45% more energy efficient than modern monoculture systems, although the yields are frequently lower (Heichel, 1978). If advances in our understanding of plant and microbial processes are applied to the design and management of rotations, then their outputs might become competitive with those of monoculture systems, and they could provide an important low energy option for the future.

ACKNOWLEDGMENTS

This work was supported by an Operating Grant from the Natural Sciences and Engineering Research Council of Canada. N. Hill was supported by the Nova Scotia Department of the Environment. We are grateful to Gillian Bidwell, Arnold Mosher and Jim Helmke for assistance and stimulating discussion, and to Robyn Warren and Basil Aldhouse and their families for facilitating field work and for sharing with us many of the concepts explored in this paper.

REFERENCES

Bizell, J. A., and Lyon, T. L., 1927, Composition of drainage waters from lysimeters at Cornell University, Proc. Intern. Cong. Soil Sci., 2:342.
Blake, G. R., 1965, Bulk density, in: "Methods of Soil Analysis," C. A. Black, ed., American Society of Agronomy, Madison, WI.
Brouzes, R., Lasik, J., and Knowles, R., 1969, The effect of organic amendment, water content, and oxygen on the incorporation of $^{15}N_2$ by some agricultural and forest soils, Can. J. Microbiol., 15:899.
Bunting, A. H., 1959, Some reflections on the ecology of weeds, in: "The Biology of Weeds," J. L. Harper, ed., Blackwell Scientific Publications, Oxford.
Candlish, E., and Clark, K. W., 1975, Preliminary assessment of

small faba beans grown in Manitoba, Can. J. Plant Sci., 55:89.
Cataldo, D. A., Haroon, M., Schrader, L. E., and Youngs, V. L., 1975, Rapid colorimetric determination of nitrate in plant tissue by nitration of salicylic acid, Commun. Soil Sci. Plant Anal., 6:71.
Dagnelie, P., 1960, Contribution à l'étude des communautes végétales par l'analyse factorielle, Bull. Serv. Carte Phytogeogr. CNRS, B-5:7.
Edmeades, D. C., and Goh, K. M., 1978, Symbiotic nitrogen fixation in a sequence of pastures of increasing age measured by a ^{15}N dilution technique, N.Z. J. Agr. Res., 21:623.
Franco, A. A., Pereira, J. C., and Neyra, C. A., 1979, Seasonal patterns of nitrate reductase and nitrogenase activities in Phaseolus vulgaris L., Plant Physiol., 63:421.
Hardy, R. W. F., Holsten, R. D., Jackson, E. D., and Burns, R. C., 1968, The acetylene-ethylene assay for N_2 fixation: Laboratory and field evaluation, Plant Physiol., 43:1185.
Harmsen, G. W., and van Schreven, D. A., 1955, Mineralization of organic nitrogen in soil, Adv. Agron., 7:299.
Heichel, G. H., 1978, Stabilizing agricultural energy needs: Role of forages, rotations, and nitrogen fixation, J. Soil and Water Conserv., 33:279.
Hudd, G. A., Lloyd-Jones, C. P., and Hill-Cottingham, D. G., 1980, Comparison of acetylene-reduction and nitrogen-15 techniques for the determination of nitrogen fixation by field bean (Vicia faba) nodules, Physiol. Plant., 48:111.
Johnson, J. W., Welch, L. F., and Kurtz, L. T., 1975, Environmental implications of N fixation by soybeans, J. Environ. Qual., 4:303.
Kepner, R. A., Bainer, R., and Barger, E. L., 1978, "Principles of farm machinery," AVI Publ. Co., Westport, CT.
Macdougall, J. I., Nowland, J. L., and Hilchey, J. D., 1969, Soil survey of Annapolis County, Nova Scotia, Nova Scotia Soil Survey (Truro, Nova Scotia), Rep. No. 16.
Magdoff, F. R., 1978, Influence of manure application rates and continuous corn on soil-N, Agron. J., 70:629.
Mathers, A. C., and Goss, D. W., 1979, Estimating animal waste applications to supply crop nitrogen requirements, Soil Sci. Soc. Am. J., 43:364.
National Academy of Sciences, 1978, "Nitrates: An Environmental Assessment," National Academy of Sciences, Washington, DC.
Patriquin, D. G., MacKinnon, J. C., and Wilkie, K. L., 1978, Seasonal patterns of denitrification and leaf nitrate reductase activity in a corn field, Can. J. Soil Sci., 58:283.
Peschken, D. P., and Harris, P., 1975, Host specificity and biology of Urophora cardui (Diptera: Tephritidae). A Biological control agent for Canada thistle (Cirsium arvense), Can. Ent., 107:1101.
Reddy, K. R., Khaleel, R., Overcash, M. R., and Westerman, P. W., 1979, A nonpoint source model for land areas receiving animal

wastes: I. Mineralization of organic nitrogen, Trans. Am. Soc. Agr. Engin., 1979:863.

Roberts, I. P., 1897, "The Fertility of the Land," MacMillan Co., NY.

Roland, A. E., and Smith, E. C., 1969, "The Flora of Nova Scotia," Nova Scotia Museum, Halifax.

Shivashankar, K., and Vlassak, K., 1978, Influence of straw and CO_2 on N_2-fixation and yield of field-grown soybeans, Plant Soil, 49:259.

Strickland, J. D. H., and Parsons, T. R., 1972, "A Practical Handbook of Seawater Analysis," Fisheries Research Board of Canada, Ottawa.

Yoshinari, T., Hynes, R., and Knowles, R., 1977, Acetylene inhibition of nitrous oxide reduction and measurement of denitrification and nitrogen fixation in soil, Soil Biol. Biochem., 9:177.

LIST OF PARTICIPANTS

AMES, Madeline
Dept. of Agronomy
University of California
Davis, CA 95616

ANDERSEN, Kjell
Plant Growth Laboratory
University of California
Davis, CA 95616

AUSUBEL, Frederick
Biology Department
Harvard University
Cambridge, MA 02138

BAHR, James T.
Mobil Chemical Company
Research & Development Labs
Edison, NJ 08817

BAIRD, Lisa Morrison
Dept. of Agronomy
University of California
Davis, CA 95616

BARNES, Donald K.
Dept. of Agronomy & Plant Genetics
University of Minnesota
St. Paul, MN 55108

BERENBAUM, Morris
Allied Chemical Company
Morristown, NJ 07960

BERINGER, John
Rothamsted Experimental Station
Harpenden, Herts
England

BREWIN, Nicholas
Dept. of Genetics
John Innes Institute
Norwich, England

BUCHANAN, Bob
Dept. of Cell Physiology
University of California
Berkeley, CA 94720

CANGELOSI, Gerry
Plant Growth Laboratory
University of California
Davis, CA 95616

CANTRELL, Michael A.
Lab for Nitrogen Fixation
Oregon State University
Corvallis, OR 97330

CARLSON, Curt
Bacteriology Department
University of California
Davis, CA 95616

CASSMAN, Ken
Land, Air, & Water Resources Dept.
University of California
Davis, CA 95616

CORBIN, David
Biology Department
University of California
San Diego, CA 92093

CSONKA, Laszlo
Plant Growth Laboratory
University of California
Davis, CA 95616

DeJONG, Ted M.
Dept. of Agronomy
University of California
Davis, CA 95616

DITTA, Gary
Biology Department
University of California
San Diego, CA 92093

DUNCAN, Margaret
Biology Department
M.I.T.
Cambridge, MA 02139

EADY, Scott
Dept. of Agronomy
University of California
Davis, CA 95616

EVANS, Harold J.
Lab for Nitrogen Fixation
Oregon State University
Corvallis, OR 97330

FIRESTONE, Mary
Soils & Plant Nutr. Dept.
University of California
Berkeley, CA 94720

FISHBECK, Kathy A.
Dept. of Agron. & Plant. Genet.
University of Minnesota
St. Paul, MN 55108

FISHER, Richard A.
Biophys. & Theoret. Biol. Dept.
University of Chicago
Chicago, IL 60637

FOCHT, Dennis D.
Soil & Environ. Sci. Dept.
University of California
Riverside, CA 92521

FOSTER, Ken
Dept. of Agronomy
University of California
Davis, CA 95616

FRIEDRICH, James W.
Allied Chemical Company
Syracuse, NY

GARCIA, Ramon L.
Allied Chemical Company
Solvay, NY 13209

GOYAL, Sham S.
Plant Growth Laboratory
University of California
Davis, CA 95616

GRANSTEDT, Robert C.
Castle & Cooke (Honduras)
P. O. Box 11281
San Francisco, CA 94101

HAGEMAN, R. H.
Dept. of Agronomy
University of Illinois
Urbana, IL 61801

HEATH-PAGLIUSO, Sharon
Plant Growth Laboratory
University of California
Davis, CA 95616

HEICHEL, Gary H.
Dept. of Agron. & Plant Genet.
University of Minnesota
St. Paul, MN 55108

HEISEY, Rod
Land, Air & Water Res. Dept.
University of California
Davis, CA 95616

HENNECKE, Hauke
Lehrstuhl fur Mikrobiologie
Universitat Munchen
D-8000 Munchen, W. Germany

HEYTLER, Peter G.
E. I. DuPont de Nemours & Co.
Wilmington, DE 19898

HICKOK, Ruth
Lab for Nitrogen Fixation
Oregon State University
Corvallis, OR 97330

HOBBS, Shaun
Prairie Regional Laboratory
University Campus
Saskatoon, Sask., Canada

HOM, Sui-Ming
Plant Growth Laboratory
University of California
Davis, CA 95616

HUA, Sui-Sheng Sylvia
USDA-SEA
Western Regional Res. Center
Berkeley, CA 94710

HUANG, H. T.
National Science Foundation
Wahington, DC 20550

HUFFAKER, Ray C.
Plant Growth Laboratory
University of California
Davis, CA 95616

INGRAHAM, John L.
Bacteriology Department
University of California
Davis, CA 95616

IYER, V. N.
Biology Department
Carleton University
Ottowa, Ontario, Canada

JAWORSKI, Ernest
Monsanto Agricultural Products
800 N. Lindbergh Blvd.
St Louis, MO 63166

JACKSON, W. A.
Soil Science Department
North Carolina State Univ.
Raleigh, NC 27650

JETER, Randy
Bacteriology Department
University of California
Davis, CA 95616

KERN, Roger
Plant Growth Laboratory
University of California
Davis, CA 95616

LEPO, Joseph E.
Lab for Nitrogen Fixation
Oregon State University
Corvallis, OR 97330

LeRUDULIER, Daniel
Les Quatre Vents
La Chapelle des Fougeretz
35520 Melesse, France

LONG, Sharon R.
Biological Laboratories
Harvard University
Cambridge, MA 02138

LOOMIS, Robert S.
Dept. of Agronomy
University of California
Davis, CA 95616

LYONS, James M.
Associate Dean
College of Ag. & Environ. Sci.
University of California
Davis, CA 95616

MAHAN, Michael
Plant Growth Laboratory
University of California
Davis, CA 95616

MARCUS-WYNER, Lynn
Dept. of Agronomy
University of California
Davis, CA 95616

MAZE, John
California Department of
Food and Agriculture
Sacramento, CA 95814

MAZUR, Barbara
E. I. DuPont de Nemours & Co.
Wilmington, DE 19898

McDERMITT, Dayle
Monsanto Agricultural Products
800 N. Lindbergh Blvd.
St. Louis, MO 63166

McNEIL, David L.
NifTAL Project
Paia, Hawaii 96779

MILLER, Bruce
Plant Growth Laboratory
University of California
Davis, CA 95616

MORRISON, Steven L.
Plant Growth Laboratory
University of California
Davis, CA 95616

MUELLER, Barbara
Argonne National Laboratory
Argonne, IL 60439

MUNNS, Don N.
Land, Air & Water Resources
Dept.
University of California
Davis, CA 95616

NEAL, John L.
Biology Department
Microbiology Section
Virginia Polytechnic Institute
Blacksburg, VA 24061

NEYRA, Carlos A.
Biochemistry & Microbiology
Dept.
Rutgers University
Cook College
New Brunswick, NJ 08903

ONG, Sally
Biology Department
University of California
San Diego, CA 92093

OSBURNE, Marcia S.
Biology Department
M.I.T.
Cambridge, MA 02139

PADWA, David
AgriGenetics Corporation
P.O. Box 1596
Santa Fe, NM 87501

PATRIQUIN, David G.
Biology Department
Dalhousie University
Halifax, NS, Canada

PAYNE, W. J.
Microbiology Department
The University of Georgia
Athens, GA 30602

PETERS, Jerry
C. F. Kettering Res. Lab.
Yellow Springs, OH 45387

PHILLIPS, Donald A.
Dept. of Agronomy
University of California
Davis, CA 95616

RAINS, D. William
Plant Growth Laboratory
University of California
Davis, CA 95616

RALEIGH, Elisabeth A.
M.I.T.
Cambridge, MA 02139

REPORTER, Minocher
C. F. Kettering Res. Lab.
Yellow Springs, HO 45387

RIGHETTI, Timothy
Land, Air & Water Resources Dept.
University of California
Davis, CA 95616

ROTHSTEIN, David
Biology Department
M.I.T.
Cambridge, MA 02139

RUSSELL, Sterling A.
Lab for Nitrogen Fixation
Oregon State University
Corvallis, OR 97330

SANCHEZ, Federico
Dept. Biochemistry & Biophys.
University of California
San Francisco, CA 94143

SATOH, Toshio
Dept. of Biology
Tokyo Metropolitan University
Fukazawa 2-1-1, Setagayaku
Tokyo, Japan

SCHLEGEL, Hans G.
Institut fur Mikrobiologie der Universitat Gottingen
Gottingen, West Germany

SILVER, Warren S.
Dept. of Biology
University of South Florida
Tampa, FL 33620

SHANMUGAM, K. T.
Microbiology & Cell Science Dept.
University of Florida
Gainesville, FL 32611

SKUJINS, J.
Biology Department
Utah State University
Logan, UT 84322

SPILLER, Hart
Plant Growth Laboratory
University of California
Davis, CA 95616

SU-SHENG, Yang
Plant Growth Laboratory
University of California
Davis, CA 95616

SZETO, Wynne
Biological Laboratories
Harvard University
Cambridge, MA 02138

TAIT, Robert C.
Plant Growth Laboratory
University of California
Davis, CA 95616

TALLEY, Steven N.
Dept. of Agronomy
University of California
Davis, CA 95616

TIEDJE, James M.
Crop & Soil Science Dept.
Michigan State University
East Lansing, MI 48824

THAYER, James R.
Crocker Nuclear Laboratory
University of California
Davis, CA 95616

TRANG, Kry
North American Plant Breeders
Princeton, IL 61356

URATSU, Sandie
Plant Growth Laboratory
University of California
Davis, CA 95616
VALENTINE, Raymond C.
Plant Growth Laboratory
University of California
Davis, CA 95616
VAN KEULEN, H.
Dept. of Agriculture
Centre for Agrobiological Research
Wageningen, The Netherlands
WALDON, Hollis
Waldon Laboratories
Ripon, CA 95366
WILCOX, Gary
Microbiology Department
University of California
Davis, CA 95616
WILLIAMS, Larry
Dept. of Agronomy
University of California
Davis, CA 95616
WOLK, C. Peter
MSU-DOE Plant Research Lab
Michigan State University
East Lansing, MI 48824
WONG, Po Keung
Plant Growth Laboratory
University of California
Davis, CA 95616
Yang, Su-Sheng
Plant Growth Laboratory
University of California
Davis, Ca 95616

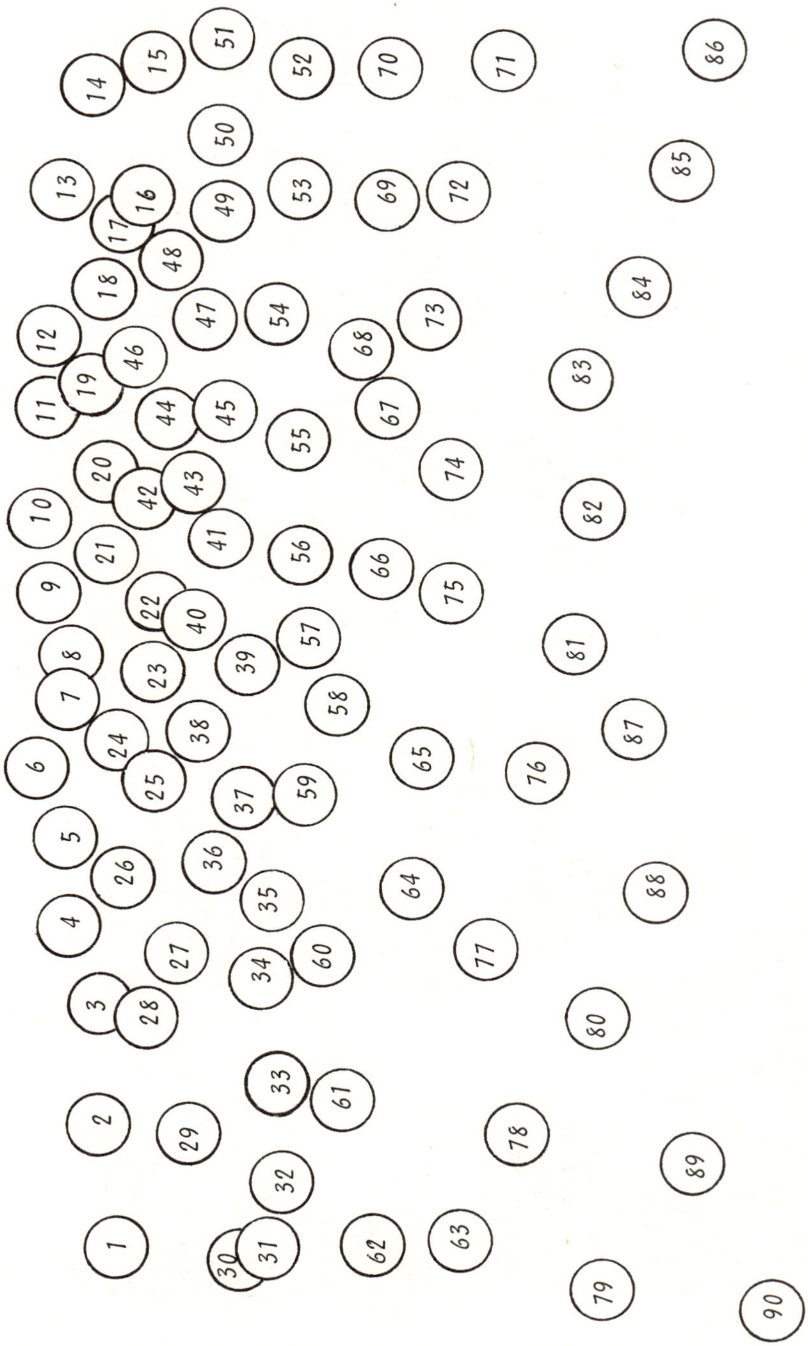
1
2
3
4
5
6
7
8
9
10
11
12
13
14
15
16
17
18
19
20
21
22
23
24
25
26
27
28
29
30
31
32
33
34
35
36
37
38
39
40
41
42
43
44
45
46
47
48
49
50
51
52
53
54
55
56
57
58
59
60
61
62
63
64
65
66
67
68
69
70
71
72
73
74
75
76
77
78
79
80
81
82
83
84
85
86
87
88
89
90

1. Dayle McDermitt
2. Ted DeJong
3. Shaun Hobbs
4. James Tiedje
5. Larry Williams
6. Jack Payne
7. Federico Sanchez
8. Bill Jackson
9. Don Phillips
10. Gerry Cangelosi
11. Scott Eady
12. Hermannes van Keulen
13. Richard Fisher
14. Bob Loomis
15. C. Peter Wolk
16. Jerry Peters
17. Madeline Ames
18. Ken Foster
19. Ken Cassman
20. Laszlo Csonka
21. David Rothstein
22. John Ingraham
23. Dick Hageman
24. Tim Righetti
25. Bob Buchanan
26. Gary Heichel
27. David Padwa
28. Don Barnes
29. Morris Berenbaum
30. Lynn Marcus-Wyner
31. Daniel Le Rudulier
32. Curt Carlson
33. Harold Evans
34. Yang Su-Sheng
35. J. Skujins
36. Warren Silver
37. Jim Bahr
38. Gary Ditta
39. Sally Ong
40. David Corbin
41. Steve Talley
42. Randy Jeter
43. Hauke Hennecke
44. Sui-Ming Hom
45. Roger Kern
46. Michael Mahan
47. Hart Spiller
48. Dennis Focht
49. Po Keung Wong
50. Jim Friedrich
51. Carlos Neyra
52. Kjell Andersen
53. Bruce Miller
54. Barbara Mueller
55. Bob Tait
56. K. T. Shanmugam
57. Toshio Satoh
58. Hans Schlegel
59. David Patriquin
60. Peter Heytler
61. John Beringer
62. Bill Rains
63. H. T. Huang
64. John Neal
65. Nicholas Brewin
66. Kry Trang
67. Sham Goyal
68. Minocher Reporter
69. Mike Cantrell
70. Joe Lepo
71. Wynne Szeto
72. Ernie Jaworski
73. Ray Valentine
74. Fred Ausubel
75. Elisabeth Raleigh
76. David McNeil
77. Steriling Russell
78. Mary Firestone
79. V. N. Iyer
80. Ruth Hickok
81. Kathy Fishbeck
82. Sharon Heath-Pagliuso
83. Sharon Long
84. Annalisa Valentine
85. Nelinia Henry
86. Suzanne Epperly
87. Sylvia Hua
88. Barbara Mazur
89. Marcia Osburne
90. Sandie Uratsu

KEY TO PHOTOGRAPH

INDEX